FLORA ZAMBESIACA

Flora terrarum Zambesii aquis conjunctarum

VOLUME SIX: PART ONE

FLORA ZAMBESIACA

MOZAMBIQUE

MALAWI, ZAMBIA, ZIMBABWE

BOTSWANA

VOLUME SIX: PART ONE

Edited by
G. V. POPE

on behalf of the Editorial Board:

S. BLACKMORE
Natural History Museum

G. Ll. LUCAS
Royal Botanic Gardens, Kew

I. MOREIRA
*Centro de Botânica, Instituto de Investigação
Científica Tropical, Lisboa*

G. V. POPE
Royal Botanic Gardens, Kew

Published by the Managing Committee on behalf of
the contributors to Flora Zambesiaca
1992

Typeset at the Royal Botanic Gardens, Kew, by
Pam Arnold, Christine Beard, Brenda Carey,
Margaret Newman, Pam Rosen
and Helen Ward

Printed in Great Britain by
Whitstable Litho Printers Ltd., Whitstable, Kent.

ISBN   0 947643 49 4

CONTENTS

FAMILY 97.   COMPOSITAE

LIST OF TRIBES INCLUDED IN VOLUME 6 PART 1

1. Mutisieae
2. Cardueae
3. Vernonieae
4. Lactuceae
5. Arctotideae

LIST OF NEW NAMES PUBLISHED IN THIS WORK

97. COMPOSITAE

By G.V. Pope

Annual, biennial or perennial herbs (often suffrutescent in the Flora Zambesiaca area with annual leafy stems or scapes from woody rootstocks and/or root tubers), or subshrubs, shrubs, or occasionally scramblers or lianes, sometimes trees, rarely aquatic or epiphytic, sometimes succulent, sometimes spinescent; tissues with schizogenous resin-ducts or articulated lacticifers. Leaves cauline and alternate or opposite, sometimes whorled, or radical and rosulate, exstipulate or sometimes with stipuliform appendages, sessile or petiolate, usually simple, entire, toothed, lobed or variously dissected. Inflorescence a capitulum with individual flowers ± sessile and aggregated on a common receptacle and surrounded by an involucre of 1–many series of bracts (phyllaries). Capitula solitary and terminal on scapes or leafy stems, or few to very numerous in lax or ± clustered cymose, often corymbiform synflorescences, occasionally scorpioidly cymose (a reduced cymose arrangement in which the subtending bracts are alternate), or spicate, racemose or paniculate, or aggregated into secondary capitula (glomerules). Common receptacle with scales (paleae) or setae subtending the florets, or epaleate and the surface smooth areolate or shallowly to deeply honeycombed (alveolate); alveolae often fimbriate or setose. Phyllaries in 2–many series, free and imbricate or ± connate, or sometimes 1-seriate and united or with cohering overlapping margins, persistent or rarely caducous, occasionally accrescent, sometimes apically appendaged. Flowers (florets) small, 1–500 or more per capitulum, hermaphrodite or unisexual (female, male or functionally male), or neuter (sterile); ovary inferior, of 2 united carpels, unilocular with 1 erect basal ovule; perianth epigynous. Calyx absent, represented by the pappus borne apically on the ovary; pappus consisting of persistent or caducous, 1–many-seriate hairs bristles awns or scales, or pappus elements ± fused to form an annular or ± cup-shaped or ear-shaped corona, or pappus absent; pappus setae barbellate or ± plumose. Corolla gamopetalous, of (3-)5 united petals, rarely absent; corolla ± regular and (3-)5-lobed (filiform or infundibuliform disk-florets), or bilabiate with a 2-lobed inner lip and a 3-dentate outer lip, or radiate with an abaxial strap-shaped limb (ray) 0–3(4)-dentate at the apex (ray-florets), or ligulate with a strap-shaped limb (ligule) 5-dentate at the apex. Stamens (3-)5, filaments free, inserted on the corolla tube, contractile; anthers introrse, usually apically appendaged, usually laterally connate into a cylinder around the style, thecae rounded sagittate or tailed at the base; pollen usually echinate, sometimes echinolophate or lophate. Style of hermaphrodite or functionally male florets elongating within the anther tube, shallowly to deeply bifid, occasionally undivided in functionally male florets, style arms with stigmatic areas on their inner sides and acute rounded or truncate at the apex, or the arms produced beyond the stigmatic surfaces as triangular, subulate or clavate appendages, variously papillate or hairy, usually with a brush of collecting hairs that sweep the pollen from the anther tube; style of female florets simpler, with acute to rounded style arms and without sweeping hairs. Fruit unilocular, 1-seeded, indehiscent (usually an achene), rarely fleshy with the single seed enclosed in a hard endocarp (drupe), sometimes produced apically into a beak (rostrum), crowned by the persistent or caducous pappus, or epappose; endosperm absent or vestigial.

The floret sexual state, its corolla form, and the combination of floret types on the common receptacle distinguish capitula as follows:
Capitula homogamous discoid – all florets of one sexual state, all hermaphrodite (or all female or functionally male); all corollas of the same form and ± regular.
Capitula homogamous ligulate – all florets hermaphrodite; all corollas ligulate.
Capitula heterogamous disciform – florets of 2(3) sexual states, the inner florets hermaphrodite or functionally male and the outer ones filiform and female; all corollas regular.
Capitula heterogamous radiant – inner florets hermaphrodite or functionally male, marginal florets larger and neuter; all corollas regular.

Capitula heterogamous radiate – central florets hermaphrodite or female or functionally
  male, outer florets female or neuter, occasionally hermaphrodite; central floret
  corollas regular, outer or marginal floret corollas radiate, or occasionally bilabiate
  with a strap-shaped outer lip.
Capitula heterogamous bilabiate – inner florets functionally male, outer florets female;
  corollas bilabiate, in the Flora Zambesiaca area the inner corollas are all equally
  2-lipped, while the outer corollas are of 2 kinds, i) 2–3 series of submarginal corollas
  equally 2-lipped, ii) a marginal series in which the corolla outer lip is strap-shaped
  (apparently radiate) and the inner lip smaller and 2-lobed (eg. *Gerbera*, Tab. **9**).
Occasionally all florets male or all female (plants monoecious or dioecious), or the
disk-florets functionally female and the ray-florets functionally male.

The Compositae, comprising approximately 1500 genera and some 25000 species, are
the largest family of the flowering plants. They are distributed throughout the world,
being absent only from Antarctica. They are especially well represented in grassland and
montane vegetation, but less common in the humid tropical forests.

The family is well defined by its combination of specialised floral characters and is
taken to be monophyletic in origin. However, the circumscription and relationships of
the major groupings within it remain uncertain. Most recent classifications, relying much
on morphological characters, recognise two subfamilies; the *Lactucoideae* and the
Asteroideae. However, a more satisfactory interpretation of phylogenetic relationships
within the family is provided by Bremer's cladistic analysis of tribal relationships [in
Cladistics **3**: 210–253 (1987)] and by the molecular studies (mainly of chloroplast DNA
evolution) by Palmer et al. [in Ann. Miss. Bot. Gard. **75**: 1180–1206 (1988)] and Jansen et
al. [in Syst. Bot. **16**: 98–115 (1991)]. These studies reveal that while the *Asteroideae*, with the
tribes *Inuleae, Astereae, Anthemideae, Senecioneae, Calenduleae, Heliantheae* and *Eupatorieae*
are a natural group, the *Lactucoideae* with the tribes *Mutisieae, Lactuceae, Vernonieae, Liabeae,
Cardueae* and *Arctotideae* may be paraphyletic.

In keeping with the findings of Palmer et al. (1988) and Jansen et al. (1991) therefore, a
linear arrangement of the tribes is adopted here, with no division into subfamilies. The
tribes accepted in this account are based on Cronquist's tribal concepts [in Amer. Midl.
Nat. **53**: 478–511 (1955)], and are arranged as follows:

volume 6 part 1	*volume 6 part 2*	*volume 6 part 3*
1. Mutisieae	6. Inuleae	9. Senecioneae
2. Carducae	7. Astereae	10. Calenduleae
3. Vernonieae	8. Anthemideae	11. Heliantheae
4. Lactuceae		12. Eupatorieae
5. Arctotideae		

The tribe *Liabeae*, closely related to the tribe *Vernonieae*, is neotropical and is not
represented in the Flora Zambesiaca area.

The foundations of this account were laid by Professor H. Wild in his series of articles
in Kirkia **6**,1: 1–62 (1967); **7**,1: 121–135 (1969); **8**,2: 167–207 (1972); **10**,1: 1–72 (1975); **11**,1:
1–23, 31–127 (1978), which we gratefully acknowledge.

Keys
Note: The following key to the tribes is specific to plants of the Flora Zambesiaca area. It is
complicated and both couplets and all alternatives contained in them should be
considered in detail and carefully compared with the plant specimen.
  Keys to the genera are provided under each tribe.

Key to the tribes *

1. Corollas all strap-shaped, equally 5-toothed at the apex   -   -   - 4. LACTUCEAE
                                                                              part 1, page 189
 – Corollas not strap-shaped, or if some strap-shaped then these with 4 or fewer equal or unequal
   apical teeth   -   -   -   -   -   -   -   -   -   -   -   - 2

* By C. Jeffrey and G.V. Pope

2. Style arms with conspicuous papillose appendages; capitula homogamous discoid, with florets all hermaphrodite, not yellow; corollas with 5(4) relatively short broad apical lobes; mature achenes black* - - - - - - - - - - - - - 12. EUPATORIEAE
– Style arms, capitula, corollas and mature achenes not showing the above combination of character-states - - - - - - - - - - - - - - - 3
3. Phyllaries uniseriate, cohering by their overlapping margins, or partly or wholly connate, with or without a calyculus; pappus present - - - - - - - - - - - 4
– Phyllaries imbricate, in 2 or more series, free or connate, if uniseriate then not cohering or pappus absent or capitula unisexual or achenes densely villous with long hairs from the base 5
4. Phyllaries with evident elongated oil-glands; achenes black* when mature
11. HELIANTHEAE
– Phyllaries without elongated oil-glands; achenes not black* when mature 9. SENECIONEAE
5. Style arms long, gradually attenuate acute, shortly hairy on the outside, style shaft similarly hairy on the upper part; capitula homogamous, all florets hermaphrodite 3. VERNONIEAE
part 1, page 56
– Styles and capitula not exhibiting the above combination of character-states, or capitula unisexual - - - - - - - - - - - - - - 6
6. Capitula with all or only the outer florets bilabiate, the corollas with 3-toothed outer lips and 2-lobed inner lips - - - - - - - - - - - - 1. MUTISIEAE
part 1, page 4
– Capitula without any such bilabiate florets - - - - - - - - 7
7. Capitula homogamous and discoid, the phyllaries without scarious unlobed appendages or apices, and without scarious margins; or capitula heterogamous and radiant with the outer florets ± enlarged and sterile; or capitula unisexual and plants dioecious; phyllary appendages if present spiniform and/or pinnately divided - - - - - - - - 8
– Capitula heterogamous and radiate or disciform but not radiant as above; or capitula homogamous and phyllaries with unlobed scarious often white or coloured appendages or apices, or with scarious usually brownish margins, but not with spiniform and/or pinnately divided appendages; or capitula unisexual and plants monoecious - - - - 12
8. Leaves spiny or bristly-spiny at least towards the base - - - - - - 9
– Leaves not spiny or bristly-spiny - - - - - - - - - - 10
9. Anthers tailed; receptacle densely setose - - - - - - - 2. CARDUEAE
part 1, page 49
– Anthers not tailed; receptacle alveolate with fringed alveolae 5. ARCTOTIDEAE
part 1, page 229
10. Corolla lobes of tubular florets much longer than broad; leaves alternate; capitula of more than 1 floret, not aggregated into glomerules - - - - - - - - - 11
– Corolla lobes ± as long as broad, or if much longer than broad then leaves opposite; capitula sometimes of only 1 floret, or capitula aggregated into glomerules - - - 13
11. Receptacle densely setose; the inner florets hermaphrodite, the outer sterile 2. CARDUEAE
part 1, page 49
– Receptacle naked or scaly, not setose; all florets hermaphrodite or unisexual 1. MUTISIEAE
part 1, page 4
12. Style arms of disk-florets connate in the lower part, the connate part thicker than the style shaft and abruptly marked off from it by a ring of short hairs at its base; capitula always bisexual
5. ARCTOTIDEAE
part 1, page 229
– Style arms of disk-florets long to short or absent but the connate lower part of the style arms if present not as above, or capitula unisexual and plants monoecious - - - - 13
13. Receptacle scaly and leaves opposite, or leaves not opposite and mature achenes black* and/or capitula glomerulate; or receptacle not scaly and pappus of scales and/or leaves opposite and/or mature achenes black and/or capitula unisexual and/or capitula glomerulate; pappus not of slender bristles, or if of bristles then bristles plumose 11. HELIANTHEAE
– Receptacle not scaly (although sometimes hairy or fimbrillate), or scaly and leaves not opposite (although sometimes opposite on upper stem) and achenes not black* when mature, or capitula not unisexual; capitula glomerulate or not; if lower leaves opposite then pappus of slender barbellate bristles, at least in disk-florets, or pappus absent, receptacle naked and capitula not glomerulate - - - - - - - - - - - - - - 14
14. Style arms of hermaphrodite or functionally male florets each with a subulate to triangular papillose appendage - - - - - - - - - - - - - 15
– Style arms of hermaphrodite or functionally male florets acute to rounded, or truncate and fringed with short hairs or papillae, or shortly conical at the apex with a subdistal fringe of hairs, not appendaged; or style undivided - - - - - - - - - - 16

---

* consider achenes to be black if the body of the achene when mature is shiny-black, at least in part; paler striations, ribs or wings may be present.

15. Receptacle alveolate; pappus a deeply and unequally laciniate corona; capitula disciform
      11. HELIANTHEAE
– Receptacle, pappus and capitula not showing the above combination of character- states
      7. ASTEREAE
16. Phyllaries with scarious usually brown often erose margins, not appendaged; leaves often pinnatipartite; style arms apically truncate and fringed, or style undivided and truncate; pappus absent, or if present then a short lacerate crown or lobed auricle, or leaves pinnatipartite and the outer pappus of 5 white petaloid scales - - - - - - 8. ANTHEMIDEAE
– Phyllaries green and herbaceous (though often with hyaline margins), or with scarious membranous often white or coloured appendages or apices, or rarely appendages foliose; leaves never pinnatipartite; a pappus of hairs or scales present in at least some florets, or if absent then style arms not truncate and fringed - - - - - - - - - 17
17. Pappus absent; fruits large curved or angular or winged achenes, or smooth drupes; capitula radiate - - - - - - - - - - - - 10. CALENDULEAE
– Pappus present or absent; fruits small achenes and not as above; capitula radiate or disciform
      18
18. Capitula radiate; ray-florets neuter, or if fertile then style arms shortly conical at the apex with a subdistal fringe of hairs - - - - - - - - - 1. MUTISIEAE
      part 1, page 4
– Capitula radiate or disciform; ray-florets if present female; style arms not as above
      6. INULEAE

Tribe 1. M U T I S I E A E Cass.

Mutisieae Cass. in Journ. Phys. **88**: 199 (1819). —C. Jeffrey in Kew Bull. **21**: 177–223, figs. 1–11 (1967). —Cabrera, *Mutisieae*, Syst. Rev. in Heywood et al., Biol. & Chem. Comp.: 1039–1066 (1977).

Suffrutescent herbs with annual stems from woody or tuberous rootstocks often with thong-like roots, sometimes annual herbs, or dioecious shrubs or trees. Leaves alternate or radical, simple, dentate occasionally deeply-lobed (pinnate and tendriliferous in *Mutisia*, a South American genus cultivated in East Africa). Capitula solitary or cymose, sometimes paniculate, 1–many-flowered; homogamous and discoid, bisexual or unisexual; or heterogamous and radiate or bilabiate, or disciform with outer florets neuter. Corollas purple or red, or less often whitish, yellow or creamy-brown. Outer florets, when differentiated, female or neuter; corollas isomorphic or dimorphic (*Gerbera*), bilabiate with outer lips 3-denticulate sometimes strap-shaped, and inner lips 2-lobed, or corolla radiate. Disk-florets bisexual, sometimes male; actinomorphic and regularly 5-lobed or deeply 5-cleft, or bilabiate. Phyllaries many-seriate, sometimes 1–2-seriate, imbricate. Receptacle epaleate or paleate, smooth, alveolate or ciliate. Anther bases sagittate with entire or ciliate tails. Style apically divided or shallowly cleft (minutely bifid or undivided and terete in male flowers of *Tarchonanthus*); style branches obtuse, rounded or truncate at the tips, not appendaged. Achenes subcylindric, turbinate or flattened flask-shaped or obovoid, truncate-attenuate or rostrate at the apex. Pappus 1–several-seriate, of barbellate or subplumose setae or narrow scales, or absent.

Keeley & Jansen, in Systematic Botany **16**, 1: 173–181 (1991), recognise the new tribe *Tarchonantheae* (Cass.) Keeley & Jansen for the genera *Brachylaena* R. Br. and *Tarchonanthus* L. However, in this flora account these genera are kept in the *Mutisieae*.

The tribe *Mutisieae* comprises approximately 91 genera in tropical and subtropical America, Africa, Asia, Australia and Hawaii.

Key to the genera

1. Plants dioecious; shrubs or trees - - - - - - - - - - - - - 2
– Plants monoecious; suffrutices with annual stems from woody rootstocks, or annual herbs   3
2. Pappus of scabrid bristles - - - - - - - - - **1. Brachylaena**
– Pappus wanting - - - - - - - - - - **2. Tarchonanthus**
3. Receptacular paleae present, resembling the inner phyllaries   **4. Erythrocephalum**
– Receptacular paleae absent - - - - - - - - - - - 4
4. Plants acaulescent, scapigerous (*Dicoma plantaginifolia* sometimes apparently acaulescent but never scapigerous) - - - - - - - - - - **7. Gerbera**
– Plants caulescent, not scapigerous; stems occasionally much abbreviated with capitula borne at ground level - - - - - - - - - - - - - 5

5. Capitula radiate; ray-floret limbs strap-shaped above, erect; pappus of narrow scales
   - **5. Pasaccardoa**
- Capitula discoid or disciform; outer-florets sometimes filiform, neuter, bilabiate but not strap-shaped; pappus of barbellate or plumose setae - - - - - - - - - 6
6. Phyllary apices rounded or narrowly obtuse; corolla limb well exserted above the pappus and involucre; leaves usually stem-sheathing at the base - - - - - **3. Pleiotaxis**
- Phyllary apices tapering-pungent; corolla hardly visible above the pappus; leaves not stem-sheathing at the base, or if so then innermost phyllaries shorter and broader than outer phyllaries - - - - - - - - - - - - - - - **6. Dicoma**

1. BRACHYLAENA R. Br.

Brachylaena R. Br. in Trans. Linn. Soc. **12**: 115 (1817). —E. Phillips & Schweick. in Bothalia **3**: 205 (1937). —Wild in Kirkia **7**: 122–125 (1969). —Paiva in Bol. Soc. Brot. sér. 2, **46**: 368–384 (1972). —Hilliard, Compos. Natal: 105–110 (1977).

Dioecious trees or shrubs. Leaves alternate, petiolate or subsessile, persistently tomentose or tomentellous on lower surface (in the Flora Zambesiaca area), entire or spinulose-denticulate or repand or coarsely dentate towards the apex. Synflorescences consisting of numerous capitula in terminal or axillary panicles or racemes, or capitula arranged in clusters or glomerules, capitula stalked or sessile. Capitula unisexual, disciform, several–many-flowered in the Flora Zambesiaca area; receptacle epaleate; phyllaries imbricate, the outermost series often extending onto the capitulum stalk. Male capitula smaller than the female, corollas funnel-shaped, 5-lobed; anther thecae exserted, bases tailed; style terete, thickened above, bifid, exserted; ovary abortive or rudimentary; pappus sparse, 1-seriate, barbellate or subplumose. Female capitula: corollas filiform, lobes 5 short; anthers 0, staminodes occasionally present; style branches exserted, short flat ovate-lanceolate; achenes subcylindric-fusiform, c. 8-ribbed, ± pubescent; pappus copious, consisting of barbellate setae in several series.

A genus of about 15 species, natives of Africa and Madagascar.

1. Twigs with leaves crowded at the ends; leaves usually mucronate; outer phyllaries not extending onto the capitulum stalk, capitula sessile or not - - - - - - - 3. *huillensis*
- Twigs leafy throughout; leaf apices not distinctly mucronate; outer phyllaries decreasing in size and extending onto the capitulum stalk for part or all of its length - - - - 2
2. Most capitula of a synflorescence apparently stalked; outer phyllaries extending onto the upper part of the stalk (absent from near the base), or the lower part of the stalk with 1–several scattered bracts or buds; inner phyllaries shortly- or long-tapering to an acute apex; petioles mostly more than 5 mm. long - - - - - - - - - - - - - - - 1. *discolor*
- Most capitula apparently sessile; outer phyllaries closely imbricate, extending to the base of the stalk; inner phyllaries tapering-attenuate to a narrowly obtuse or blunt apex; petioles mostly less than 5 mm. long - - - - - - - - - - - - - 2. *rotundata*

1. **Brachylaena discolor** DC., Prodr. **5**: 430 (1836). —Harv. in Harv. & Sond., F.C. **3**: 117 (1865). —Sim, For. Fl. Port. E. Afr.: 77, t. 73 (1909). —E. Phillips & Schweick. in Bothalia **3**: 219 (1937). —Mogg in Macnae & Kalk, Nat. Hist. Inhaca Isl.: 154 (1958). —Gomes e Sousa, Dendrol. Moçamb. Estudo Geral **2**: 682 (1967). —Wild in Kirkia **7**: 123 (1969). —Hilliard & Burtt in Notes Roy. Bot. Gard. Edinb. **31**: 3 (1971). —Paiva in Bol. Soc. Brot. sér. 2, **46**: 375 (1972). —Hilliard, Compos. Natal: 109 (1977). —K. Coates Palgrave, Trees Southern Afr.: 906 (1977). TAB. **1** fig. A. Types: Mozambique, Delagoa Bay (sphalm. Algoa), *Forbes* s.n.; and from South Africa (Cape Province and Natal).
   Brachylaena natalensis Sch. Bip. in Walp., Repert. **2**: 972 (1843). —Harv. in Harv. & Sond., F.C. **3**: 117 (1865). Type from South Africa (Natal).
   Brachylaena rhodesiana S. Moore in Journ. Linn. Soc., Bot. **40**: 108 (1911) pro parte quoad specim. mossamb.
   Brachylaena discolor var. *mossambicensis* Paiva in Bol. Soc. Brot. sér. 2, **46**: 379 (1972). Type: Mozambique, Maputo, Santaca, *Gomes e Sousa* 3611 (COI, holotypus; K; LISC; LMU; SRGH).

A shrub or tree to 10(12) m. tall, usually evergreen; branches brown to dark-purplish, lenticellate and narrowly sulcate, at first greyish tomentellous soon glabrescent, ± leafy throughout. Leaves with petioles mostly 5–16 mm. long; lamina mostly 5–12 × 1.3–5 cm., larger in coppice growth, oblanceolate to obovate, rarely oblong-elliptic, rounded sometimes briefly acuminate at apex, ± narrowly cuneate and usually tapering to a long petiole, entire and faintly revolute, or repand-dentate to serrate-or coarsely-dentate in

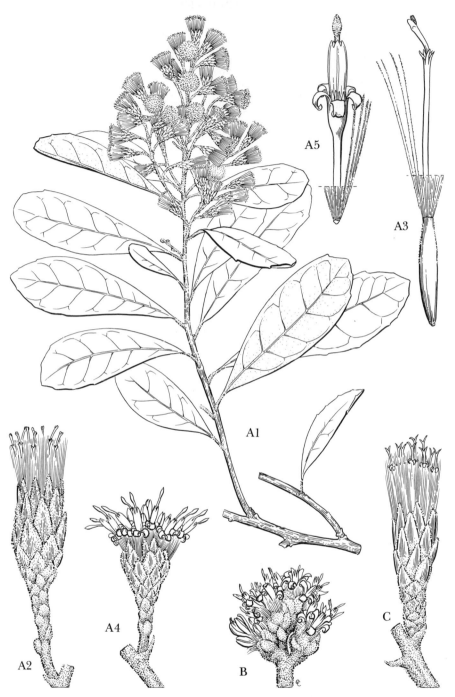

Tab. 1. A. —BRACHYLAENA DISCOLOR. A1, twig with female synflorescence (×⅔); A2, capitulum of female florets (× 3); A3, female floret (× 8), A1–A3 from *Watmough* 326; A4, capitulum of male florets (× 3); A5, male floret (× 8), A4 & A5 from *Barbosa & Lemos* 8665. B. —BRACHYLAENA HUILLENSIS, synflorescence, glomerule of male capitula (× 3), from *Gomes e Sousa* 3864. C. —BRACHYLAENA ROTUNDATA, capitulum of female florets (× 3), from *Chase* 4276. Drawn by Eleanor Catherine.

the apical part, sometimes spinescent-denticulate; upper surface thinly araneose when young, or glabrous, shiny dull-green with subprominent reticulate venation; lower surface greyish tomentellous with prominent veins; lateral veins forked near the margin. Capitula numerous in dense terminal panicles 4–40 cm. long, or in short raceme-like panicles in leaf axils. Involucres mostly narrowly obconic, somewhat cyathiform in male capitula. Phyllaries thinly coriaceous, acute, lanate on the margins otherwise glabrous; from c. 1 mm. long outside, the inner increasing to c. 4 mm. long in male capitula and to 6(9) mm. long in female capitula, becoming lanceolate or tapering-lorate; the outermost 5–11 series decreasing in size and extending onto upper part of the 1–10 mm. long capitulum stalk, the lower part of the capitulum stalk often with scattered small bracts or buds; bracts below the involucre lanate. Male flowers: corollas dull-yellow, 3–5 mm. long, lobes c. 1 mm. long and ± recurved; pappus sparse, uniseriate, setae 3–4 mm. long with seta barbs ± equalling the seta axis in width. Female flowers: corollas 3–6 mm. long, terete narrowing to the apex, lobes erect up to c. 0.5 mm. long; achenes pale-brownish to purplish, 3–4 mm. long, subcylindric-subfusiform, narrowly c. 8-ribbed, sparsely puberulous; pappus several-seriate, setae 4–5(7) mm. long, ± terete, barbellate.

Mozambique. GI: Bazaruto Isl., ♀ fl. viii.1936, *Gomes e Sousa* 1823 (COI; K; LISC). M: Maputo, Marracuene, Costa do Sol, ♂ fl. 10.viii.1959, *Barbosa & Lemos* 8665 (COI; K; LISC; SRGH).
Also in South Africa (Cape Province and Natal) and Swaziland. Abundant in dune and coastal forests, coastal woodlands, river mouths and mangrove margins.
Brachylaena transvaalensis E. Phillips & Schweick. is often included in *Brachylaena discolor* sens. lat. However, this large tree of inland medium altitude forests (1000–1600 m.) in South Africa and Swaziland is sufficiently distinct as to be treated as a separate species. It differs from *B. discolor* in its leaves which are characteristically narrowly elliptic, acute and ± acuminate at the apex, and in its smaller capitula (involucres 3–5 × 2–4 mm.) which are mostly sessile or subsessile with phyllaries extending completely to the base of the stalk (capitula of *B. discolor* usually have distinct stalks with only a few scattered bracts and buds).

2. **Brachylaena rotundata** S. Moore in Journ. Bot. **41**: 131 (1903). —Eyles in Trans. Roy. Soc. S. Afr. **5**: 508 (1916). —E. Phillips & Schweick. in Bothalia, **3**, 3: 218 (1937). —O.B. Mill. in Journ. S.Afr. Bot. **18**: 86 (1952). —Wild in Kirkia **7**: 123 (1969). —Paiva in Bol. Soc. Brot. sér. 2, **46**: 372 (1972). —K. Coates Palgrave, Trees Southern Afr.: 909 (1977). TAB. 1 fig. C. Type from South Africa (Transvaal).
Brachylaena rhodesiana S. Moore in Journ. Linn. Soc., Bot. **37**: 448 (1906); in op. cit. **40**: 108 (1911) pro parte excl. specim. mossamb. —Eyles in Trans. Roy. Soc. S. Afr. **5**: 508 (1916). —Steedman, Trees, Shrubs & Lianes S. Rhod.: 83 (1933). Type: Zimbabwe, Matopo Hills, Maleme (Malami) Valley, *Gibbs* 72 (BM, holotype; K, isotype; LISC, photo).
Brachylaena discolor sensu Monro in Proc. Rhod. Sci. Ass. **8**, 2: 90 (1908). —Eyles in Trans. Roy. Soc. S. Afr. **5**: 508 (1916) non DC. (1836).

A shrub or small tree to 7(10) m. tall, deciduous; branches brown to dark-purplish, lenticellate and narrowly sulcate, at first greyish- or brown-tomentose later glabrescent. Leaves with petioles mostly 2–5 mm. long, sometimes to 7 mm. long; lamina mostly 4–10(15) × 2.5–6 cm., larger in coppice growth, oblong-elliptic to broadly oblanceolate or elliptic, obtuse to rounded rarely subacute at the apex, cuneate to ± rounded at the base, entire, sometimes repand-dentate to coarsely dentate towards the apex, sometimes spinescent-denticulate; upper surface thinly araneose especially when young, or glabrescent; lower surface pale-brown or greyish tomentellous with prominent veins. Capitula sometimes appearing before the leaves, young synflorescences with buds in axillary and terminal spikes, mature synflorescences with numerous capitula in dense terminal panicles 4–40 cm. long, or in short raceme-like panicles in axils of old leaves. Involucres cyathiform to obconic. Phyllaries minutely glandular outside, subobtuse, margins ciliolate, the outer phyllaries from c. 1 mm. long and ovate, the inner to c. 5 mm. long becoming lorate-lanceolate, narrowly obtuse or blunt at the apex; the outermost 5–8 series decreasing in size and extending down to the base of the capitulum stalk. Male flowers: corollas dull-yellow, 3–5 mm. long, lobes c. 1.5 mm. long and ± recurved; pappus uniseriate, setae 3–4 mm. long, subplumose, the seta barbs exceeding the seta axis in width. Female flowers: corollas dull-yellow, 3–5 mm. long, filiform, lobes erect up to c. 0.5 mm. long; achenes c. 4 mm. long, subcylindric-fusiform, narrowly c. 8-ribbed, pubescent; pappus 2–several-seriate, setae 4–5 mm. long, ± terete or flattened, seta barbs ± equalling the seta axis in width.

Botswana. SE: Lobatse, railway embankment, c. 1066 m. st. 29.ii.1976, *Mott* 883 (K; SRGH).
Zimbabwe. W: Matopos, ♂ fl. 6.ix.1947, *Wild* 1989 (K; SRGH). C: Gweru, ♀ fl. 28.ix.1958, *Loveridge*

220 (K; LISC; SRGH). E: Mutare, Darlington, ♀ fl. 11.ix.1951, *Chase* 4276 (BM; COI; K; LISC; SRGH). S: Chibi Distr., Nyoni Hills, Tokwe Dam, ♂ fl. 21.ix.1967, *Müller* 682 (COI; K; SRGH). **Mozambique**. MS: base da Serra da Gorongosa, ♂ fl. 8.x.1944, *Mendonça* 2387 (LISC).

Also in South Africa (Natal and Transvaal). Of scattered distribution in mixed deciduous woodland and gully forests, often on rocky outcrops and hillsides at low to medium altitudes, in sandy soils and sometimes on termitaria.

A variant with narrowly elliptic leaves, acute at the apex and gently tapering to petioles up to c. 8 mm. long, occurs in central Mozambique from Dombe to Gorongosa Mt., eg. *Mendonça* 2387 (LISC), *Pedro* 4404 (K) and *Tinley* 2698 (K; SRGH). Similar leaf shapes are found in specimens from the Nyoni Hills area of SE. Zimbabwe, eg. *Müller* 682 (K; SRGH) and *Harvie* 1/53 (K; SRGH), these intergrading here with the more typical *B. rotundata* leaf.

This narrow-leaved, long-petioled variant approaches *Brachylaena transvaalensis* in leaf shape but has the capitula characteristic for *B. rotundata* with larger obconic not cyathiform involucres, and with narrow inner phyllaries blunt at the apex, not broadly acute.

3. **Brachylaena huillensis** O. Hoffm. in Engl., Bot. Jahrb. **32**: 149 (1902). —E. Phillips & Schweick. in Bothalia **3**: 212 (1937). —Mendonça, Contrib. Conhec. Fl. Angol., **1** Compositae: 54 (1943). —Wild in Kirkia **7**: 124 (1969). —Paiva in Bol. Soc. Brot. sér. 2, **46**: 369 (1972). —Hilliard, Compos. Natal: 107 (1977). —K. Coates Palgrave, Trees Southern Afr.: 907 (1977). TAB. **1** fig. B. Type from Angola.

   Tarchonanthus camphoratus sensu Hiern, Cat. Afr. Pl. Welw. **1**, 3: 554 (1898) pro parte quoad specim. *Welwitsch* 6745, non L. (1753).

   Brachylaena hutchinsii Hutch. in Bull. Misc. Inf., Kew **1910**: 126 (1910); in Hook., Ic. Pl. ser. 4, **10**: t.2928 (1911). Types from Kenya.

   Brachylaena sp. —Mendonça, Contrib. Conhec. Fl. Angol., **1** Compositae: 54 (1943).

A shrub or small tree to 7(30) m. tall, deciduous in dry areas; bark dark-grey, finely fissured, flaking in strips; branches becoming dark-purplish, lenticellate and narrowly sulcate, at first greyish- or brown-tomentellous later glabrescent; leaves crowded at the ends of twigs. Leaves with petioles 2–14 mm. long; lamina flat or revolute, mostly 3–10(12) × 1.2–3(4) cm., larger in coppice growth, narrowly elliptic-oblanceolate to obovate or elliptic to narrowly oblong-elliptic, acute to rounded and distinctly mucronate at the apex, cuneate below, entire, slightly to strongly revolute, occasionally in obovate leaves ± strongly spinescent-denticulate, particularly towards the apex; upper surface thinly araneose, soon glabrescent and shiny dull-green with reticulate venation subprominent; lower surface greyish- or silvery-tomentellous with prominent veins. Capitula few to many in short spikes, racemes or subglobose clusters; male synflorescences usually contracted into dense, stalked, axillary or terminal glomerules c. 1 cm. in diam., glomerules sometimes racemosely arranged on branchlets; female synflorescences composed of numerous, stalked, 2–6-capitulate clusters in leaf axils, or clusters racemosely arranged on numerous branchlets to c. 7 cm. long. Male capitula sessile or sometimes shortly stalked, 6–7-flowered; involucres to c. 3 mm. long, cyathiform; phyllaries increasing from c. 1 mm. long and ovate outside to c. 3 mm. long inside becoming oblong, tomentose; corollas creamy-white, 3.5–5.5 mm. long, narrowly funnel-shaped with lobes c. 2 mm. long and ± recurved; pappus sparse, setae 2–3 mm. long with barbs exceeding the seta axis in width. Female capitula stalked, 5–6-flowered; involucres to c. 6 mm. long, cylindric-obconic; phyllaries increasing from c. 1 mm. long and ovate outside to c. 5 mm. long inside becoming oblong, tomentose, outer phyllaries not extending onto the capitulum stalk; corollas 4–6 mm. long, filiform, lobes erect up to c. 0.5 mm. long; achenes 3–4 mm. long, subcylindric, obscurely 6–8-ribbed, pubescent; pappus several-seriate, setae 4–5 mm. long ± terete, seta barbs ± equalling the seta axis in width.

Zimbabwe W: Matobo, lower S. slopes of Tebase, st. 26.viii.1972, *Müller* 2060 (COI;SRGH). S: Gwanda, Doddieburn Ranch, Makoli Kopje, c. 800 m., st. 11.v.1972, *Pope* 740 (COI; K; LISC: SRGH). **Mozambique**. N: Mepanga, near the village Mocimboa da Praia, st. 10.xi.1960, *Gomes e Sousa* 4616 (K). GI: Guijá, proximo de Massingir, ♂ fl. 1.xii.1944, *Mendonça* 3211 (LISC). M: Reserva de Caca do Maputo (Bela Vista), ♂ fl. 20.xi.1940, *Torre* 2095 (LISC).

Also in Uganda, Kenya, Tanzania, Angola and South Africa (Transvaal and Natal). Locally common in coastal and dune forest, also in inland deciduous woodland, gully and kloof forests at low to medium altitudes. On rocky granite or quartzite outcrops.

An evergreen forest tree with fluted stems in Kenya.

On rocky, granite outcrops of S. and W. Zimbabwe this species grades into a variant with obovate leaves often spinescent-dentate on the margins. These leaves remain greenish on the upper surface and do not dry brownish or become strongly revolute as often happens in this species. Specimens belonging to this variant are; *Biegel & Pope* 3258 (COI, K, SRGH), *Müller* 2060, 2061 (COI, SRGH) and *Pope* 740 (COI, K, SRGH).

2. TARCHONANTHUS L.

Tarchonanthus L., Sp. Pl. **2**: 842 (1753); Gen. Pl. ed. 5: 365 (1754). —Wild in Kirkia **7**: 121–122 (1969). —Paiva in Bol. Soc. Brot. sér. 2, **46**: 355–368 (1972). —Hilliard, Compos. Natal: 110–112 (1977).

Dioecious trees or shrubs, often aromatic. Leaves alternate, petiolate or subsessile, discolorous, persistently tomentose or lanate on the lower surface, entire to coarsely irregularly serrate towards the apex (sometimes 3-lobed at the apex in South Africa). Synflorescence of numerous heads in terminal or axillary panicles, the whole synflorescence usually tomentose-lanate. Capitula unisexual, disciform; receptacle epaleate or sometimes with long silky hairs; phyllaries 1–2-seriate. Male capitula many-flowered, corollas tubular or funnel-shaped, longer than in the female florets, villous or lanate outside, lobes 5 recurved; anther thecae exserted, bases tailed; style narrowly terete, simple or minutely bifid, exserted; ovary abortive or rudimentary; pappus 0. Female capitula 1- or 2–3-flowered; corollas shorter than the ovary, villous or lanate outside, lobes 4–5 equalling the tube in length; anthers 0; style branches exserted, short flat ovate; achenes ± flattened obovoid or ellipsoid, densely lanate; pappus 0.

A genus of 2 species in Africa.

Leaf upper surface smooth or minutely tessellate-reticulate, glandular, margins sub-entire, lamina up to 7 cm. long (except in coppice shoots); petiole 0 or 1–6 mm. long; female capitula c. 3-flowered - - - - - - - - - - - - - 1. *camphoratus*
Leaf upper surface bullate and eglandular, margins often irregularly dentate towards the apex, 3(5)-lobed in South Africa, lamina mostly more than 7 cm. long; petiole mostly 10 mm. or more long; female capitula 1-flowered - - - - - - - - - - - 2. *trilobus*

1. **Tarchonanthus camphoratus** L., Sp. Pl. **2**: 842 (1753). —DC., Prodr. **5**: 431 (1836). —Harv. in Harv. & Sond., F.C. **3**: 118 (1894). —Oliv. & Hiern in F.T.A. **3**: 321 (1877). —O. Hoffm. in Bol. Soc. Brot. **7**: 231 (1889); in Engl. & Prantl, Nat. Pflanzenfam. **4**, 5: 174 (1890); in Warb., Kunene-Samb.-Exped. Baum: 409 (1903). —Hiern, Cat. Afr. Pl. Welw. **1**, 3: (1898) pro parte excl. specim. Welwitsch 6745. J.M. Wood, Handb. Fl. Nat. 69 (1907). —Sim, For. Fl. Port. E. Afr.: 77 (1909). —Eyles in Trans. Roy. Soc. S. Afr. **5**: 508 (1916). —Mendonça, Contrib. Conhec. Fl. Angol., **1** Compositae: 54 (1943). —Brenan, Check-list For. Trees Shrubs Tang. Terr.: 160 (1949). —Compton in Journ. S. Afr. Bot. **6**, Suppl.: 74 (1966). —Cufod. in Bull. Jard. Bot. Brux. **36**, 3: 1091 (1966). —Merxm., Prodr. SW. Afr. 139: 176 (1967). —Wild in Kirkia **7**, 1: 121 (1969). —Paiva in Bol. Soc. Brot. sér. 2, **46**: 360 (1972). —Hilliard, Compos. Natal: 111 (1977). TAB. **2** fig. A. Type from South Africa.

   Tarchonanthus minor Less., Syn. Comp.: 208 (1832). —Harv. in Harv. & Sond., F.C. **3**: 118 (1865). Type – no specimen cited.

   Tarchonanthus litakunensis DC., Prodr. **5**: 431 (1836). Syntypes from South Africa.

   Tarchonanthus obovatus DC., Prodr. **5**: 431 (1836). Type from South Africa.

   Tarchonanthus angustissimus DC., Prodr. **5**: 431 (1836). Syntypes from South Africa.

   Tarchonanthus abyssinicus Sch. Bip. in Schweinf. & Aschers., Beit. Fl. Aethiop.: 287 (1867) nom. nud.

Dense shrubs or small trees up to c. 8 m. tall, with a strong camphor odour. Trunk up to c. 40 cm. in diam., bark greyish, fissured; twigs leafy, closely greyish- or pale brown-felted. Leaves subsessile or with a petiole up to c. 6 mm. long and tomentellous; lamina very variable in size c. 2 × 0.5 to c. 8.5(12) × 3.5(5) cm., narrowly oblong-elliptic, elliptic or oblanceolate to narrowly obovate, apex subacute obtuse or rounded, base cuneate, margins entire; upper surface glabrescent, finely tessellate-reticulate with numerous golden glandular-globules along deeply depressed veins; under surface white-felted, the tomentum ± obscuring minute golden glands, midrib and nerves prominent beneath. Heads numerous in large leafy terminal panicles. Male capitula; involucres 2.5–3.5 × 3.5–5 mm., broadly turbinate to shallowly cupuliform, phyllaries 5–6, ovate, connate below, sometimes also with several free shorter narrowly oblong-ovate phyllaries, tomentose-araneose outside; florets (3)10–25, corollas c. 2.5 mm. long broadly funnel-shaped, densely woolly outside, anther thecae 1.5–2 mm. long, exserted, style linear exserted by c. 1.5 mm. above the anther tube, ovary rudimentary. Female capitula; involucres 3–3.5 × 6 mm., broadly turbinate to subglobose; phyllaries 2–3-seriate, imbricate, rotund to narrowly ovate, shortly connate at base or free, tomentose-araneose outside and often long ciliate about the upper margins; florets 4–5, corollas c. 1 mm. long, shortly funnel-shaped, lanate outside, stigma shortly bifid exserted; achene c. 2.5–3(4) ×

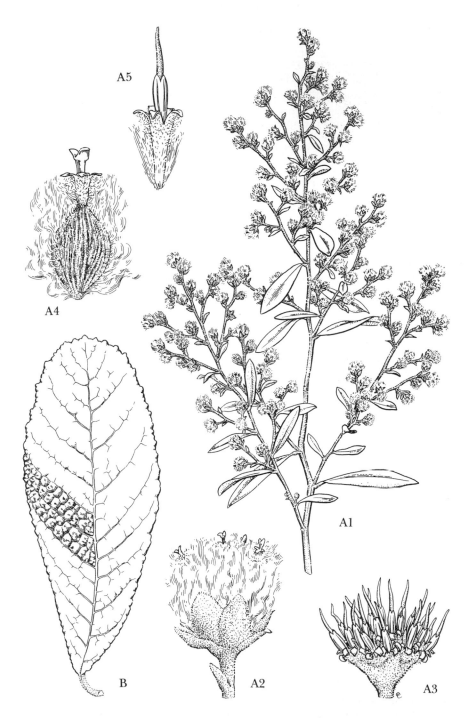

Tab. 2. A. —TARCHONANTHUS CAMPHORATUS. A1, twig (× ⅔); A2, female capitulum (× 4); A3, male capitulum (× 4); A4, female floret (× 8); A5, male floret (× 8), A1,A2 & A4 from *Hansen* 3141, A3 & A5 from *Norrgrann* 160. B. —TARCHONANTHUS TRILOBUS var. GALPINII, leaf (× ⅔), from *Chase* 8373. Drawn by Eleanor Catherine.

1.5 mm., somewhat compressed ellipsoid, densely covered in a mass of long sericeous-woolly hairs.

Botswana. N: Boteti R. above Rakops, ♀ fl. 10.xii.1978, *P.A. Smith* 2591 (K; PRE; SRGH). SE: Outside Content Farm, 990 m., 24°34'S, 25°57'E, ♀ fl. 18.vi.1977, *Hansen* 3141 (C; GAB; K; PRE; SRGH; UPS). SW: 30 km. NW. of Ghanzi, 1100 m., ♀ fl. 25.vii.1976, *Mott* 985 (K; SRGH; UBLS). **Zambia**. S: Choma, 7.viii.1961, *Bainbridge* 527 (LISC). **Zimbabwe**. W: Matopos, ♂ fl. viii.1930, *Hutchinson* 4131 (K). C: Hunter's Road, c. 32 km. NNE. of Gweru, ♀ fl. 16.iv.1967, *Biegel* 2073 (COI; K; SRGH). E: Mozambique border, ♂ fl. 20.iii.1948, *Chase* 751 (K; SRGH).
Also in Saudi Arabia, Ethiopia, Uganda, Kenya, Tanzania, Angola, Namibia, South Africa, Swaziland and Lesotho. In riverine vegetation, mixed deciduous woodland and scrubland on Kalahari Sand, on calcrete and granite outcrops.

2. **Tarchonanthus trilobus** DC., Prodr. **5**: 432 (1836). —Harv. in Harv. & Sond., F.C. **3**: 118 (1865). —J.M. Wood, Handb. Fl. Nat.: 69 (1907). Type from South Africa.

Small trees or shrubs up to c. 8(10)m. tall. Trunk up to c. 35 cm. in diam.; bark dark-grey, rough, flaking longitudinally; branches grey-felted; twigs robust, pale-brown, woolly-felted. Leaves crowded towards the apex of twigs; petiole up to c. 2.3 cm. long, woolly-felted, canaliculate; lamina thinly coriaceous, up to c. 18 × 7 cm., obovate-elliptic to oblong-obovate, ± rounded or distinctly 3(5)-lobed at the apex, cuneate at the base, margins subentire to irregularly ± coarsely dentate towards the apex; upper surface glabrescent, ± bullate; under surface with prominent midribs and veins and a whitish or pale-buff, lanate indumentum. Capitula numerous in axillary, spiciform panicles amongst the leaves; panicle branches up to c. 3 cm. long, subtended by linear-spathulate bracts up to c. 1.5 mm. long. Male capitula: involucres 2.5–3.5 × 3–4.5 mm., obconic-turbinate, phyllaries linear-lanceolate, connate below, densely lanate outside, glabrous inside; florets 10–16, corollas 3–4.5 mm. long, funnel-shaped, densely lanate outside; anther thecae 1.5–2 mm. long, exserted; style linear with a glandular simple apex exserted by up to c. 1.5 mm. from the anther tube; ovary rudimentary. Female capitula: involucres 4–5 × 2–5 mm., broadly cylindric-campanulate to ellipsoid or subglobose; phyllaries 4–8, 1–2-seriate, imbricate, coriaceous, ovate-lanceolate, attenuate-acute to apex, shortly connate at base or free, densely lanate outside, the indumentum totally obscuring the capitulum; florets 1, corolla tube c. 1.5 mm. long, lanate outside, stigma shortly spreading bifid, exserted; achene c. 2.5–3(5) × 1.5(4) mm., somewhat flattened obovoid, densely covered in a mass of long sericeous-woolly hairs.

Var. **galpinii** (Hutch. & Phillips) Paiva in Bol. Soc. Brot., sér. 2, **46**: 358 (1972). —Hilliard, Compos. Natal: 112 (1977). TAB. **2** fig. B. Type from South Africa.
    Tarchonanthus galpinii Hutch. & Phillips in Bull. Misc. Inf., Kew **1936**: 85 (1936). —Compton in Journ. S. Afr. Bot. **6**, Suppl.: 74 (1966). —Wild in Kirkia **7**: 122 (1969). Type as above.

Leaves rounded to obtuse at the apex.

Zimbabwe. E: Nyanga Distr., near Mt. Nyahokwe summit, Ziwa Farm, c. 22 km. N. of Nyanga (Inyanga) Village, c. 1372 m. ♂ fl. 27.ii.1966, *Chase* 8384 (COI; K; LISC; SRGH). S: Bikita Distr., S. slope of Mt. Horzi, c. 1097 m., 10.v.1969, *Pope* 148 (COI; K). **Mozambique**. M: Namaacha, ♂ fl. 8.vi.1951, *Barbosa & Myre* 148 (LMU; SRGH).
Also in South Africa (Transvaal, Natal) and Swaziland. Usually on dry, submontane, rocky hillsides.
Var. *trilobus* with leaves distinctly 3(5)-lobed at the apex intergrades with this variety in the Lebombo Mountains and extends through Natal to the Transkei.

3. PLEIOTAXIS Steetz

Pleiotaxis Steetz in Peters, Reise Mossamb., Bot. **2**: 499 (1864). —C. Jeffrey in Kew Bull. **21**: 180 (1967). —Wild in Kirkia **8**: 194 (1972).

Perennial herbs or suffrutices with annual stems from woody rootstocks. Stems 1–several, simple or branching near the apex, sericeous-lanate. Leaves alternate, ± sessile, usually stem-sheathing at the base, lamina discolorous, plane or bullate, membranous to subcoriaceous, finely araneose-lanate to glabrescent, indumentum densest on the lower surface, nerves often impressed on upper surface and strongly prominent beneath. Capitula homogamous, discoid; in few to many-headed racemose or narrowly subpaniculate arrangements, or capitula large terminal and solitary, sometimes with 2–4 subsidiary heads. Involucres subcylindric, turbinate or campanulate. Phyllaries many-seriate, closely appressed imbricate, progressively larger to the inside; blades often purple or purple-tinged. Receptacle epaleate, shallowly alveolate. Florets deep- or bright-red or purple, cream or white, hermaphrodite, corollas actinomorphic, long slender tubular below and abruptly dilated into a short cylindric deeply linear-lobed limb exserted beyond the pappus and involucre; anther thecae appendiculate, the appendages tapering claw-shaped or stipitate-glandular, bases sagittate with elongate tails which are long-haired or fimbriate on their margins; style shallowly or deeply 2-fid, the branches somewhat swollen-triangular apically and hairy outside. Achenes narrowly subcylindric, obscurely narrowly many-ribbed, sometimes somewhat angular, densely hispid in the Flora Zambesiaca area; pappus of copious barbellate or subplumose setae.

A genus of about 26 species confined to tropical Africa.

1. Corollas creamy-white; involucres subcylindric ± narrowed to the apex, hardly spreading, mostly less than 10 mm. wide   -   -   -   -   -   -   -   -   -   -   -   -   -   2
- Corollas deep-red to purplish-red; involucres campanulate or turbinate, spreading, mostly more than 10 mm. wide (sometimes less than 10 mm. wide but then leaves bullate)   -   -   3
2. Leaves broadly elliptic-obovate, 2–2.5 times as long as wide, rounded to obtuse at the apex, ± crowded on the lower stem; phyllaries ± obtuse with broad hyaline margins somewhat lacerate about the apex   -   -   -   -   -   -   -   -   -   -   -   -   10. racemosa
- Leaves narrowly oblong-oblanceolate, 3.5–4 times as long as wide, acute at the apex, not crowded at the stem base; phyllaries ± acute, margins obscurely narrowly hyaline   9. oxylepis
3. Leaf flat, membranous, upper surface glabrous or glabrescent with only the main lateral nerves sometimes impressed; leaf-sheath longer than wide (lamina sometimes finely araneose or somewhat bullate in P. dewevrei but then the petiole-like midrib and sheath apex with long marginal teeth)   -   -   -   -   -   -   -   -   -   -   -   -   -   -   4
- Leaf bullate, subcoriaceous, upper surface sparsely finely araneose-lanate with all nerves strongly impressed; leaf-sheath length ± equalling width   -   -   -   -   -   6
4. Leaf upper surface with a raised finely reticulate tertiary venation, lamina lanceolate-elliptic, 2–4 times as long as wide   -   -   -   -   -   -   -   -   -   1. pulcherrima
- Leaf upper surface plane, reticulate venation not visible, lamina narrowly oblong-lanceolate to linear, 4–15 times as long as wide   -   -   -   -   -   -   -   -   -   5
5. Involucres more than 22 mm. long and (14)18 mm. wide; pappus setae appressed-barbellate, barb length up to 1.5 times the seta axis width; achenes 10–15 mm. long; leaves narrowly oblong-lanceolate   -   -   -   -   -   -   -   -   -   -   -   3. dewevrei
- Involucres 14–20 mm. long and 8–14 mm. wide; pappus setae spreading-subplumose, bristles more than twice the seta axis width; achenes 7–9 mm. long; leaves linear   2. rogersii
6. Leaves linear, the largest to 19 cm. long and less than 1 cm. wide; phyllaries glabrous
                                                                                6. angustirugosa
- Leaves elliptic, lanceolate or oblanceolate, the largest more than 1 cm. wide; phyllaries araneose to sparsely tomentose, at least the outer   -   -   -   -   -   -   -   7
7. Leaves mostly less than 6(8) × 3 cm.; capitula small, usually 3–many on each stem; involucre less than 20 × 20 mm.; largest phyllaries 2.5–4 mm. wide   -   -   -   -   -   -   8
- Leaves mostly more than (8)10 × 3 cm.; capitula large, usually solitary and terminal on each stem; involucre more than 20 × 20 mm.; largest phyllaries up to 7 mm. wide   -   -   -   9
8. Leaves mostly tapering into a petiole-like base 1–5 mm. long; phyllaries light to dark-purple, not completely obscured by a sparse whitish tomentum, margins hyaline glabrous; plants usually long-stemmed lax subshrubs, from northern Zambia   -   -   -   -   -   7. argentea
- Leaves sessile, subauriculate at the base; phyllary surface completely obscured by a dense araneose-lanate indumentum; plants suffruticose, from western Zambia, Angola and Zaire
                                                                                8. fulva
9. Leaves tapering to a distinct petiole up to c. 3 cm. long; stems leafy to near the apex with leaves hardly diminishing in size up the stem   -   -   -   -   -   -   -   4. eximia
- Leaves sessile, semi-amplexicaul and subauriculate at the base; lower stem crowded leafy, upper-stem leaves 0 or few, diminishing in size   -   -   -   -   -   5. antunesii

1. **Pleiotaxis pulcherrima** Steetz in Peters, Reise Mossamb., Bot. **2**: 500, t.51 (1864). —Oliv. & Hiern in F.T.A. **3**: 440 (1877). —O. Hoffm. in Engl., Bot. Jahrb. **15**: 536 (1893); in Engl., Pflanzenw. Ost-Afr. **C**: 420 (1895). —Hiern, Cat. Afr. Pl. Welw. **1**, 3: 610 (1898) pro parte. —S. Moore in Journ. Bot. **65**, Suppl. 2: 64 (1927). —De Wild., Pl. Bequaert. **5**: 157 (1929). —Mendonça, Contrib. Conhec. Fl. Angol., **1** Compositae: 133 (1943). —C. Jeffrey in Kew Bull. **21**: 182 (1967). —Wild in Kirkia **8**: 197 (1972). —Maquet in Fl. Rwanda, Spermat. **3**: 682, fig. 213(2), (1985). —Lisowski, Aster. Fl. Afr. Centr. [in Fragm. Flor. Geobot. **36**, Suppl. 1]: 507 (1991). TAB. **3** fig. A. Syntypes: Mozambique, Boror, *Peters* (B†); Rios de Sena, *Peters* (B†).

Tab. 3. A.—PLEIOTAXIS PULCHERRIMA. A1, habit (×⅓); A2, floret (×2), A1 & A2 from *Brummitt & Little* 9490. B.—PLEIOTAXIS ARGENTEA, leaf, note bullate upper surface and short stem-sheath (×⅔), from *Fanshawe* 8577. C.—PLEIOTAXIS ROGERSII, leaf, note long stem-sheath fimbriate-toothed on margin (×⅔), from *Mutimushi* 3495. Drawn by Eleanor Catherine.

Pleiotaxis pulcherrima var. *poggeana* O. Hoffm. in Engl., Bot. Jahrb. **15**: 537 (1893). —De Wild., Pl. Bequaert. **5**: 159 (1929). —Mendonça, Contrib. Conhec. Fl. Angol., 1 Compositae: 134 (1943). Type from Angola.
Pleiotaxis sapinii S. Moore in Journ. Bot. **63**: 45 (1925). —De Wild., Pl. Bequaert. **5**: 160 (1929). Type from Zaire.
Pleiotaxis latisquama S. Moore in Journ. Bot. **63**: 46 (1925). Type: Zambia, Bwana Mkubwa, *Rogers* 8381 (BM, holotype; BOL; K).

An erect suffrutex 18–80 cm. tall, from a woody rootstock. Stems annual, 1–several, simple or sometimes branching at the apex, leafy, white sericeous-lanate. Leaves ± discolorous, ± ascending, stem-sheathing at the base; mostly 6–24 × 2–7 cm., lanceolate-elliptic, acute at the apex, rounded to cuneate below and usually narrowing to a winged petiole-like midrib 1–25 mm. long, before grading into a cylindrical, membranous, sericeous-lanate sheath 10–20 mm. long, or lamina base amplexicaul passing directly into a deep funnel-shaped sheathing base; margins finely sharply serrulate; upper surface green and glabrous, lower surface white araneose-sericeous; main nerves 6–10 on each side of the midrib, impressed on upper surface, prominent beneath, tertiary venation finely reticulate and prominent on both surfaces though ± obscured by indumentum beneath. Capitula solitary, terminal, occasionally 2–4, on short to long stout stalks, sometimes subsessile in one or more leaf axils. Involucres mostly 20–35 × 12–20 mm., spreading to c. 55 mm. wide, obconic-campanulate. Phyllaries numerous, coriaceous, glabrous, finely striate, margins narrowly hyaline, from 4 mm. long and triangular-ovate outside, increasing to c. 24 mm. long and oblanceolate-lorate inside; middle phyllaries up to c. 8 mm. wide, oblong and ± expanded in the upper part, rounded at the apex; inner phyllaries narrower. Florets very numerous; corollas bright- or deep-red, up to c. 25 mm. long, with a long slender tube abruptly dilated into a short cylindric deeply narrowly-lobed limb which exceeds the pappus and involucre, glabrous. Achenes 7–11 mm. long, narrowly cylindric, obscurely narrowly many-ribbed, hispid; pappus of copious barbellate setae 7–17 mm. long.

Zambia. N: c. 95 km. from Mbala on Nakonde road, 23.iv.1986, *Philcox, Pope & Chisumpa* 10093 (BR; K; NDO; SRGH). W: Mwinilunga Distr., between Ikelenge and Kalene, c. 1625 m., 15.v.1986, *Philcox, Pope & Chisumpa* 10311 (BR; K; SRGH). E: Petauke Distr., Kachalola Protected For., 30°41'E, 14°44'S, c. 1000 m., 4.iii.1973, *Kornaś* 3409 (K). S: Mumbwa, *Macaulay* 787 (K). **Malawi**. N: Rumphi Distr., Nyika, c. 1371 m., 26.iv.1973, *Pawek*, 6571 (K; MAL; MO). C: Lilongwe Distr., Dzalanyama For. Res., c. 4 km. from Sinyala Gate, c. 1125 m., 24.ii.1982, *Brummitt, Polhill & Banda* 16068 (K; MAL; SRGH). S: Blantyre, c. 1100 m., 17.vi.1946, *Brass* 16337 (K; MO). **Mozambique**. N: Mutuáli, Lurio road, 7 km. from railway line, 3.iii.1954, *Gomes e Sousa* 4237 (COI; K; LISC; SRGH). Z: entre Mocuba & Ile, 18.vi.1943, *Torre* 5505A (LISC).
Also in Angola, Zaire, Rwanda, Burundi and Tanzania. Miombo woodland on rocky hillsides or on sandy soil, and in grassland.
Typical *Pleiotaxis pulcherrima*, as depicted in the illustration of the type, has somewhat narrowly elliptic leaves which taper into a petiole-like narrowly-winged midrib before passing into an appressed, cylindrical, stem-sheathing base. Occurring sympatrically with this is a variant which is characterised by a larger and more robust habit, and larger leaves which are amplexicaul with broad, deeply funnel-shaped stem-sheathing bases. The latter variant occurs in western Zambia and adjacent Zaire and in northern Malawi and adjacent Zambia and Tanzania; *Philcox, Pope & Chisumpa* 10311, *Mutimushi* 3199 and *Edwards* 719 from the Mwinilunga area, *Fanshawe* 10088 and *Mutimushi* 1825 from the Kitwe area, *Carson* 49 from northern Zambia and *Pawek* 6571 from northern Malawi.
S. Moore described *Pleiotaxis sapinii* for the Zairian material of this variant, and *Pleiotaxis latisquamea* for the Zambian Copperbelt material. However, the discontinuity of variation between these plants is not satisfactorily distinct, and so for the time being it is considered best to follow Wild's broad concept of this species and refer all this material to *Pleiotaxis pulcherrima*.

2. **Pleiotaxis rogersii** S. Moore Journ. Bot. **63**: 50 (1925). —De Wild., Pl. Bequaert. **5**: 159 (1929). —C. Jeffrey in Kew Bull. **21**: 188 (1967). —Wild in Kirkia **8**: 197 (1972). —Lisowski, Aster. Fl. Afr. Centr. [in Fragm. Flor. Geobot. **36**, Suppl. 1]: 512 (1991). TAB. 3 fig. C. Type from Zaire.
Pleiotaxis davyi S. Moore Journ. Bot. **63**: 50 (1925). —De Wild., Pl. Bequaert. **5**: 153 (1929). Type from Zaire.
Pleiotaxis pulcherrima var. *angustifolia* sensu De Wild., Pl. Bequaert. **5**: 158 (1929) non O. Hoffm.

An erect slender suffrutex 30–135 cm. tall, from a woody rootstock. Stems annual, 1–several, simple or sometimes branching above, leafy, sericeous-lanate. Leaves ± discolorous, stem-sheathing at the base, mostly 5–16 × 0.4–2 cm., linear-elliptic to narrowly-lanceolate, tapering-acute to somewhat attenuate at the apex, cuneate and subsessile to subauriculate at the base, or narrowing to a winged petiole-like midrib to c.

10 mm. long, before grading into a cylindrical or very narrowly infundibuliform, membranous sericeous sheath up to c. 20 mm. long; margins finely sharply serrate, sometimes remotely so, the serrations becoming linear teeth c. 3 mm. long towards the base of the winged midrib and upper margins of the sheath; upper surface green, glabrous; lower surface greyish araneose-sericeous; main nerves 5–12 on each side of the midrib, prominent beneath; reticulate tertiary venation inconspicuous. Capitula solitary and terminal, or 2–4 on short or long stalks. Involucres mostly 14–20 × 8–14 mm., obconic-campanulate. Phyllaries appressed imbricate, the outer sometimes extending briefly onto the stalk, scarious-coriaceous, tomentellous to glabrescent outside, margins narrowly reddish-tinged hyaline, from c. 3 mm. long and narrowly ovate outside, increasing to c. 18 mm. long and becoming lorate-lanceolate inside; middle phyllaries up to c. 4.5 mm. wide, narrowly ovate and rounded to obtuse at the apex; inner phyllaries narrower. Florets numerous; corollas bright- to deep-red, up to c. 18 mm. long with a long slender tube abruptly dilated into a short cylindric, deeply narrowly-lobed limb which exceeds the pappus and involucre, glabrous. Achenes 7–8.5 mm. long, narrowly cylindric, obscurely narrowly many-ribbed, densely hispid; pappus of copious, subplumose setae, 9–12 mm. long.

Zambia. W: c. 22 km. on Solwezi to Mwinilunga road, c. 1610 m., 14.v.1986, *Philcox, Pope, Chisumpa & Ngoma* 10247 (BR; K; LISC; MO; NDO; SRGH).
Also in Zaire. Miombo and *Marquesia* woodlands, sometimes in mushitu (swamp forest).
P. decipiens C. Jeffrey (Angola) and *P. lawalreeana* Lisowski (Zaire) closely resemble this species but may be readily distinguished by their glabrous achenes and the absence of linear teeth at the base of the winged midrib and on the leaf sheath upper margins.

3. **Pleiotaxis dewevrei** O. Hoffm. ex Dur. & De Wild. in Compt. Rend. Soc. Bot. Belg. **39**: 34 (1900).
—De Wild., Pl. Bequaert. **5**: 153 (1929). —C. Jeffrey in Kew Bull. **21**: 188 (1967). —Wild in Kirkia **8**: 196 (1972). —Lisowski, Aster. Fl. Afr. Centr. [in Fragm. Flor. Geobot. **36**, Suppl. 1]: 514 (1991). Type from Zaire.
Pleiotaxis vernonioides S. Moore in Journ. Linn. Soc., Bot. **35**: 365 (1902). —De Wild., Pl. Bequaert. **5**: 160 (1929). Type from Tanzania.
Pleiotaxis sciaphila S. Moore in Journ. Bot. **63**: 49 (1925). —De Wild., Pl. Bequaert. **5**: 160 (1929). Type from Zaire.
Pleiotaxis antunesii var. *planifolia* S. Moore in Journ. Bot. **63**: 49 (1925). —De Wild., Pl. Bequaert. **5**: 152 (1929). Type: Zambia, Luwingu, *Jelf* 46 (BM, holotype).

An erect suffrutex, 20–90 cm. tall, from a woody rootstock. Stems annual, 1–several, simple or sometimes branching near the apex, leafy to near the apex, white sericeous-lanate. Leaves ± discolorous, membranous, stem-sheathing at the base, mostly 4.5–16 × 1–4 cm., narrowly oblong-lanceolate to narrowly elliptic, ± tapering acute at the apex, cuneate to rounded below narrowing to a winged petiole-like midrib before grading into a cylindric, partially or deeply cleft, membranous sericeous-lanate sheath 5–20 mm. long, or sometimes lamina subsessile and subauriculate; margins finely sharply irregularly serrate, the serrations becoming linear teeth to c. 5 mm. long on the winged midribs and upper margins of the sheath; upper surface green often drying brown and sparsely finely appressed lanate or glabrescent; lower surface white araneose-sericeous; main nerves 10–16 on each side of the midrib, prominent beneath; reticulate tertiary venation inconspicuous. Capitula solitary, terminal, occasionally 2–4 on long stout stalks. Involucres mostly 22–30 × 18–28 mm., spreading to c. 45 mm. wide, obconic-campanulate. Phyllaries numerous, appressed imbricate, coriaceous or ± cartilaginous, sparsely finely puberulous to glabrescent, margins very narrowly hyaline; the outer extending briefly onto the capitulum stalk, from c. 4 mm. long and narrowly ovate, increasing to c. 23 mm. long becoming lorate-lanceolate inside; middle phyllaries up to c. 6 mm. wide, obtuse at apex; inner phyllaries narrower. Florets very numerous; corollas bright- or deep-red, up to c. 25 mm. long, with a long slender tube abruptly dilated into a short cylindric, deeply narrowly-lobed limb which exceeds the pappus and involucre, glabrous. Achenes 10–15 mm. long, narrowly cylindric, obscurely narrowly many-ribbed, densely hispid; pappus of copious barbellate setae 10–17 mm. long.

Zambia. N: Mbala, Mpulungu near harbour, 24.iv.1986, *Philcox, Pope & Chisumpa* 10098 (BR; K; LISC; MO; NDO; SRGH). W: Mufulira, 9.vi.1934, *Eyles* 8177 (BM; K; SRGH).
Also in Zaire, Burundi and Tanzania. In deciduous woodlands, usually miombo woodland, in good understorey or amongst tall grasses, in sandy soil or on rocky hillsides.

4. **Pleiotaxis eximia** O. Hoffm. in Engl., Bot. Jahrb. **15**: 539 (1893). —Hiern, Cat. Afr. Pl. Welw. **1**, 3: 610 (1898). —S. Moore in Journ. Bot. **65**, Suppl. 2: 65 (1927). —De Wild., Pl. Bequaert. **5**: 154 (1929). —Mendonça, Contrib. Conhec. Fl. Angol., 1 Compositae: 137 (1943). —C. Jeffrey in Kew Bull. **21**: 193 (1967). —Wild in Kirkia **8**: 200 (1972). —Lisowski, Aster. Fl. Afr. Centr. [in Fragm. Flor. Geobot. **36**, Suppl. 1]: 524 (1991). Syntypes from Angola.

An erect robust silvery-grey suffrutex, 15–80 cm. tall, from a woody rootstock. Stems annual, 1–many, simple or sometimes branching near the apex, leafy usually to apex, densely greyish lanate. Leaves stiff, silvery-grey, concolorous, sometimes paler beneath, bullate, stem-sheathing briefly at the base, mostly 8–25 × 2–7.5 cm., elliptic, tapering acute at the apex; cuneate below, tapering into a narrowly winged petiole-like midrib up to c. 30 mm. long before widening into a short membranous lanate sheathing-base to c. 10 mm. long; margins irregularly serrate, sometimes obscurely so; finely araneose-lanate on both surfaces often more densely so beneath; main nerves 12–20 on each side of the midrib, impressed on the upper surface and prominent beneath, tertiary venation reticulate and prominent beneath. Capitula large, usually solitary and terminal on each stem, occasionally 2–4 subsidiary heads in axils below, capitula stalks often somewhat thickened and densely grey- or fulvous-lanate below the capitula. Involucres mostly 20–40 × 25–40 mm., spreading to c. 60 mm. wide, turbinate. Phyllaries numerous, stiffly coriaceous, at first appressed imbricate soon loosely spreading, usually persistently close felted-lanate outside, sometimes glabrescent, margins often narrowly hyaline purple-tinged and glabrous; the lowermost phyllaries scattered on the capitulum stalk, mostly from 5–12 mm. long, sometimes to c. 15 mm. long, narrowly triangular to linear-lanceolate increasing to c. 25 mm. long, sometimes to 30 mm. long, and becoming linear inside; middle phyllaries up to 8 mm. wide, lorate-lanceolate or oblong; inner phyllaries narrower. Florets numerous, corollas bright- or deep-red, up to c. 30 mm. long, with a long slender tube abruptly dilated into a short cylindric deeply-lobed limb which exceeds the pappus and involucre, glabrous; corolla lobes 6–7 mm. long, linear. Achenes 9–14 mm. long, narrowly cylindric, obscurely many-ribbed, densely appressed hispid; pappus of copious barbellate setae to c. 18 mm. long.

Phyllaries all lorate-lanceolate, gently tapering in the upper half to a narrowly obtuse apex, usually persistently felted-lanate outside (usually fulvous), the largest 3–6 mm. wide; involucre usually up to c. 35 × 30 mm., spreading to c. 50 mm. wide -   -   -   -   -   subsp. *eximia*

Middle phyllaries distinctly broader than the others, pronounced cucullate and shortly tapered or rounded at the apex, the middle and inner phyllaries whitish pubescent soon glabrescent outside, the largest usually 6–8 mm. wide; involucre usually 35–60 × 35 mm., spreading to c. 60 mm. wide -   -   -   -   -   -   -   -   -   -   -   subsp. *kassneri*

Subsp. **eximia**

    Pleiotaxis amoena R.E. Fr., Wiss. Ergebn. Schwed. Rhod.-Kongo-Exped. **1**, 2: 347, t. 22, fig. 2 (1916). —De Wild., Pl. Bequaert. **5**: 152 (1929). Type: Zambia, Katwe, *Fries* 1209 (UPS, holotype).

    Pleiotaxis antunesii sensu Eyles in Trans. Roy. Soc. S. Afr. **5**: 521 (1916) non O. Hoffm. (1893).

    Botswana. N: Chobe Distr., Panda Forest, v.1966, *Mutakela* 64 (SRGH). **Zambia**. B: c. 85 km. from Zambezi (Balovale) on Kabompo road, 26.iii.1961, *Drummond & Rutherford-Smith* 7376 (K; SRGH). N: c. 32 km. S. of Mpika, 24.vii.1930, *Pole Evans* 3053 (PRE; SRGH). W: Kitwe, 16.vi.1967, *Mutimushi* 1973 (K; M; NDO). C: c. 64 km. E. of Lusaka, c. 1220 m., 10.iv.1955, *Best* 78 (K; SRGH). S: Mazabuka, Kafue floodplain between Great North Road and Kabanje Village, 9.iii.1963, *van Rensburg* 1667 (K; M; SRGH). **Zimbabwe**. N: Charama Escarpment, near turnoff on Gokwe-Charama road, 27.iii.1962, *Bingham* 181 (K; M; SRGH). W: Lupane, 4.iv.1972, *Chiparawasha* 435 (K; LISC; SRGH). C: Montebello, c. 40 km. NW. of Gweru, 5.iv.1967, *Biegel* 2049 (K; SRGH).

    Also in Angola and Zaire. Miombo woodland and mixed deciduous woodland on sandy soil (usually Kalahari Sand), in wooded tall-grassland, sometimes on rocky hillsides.

Subsp. **kassneri** (S. Moore) G.V. Pope comb. et stat. nov. Type from Zaire.

    Pleiotaxis kassneri S. Moore in Journ. Bot. **63**: 48 (1925). —De Wild., Pl. Bequaert. **5**: 155 (1929). Type as above.

    Pleiotaxis kassneri var. *angustifolia* S. Moore in Journ. Bot. **63**: 48 (1925). Type from Zaire.

    Zambia. N: Mbala Distr., Howser's Farm, c. 1828 m., 23.iii.1955, *Richards* 5068 (BR; K).

    Also in Tanzania and Zaire. Wooded grassland with tall grasses, on escarpments and rocky hillsides or in sandy soil on flat ground.

5. **Pleiotaxis antunesii** O. Hoffm. in Engl., Bot. Jahrb. **15**: 539 (1893); in Bol. Soc. Brot. **10**: 183 (1893). —Hiern, Cat. Afr. Pl. Welw. **1**, 3: 611 (1898). —S. Moore in Journ. Bot. **65**, Suppl. 2: 65 (1927). —De Wild., Pl. Bequaert. **5**: 152 (1929). —Mendonça, Contrib. Conhec. Fl. Angol., 1

Compositae: 137 (1943). —C. Jeffrey in Kew Bull. **21**: 193 (1967). —Lisowski, Aster. Fl. Afr. Centr. [in Fragm. Flor. Geobot. **36**, Suppl. 1]: 523 (1991). Type from Angola.

An erect silvery-grey suffrutex 20–60 cm. tall, from a woody rootstock with thong-like roots. Stems annual, 1–many, usually simple, leafy with leaves crowded on the lower stem, usually leafless near the apex, densely greyish araneose-lanate. Leaves ± discolorous, bullate, sessile, not or hardly stem-sheathing at the base, mostly 5–15 × 2–5 cm., elliptic, acute at the apex, not or slightly narrowing below to a semi-amplexicaul subauriculate base; margins erose-serrulate, sometimes obscurely so; finely silvery-grey araneose-lanate on both surfaces, more densely so and paler beneath; main nerves 12–15 on each side of the midrib, impressed on the upper surface and prominent beneath, venation prominently reticulate beneath. Capitula large, usually solitary and terminal on the stem, borne well above the foliage on long stout, densely grey or fulvous araneose-lanate stalks. Involucres mostly 25–30 × 20–25 mm., spreading to c. 50 mm. wide, turbinate. Phyllaries numerous, stiffly subcoriaceous, at first appressed-imbricate soon loosely spreading, sparsely felted-lanate, margins narrowly hyaline glabrous, sometimes lacerate or shallowly torn about the apex, the lowermost sometimes extending briefly onto the capitulum stalk, from 4 mm. long and narrowly triangular outside, increasing to c. 25 mm. long becoming linear inside; middle phyllaries up to c. 7 mm. wide and lorate-lanceolate, inner phyllaries narrower. Florets numerous; corollas brilliant- to pale-red, up to c. 30 mm. long with a long slender tube abruptly dilated into a short cylindric deeply lobed limb which exceeds the pappus and involucre, glabrous; corolla lobes 6–7 mm. long, linear. Achenes c. 9 mm. long, narrowly cylindric, obscurely many-ribbed, densely appressed hispid; pappus of copious subplumose-barbellate setae c. 16 mm. long.

Zambia. B: c. 103 km. SE. of Senanga on Sesheke road, c. 1050 m., 1.ii.1975, *Brummitt, Chisumpa & Polhill* 14223 (K).
Also in Angola and Zaire. *Baikiaea* woodland on Kalahari Sand.

6. **Pleiotaxis angustirugosa** C. Jeffrey in Kew Bull. **21**: 195, fig. 7 (1967). —Wild in Kirkia **8**: 198 (1972).
Type: Zambia, Barotseland, Chavuma, *Gilges* 398 (K; PRE; SRGH, holotype).

An erect suffrutex, to c. 1 m. tall, from a ?woody rootstock. Stems annual, 1–several, sericeous-lanate, sparsely branched and leafy near the apex, simple and leafless below, becoming woody. Leaves sessile discolorous, markedly bullate, briefly stem-sheathing at the base, mostly 4.5–19 × 0.4–0.7 cm., linear; acute at the apex, semi-amplexicaul and subauriculate below from a short membranous, sericeous-lanate sheath ± as long as wide; margins revolute, serrate; upper surface bullate and ± sparsely lanate-pubescent, under surface densely greyish or brownish araneose-sericeous; main nerves usually more than 20 on each side of the midrib, impressed on the upper surface and prominent beneath, venation reticulate. Capitula many, usually terminal on branches with 2–3 subsidiary capitula in axils below, on short or long stalks. Involucres mostly 20–25 × 15–20 mm., obconic-campanulate. Phyllaries green often purple-tinged, appressed imbricate, thinly coriaceous, glabrous outside, margins hyaline often purple-tinged, from 2 mm. long and ovate outside increasing to c. 23 mm. long and becoming narrowly lorate inside; middle phyllaries to c. 3.5 mm. wide, ovate-oblong, rounded at the apex. Corollas red, up to c. 24 mm. long, with a long slender tube abruptly dilated into a short cylindric deeply narrowly lobed limb which exceeds the pappus and involucre, glabrous; corolla lobes up to c. 6 mm. long, linear. Achenes ?6–10 mm. long, narrowly-cylindric, obscurely many-ribbed, densely appressed hispid; pappus of copious barbellate setae to c. 15 mm. long.

Zambia. B: Sesheke, i.1925, *Borle* (PRE).
Endemic, as far as is known at present, to Barotseland in Western Province of Zambia. Ecology not known.

7. **Pleiotaxis argentea** M.R.F. Taylor in Bull. Misc. Inf., Kew **1940**: 61 (1940). —C. Jeffrey in Kew Bull. **21**: 197 (1967). —Wild in Kirkia **8**: 199 (1972). —Lisowski, Aster. Fl. Afr. Centr. [in Fragm. Flor. Geobot. **36**, Suppl. 1]: 518 (1991). TAB. **3** fig. B. Type: Zambia, Lunzua Escarpment, *Burtt* 6437 (BM; K, holotype).

A lax perennial herb or subshrub to 0.5–2.5 m. tall, from a woody rootstock. Stems 1–many, sericeous-lanate, simple and leafless in the lower two thirds, becoming woody; ± diffusely branched and leafy in the upper part, with leaves clustered towards the ends of branches. Leaves discolorous, markedly bullate, briefly stem-sheathing at the base, the midribs often somewhat recurved; lamina mostly 3–10 × 1–4 cm., ovate to oblong-ovate, acute at the apex, broadly cuneate below, tapering into a narrowly winged petiole-like midrib to c. 5 mm. long, before widening into a short membranous sericeous-lanate sheath ± as long as wide, or lamina base somewhat rounded and subpetiolate; margins sharply serrulate; upper surface bullate, silvery- or greyish-green thinly lanate-pubescent, sometimes sparsely so; lower surface densely greyish or brownish araneose-sericeous; main nerves 7–11 on each side of the midrib, tertiary venation reticulate, nerves and veins ± impressed on upper surface, prominent beneath. Capitula numerous, usually terminal on branches, with 2–7 subsidiary capitula in axils below, on short or long, stout stalks. Involucres mostly 15–20 × 8–20 mm., spreading to c. 30 mm. wide, cylindric-campanulate. Phyllaries usually purple, chartaceous, araneose-tomentellous outside, margins glabrous, broadly hyaline, from 2 mm. long and ovate outside, increasing to c. 20 mm. long and becoming narrowly lanceolate-lorate inside; middle phyllaries up to c. 4 mm. wide, ovate, obtuse to ± rounded; inner phyllaries narrower. Florets numerous; corollas bright- or deep-red, up to c. 18 mm. long with a long slender tube abruptly dilated into a short cylindric, deeply narrowly-lobed limb which exceeds the pappus and involucre, glabrous. Achenes 6–9 mm. long, narrowly cylindric, obscurely many-ribbed, densely appressed hispid; pappus of copious barbellate setae to c. 14 mm. long.

Zambia. N: Namkungu Ridge, c. 16 km. E. of Kasama, 29.iv.1986, *Philcox, Pope & Chisumpa* 10201 (BR; K; LISC; NDO; SRGH).
Also in Zaire. Escarpment miombo, deciduous woodland on rocky hillsides and rocky outcrops.

8. **Pleiotaxis fulva** Hiern, Cat. Afr. Pl. Welw. **1**, 3: 611 (1898). —De Wild., Pl. Bequaert. **5**: 155 (1929). —Mendonça, Contrib. Conhec. Fl. Angol., 1 Compositae: 138 (1943). —C. Jeffrey in Kew Bull. **21**: 197 (1967). —Wild in Kirkia **8**: 199 (1972). —Lisowski, Aster. Fl. Afr. Centr. [in Fragm. Flor. Geobot. **36**, Suppl. 1]: 519 (1991). Type from Angola.
Pleiotaxis arenaria Milne-Redhead in Bull. Misc. Inf., Kew **1937**: 424 (1937). Type: Zambia, Mwinilunga, *Milne-Redhead* 970 (K, holotype).

An erect tomentose suffrutex, 30–90 cm. tall, from a woody rootstock. Stems annual, becoming woody at the base, 1–several, simple below and sparingly branched near the apex, leafy throughout, tawny sericeous-lanate in upper part, whitish araneose-lanate below. Leaves discolorous, markedly bullate, briefly stem-sheathing at the base, mostly 3–6 × 1–2 cm., oblong-lanceolate, obtuse at apex, subauriculate below, from a membranous sericeous-lanate stem-sheath ± as long as wide; margins erose-serrulate; upper surface grey-green, thinly lanate-pubescent, under surface densely tawny or greyish araneose-sericeous; main nerves 11–16 on each side of the midrib, tertiary venation reticulate, venation impressed on upper surface and prominent beneath. Capitula many, 1–2 on each branch, subsidiary heads at first subsessile in leaf or bract axils, stalks to c. 6 cm. long. Involucres mostly 12–15 × 7–10 mm., spreading to c. 15 mm. wide, narrowly cylindric-campanulate. Phyllaries appressed imbricate, chartaceous, margins purple-tinged hyaline, obtuse to attenuate-acute at the apex, tawny araneose-lanate, the indumentum usually obscuring the surface, from c. 1.5 mm. long and narrowly triangular outside, increasing to c. 14 mm. long becoming very narrowly lanceolate inside; middle phyllaries up to c. 2.5 mm. wide, lanceolate. Florets numerous; corollas bright- to deep-red, up to c. 16 mm. long with a long slender tube abruptly dilated into a short cylindric deeply-lobed limb which exceeds the pappus and involucre, glabrous; corolla lobes linear, eventually recurved or spreading to expose the anther tube. Achenes c. 5 mm. long (immature), narrowly subfusiform-cylindric, stramineous strigose-hispid; pappus of copious barbellate setae 9–11 mm. long.

Zambia. W: c. 27 km. from Mwinilunga on Kabompo Road, 6.vi.1963, *Edwards* 646 (K; M; SRGH).
Also in Angola and Zaire. *Brachystegia-Isoberlinia* woodland, dry evergreen *Cryptosepalum* forest on sand, amongst understorey grasses, and also in grassland fringing dambos.

9. **Pleiotaxis oxylepis** C. Jeffrey in Kew Bull. **21**: 197, fig. 8 (1967). —Wild in Kirkia **8**: 201 (1972). Type: Zambia, Mbala Distr., Kalambo Falls, *Exell, Mendonça & Wild* 1283 (BM; LISC; SRGH, holotype).

An erect suffrutex to c. 80 cm. tall, from a woody rootstock. Stems annual, usually solitary, simple, leafy except for the apical part, somewhat compressed-angular, thinly araneose-lanate. Leaves concolorous, membranous, sessile, not stem-sheathing at the base, mostly 11–25 × 2–9 cm., oblong-elliptic, acute at the apex, cuneate or rounded subauriculate below, ± stem-clasping, margin coarsely sharply serrate; very sparsely finely

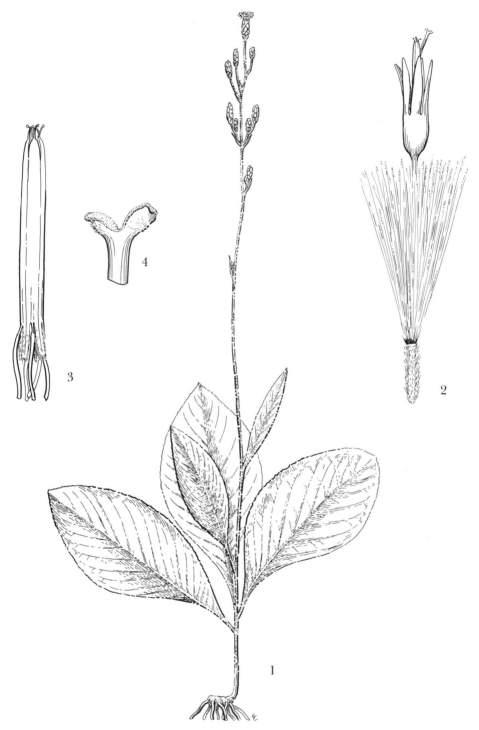

Tab. 4. PLEIOTAXIS RACEMOSA. 1, habit (× ⅓); 2, floret (× 4); 3, anther tube, note stipitate apices and tailed bases (× 10); 4, style apex (× 20), 1–4 from *Richards* 19710. Drawn by Eleanor Catherine.

araneose-lanate on both surfaces, or glabrescent; main nerves 9–17 on each side of the midrib, prominent beneath, veins usually faintly loosely reticulate. Synflorescence cylindric subpaniculate, 11–25-capitulate; capitula 1–4 on short branches from bract axils, capitula stalks 1–50 mm. long. Involucres mostly 15–23 × 4–7 mm., hardly spreading, narrowly cylindric. Phyllaries dark-purple, appressed imbricate, rarely extending onto the stalk, scarious-membranous, margins obscurely narrowly hyaline, sparsely araneose soon glabrescent, from c. 3 mm. long and triangular-ovate outside, increasing to c. 22 mm. long becoming linear inside; middle phyllaries up to c. 4 mm. wide and oblong-ovate, inner phyllaries narrower. Florets many; corollas yellowish-white, up to c. 20 mm. long with a long slender tube abruptly dilated into a short cylindric deeply-lobed limb which exceeds the pappus and involucre, glabrous; corolla lobes c. 4 mm. long, linear. Achenes to c. 8 mm. long, narrowly cylindric, densely appressed hispid; pappus of copious fine barbellate setae c. 13 mm. long.

Zambia. N: Mbala Distr., Kalambo Falls, c. 900 m., 15.ii.1964, *Richards* 19023 (K; M; SRGH).
Also in Tanzania, apparently endemic in this area. Miombo woodland, often on steep slopes.

10. **Pleiotaxis racemosa** O. Hoffm. in Engl., Bot. Jahrb. **15**: 540 (1893); in Engl., Pflanzenw. Ost-Afr. **C**: 420 (1895). —De Wild., Pl. Bequaert. **5**: 159 (1929). —C. Jeffrey in Kew Bull. **21**: 199 (1967). —Wild in Kirkia **8**: 201 (1972). TAB. **4**. Type from Tanzania.

An erect often subscapose suffrutex 50–80 cm. tall, from a woody rootstock with many thong-like roots. Stems annual, usually solitary, simple, ± densely leafy towards the base, leafless above, somewhat compressed-angular, araneose-lanate. Leaves concolorous, chartaceous-membranous, not stem-sheathing at the base; mostly 6–27 × 3.5–10 cm., broadly elliptic to broadly oblong-obovate, rounded to obtuse at the apex, sometimes subacute-apiculate, broadly cuneate below ± narrowing into a winged petiole-like midrib 0–30 mm. long, narrowly semi-amplexicaul sometimes subauriculate at the base; margins serrate to coarsely irregularly serrate, sometimes subentire; sparsely finely araneose-lanate on both surfaces, sometimes with scattered patent hairs on the upper surface; main nerves 10–12 on each side of the midrib. Capitula 3–31 in a racemose to subpaniculate arrangement, 1–5 on short racemose branches from bract axils, capitula stalks stout 1–50 mm. long. Involucres mostly 13–20 × 5–12 mm., hardly spreading, narrowly cylindric. Phyllaries dark-purple sometimes stramineous, appressed imbricate, rounded obtuse, scarious-membranous, margins broadly hyaline ± torn, glabrous, from c. 2 mm. long and ovate outside, increasing to c. 19 mm. long becoming linear inside; middle phyllaries up to c. 6 mm. wide and oblong-ovate; inner phyllaries narrower. Florets many; corollas yellowish-white, up to c. 19 mm. long with a long slender tube abruptly dilated into a short cylindric deeply-lobed limb which exceeds the pappus and involucre, glabrous; corolla lobes c. 4 mm. long, linear. Achenes to c. 7 mm. long, subfusiform-cylindric, obscurely ribbed, densely appressed hispid; pappus of copious fine barbellate setae c. 13 mm. long.

Zambia. N: Mbala Distr., Mpulungu, c. 1432 m., 10.iv.1961, *Phipps & Vesey-FitzGerald* 3012 (K; LISC; M; S; SRGH).
Also in Tanzania. Open escarpment woodland or mixed deciduous woodlands, and short open or wooded grasslands, mostly on rocky hillsides.

4. ERYTHROCEPHALUM Benth.

Erythrocephalum Benth. in Benth. & Hook. f., Gen. Pl. **2**: 488 (1873). —C. Jeffrey in Kew Bull. **21**, 2: 201 (1967). —Wild in Kirkia **8**: 173 (1972).

Annual herbs, or suffrutices with annual stems from woody rootstocks. Vegetative indumentum of appressed fine ± matted hairs usually intermixed with large patent many-celled uniseriate hairs each bearing a fine filamentous terminal cell. Stems 1– several, simple or branching near the apex, araneose-lanate to coarsely pubescent. Leaves alternate, semi-amplexicaul, usually discolorous, indumentum as for stem but denser on the lower surface, sometimes ± scabrid. Capitula homogamous and discoid or heterogamous with inner florets regular and outer florets bilabiate-radiate, 1–many, solitary and terminal on stem and branches. Involucres ± broadly campanulate, ± truncate at the base. Phyllaries many-seriate, appressed imbricate, the innermost grading into the paleae of the common receptacle, fimbriate-denticulate on the margins or sometimes denticulate about the apex only; apices sometimes long caudate, or in inner phyllaries expanded ovate and appendage-like. Receptacle paleate. Ray-florets when present functionally male, with deep-red bilabiate-radiate corollas, outer lip (ray) large 3-fid, inner lip shorter consisting of 2 linear lobes. Disk-florets deep-red or creamy-white, occasionally orange-tinged, hermaphrodite and actinomorphic, abruptly dilated into a deeply lobed limb, lobes erect linear. Anther bases produced into oblong fimbriate or ciliolate tails. Styles deeply 2-fid, branch apices shortly conical with a sub-distal fringe of hairs. Achenes doliform-cylindric, c. 5-angled or ribbed, usually with a ± swollen or lobed cartilaginous carpopodium at the base, minutely puberulous or glabrous; pappus of 4–5 caducous barbellate setae.

A tropical African genus of about 12 species.

1. Capitula radiate; rays large red - - - - - - - - - - - 2
 - Capitula discoid; corollas creamy-white or deep-red - - - - - - - 3
2. Leaves up to 12 mm. wide and more than 11.5 cm. long, linear or linear-lanceolate, more than 9 times longer than wide - - - - - - - - - - - 2. *decipiens*
 - Leaves more than 12 mm. wide, elliptic to narrowly lanceolate-elliptic, or broadly elliptic, up to 9 times as long as wide - - - - - - - - - - - 1. *zambesianum*
3. Plants annual; receptacular paleae densely persistently araneose outside; involucres 7–9 mm. long; leaves thinly chartaceous - - - - - - - - - 6. *microcephalum*
 - Plants perennial with numerous spreading stout roots from a woody rootstock; receptacular paleae glabrescent outside; involucres more than 9 mm. long; leaves not thinly chartaceous 4
4. Corollas deep-red; leaves ± concolorous, sparsely finely appressed araneose on both surfaces, reticulate nervation clearly visible and subprominent on both surfaces 3. *dictyophlebium*
 - Corollas creamy-white; leaves discolorous, upper surface sparsely appressed araneose or ± scabrid, lower surface densely appressed araneose-lanate, reticulate nervation not clearly visible - - - - - - - - - - - - - - - - - 5
5. Leaf bases ± auriculate; capitula subtended and overtopped by 1–4 leaves or large leaf-like bracts; leaf upper surface scabrid or scabridulous - - - - - 4. *scabrifolium*
 - Leaf bases narrowly tapering or narrowly rounded semi-amplexicaul; capitula without subtending leaf-like bracts, or if bracts present then small and ± equalling the capitulum in length; leaf upper surface seldom scabrid - - - - - 5. *albiflorum*

1. **Erythrocephalum zambesianum** Oliv. & Hiern in F.T.A. **3**: 441 (1887). —O. Hoffm. in Engl., Pflanzenw. Ost-Afr. **C**: 420 (1895). —S. Moore in Journ. Linn. Soc., Bot. **40**: 123 (1911). —Eyles in Trans. Roy. Soc. S. Afr. **5**: 522 (1916). —C. Jeffrey in Kew Bull. **21**: 203 (1967). —Wild in Kirkia **8**: 174 (1972). —Lisowski, Aster. Fl. Afr. Centr. [in Fragm. Flor. Geobot. **36**, Suppl. 1]: 529 (1991). TAB. **5**. Syntypes: Malawi, Shire R., *Waller* (K); Malawi, Manganja Hills, *Kirk* s.n. (K).
 Erythrocephalum zambesianum var. *angustifolium* S. Moore in Journ. Linn. Soc., Bot. **35** (1904). Type: Malawi, *Simons* (BM, holotype).
 Erythrocephalum castellaneum Busc. & Muschl. in Engl., Bot. Jahrb. **49**: 509 (1913). Type: Zambia, Luapula, *Aosta* 1040 (B†).
 Erythrocephalum helenae Busc. & Muschl. in Engl., Bot. Jahrb. **49**: 508 (1913). Type: Zambia, between Kabwe and Bwana Mkubwa, *Aosta* 191 (B†).
 Erythrocephalum aostae Busc. & Muschl. in Engl., Bot. Jahrb. **49**: 508 (1913). Type: Mozambique, *Aosta* 6 (B†).
 Erythrocephalum bicolor Merxm., Mitt. Bot. Staatss. Münch., **1**, 6: 207 (1953). Type: Zimbabwe, Rusape, *Dehn* S.43/52 (M, holotype).
 Erythrocephalum niassae Wild in Kirkia **8**: 169; 176 (1972). Type: Mozambique, Niassa, Lichinga (Vila Cabral), *Torre* 8 (COI, holotype; LISC).
 Erythrocephalum longifolium Benth. ex Oliv. in Trans. Linn. Soc. Lond. **29**: 102 (1873). —C. Jeffrey in Kew Bull. **21**: 203 (1967). —Lisowski, Aster. Fl. Afr. Centr. [in Fragm. Flor. Geobot. **36**, Suppl. 1]: 530 (1991). Type from Tanzania.
 Erythrocephalum nutans Benth. ex Oliv. in Trans. Linn. Soc. Lond. **29**: 102 (1873). Type from Tanzania.

An erect suffrutex 12–80 cm. tall, from a woody rootstock with numerous spreading thong-like roots. Stems annual, usually single, simple or branching above, somewhat zigzag above, leaves ± crowded in upper part, whitish-grey araneose-lanate with long fine matted-appressed hairs usually interspersed with few to numerous purplish patent larger hairs. Leaves discolorous, sessile, mostly 7–22 × 1.5–7.5 cm., elliptic to lanceolate, ± tapering-acute to the apex, semi-amplexicaul and subauriculate below, margins sharply serrulate often bi-serrulate; upper surface green, thinly araneose, glabrescent; lower surface persistently appressed greyish araneose-lanate, sometimes with scattered large brownish hairs on both surfaces. Capitula solitary and terminal on the stem, or 2–6

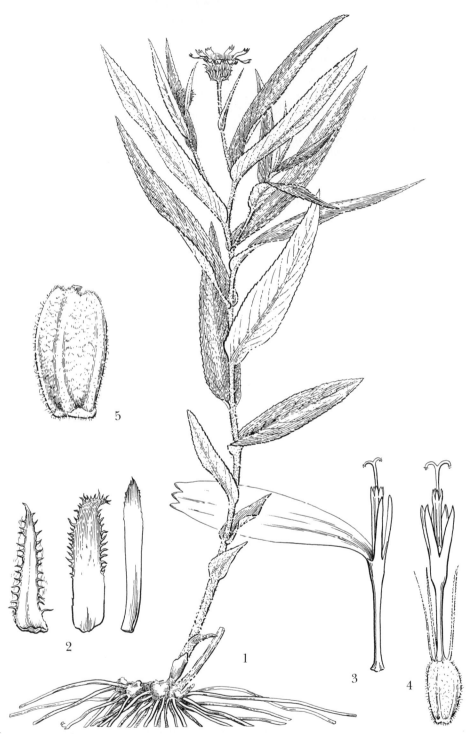

Tab. 5. ERYTHROCEPHALUM ZAMBESIANUM. 1, habit (× ⅓); 2, phyllaries, outer middle and
inner (× 3), 1 & 2 from *Robson* 1037; 3, outer floret, bilabiate with strap-shaped outer lip and
2-lobed inner lip (× 3); 4, inner floret, regular (× 3), 3 & 4 from *Brummitt & Banda* 8415; 5, achene
(× 6), from *Brummitt* 8448. Drawn by Eleanor Catherine.

terminal on branches. Involucres mostly 10–22 × 12–35 mm., broadly campanulate, truncate at the base, araneose-lanate to glabrescent. Outer phyllaries 6–30(38) mm. long, lanceolate, fimbriate-denticulate on the margins, attenuate above or with a caudate appendage-like apical portion 3–22 mm. long, this sometimes elongating to c. 30 mm. long and then greatly exceeding inner phyllaries; inner phyllaries up to c. 25 mm. long, tapering-lorate, grading into receptacular paleae; phyllary margins narrowly membranous often purple, fimbriate-denticulate with linear teeth 0.5–3 mm. long, sometimes inner phyllary apices expanded. Receptacular paleae 10–15 mm. long, triangular-lorate, reddish-purple-tipped and denticulate about the apex. Ray-florets deep-red, 14–30 mm. long, with exserted stamens and style, ovary rudimentary; rays 9–22 mm. long widening to apex, 2.5–9 mm. at broadest part, 3-fid with lobes c. 2 mm. long, inner lip of 2 linear lobes 4.5–6 mm. long; pappus of a few caducous setae 4–5 mm. long. Disk-florets deep-red, 11–14 mm. long, narrowly tubular below with a deeply lobed campanulate limb 5–6.5 mm. long, lobes 4–5 mm. long linear, ovary fertile. Achenes 5–6 × 2.5–3 mm., angular-barrel-shaped, 5-ribbed, minutely puberulous, carpopodium swollen; pappus of caducous barbellate setae 6–9 mm. long.

Zambia. N: Kawimbe Rocks, c. 1740 m., 17.i.1964, *Richards* 18823 (K; M; MO; SRGH). C: Kabwe, 6.ii.1965, *Fanshawe* 9181 (K; NDO; SRGH). E: Kapatamoyo, near Chipata, c. 1200 m., 5.i.1959, *Robson* 1037 (BM; K; LISC; PRE; SRGH). **Zimbabwe**. C: Makoni Distr., c. 15 km. W. of Rusape, 12.ii.1974, *Pope* 1167 (K; SRGH). E: Mutare Distr., Kukwanisa, c. 1492 m., 5.i.1966, *Biegel* 1641 (K; SRGH). **Malawi**. N: Mzimba Distr., Mzuzu, Marymount, c. 1371 m., 31.i.1969, *Pawek* 1671 (K). C: Kasungu Distr., Chimaliro Forest, Phaso Road, 10.i.1975, *Pawek* 8877 (K; MAL; MO). S: Matenje Road, c. 2 km. N. of Limbe, c. 1190 m., 7.ii.1970, *Brummitt* 8448 (K; LISC; MAL; PRE; SRGH). **Mozambique**. N: near Mandimba, 18.xii.1941, *Hornby* 3521 (K). Z: Quelimane Distr., Munguluni Mission, ii.1949, *Faulkner in Kew* 110 (COI; K). T: Moatize, 11.i.1966, *Correia* 368 (LISC). MS: Chimoio, Gondola, 24.i.1948, *Mendonça* 3693 (LISC).

Also in Tanzania and Zaire. High rainfall miombo woodland, *Brachystegia/Uapaca* woodland, and grassland. Often in sandy soil or rocky hillsides.

In addition to typical *Erythrocephalum zambesianum*, with leaves elliptic and 7–12 × 2–3(4) cm., 4 variants based on leaf shape, size and indumentum can be recognised:

Variant 1, from the Lichinga Plateau in northern Mozambique, has leaves more narrowly lanceolate than those of typical *Erythrocephalum zambesianum* from further south. These leaves often have somewhat revolute margins and an impressed nervation on the upper surface which gives them a slightly bullate appearance (especially towards the apex, and at least when dry). Material of this form was described as *Erythrocephalum niassae* by Wild. However, no satisfactory discontinuity in variation in these or other features is seen and the two taxa cannot be kept separate at the species level. Examples of specimens intermediate between the taxa are; in Zambia, *Brummitt et al.* 16968 (K) from the Kundalila Falls, *Hooper & Townsend* 645 (K) from Mkushi and *Wright* 288 (K) from Chipata; in Malawi, *Lawrence* 538 (K) from Soche Mt. (Blantyre).

Variant 2, occurs in central and eastern Zambia and adjacent parts of Malawi. It is a large-leaved form with leaves up to c. 20 × 5 cm. and oblong-lanceolate in shape. Examples of this are; in Zambia, *Fanshawe* 1816 (K) and *Robson* 1037 (K) from Kapiri Mposhi and Chipata respectively; and in Malawi, *Pawek* 8877 (K) and *Richards* 10544 (K) from Chimaliro Forest (Kasungu) and Kaziwizwi River (Rumphi) respectively.

Variant 3, from the Isoka–Mbala area of northern Zambia, represented by *Richards* 216 & 17518 (K), and adjacent areas in Tanzania, is a large-leaved somewhat robust plant. Leaves are elliptic, up to c. 20 × 7 cm., with sharply irregularly serrate-dentate margins and involucres up to c. 3.5 cm. in diam.

Variant 4 is characterised by all the Zimbabwean material seen so far, as well as some specimens from adjacent Mozambique. It has leaves which are persistently thinly araneose on the upper surface and which tend to be broader and more elliptic-lanceolate in shape than typical *E. zambesianum*. Similar material is found in the Iringa and Mbeya Districts of Tanzania.

2. **Erythrocephalum decipiens** C. Jeffrey in Kew Bull. **21**: 201, fig. 9 (1967). —Wild in Kirkia **8**: 175 (1972). —Lisowski, Aster. Fl. Afr. Centr. [in Fragm. Flor. Geobot. **36**, Suppl. 1]: 528 (1991). Type from southern Tanzania.

An erect slender suffrutex 14–60 cm. tall, from a woody rootstock with numerous spreading thong-like roots. Stems annual, usually single, simple, or branching above, leafy with leaves ± crowded in the upper part, whitish-grey araneose-lanate with fine matted-appressed hairs. Leaves discolorous, sessile, mostly 7–15 × 0.4–1.5(2.5) cm., linear-elliptic, tapering-acute to the apex, narrowly cuneate and semi-amplexicaul below, margins minutely sharply serrulate ± revolute; upper surface green, glabrous, lower surface appressed greyish araneose-lanate. Capitula solitary and terminal on the stem, less often 2–5 terminal on branches. Involucres mostly 10–14 × 9–15 mm., broadly campanulate, truncate at the base, araneose-lanate to glabrescent. Outer phyllaries 6–16 mm. long, lanceolate-lorate, fimbriate-denticulate on the margins, acute or with a narrow attenuate or caudate appendage-like apical portion 2–9 mm. long; inner phyllaries up to c. 14 mm. long, tapering-lorate, grading into similar but shorter receptacular paleae; margins narrowly membranous, often purple and fimbriate-denticulate above, with linear teeth to c. 1 mm. long, sometimes apex expanded-ovate and appendage-like. Receptacular paleae flat, to c. 12 mm. long, lorate, reddish-purple-tipped and denticulate about the apex. Ray-florets deep-red, up to c. 17 mm. long, with exserted stamens and style but rudimentary ovary; rays up to c. 12 mm. long, widening to apex, c. 2.5 mm. at broadest part, 3-fid with lobes 0.5–2(4) mm. long, inner lip composed of 2 linear lobes c. 7 mm. long, pappus of a few caducous setae 5–6 mm. long. Disk-florets deep-red, 8–14 mm. long, narrowly tubular below with a deeply lobed campanulate limb c. 5 mm. long, lobes c. 4 mm. long linear, ovary fertile. Achenes 5–7 × 2 mm., subcylindric, narrowly 5-ribbed, minutely puberulous; carpopodium swollen, exceeding the ribs in thickness; pappus of caducous barbellate setae c. 8 mm. long.

Mozambique. N: Niassa Distr., Marrupa, 28.i.1981, *Nuvunga* 415 (K; LMU).
Also in Tanzania and Zaire. Miombo woodland and open grassland.
This species approaches the Angolan plant *E. foliosum* (Klatt) O. Hoffm. in its linear leaves, but the latter has discoid capitula and shorter, wider achenes.

3. **Erythrocephalum dictyophlebium** Wild in Kirkia 8: 170; 178 (1972). Type: Zambia, Kawambwa-Mambilima Falls (Johnson Falls) road, *Richards* 15502 (BM; K, holotype; M).

An erect suffrutex to c. 40 cm. tall, from a woody rootstock with numerous spreading, thong-like roots. Stems annual, 1–several, simple or branching above, leafy to the apex, coarsely pilose-pubescent, the indumentum a mixture of fine matted-appressed whitish hairs and numerous large brownish patent hairs. Leaves concolorous, sessile, mostly 6–11 × 1.5–4 cm., elliptic, acute at the apex, broadly cuneate to narrowly rounded below, semi-amplexicaul, sparsely denticulate to subentire on the margin, green and sparsely finely appressed white araneose on both surfaces; nervation clearly visible, subprominent and reticulate on both surfaces. Capitula solitary and terminal on stem and apical branches. Involucres mostly c. 9 × 12 mm., increasing to c. 15 mm. wide when mature, becoming broadly campanulate and truncate at the base, araneose-lanate. Phyllaries 8–10 mm. long, tapering lorate; margins reddish-purple and somewhat membranous, ± expanded and fimbriate-denticulate about the apex; inner phyllaries grading into the receptacular paleae. Ray-florets absent. Disk-floret corollas deep-red, mostly c. 14 mm. long, narrowly tubular below and abruptly dilated into a subcylindric-campanulate, deeply lobed limb c. 5 mm. long, lobes c. 4 mm. long, linear. Achenes (immature) subcylindric, minutely puberulous, pappus of caducous barbellate setae c. 7 mm. long.

Zambia. N: Kawambwa-Mambilima Falls (Johnston Falls) road, c. 900 m., 4.xii.1961, *Richards* 15502 (K; M).
So far recorded only from this locality. Grassland.

4. **Erythrocephalum scabrifolium** C. Jeffrey in Kew Bull. 21: 206, fig. 11 (1967). —Wild in Kirkia 8: 177 (1972). —Lisowski, Aster. Fl. Afr. Centr. [in Fragm. Flor. Geobot. 36, Suppl. 1]: 533 (1991). Type: Zambia, Mbala Distr., road to Iyendwe Valley from Kambole, *Richards* 10814 (K, holotype).

An erect robust suffrutex to c. 60 cm. tall, from a woody rootstock with numerous spreading, thong-like roots. Stems annual, 1–several, simple or branching above, leafy to the apex, leaves diminishing in size and number and drying early on the lower stem, coarsely pilose-pubescent. Leaves discolorous, sessile, mostly 5–15 × 0.5–2.8 cm., linear-elliptic to narrowly oblong-lanceolate or narrowly panduriform-lanceolate, attenuate-acute or shortly tapering obtuse to the apex, the lamina narrowing slightly from below the middle before expanding briefly into a semi-amplexicaul auriculate base, margins serrulate to subentire; upper surface dark-green and scabrous or scabridulous-araneose, the indumentum a mixture of stiff patent hairs and sparse appressed fine white araneose hairs; lower surface densely brownish, araneose-lanate, the indumentum of appressed fine matted hairs interspersed with few to numerous patent-pilose hairs. Capitula 1–7, solitary and terminal on the stem and short apical branches, apparently subsessile being subtended by 1–4 large leaf-like bracts up to c. 5 cm. long which overtop the capitula. Involucres mostly 10–15 × 7–12 mm., spreading to c. 25 mm. wide when mature, becoming broadly campanulate and truncate at the base, araneose-lanate. Outer phyllaries 5–11 mm. long including an attenuate or caudate-aristate apex 2–6 mm. long, ovate-lanceolate or oblong with a caudate apical appendage, fimbriate-denticulate on the margins especially in the upper part; inner phyllaries up to c. 12 mm. long, tapering lorate, grading into the receptacular paleae, margins membranous, often purplish and fimbriate-denticulate about the apex, sometimes the inner phyllary apices expanded-ovate and appendage-like. Receptacular paleae similar to the inner phyllaries. Ray-florets absent. Disk-floret corollas creamy-white, mostly 8–12 mm. long, narrowly tubular below, abruptly dilated into a subcylindric-campanulate deeply lobed limb 5–6 mm. long, limb somewhat swollen near the base, lobes 3.5–5 mm. long, linear. Achenes 5–7 × 2–3.5 mm., subcylindric, 5-angled, obscurely ribbed on the angles, minutely puberulous, carpopodium swollen; pappus of caducous scale-like barbellate setae 6–8 mm. long.

Zambia. N: c. 20 km. from Chinsali on Kasama road, 13.iv.1986, *Philcox, Pope & Chisumpa* 9920 (BR; K; LISC; SRGH). W: Kitwe, 16.ii.1967, *Mutimushi* 1824 (K; NDO). Malawi. N: Chitipa Distr., Nthalire, 10°19'S, 33°38'E, 23.v.1989, *Pope, A. Radcl.-Sm. & Goyder* 2320 (BR; K; LISC; MAL).
Also in Tanzania, Zaire and Angola. High rainfall miombo woodland.
Material from the Zambian Copperbelt (*Fanshawe* 778 and *Mutimushi* 1824, 2416, all K) with leaves shortly tapering subobtuse, araneose-scabridulous on the upper surface and obscurely serrulate probably represents a separate infra-specific taxon.

5. **Erythrocephalum albiflorum** Wild in Kirkia **8**: 170; 177 (1972). Type: Zambia, Solwezi, Mulenga Protected Forest Area, *Drummond & Rutherford-Smith* 7034 (K; LISC, isotypes; SRGH, holotype).

An erect suffrutex to c. 60 cm. tall, from a woody rootstock with numerous spreading, thong-like roots. Stems annual, 1–several, simple or branching above, leafy to the apex with leaves diminishing in size and number below, somewhat stem-sheathing on the lower stem, araneose-lanate sometimes becoming coarsely pilose-pubescent; the indumentum of fine matted-appressed whitish hairs sometimes intermixed with few to numerous large brownish patent hairs. Leaves discolorous, sessile, mostly 5–14 × 0.5–25 cm., linear-elliptic to narrowly oblong-lanceolate, attenuate acute to the apex, the lower part cuneate tapering to a semi-amplexicaul base, margins slightly revolute subentire; upper surface dark-green and sparsely araneose-lanate with appressed fine white hairs or sometimes these intermixed with stiff patent hairs and the indumentum scabridulous-araneose; lower surface densely greyish, sometimes brownish, araneose-lanate. Capitula 1–5, solitary and terminal on the stem and short apical branches, subtending leaf-like bracts wanting or small and ± equalling the capitula in length. Involucres mostly 11–15 × 10–20 mm., increasing to c. 25 mm. wide when mature, becoming broadly campanulate and truncate at the base, araneose-lanate. Outer phyllaries 5–11 mm. long, ovate or linear-lanceolate, acute or with a caudate appendage-like apical portion elongating to c. 20 mm. long, fimbriate-denticulate on the margins especially in the upper part; inner phyllaries up to c. 16 mm. long, tapering lorate, grading into the receptacular paleae, margins membranous often purplish and fimbriate-denticulate about the apex, the apex sometimes expanded and appendage-like. Receptacular paleae similar to the inner phyllaries. Ray-florets absent. Disk-floret corollas creamy-white, mostly c. 11 mm. long, narrowly tubular below and abruptly dilated into a subcylindric-campanulate, deeply lobed limb c. 5 mm. long, lobes c. 4 mm. long, linear. Achenes c. 5 mm. long, subcylindric, 5-angled, obscurely ribbed on the angles, minutely puberulous; carpopodium thin; pappus of caducous linear scale-like barbellate setae 6–8 mm. long.

Zambia. W: Chingola, 18.i.1956, *Fanshawe* 2747 (K; NDO; SRGH). Malawi. ?S: *Buchanan* 1269 (K).
Known so far only from Zambia W and ? Malawi S. *Brachystegia* and mixed deciduous woodland.

6. **Erythrocephalum microcephalum** Dandy in Bull. Misc. Inf., Kew **1927**: 75 (1927). —C. Jeffrey in Kew Bull. **21**: 206 (1967). Type from Tanzania.

An erect slender to bushy annual herb to c. 70 cm. tall. Stems sparsely to much-branched, leafy, araneose-lanate or often sparsely so; indumentum a mixture of appressed fine whitish hairs and scattered large brownish patent hairs. Leaves discolorous, sessile, mostly 4–10 × 0.8–2.5 cm., narrowly elliptic-lanceolate, subacute at the apex, broadly cuneate to narrowly rounded sub-cordate below, semi-amplexicaul or sometimes narrowing slightly from below the middle before expanding into a subauriculate base; margins serrulate to subentire; upper surface dark-green, thinly araneose; lower surface appressed greyish araneose-lanate on both surfaces, often with scattered large brownish many-celled simple hairs. Capitula 3–many, solitary and terminal on leafy branches, stalks to c. 8 cm. long, ebracteate. Involucres mostly 7–9 ×

9–12 mm., increasing to c. 15 mm. wide, broadly campanulate, truncate at the base, densely araneose-lanate. Outer phyllaries from c. 4 mm. long, very narrowly triangular, spaced denticulate-pectinate on the margins; inner phyllaries to c. 7 mm. long, broadly triangular-ovate with membranous margins fimbriate denticulate about the apex. Receptacular paleae similar to the inner phyllaries, araneose-lanate outside. Ray-florets absent. Disk-floret corollas pale-yellowish to orange-coloured, mostly 5–7 mm. long, narrowly tubular below and abruptly dilated into a subcylindric-campanulate, deeply lobed limb, 2–3 mm. long, lobes c. 2 mm. long, linear. Achenes to c. 5 mm. long, c. 2 mm. in diam., cylindric, obscurely 5-ribbed, glabrous sometimes sparsely finely puberulous at first, irregularly minutely pitted, carpopodium 5-lobed with a lobe at the base of each rib; pappus of caducous barbellate setae 4–5 mm. long.

Zimbabwe. W: Hwange Nat. Park, c. 6 km. W. of Robins Camp, Little Toms River, c. 960 m., 28.xi.1974, *Rushworth* 2585 (K; SRGH).
Also in Uganda, Kenya and Tanzania. Grassland on basaltic (black cotton) clay soils.

5. PASACCARDOA Kuntze

Pasaccardoa Kuntze, Rev. Gen. Pl. **1**: 354 (1891). —C. Jeffrey in Kew Bull. **21**: 208 (1967).
—Wild in Kirkia **8**: 192 (1972).
Phyllactinia Benth. in Benth. & Hook. f., Gen. Pl. **2**: 488 (1873), non J.H. Lev. (1851), nom. illegit.

Annual herbs, or suffrutices with annual stems from woody rootstocks. Vegetative indumentum of fine flagelliform hairs each consisting of an appressed filamentous terminal cell borne on a uniseriate many-celled stalk. Leaves alternate, narrowly oblanceolate to elliptic, somewhat discolorous, densely araneose-lanate on the under surface, gland-dotted. Capitula heterogamous, radiate, few–many, solitary and terminal on branches sometimes also subsessile along branches, subtended by leaf-like bracts. Phyllaries numerous, many-seriate, araneose-lanate to puberulous, at least the outer subulate-aristate with stiff squarrose or recurved apices, or lamina straight and pungent. Receptacle deeply alveolate, paleae absent. Ray-florets neuter, stamens and style absent or rudimentary, corollas deep-red or creamy-white; ray erect, 3-fid, inner lip wanting; ovary rudimentary, pappus of many narrow scales. Disk-florets deep-red or creamy-white, hermaphrodite and actinomorphic. Anther apical appendages narrowly triangular, theca bases sagittate with tapering-acute fimbriate or ciliolate tails. Style deeply 2-fid, apices subobtuse with a subdistal fringe of hairs. Achenes subcylindric, 10-ribbed with numerous stiff white hairs; pappus several-seriate of numerous spreading narrowly lanceolate scales.

A tropical African genus of c. 4 species.

1. Plants annual; achenes 6–8 mm. long; capitula terminal and often also subsessile along the branches; capitula stalks shorter than the involucre - - - - - - 3. *grantii*
- Plants perennial; achenes 5–6 mm. long; capitula terminal only, on branches or stalks longer than the involucre - - - - - - - - - - - - - - 2
2. Rays deep-red, 3.5–6 mm. wide, ± exserted; branches ascending; stems mostly erect 1. *jeffreyi*
- Rays yellowish or cream-coloured, 1.5–3 mm. wide, hardly exserted beyond the involucre; branches short, strongly divaricate; stems prostrate or decumbent - - 2. *procumbens*

1. **Pasaccardoa jeffreyi** Wild in Kirkia **8**: 168; 193 (1972). —Lisowski, Aster. Fl. Afr. Centr. [in Fragm. Flor. Geobot. **36**, Suppl. 1]: 538 (1991). Type from Angola.

An erect suffrutex, or one or more stems decumbent, from a woody rootstock. Stems annual, single or 2–4, 25–85 cm. tall, becoming woody below, branching above or sometimes throughout its length, leafy, angular-striate, thinly-lanate becoming puberulous; branches ascending. Leaves subsessile, mostly 3–10(13) × 0.5–3 cm., oblanceolate, obtuse or rounded at the apex and narrowly cuneate below, margins serrulate with minute callose teeth, upper surface thinly araneose-lanate soon roughly puberulous, lower surface greyish araneose-lanate, both surfaces densely gland-dotted. Capitula many, solitary and terminal on leafy short ascending branches, capitula subtended and usually exceeded by 1–several small leaf-like bracts. Involucre mostly 16–22(25) × 18–30 mm., spreading to c. 40 mm. wide, broadly campanulate to obconic.

Phyllaries numerous; the outer straight or ± squarrose, from c. 5 mm. long, pungent-subulate, aristate, araneose-lanate; the inner straight, up to c. 23 mm. long, linear-lanceolate, tapering to a sharply acute sometimes aristate apex, tomentellous-puberulous, margins sharply minutely toothed. Receptacle deeply alveolate with pits to c. 2 mm. deep, paleae absent. Ray-florets neuter; corollas deep-red, up to c. 18 mm. long including the ray, deeply cleft; ray erect, c. 17 × 3.5–6 mm. long, oblanceolate, 3-fid, sparsely glandular outside; ovary rudimentary 5–7 mm. long, slightly swollen in lower part; pappus of many chartaceous narrowly lanceolate scales 0.5–3 mm. long. Disk-florets deep-red, tubular in the lower 2–3 mm. and abruptly dilated into a deeply lobed cylindric to conic limb 5–7 mm. long; lobes to c. 6 mm. long, linear, recurved at apex, glandular outside; anther-tube red-purple, the conical tip exserted. Achenes 5–6 mm. long, subcylindric, 10-ribbed with numerous long stiff white hairs ± appressed in the grooves, those near the achene base shorter, more robust and spreading; pappus several-seriate of numerous spreading, narrowly lanceolate, chartaceous scales 0.5–3 mm. long.

Zambia. N: Mwinilunga Distr., Lisombo R. tributary, 14.vi.1963, *Loveridge* 984 (K; SRGH).
Also in Angola and Zaire. Miombo woodland and wooded grassland, on hillsides or on Kalahari Sand.
Pasaccardoa jeffreyi subsp. *kasaiensis* Lisowski, recorded for Angola and Zaire, is recognised by its distinctive capitula with strongly recurved phyllaries.

2. **Pasaccardoa procumbens** (Lisowski) G.V. Pope stat. nov. Type from Zaire.
    Pasaccardoa jeffreyi subsp. *procumbens* Lisowski in Bull. Jard. Bot. Nat. Belg. **57**: 258 (1987); Aster. Fl. Afr. Centr. [in Fragm. Flor. Geobot. **36**, Suppl. 1]: 540 (1991). Type as above.

A decumbent suffrutex from a woody rootstock. Stems annual, 1–several, up to c. 50 cm. long, trailing or some at first erect, branching towards the apex, leafy, thinly araneose-lanate to coarsely puberulous; branches to ?c. 15 cm. long, ± strongly divaricate particularly on procumbent stems. Leaves subsessile or with a petiole to c. 5 mm. long; lamina mostly 3–8 × 0.5–1.2 cm., ± narrowly elliptic, sometimes obovate-elliptic, acute to rounded at the apex and narrowly cuneate below, margin serrulate with minute callose teeth, upper surface sparsely thinly araneose-lanate to roughly puberulous, lower surface greyish araneose-lanate, both surfaces densely gland-dotted. Capitula solitary and terminal on divaricate, leafy branches, subtended by leaf-like bracts or the outermost phyllaries green and narrowly leaf-like. Involucres mostly 10–17 × 15 mm., spreading to c. 25 mm. wide, broadly campanulate. Phyllaries numerous; the outer from c. 4 mm. long, lanceolate-subulate tapering to a sharp recurved apical bristle, sparsely araneose to puberulous; the inner to c. 15 mm. long, linear-lanceolate, sharply acute and ± aristate at the apex, margins minutely toothed near the apex. Receptacle deeply alveolate with pits to c. 2 mm. deep, paleae absent. Ray-florets neuter, corollas creamy-white to yellowish, up to c. 14 mm. long, cleft almost to the base; ray erect, 1.5–3 mm. wide, narrowly oblanceolate, 3-fid, sparsely glandular outside, hardly exceeding the involucre; ovary rudimentary, 4–5 mm. long, narrowly terete; pappus of many chartaceous narrowly lanceolate scales 0.2–3 mm. long. Disk-floret corollas creamy-white, tubular in the lower c. 2 mm., abruptly dilated into a deeply-lobed cylindric to conical limb c. 5 mm. long; lobes c. 4 mm. long linear, recurved at the apex, glandular outside; anther tube purplish-brown, the conical tip exserted. Achenes c. 5 mm. long, subcylindric, 10-ribbed with numerous long stiff white hairs ± appressed in the grooves, the hairs shorter near the achene base and spreading; pappus several-seriate of numerous spreading narrowly lanceolate chartaceous scales 0.5–3 mm. long.

Zambia. W: Mwinilunga airstrip, c. 1625 m., 16.v.1986, *Philcox, Pope, Chisumpa & Ngoma* 10315 (BR; K; LISC; NDO; SRGH).
Also in Zaire. Wooded grassland on sandy soil.
The trailing stems with short strongly divaricate branches, and the creamy-white or yellowish rays which hardly exceed the involucre set this species apart from *P. jeffreyi*.

3. **Pasaccardoa grantii** (Benth. ex Oliv.) Kuntze, Rev. Gen. Pl. **1**: 355 (1891). —O. Hoffm. in Engl., Pflanzenw. Ost-Afr. **C**: 420 (1895). —C. Jeffrey in Kew Bull. **21**: 208 (1967). —Wild in Kirkia **8**: 193 (1972). —Lisowski, Aster. Fl. Afr. Centr. [in Fragm. Flor. Geobot. **36**, Suppl. 1]: 536, fig. 108 (1991). TAB. **6**. Type from Tanzania.
    Phyllactinia grantii Benth. ex Oliv. in Trans. Linn. Soc., Lond. **29**: 102, t. 68 (1873). —Oliv. & Hiern in F.T.A. **3**: 442 (1877). Type as above.
    Pasaccardoa dicomoides De Wild. & Muschl. in Bull. Soc. Bot. Belg. **49**: 223 (1913). Type from Zaire.

Tab. 6. PASACCARDOA GRANTII. 1, habit (×½), from *Philcox, Pope & Chisumpa* 9934; 2, phyllaries, outer and inner (× 2); 3, marginal floret, sterile, ray 3-dentate and ovary rudimentary (× 5); 4, inner floret corolla (× 5); 5, achene, coarsely hairy at base otherwise finely setose (× 5), 2–5 from *Vesey FitzGerald* 3215. Drawn by Eleanor Catherine.

Pasaccardoa kassneri De Wild. & Muschl. in Bull. Soc. Bot. Belg. **49**: 225 (1913). Type from Zaire.

An erect slender to laxly bushy annual herb, 20–35 cm. tall. Stems leafy, araneose-lanate to roughly puberulous; branches ascending-divaricate, to c. 20 cm. long. Leaves with petioles to c. 1.2 cm. long; lamina sub-membranous mostly 3.5–7 × 1–2.4 cm., elliptic-oblanceolate, ± obtuse at the apex and cuneate below, margins serrulate with minute callose teeth, upper surface sparsely thinly araneose-lanate to glabrescent, lower surface greyish araneose-lanate, both surfaces densely gland-dotted. Capitula solitary and terminal on branches, usually also 1–3 arranged along the branches on axillary stalks shorter than the involucre; capitula subtended by a whorl of 3–6 large leaf-like bracts, bracts up to c. 5.5 × 1.7 cm., but usually smaller, exceeding the capitulum. Involucres mostly 8–13 × c. 15 mm., spreading to c. 25 mm. wide, obconic-campanulate. Phyllaries numerous; the outer from c. 4 mm. long, subulate-aristate, tapering into a long stiff squarrose or recurved bristle, araneose-lanate; the inner to c. 12 mm. long, linear-lanceolate, tapering to a sharply acute or somewhat aristate apex, margins minutely toothed, particularly near the apex. Receptacle deeply alveolate, the pits to c. 2.5 mm. deep, paleae absent. Ray-florets neuter, corollas deep-red or creamy-white, up to c. 14 mm. long, cleft almost to the base, ray erect, narrowly oblanceolate, 3-fid, sparsely glandular outside; ovary rudimentary, 4–5 mm. long, subterete or somewhat swollen near the base; pappus of many spreading chartaceous narrowly lanceolate scales 0.5–4 mm. long. Disk-floret corollas deep-red or creamy-white, tubular in the lower c. 2 mm., abruptly dilated into a deeply lobed cylindric to conical limb 4–5 mm. long, lobes to c. 4 mm. long, linear, recurved at the apex, glandular outside; anther tube reddish-purple, the conical tip exserted. Achenes 6–8 mm. long, subcylindric, 10-ribbed with numerous stiff white hairs ± appressed in the grooves, hairs near the base of the achene shorter more robust and spreading; pappus several-seriate of numerous spreading narrowly lanceolate chartaceous scales 0.5–4 mm. long.

Zambia. N: near Muzombwe, W. side of Mweru Wantipa, c. 1066 m., 15.iv.1961, *Phipps & Vesey-FitzGerald* 3215 (K; LISC; M; SRGH), fls. red; c. 5 km. E. of Kampumbo School on road between Isoka and Muyombe, 15.iv.1986, *Philcox, Pope & Chisumpa* 9934 (BR; K; LISC; MO; NDO; SRGH), fls. creamy-white.

Also in Tanzania and SE. Zaire. Deciduous woodland and wooded grassland, on alluvium and sandy soils.

6. DICOMA Cass.

Dicoma Cass. in Bull. Sci. Soc. Philom. Paris **1817**: 12 (1817). —F.C. Wilson in Bull. Misc. Inf., Kew **1923**: 377 (1923). —C. Jeffrey in Kew Bull. **21**, 2: 208 (1967). —Wild in Kirkia **8**: 178 (1972). —Hilliard, Compos. Natal: 582 (1977). —Lawalrée & Mvukiyumwami in Bull. Jard. Bot. Nat. Belg. **52**: 151 (1982).

Annual herbs, or suffrutices with annual stems from woody taproots or from woody rootstocks with thong-like roots, or sometimes from semi-woody tuberous rootstocks. Stems 1–several, erect or decumbent, sometimes caespitose, simple or branching, sericeous-araneose to lanate or glabrescent, occasionally abbreviated or absent. Leaves alternate, occasionally subrosulate, sessile and ± stem-clasping occasionally stem-sheathing at the base, usually discolorous, entire or serrulate on the margins or revolute, indumentum as for stems and usually more dense on the leaf lower surface, 3–5-veined from the base or obscurely so. Capitula homogamous and discoid, all florets hermaphrodite, or capitula heterogamous and disciform (in the Flora Zambesiaca area) with the outer florets neuter, 1–numerous, sessile or shortly stalked, solitary and terminal or spiciform to subpaniculate. Phyllaries many-seriate, imbricate, appressed or spreading, the outer sometimes squarrose or reflexed, sometimes extending onto the capitulum stalk, the innermost sometimes shorter than the rest, chartaceous to stiffly coriaceous, pungent or subulate. Receptacular paleae absent. Ray-florets usually absent, or when present neuter with filiform sub-bilabiate corollas, and a pappus of several-seriate barbellate setae. Disk-floret corollas yellowish to creamy-brown or purplish to whitish, hermaphrodite and regularly deeply 5-lobed, the lobes ± equalling the tube in length, corollas glabrous. Anther apical appendage lorate-lanceolate, thecae bases sagittate produced into long retrorse-ciliate tails. Style shortly bifid, the branches not separating, tips obtuse with minute sweeping hairs. Achenes ± compressed turbinate, obscurely 2–5-ribbed or strongly 10-ribbed, usually glandular or glutinous, densely long-ascending hispid; disk-floret pappus copious, 2–several-seriate of slender plumose or barbellate setae, the setae all similar increasing in length to the inside, or somewhat dimorphic with the outer setae slender and the inner stouter narrowly scale-like and often ± scarious-winged in the lower part.

A genus of about 32 species in tropical and southern Africa, 1 species extending to Socotra and India, and 4 endemic in Madagascar.

1. Plants annual - - - - - - - - - - - 18. *tomentosa*
  – Plants perennial - - - - - - - - - - - - - 2
2. Pappus setae barbellate or ciliate; innermost phyllaries similar to but longer than the outer, often exceeding the pappus in length; rootstock without numerous spreading, thong-like roots 3
  – Pappus setae all distinctly plumose; innermost phyllaries shorter and broader than the outer, ± equalling the pappus in length; rootstock small with numerous, spreading thong-like roots 11
3. Capitula heterogamous, disciform; outer floret pappus exceeding that of inner florets, outer florets sterile; inner floret pappus 2-seriate; plants with a slender taproot; leaf upper surface appressed araneose-lanate, sometimes thinly so - - - - - - - - 4
  – Capitula homogamous, discoid; pappus of all florets subequal, or slightly larger and more robust in inner florets; all florets fertile; pappus elements many-seriate; plants with tuberous rootstocks or stout woody taproots; leaves glabrous above - - - - - - - - 8
4. Leaf ± abruptly narrowed and petiole-like in the lower part, or strongly spathulate, the expanded portion of the lamina elliptic or rotund to very broadly ovate, apex rounded-obtuse 5
  – Leaf tapering gradually to the base, or narrowing to a broadly-winged midrib, the expanded portion of the lamina linear to narrowly elliptic-lanceolate, apex tapering acute, (leaves somewhat petiolate in *D. capensis* but then lamina mostly less than 5 mm. wide with undulate margins) - - - - - - - - - - - - - - 6
5. Leaves concolorous, the indumentum similar and ± equally dense on leaf upper and lower surfaces; indumentum often brownish-tinged, at least on young growth, more obvious on lamina margins and midrib; expanded portion of lamina ± elliptic, margins not undulate 13. *schinzii*
  – Leaves discolorous, distinctly more densely greyish-lanate on lower surface, the upper surface showing green through the indumentum; indumentum of margins and midrib never brownish-tinged; expanded portion of lamina rotund to very broadly ovate, margins usually undulate
  14. *arenaria*
6. Leaves mostly less than 5(7) mm. wide, often linear-elliptic; leaf margins crispate-undulate; pappus up to c. 11 mm. long, overtopped by the involucre - - - - 15. *capensis*
  – Leaves mostly more than 7 mm. wide, elliptic, margins seldom undulate; pappus usually 20 mm. or more long; erect subshrubs to c. the involucre - - - - - - - - - 7
7. Leaves subspathulate-oblanceolate widest above the middle, narrowing into a winged midrib below; involucre 15–20 mm. long; pappus c. 20 mm. long; inner pappus setae of central florets narrowly scarious-winged towards the base - - - - - - - 16. *niccolifera*
  – Leaves elliptic, widest about the middle; involucre 23–33 mm. long; pappus to c. 25 mm. long; inner pappus setae of central florets broadly membranous undulate-winged towards the base
  17. *macrocephala*
8. Leaves elliptic or oblanceolate, those subtending the capitula few and equal to or shorter than the involucre; erect subshrubs to c. 1 m. tall - - - - - - - 9. *galpinii*
  – Leaves filiform to linear-lanceolate or narrowly lanceolate, mostly more than 4 times as long as wide, (or if less and lamina elliptic then plants decumbent with the subtending leaves usually exceeding the capitula); decumbent or erect-caespitose herbs to c. 30(50) cm. tall 9
9. Inner phyllaries with pungent apices, ± equalling the pappus in length; capitula mostly 10–15 mm. long, obconic; stems erect - - - - - - - - - - 12. *gerrardii*
  – Inner phyllaries numerous with subulate apices exceeding the pappus; capitula 15–30 mm. long, broadly campanulate, the base truncate-concave; stems decumbent, or erect and then leaves filiform - - - - - - - - - - - - - - - - - - 10
10. Leaves filiform, revolute, many subtending the capitula; capitula terminal, solitary; stems simple, strict, erect - - - - - - - - - - - - - 11. *attenuata*
  – Leaves linear-elliptic to narrowly-elliptic, not or hardly revolute, few–many subtending the capitula; capitula 2–many per stem; stems usually branching near the apex, decumbent sometimes laxly erect - - - - - - - - - - - 10. *anomala*
11. Plants acaulescent, or stems abbreviated up to c. 5 cm. tall - - - - - - 12
  – Plants caulescent; stems more than 5 cm. tall - - - - - - - 13
12. Leaves strongly 5-nerved from near the base, margins usually subentire to serrulate; phyllaries appressed-ascending, the outer usually tapering-attenuate - - - 4. *plantaginifolia*
  – Leaves more markedly pinnately-nerved, margins sharply serrate; outermost phyllaries patent and extending onto the capitulum stalk, the outer ovate-lanceolate, shortly acuminate
  5. *humilis*

13. Largest phyllaries 6–10 mm. wide, less than 3 times as long as wide, narrowly ovate, acute to shortly acuminate - - - - - - - - - - - - 6. *zeyheri*
– Largest phyllaries up to 5(7) mm. wide, more than 3 times as long as wide, ± narrowly lanceolate, tapering acuminate, (sometimes to c. 7 mm. wide in *D. kirkii* subsp. *kirkii* but then upper cauline leaves widest below the middle and tapering to the apex) - - - - - - - 14
14. All phyllaries appressed ascending, the outer not patent or reflexed - - - - 15
– Outer (lowermost) phyllaries patent to reflexed - - - - - - - - 16
15. Involucre mostly less than 20 mm. long, phyllary apices exceeding the pappus by less than 6 mm.; leaves less than 20 mm. wide, not sheathing at the base - - - - 8. *angustifolia*
– Involucre more than 25 mm. long, phyllary apices exceeding the pappus by more than 8 mm.; leaves usually more than 20 mm. wide, sheathing at the base - - - - 7. *poggei*
16. Leaves widest above the middle, oblanceolate, long-tapering to the base, 12–27 × 2–6 cm.
2. *sessiliflora*
– Leaves widest about or below the middle, linear-elliptic or lorate-lanceolate and apically attenuate but not long-tapering to the base, at least the upper cauline leaves less than 12 cm. long and 1.5 cm. wide - - - - - - - - - - - - - - - 17
17. Leaves narrowly-elliptic, base tapering to a short petiole, hardly stem-clasping 3. *elliptica*
– Leaves lorate-lanceolate, base not tapering, sessile and subauriculate or narrowing only slightly and semi-amplexicaul ± stem-sheathing - - - - - - - - 1. *kirkii*

1. **Dicoma kirkii** Harv. in Harv. & Sond., F.C. **3**: 518 (1865). —Oliv. & Hiern in F.T.A. **3**: 444 (1877). —Eyles in Trans. Roy. Soc. S. Afr. **5**: 522 (1916). —F.C. Wilson in Bull. Misc. Inf., Kew **1923**: 387 (1923). Type: Zambia, Highlands of Batoka Country, *Kirk s.n.* (K, holotype).

An erect perennial herb, 18–80(100) cm. tall from a small woody rootstock; roots spreading, thong-like. Stems annual, solitary, usually branching near the apex, leafy, sericeous or araneose-lanate. Leaves discolorous; mid- to lower-cauline leaves mostly 5–15 × 1–2.5 cm., narrowly lanceolate and shortly acute, or these 8–16 × 0.5–1.7 cm., linear-lanceolate and tapering acute; lowermost leaves decreasing in size to the stem base, becoming scale-like; upper-cauline leaves linear and narrowly tapering-attenuate; bases hardly narrowing and sometimes subauriculate before becoming stem-clasping or stem-sheathing; stem-sheaths 5–25 mm. long, submembranous; margins flat and serrulate or strongly revolute; upper surface green and thinly araneose to glabrescent, lower surface white appressed araneose-lanate, sometimes sparsely so; 3(5) veined from the base. Capitula 1–numerous, clustered or subpaniculate at the stem apex, or terminal on short axillary branches and racemose. Involucres mostly 20–35 × 15–30 mm., spreading to c. 40 mm. in diam., broadly obconic. Phyllaries numerous, greenish-stramineous, stiffly chartaceous, glabrous with margins araneose-ciliate at first; the outer 6–9 series stiffly diverging or ascending, the outermost 2–4 series patent to reflexed, increasing from c. 6 mm. to c. 30 mm. long, the largest up to c. 3.5 mm. wide or 4–7 mm. wide, flat and pale scarious-winged in the lower part, long-attenuate pungent with inrolled margins in the upper part; the innermost 2–3 series of phyllaries shorter, lorate or narrowly oblong, cuspidate, white membranous. Corollas yellowish, 9–11 mm. long, infundibuliform; lobes c. 5 mm. long, linear, erect. Achenes 2–4 mm. long, turbinate, densely long-ascending hispid; pappus copious, several seriate, of plumose setae 8–10 mm. long, shorter on the outside, seta bristles c. 1 mm. long.

Mid- and lower-cauline leaves mostly 10–25 mm. wide, bases semi-amplexicaul or shortly conduplicate about the stem, clearly 3–5-nerved from the base, margins flat or hardly revolute; widest phyllaries mostly (3)4–7 mm. wide, ovate-lanceolate below becoming subulate above
subsp. *kirkii*
Mid- and lower-cauline leaves mostly up to c. 10(15) mm. wide, bases stem-sheathing at least in the lower leaves, obscurely 3-nerved from the base, margins usually revolute; widest phyllaries up to c. 4 mm. wide, gradually tapering subulate - - - - - - subsp. *vaginata*

Subsp. **kirkii**
Dicoma kirkii var. *angustifolia* S. Moore in Journ. Linn. Soc., Bot. **40**: 123 (1911). —Eyles in Trans. Roy. Soc. S. Afr. **5**: 522 (1916). Type: Zimbabwe, Chirinda, *Swynnerton 444* (BM, holotype; K; SRGH).
Dicoma sessiliflora subsp. *kirkii* (Harv.) Wild in Kirkia **8**: 184 (1972). Type: Zambia, Highlands of Batoka Country, *Kirk s.n.* (K, holotype).
Dicoma sessiliflora subsp. *sessiliflora* —Wild in Kirkia **8**: 184 (1972) pro parte, quoad specim. *Chase 5815, Leach & Bullock 13127*.
Dicoma sessiliflora sensu Eyles in Trans. Roy. Soc. S. Afr. **5**: 522 (1916). —sensu F.C. Wilson in Bull. Misc. Inf., Kew **1923**: 386 (1923) pro parte, quoad *Rogers 4058*.

Zimbabwe. N: Guruve Distr., Great Dyke, 2.vi.1978, *Nyariri* 194 (K; SRGH). W: Shangani Res., 1.iv.1951, *West* 3147 (K; SRGH). C: Chegutu (Hartley), Parks & Wild Life Land, grid ref. 1729 D3, c. 920 m., 3.vii.1974, *Miller* 2154 (K; SRGH). E: Mutare, Matika's Mt., 14.v.1950, *Chase* 2206 (BM; COI; K; LISC; SRGH). S: Chibi Distr., near Madzivire Dip, c. 6.4 km. N. of Rundi R., 4.v.1962, *Drummond* 7929 (K; SRGH). **Mozambique**. MS: c. 48 km. E. of Vila de Manica, c. 594 m., 14.vi.1959, *Leach* 9102 (K; SRGH).

Known only from the Flora Zambesiaca area. Miombo, *Brachystegia boehmii* or *Uapaca* woodlands, often on granite outcrops or hillsides.

Subsp. **vaginata** (O. Hoffm.) G.V. Pope in Kew Bull. **46**: 703 (1991). Type from Tanzania.

    Dicoma vaginata O. Hoffm. in Engl., Bot. Jahrb. **30**: 442 (1901). —F.C. Wilson in Bull. Misc. Inf., Kew **1923**: 388 (1923). Type as above.

    Dicoma sessiliflora subsp. *sessiliflora* sensu Wild in Kirkia **8**: 184 (1972) pro parte, quoad specim. *McClounie* 19 et *Fanshawe* 4506.

    Dicoma sessiliflora sensu F.C. Wilson in Bull. Misc. Inf., Kew **1923**: 386 (1923) pro parte, quoad specim. *McClounie* 19. —C. Jeffrey in Kew Bull. **21**: 210 (1967) quoad adnot.

Zambia. C: Chakwenga Headwaters, 100–129 km. E. of Lusaka, 8.ix.1963, *Robinson* 5661 (BR; K; M). E: Chipata, 2.vi.1958, *Fanshawe* 4506 (K; M). **Malawi**. N: Chitipa, c. 1300 m., 13.vii.1970, *Brummitt* 12077 (K; SRGH). C: Kongwe Forest, near Dowa, 4.v.1980, *Blackmore, Brummitt & Banda* 1429 (K; MAL). **Mozambique**. N: Niassa Distr., Unango, ix.1964, *Magalhães* 97 (COI).

Also in Tanzania and Zaire. Miombo, or sometimes *Uapaca* woodlands, usually on hillsides.

2. **Dicoma sessiliflora** Harv. in Harv. & Sond., F.C. **3**: 518 (1865). — Oliv. & Hiern in F.T.A. **3**: 444 (1877). —F.C. Wilson in Bull. Misc. Inf., Kew **1923**: 386 (1923) pro parte, quoad typum et specim. *Sharpe* 162 et *Kirk* s.n. —C. Jeffrey in Kew Bull. **21**: 210 (1967) pro parte, excl. specim. *Scott* in E.A.H. 11818, *Chandler* 933, *Stolz* 741, 1421, *Geilinger* 2598, *Semsei* 2457. —Wild in Kirkia **8**: 183 (1972) pro parte, quoad specim. *Faulkner* 64 et *Chase* 6629. TAB. **7** fig. A. Type: Malawi (Mozambique border), Manganja Hills, *Meller* s.n. (K, holotype).

A slender, erect perennial herb, to c. 1 m. tall from a small woody rootstock; roots numerous, thong-like. Stems annual, usually solitary, simple, leafy with the largest leaves somewhat clustered on the lower stem, closely appressed white araneose-lanate to glabrescent. Leaves discolorous, sessile; lower-stem leaves 12 27 × (1.5)2–6 cm., oblanceolate, broadest above the middle and tapering-attenuate to the base; upper-stem leaves fewer, becoming linear and shorter; apices subobtuse to acute, bases narrow semi-amplexicaul; margins sparsely serrulate or slightly revolute; green and glabrous on upper surface, white araneose-lanate beneath (the indumentum sometimes peeling off); venation finely reticulate and prominent on both surfaces although obscured by the indumentum below. Capitula few to numerous, spicately arranged, 1–2 sessile in upper leaf axils with others somewhat clustered at the stem apex, or less often shortly stalked and subpaniculate. Involucres mostly 20–30 × 16–30 mm., spreading to c. 40 mm., broadly obconic. Phyllaries numerous, greenish-stramineous, stiffly chartaceous, araneose-ciliate becoming pectinate-ciliate on the margins otherwise glabrous; the outer 6–9 series stiffly diverging-ascending, the outermost 2–4 series patent to reflexed, increasing from c. 6 mm. long outside to c. 30 mm. long inside, lanceolate-subulate, flat and broadly pale scarious-winged in lower part, long pungent-acuminate above with the margins inrolled; the innermost 2–3 series of phyllaries shorter, lorate or narrowly oblong, cuspidate, white scarious. Corollas yellowish, c. 10 mm. long, infundibuliform; lobes c. 5 mm. long, linear, erect; achenes to c. 4 mm. long, turbinate, densely long-ascending hispid; pappus copious, several seriate, of plumose setae c. 9 mm. long, shorter on the outside, seta bristles c. 1 mm. long.

Malawi. S: Mangochi Distr., Lingamasa, 11.vii.1963, *Salubeni* 59 (K; LISC; MAL; SRGH). **Mozambique**. N: 39 km. W. of Ribáuè, c. 670 m., 17.v.1961, *Leach & Rutherford-Smith* 10894 (K; SRGH). Z: Quelimane Distr., Munguluni (M'gulumi Mission, ix.1945, *Faulkner* 64 (COI; K; PRE; S). T: c. 10 km. N. of Tete, 26.vi.1947, *Hornby* 2776 (K; SRGH). MS: c. 30 km. N. of Dombe, c. 150 m., 4.vi.1971, *Biegel & Gordon* 3530 (K; SRGH).

Also in Tanzania and Zaire. Miombo and mixed deciduous woodland, and wooded grasslands.

3. **Dicoma elliptica** G.V. Pope in Kew Bull. **46**: 702 (1991). Type: Malawi, Nkhota Kota to Dowa, 2.x.1950, *Jackson* 196 (K, holotype).

    Dicoma sessiliflora sensu auctt. incl. F.C. Wilson in Bull. Misc. Inf., Kew **1923**: 386 (1923) pro parte, quoad specim. *Kassner* 2991, 2689; *Buchanan* 457, 738, 989; *Cameron* 120. —Brenan in Mem. N.Y. Bot. Gard. **8**, 5: 488 (1954) quoad *Brass* 16901 et *Vernay* 17093. —Wild in Kirkia **8**: 183 (1972) pro parte, quoad specim. *Jackson* 2233, *Brass* 16901, *Barbosa & Carvalho* 3303, *Mitchell* 2987.

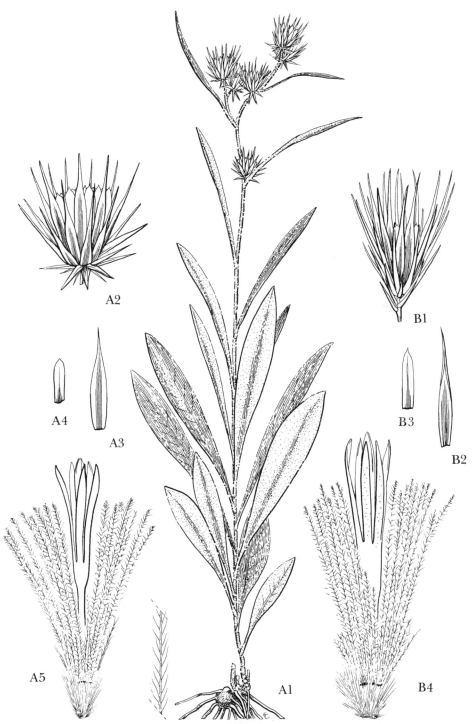

Tab. 7. A. —DICOMA SESSILIFLORA. A1, habit (×⅓), from *Salubeni* 59; A2, capitulum; A3, middle phyllary; A4, inner phyllary; A5, floret and detail of plumose pappus seta (× 6), A2–A5 from *Faulkner* Kew 64. B. —DICOMA POGGEI. B1, capitulum, all phyllaries appressed; B2, middle phyllary; B3, inner phyllary; B4, floret, B1–B4 from *Richards* 1765. (Capitula and phyllaries × 1, florets × 5). Drawn by Eleanor Catherine.

As for *Dicoma sessiliflora* except that the stems are more uniformly leafy, the lower-stem leaves are never clustered and decrease somewhat in size and number towards the stem base. The leaves differ in being mostly 8–16 × 0.8–1.5(2.5) cm., linear-elliptic or narrowly elliptic, widest about the middle or ± parallel-sided; apex tapering acute; base narrowly cuneate tapering to a petiole-like hardly stem-clasping base; margins sparsely serrulate, revolute.

Zambia. E: Chipata, E. of Machinje Hills, Lubanda Stream, 15.v.1965, *Mitchell* 2987 (K). **Malawi**. N: Karonga Distr., N. of Kayelekera, 30 km. W. of Karonga, 10.vi.1989, *Brummitt* 18476 (K). C: Ntchisi Mt., c. 1400 m., 24.vii.1946, *Brass* 16901 (K; MO; SRGH). S: Blantyre township, 14.v.1948, *Faulkner* Kew 237 (K). **Mozambique**. N: Mandimba, 1.viii.1934, *Torre* 427 (COI). T: c. 54.7 km. from Chicoa to Fingoè, 26.vi.1949, *Barbosa & Carvalho* 3303 (K; LISC).
Also in Zaire. Occasional in miombo woodland, sometimes beside streams.

4. **Dicoma plantaginifolia** O. Hoffm. in Engl., Bot. Jahrb. **15**: 546 (1893). —Hiern, Cat. Afr. Pl. Welw. **1**, 3: 614 (1898). —S. Moore in Journ. Linn. Soc., Bot. **37**: 327 (1906); op. cit. **47**: 282 (1925); Journ. Bot. **65**, Suppl. 2, Gamopet.: 66 (1927). —Mendonça, Contrib. Conhec. Fl. Angol., 1 Compositae: 146 (1943). —C. Jeffrey in Kew Bull. **21**: 210 (1967). —Wild in Kirkia **8**: 180 (1972). —Lawalrée & Mvukiyumwami in Bull. Jard. Bot. Nat. Belg. **52**: 158 (1982). —Lawalrée in Lisowski, Aster. Fl. Afr. Centr. [in Fragm. Flor. Geobot. **36**, Suppl. 1]: 546 (1991). Type from Angola.
Dicoma nana Welw. ex Hiern, Cat. Afr. Pl. Welw. **1**: 614 (1898). —F.C. Wilson in Bull. Misc. Inf., Kew **1923**: 387 (1923). —Mendonça, Contrib. Conhec. Fl. Angol., 1 Compositae: 146 (1943). —Lawalrée in Lisowski, Aster. Fl. Afr. Centr. [in Fragm. Flor. Geobot. **36**, Suppl. 1]: 546 (1991). Type from Angola.
Dicoma quinquenervia sensu F.C. Wilson, in Bull. Misc. Inf., Kew **1923**: 388 (1923) pro parte quoad specim. *Rogers* 8528 et *Macaulay* 803, non Bak.
Dicoma pygmaea Hutch., Botanist S. Afr.: 526 (1946). Type: Zambia, Chilanga, *Rogers* 8528 (K, holotype).

A perennial herb, usually acaulescent, from a woody rootstock; roots numerous spreading thong-like. Stems sometimes developed but abbreviated, erect to c. 5 cm. tall, simple or with a few short branches, and with 1–3 persistent leaf bases. Leaves few to numerous, prostrate discolorous subsessile, 6–20 × 1.5–12 cm., oblong-elliptic, oblanceolate-elliptic or ± orbicular, rounded and somewhat cuspidate at the apex or obtuse, broadly cuneate below narrowing to a winged midrib up to c. 2 cm. long, somewhat stem-clasping at the base, margins sparsely denticulate or serrulate, slightly revolute, green and glabrous on upper surface, white araneose-sericeous beneath (the indumentum sometimes peeling off), strongly 5(7)-nerved from the base, venation finely reticulate and prominent on both surfaces, or obscured by the indumentum beneath. Capitula subsessile, 1–7, solitary or densely clustered in the leaf rosette. Involucres mostly 25–36 mm. long, 20–50 mm. in diam. in the upper part, broadly obconic-campanulate. Phyllaries stiffly slightly diverging, greenish to straw-coloured, chartaceous, sharply ciliolate on upper margins otherwise glabrous, the outermost usually not extending onto the capitulum stalk; the outer phyllaries from c. 7 mm. long and linear-lanceolate, usually somewhat falcate, tapering to a pungent apex; the inner phyllaries to c. 35 mm. long and narrowly lorate-lanceolate, ± scarious-winged in the lower part, inrolled and long-acuminate finely pungent to the apex; innermost phyllaries somewhat shorter, membranous winged and sharply cuspidate. Corollas creamy to brownish usually with purplish lobes, 10–13 mm. long, infundibuliform; lobes 5–6.5 mm. long, linear, erect; achenes c. 2 mm. long, narrowly turbinate, densely long-ascending hispid; pappus copious, several seriate of plumose white setae 8–11 mm. long, shorter on the outside, seta bristles 0.5–1 mm. long.

Zambia. N: Kasama Distr., Chishimba Falls, 20.v.1962, *Robinson* 5193 (K; SRGH). W: Ndola, 30.vii.1953, *Fanshawe* 188 (K). C: Chisamba, 10.viii.1957, *West* 4022 (SRGH). S: between Choma and Pemba, 17.viii.1929, *Burtt Davy* 714 (K). **Malawi**. N: Chitipa Distr., c. 21 km. SSE. of Chisenga, c. 1900 m., 11.vii.1970, *Brummitt* 11998 (K). C: Kasungu, c. 1000 m., 26.viii.1946, *Brass* 17425 (K; MO; SRGH).
Also in Tanzania, Zaire, Burundi and Angola. A pyrophyte of miombo woodland and submontane grassland, flowers usually precocious.

Richards 5845 (K) from the Lumi R. near Mbala (in Zambia) differs from other material of this species, even from other Mbala-area material seen, in its outermost phyllaries which are shorter and ovate-lanceolate rather than linear-lanceolate and ± falcate, sharply acute not attenuate at the apex and thinner in texture.

5. **Dicoma humilis** Lawalrée in Bull. Jard. Bot. Nat. Belg. **52**: 157 (1982). Type from Zaire.

Similar to *D. plantaginifolia* but differing in leaf and phyllary characters. Leaves 9–16 × 1–4.5 cm., elliptic to oblanceolate, acute at the apex, narrowing below into a winged midrib to c. 3 cm. long, somewhat stem-clasping; margins sharply serrate; upper surfaces green and glabrous, lower surfaces white araneose-sericeous (the indumentum sometimes peeling off); ± markedly pinnately-nerved. Capitula stalked, 1–7 clustered in the leaf rosette. Involucres mostly 20–25 mm. long, 15–25 mm. in diam. in the upper part, broadly obconic-campanulate. Phyllaries numerous, stiffly slightly diverging, the outermost ± patent and extending onto the capitulum stalk, greenish-stramineous, chartaceous, sharply ciliolate on upper margins otherwise glabrous; the outer phyllaries from c. 3 mm. long, ovate-lanceolate, shortly acuminate; the middle phyllaries to c. 24 mm. long, narrowly oblong-lanceolate, ± broadly scarious-winged, inrolled and acuminate-pungent in the upper part; the inner series of phyllaries somewhat shorter, mostly white-scarious and sharply cuspidate.

Zambia. N: Felisi, c. 1036 m., 3.vi.1950 *Bullock* 2903 (K).
Also in Zaire. Wooded grassland or miombo woodland on sandy soil.

Further material needs to be seen to establish whether this is an extreme variation in *D. plantaginifolia* or a distinct species occurring sympatrically with it.

6. **Dicoma zeyheri** Sond. in Linnaea **23**: 70 (1850). —Harv., Thes. Cap. **1**: 43, pl. 68 (1859); in Harv. & Sond., F.C. **3**: 518 (1865). —F.C. Wilson in Bull. Misc. Inf., Kew **1923**: 387 (1923). —Wild in Kirkia **8**: 182 (1972). —Hilliard, Compos. Natal: 586 (1977). Type from South Africa (Transvaal).

An erect robust perennial herb, 15–40 cm. tall from a woody rootstock; roots numerous, thong-like. Stems annual, 1–several, simple or shortly branched near the apex, leafy, thinly whitish araneose. Leaves discolorous, mostly 50–130 × 10–47 mm., oblanceolate to narrowly obovate, subacute or shortly acuminate to the apex, narrowing below to a short, winged midrib, somewhat stem-clasping at the base, sparsely denticulate, slightly revolute on the margins, green and glabrous on the upper surface, white araneose-sericeous below (the indumentum sometimes peeling off); lower lateral nerves at an acute angle to the midrib, venation finely reticulate and subprominent on both surfaces or obscured by the indumentum below. Capitula solitary and terminal, or less often up to c. 12 terminal on short branches in upper leaf axils. Involucres mostly 30–40 mm. long, 30–50 mm. in diam., broadly campanulate. Phyllaries numerous large, the outer spreading-reflexed the remainder stiffly ascending-diverging, the outermost usually extending down the capitulum stalk, also reflexed, silvery-green sometimes reddish-tinged, coriaceous, glabrous, the margins narrowly membranous sometimes ± inrolled in the upper part, minutely serrulate, apices cuspidate to acuminate-pungent; from c. 10 mm. long on the outside to c. 35 mm. long inside; the middle phyllaries 6–10 mm. wide, ovate-acuminate to lanceolate-acuminate. Corollas creamy to brownish with purplish lobes, 10–11 mm. long, infundibuliform; lobes 5–6 mm. long, linear, erect; achenes c. 2 mm. long, narrowly turbinate, densely long ascending hispid; pappus copious, several seriate, of plumose white setae 7–10 mm. long, shorter on the outside, seta bristles c. 1 mm. long.

Mozambique. M: Lebombo, Mt. Mpondium, 22.ii.1955, *Exell, Mendonça & Wild* 527 (BM; SRGH).
Also in South Africa (Natal and the Transvaal) and Swaziland. Grassland or open bush-grassland on stony soil usually on hillsides.

7. **Dicoma poggei** O. Hoffm. in Engl., Bot. Jahrb. **15**: 546 (1893). —Lawalrée & Mvukiyumwami in Bull. Jard. Bot. Nat. Belg. **52**: 158 (1982). —Lawalrée in Lisowski, Aster. Fl. Afr. Centr. [in Fragm. Flor. Geobot. **36**, Suppl. 1]: 547 (1991). TAB. **7** fig. B. Type from Zaire.
Dicoma quinquenervia Bak. in Bull. Misc. Inf., Kew **1895**: 290 (1895). —F.C. Wilson in Bull. Misc. Inf., Kew **1923**: 388 (1923). —Mendonça, Contrib. Conhec. Fl. Angol., 1 Compositae: 145 (1943). —Wild in Kirkia **8**: 181 (1972). Type: Zambia, Mweru, Chama R., *Carson* 4 (K, holotype).
Dicoma superba S. Moore in Journ. Linn. Soc., Bot. **37**: 326 (1906). Type from Angola.
Dicoma quinquenervia var. *latifolia* S. Moore in Journ. Linn. Soc., Bot. **47**: 282 (1925). Type as for *D. superba* S. Moore.

An erect slender perennial herb, 25–70 cm. tall from a woody rootstock; roots numerous spreading thong-like. Stems annual, usually single, simple, leafy, thinly whitish araneose. Leaves discolorous, softly coriaceous, sessile, mostly 7–150 × 15–55 mm., narrowly to broadly elliptic, acute to obtuse at the apex, ± broadly cuneate below tapering into a ± deeply cleft stem-sheathing base 10–35 mm. long; margins sparsely callose-denticulate; upper surface green, glabrous; lower surface white araneose (the indumentum sometimes peeling off); strongly 3–5-nerved from the base, tertiary venation finely reticulate and subprominent on both surfaces or obscured by indumentum below. Capitula 2–15, subsessile or shortly stalked, 2–6 clustered at the stem apex, usually with 1–3 in the axils of upper leaves. Involucres mostly 25–40 mm. long, 20–30 mm. in diam. in upper part, obconic. Phyllaries green chartaceous glabrous, sharply serrulate-ciliolate, apparently dimorphic; the outer c. 7 series straight, loosely diverging, the outermost extending briefly onto the capitulum stalk, from c. 7 mm. long outside to c. 35 mm. long inside, subulate-lanceolate to linear-lanceolate, broadly scarious-winged in the lower part, inrolled and long-acuminate pungent above; the inner 1–2 series shorter, to c. 17 mm. long, lorate, distinctly scarious-winged and cuspidate at the apex. Corollas creamy or brownish with purplish lobes, 8–12 mm. long, narrowly infundibuliform; lobes 5–6 mm. long, linear erect; achenes 1.5–2.5 mm. long, turbinate, densely long ascending hispid; pappus copious, several seriate, of plumose white setae 9–11 mm. long, shorter on the outside, seta bristles c. 1 mm. long.

Zambia. N: Isoka Distr., Nakonde, Old Fife Coffee Res. Sta., c. 1850 m., 22.iv.1986, *Philcox, Pope & Chisumpa* 10074 (BR; K; LISC; SRGH). W: Solwezi, 13.vi.1930, *Milne-Redhead* 494 (K; PRE).
Also in W. Africa, Burundi, Zaire, Tanzania and Angola. In miombo, *Marquesia* and escarpment woodlands, mushitu (swamp forest) margins and riverine vegetation, with short grasses.

8. **Dicoma angustifolia** (S. Moore) Wild in Kirkia **8**: 181 (1972). Type from Angola.
   Dicoma superba var. *angustifolia* S. Moore in Journ. Linn. Soc., Bot. **37**: 327 (1906). Type as above.
   Dicoma quinquenervia var. *angustifolia* (S. Moore) S. Moore op. cit. **47**: 282 (1925); Journ. Bot. **65**, Suppl. 2, Gamopet: 66 (1927). Type as above.
   Dicoma sessiliflora sensu F.C. Wilson in Bull. Misc. Inf., Kew **1923**: 386 (1923) pro parte quoad specim. Zamb. *Rogers* 8115 et *Macaulay* 747. —sensu Mendonça, Contrib. Conhec. Fl. Angol., **1** Compositae: 144 (1943) quoad specim. cit. —sensu C. Jeffrey in Kew Bull. **21**: 210 (1967) pro parte.
   Dicoma saligna Lawalrée in Lawalrée & Mvukiyumwami in Bull. Jard. Bot. Nat. Belg. **52**: 160 (1982). —Lawalrée in Lisowski, Aster. Fl. Afr. Centr. [in Fragm. Flor. Geobot. **36**, Suppl. 1]: 553 (1991). Type from Zaire.

An erect slender perennial herb, 35–70 cm. tall from a woody rootstock; roots numerous thong-like. Stems annual, 1–3, simple, leafy, thinly whitish araneose. Leaves discolorous, softly sericeous, subsessile; 2.5–11(13) × 0.4–1.5(2.5) cm., lorate- or linear-elliptic, usually shortly tapered to an acute apex, cuneate or shortly subpetiolate below, margins strongly revolute, upper surface green and glabrous, lower surface white araneose-sericeous (the indumentum sometimes peeling off); midrib impressed on upper surface, prominent beneath, venation finely reticulate and prominent on both surfaces or obscured by the indumentum below. Capitula 3–21, ± spicately arranged, subsessile and usually solitary in upper leaf axils, sometimes somewhat clustered at the stem apex. Involucres mostly 15–20 mm. long, 10–20(25) mm. in diam. in upper part, obconic-campanulate, whitish araneose-lanate. Phyllaries greenish-stramineous, stiffly chartaceous, araneose-ciliate on the margins otherwise glabrous; the outer 6–8 series straight, loosely stiffly diverging (not patent or reflexed and not extending onto the capitulum stalk), from c. 3 mm. long outside to c. 20 mm. inside, lanceolate-subulate, ± scarious-winged in the lower part, flat or inrolled and long-acuminate pungent above; the inner 1–2 series of phyllaries shorter, lorate, whitish membranous and attenuate or sharply cuspidate. Corollas yellowish, ± equalling the inner phyllaries, 8–10 mm. long, infundibuliform; lobes 4–5 mm. long, linear, erect; achenes to c. 3 mm. long, turbinate, densely long ascending hispid; pappus copious, several-seriate, of plumose setae 8–9 mm. long, shorter on the outside, seta bristles c. 1 mm. long.

Zambia. B: c. 64 km. from Zambezi (Balovale) on Kabompo road, 26.iii.1961, *Drummond & Rutherford-Smith* 7361 (K; SRGH). N: c. 43.5 km. S. of Mbala, 18.vii.1930, *Hutchinson & Gillett* 3864 (BM; K; SRGH). W: Ndola, 7.ix.1967, *Mutimushi* 2072 (K; NDO). C: c. 10 km. beyond Chiwefwe, c. 1341 m., 14.vii.1930, *Hutchinson & Gillett* 3656 (BM; K; SRGH).
Also in Angola and Zaire. Occasional in partial shade of open miombo woodland and mixed deciduous woodland on Kalahari Sand.

9. **Dicoma galpinii** F.C. Wilson in Bull. Misc. Inf., Kew **1923**: 385 (1923). —Wild in Kirkia **8**: 189 (1972). Type from South Africa (Transvaal).

An erect perennial herb or subshrub, up to c. 1 m. tall from a woody taproot. Stems several, becoming somewhat woody below, branched, leafy, fulvous-lanate on young growth, whitish-lanate to glabrescent on older stems. Leaves discolorous subsessile or with the midrib petiole-like in the lower c. 10 mm.; lamina mostly 2–6 × 0.5–1.5 cm., elliptic to oblanceolate, rounded or obtuse-mucronate to cuspidate at the apex, cuneate to attenuate below, minutely callose-serrulate on the margins, dark-green gland-pitted and glabrous on the upper surface, or briefly whitish araneose at first, lower surface densely greyish lanate, margins and midrib sometimes fulvous hairy. Capitula many, solitary and terminal on branches, subtended by 1–5, usually small, leaf-like bracts. Involucres 12–15(17) × 10–15 mm., phyllaries spreading to c. 20 mm. wide, broadly obconic-campanulate. Phyllaries stiffly subcoriaceous, usually squarrose in the upper half, glabrous, usually finely purple-striped on either side of the midrib, sharply serrulate; outer phyllaries from 3 mm. long, finely subulate; the inner phyllaries to c. 10(14) mm. long, very narrowly triangular, tapering to a fine pungent apex. Corollas pale-mauve, 6–7 mm. long, tubular below widening into a deeply lobed limb, lobes linear erect but recurved at apex; achenes c. 2.5 mm. long, narrowly turbinate, strongly 10-ribbed, somewhat glutinous, densely brownish long hispid between the ribs, the hairs obscuring the surface; basal hairs shorter more numerous and ± spreading; pappus copious, dimorphic, outer setae slender subplumose-ciliolate 5–6 mm. long, inner c. 20 setae narrowly scale-like c. 7 mm. long and white scarious-winged in the lower part, subplumose-ciliolate.

Zimbabwe. S: Gonarezhou (Gona-Re-Zhou) on old track to Chitsa's Kraal, c. 4 km. from Chibiribira Falls, 1.vi.1971, *Grosvenor* 602 (K; LISC; SRGH). **Mozambique**. M: Maputo, near Changalane, 3.iv.1948, *Torre* 7928 (LISC).
Also in South Africa (Transvaal). In mixed deciduous scrub or woodland or dry riverine bush, usually on dry rocky hillsides.

10. **Dicoma anomala** Sond. in Linnaea **23**: 71 (1850). —Harv. & Sond., F.C. **3**: 517 (1865). —Oliv. & Hiern in F.T.A. **3**: 443 (1877). —O. Hoffm. in Engl., Bot. Jahrb. **15**: 545 (1893); in Engl., Pflanzenw. Ost-Afr. **C**: 420 (1895). —Hiern in Cat. Afr. Pl. Welw. **1**, 3: 614 (1898). —F.C. Wilson in Bull. Misc. Inf., Kew **1923**: 382 (1923). —S. Moore in Journ. Bot. **65**, Suppl. 2: 65 (1927). —Mendonça, Contrib. Conhec. Fl. Angol., **1**, Compositae: 142 (1943). —C. Jeffrey in Kew Bull. **21**: 209 (1967). —Wild in Kirkia **8**: 189 (1972). —Hilliard, Compos. Natal: 583 (1977). —Lawalrée & Mvukiyumwami in Bull. Jard. Bot. Nat. Belg. **52**: 155 (1982). —Maquet in Fl. Rwanda, Spermat. **3**: 680, fig. 213(1), (1985). —Lawalrée in Lisowski, Aster. Fl. Afr. Centr. [in Fragm. Flor. Geobot. **36**, Suppl. 1]: 543, fig. 109 (1991). Type from South Africa (Transvaal).
Dicoma anomala var. *sonderi* Harv. in Harv. & Sond., F.C. **3**: 517 (1865). Type as above.
Dicoma anomala var. *cirsioides* Harv. in Harv. & Sond., F.C. **3**: 517 (1865). Type from South Africa.
Dicoma karaguensis Oliv. in Trans. Linn. Soc., Lond. **29**: 103, t. 70 (1875). Type from Tanzania.
Dicoma anomala forma *karaguensis* (Oliv.) Oliv. & Hiern in F.T.A. **3**: 443 (1877). Type as above.
Dicoma anomala var. *karaguensis* (Oliv.) O. Hoffm. in Engl., Bot. Jahrb. **15**: 545 (1893); in Engl., Pflanzenw. Ost-Afr. **C**: 420 (1895); in Warburg, Kunene-Sambesi Exped., Baum: 426 (1903). —S. Moore in Journ. Bot. **65**, Suppl. 2: 65 (1927). Type as above.
Dicoma megacephala Bak. in Bull. Misc. Inf., Kew **1897**: 271 (1897). Type: Malawi, between Khondowe and Karonga, *Whyte* s.n. (K, holotype).
Dicoma nyikensis Bak. in Bull. Misc. Inf., Kew **1897**: 271 (1897). Type: Malawi, Nyika Plateau, *Whyte* s.n. (K, holotype).
Dicoma anomala var. *microphylla* O. Hoffm. in Warburg, Kunene-Sambesi Exped., Baum: 426 (1903). —Mendonça, Contrib. Conhec. Fl. Angol., **1** Compositae: 143 (1943). Type from Angola.
Dicoma anomala var. *latifolia* O. Hoffm. in Warburg, Kunene-Sambesi Exped., Baum: 426 (1903). —S. Moore in Journ. Bot. **65**, Suppl. 2, Gamopet.: 66 (1927). —Mendonça, Contrib. Conhec. Fl. Angol., **1** Compositae: 143 (1943). Type from Namibia.
Dicoma anomala var. *leptothrix* S. Moore in Journ. Bot. **65**, Suppl. 2, Gamopet.: 65 (1927). —Mendonça, Contrib. Conhec. Fl. Angol., **1** Compositae: 143 (1923). Type from Angola.
Dicoma anomala var. *megacephala* (Bak.) Mendonça, Contrib. Conhec. Fl. Angol., **1** Compositae: 143 (1943). Type as for *D. megacephala* Bak.
Dicoma anomala subsp. *cirsioides* (Harv.) Wild in Kirkia **8**: 191 (1972). —Hilliard Compos. Natal: 584 (1977). Type as for *D. anomala* var. *cirsioides* Harv.

Decumbent spreading perennial herbs. Aerial stems arise singly from woody, subterranean stems 2–18 cm. long, borne apically on aromatic turbinate semi-woody tubers. Stems annual, few to many, radiating-decumbent, 5–60 cm. long, wiry and flexuous or more stiffly robust, simple or branched, uniformly leafy, closely densely araneose. Branches when present few to many at the stem apex, or throughout the length of the stem, decumbent leafy short and stiff, or to c. 15 cm. long and flexuous. Leaves discolorous, subsessile or with the midrib petiole-like in the lower c. 10 mm.; mostly

25–110 × c. 6 mm. and linear-elliptic, sometimes linear and c. 2 mm. wide, or 20–60 × 6–15 mm. and narrowly elliptic; acute at the apex, tapering into a narrowly winged midrib or broadly cuneate at the base, sharply callose-tipped serrulate, gland-pitted and glabrous on the upper surface or evanescent finely araneose, densely greyish araneose-lanate beneath, midrib prominent and puberulous beneath or glabrescent, often with 2 distinct lateral veins from the base. Capitula from 1–30 per stem, solitary and terminal on stem and branches, sessile in a rosette of 2–many subtending leaves. Involucres mostly 18–30 × (15)20–35 mm., spreading to c. 60 mm. in diam., very broadly obconic-campanulate, truncate-concave at the base. Phyllaries 90–200 in many series, increasing from c. 2 mm. long outside to c. 30 mm. long inside, exceeding the pappus, narrowly triangular-subulate or stiffly subulate, straight, or outer ones ± curved, minutely sharply serrulate on the margins, lamina with narrow dark-green or purplish grooves on either side of a broad midrib, finely araneose-lanate or glabrescent. Florets 20–90+, corollas purplish-mauve or white, 8–12 mm. long, infundibuliform; lobes erect, linear, glandular without near the apex. Achenes to c. 2 mm. long, ± compressed-turbinate, obscurely c. 10-ribbed, densely strigose-hispid, glandular-viscid; pappus copious, many-seriate, setae 7–12 mm. long, finely barbellate, the outer slender, the innermost very narrowly scarious-winged in the lower part.

Zambia. N: escarpment above Ndundu to Kawimbe road, c. 1800 m., 11.iv.1966, *Richards* 21413 (K). W: Solwezi, 15.vi.1930, *Milne-Redhead* 514 (K; PRE). C: Lusaka, 11.vi.1955, *Fanshawe* 2330 (K). E: Mafinga, 24.v.1973, *Chisumpa* 54 (K; NDO). S: Livingstone, c. 914 m., vi.1909, *Rogers* 7151 (K; SRGH). **Zimbabwe**. N: Makonde Distr., Copper King, 28.iv.1965, *Wild* 7357 (K; SRGH). W: Bulawayo, 17.vi.1931, *Eyles* 7129 (K). C: 3 km. on Norton Road from junction with Harare-Beatrice road, c. 1800 m., 30.viii.1960, *Rutherford-Smith* 17 (K; LISC). E: Chimanimani Mts., 'Stonehenge' near Bundi River source, c. 1828 m., 8.v.1958, *Chase* 6903 (K; SRGH). S: Mberengwa Distr., Mt. Buhwa, c. 1219 m., 11.ix.1965, *Leach & Bullock* 13125 (COI; K; SRGH). **Malawi**. N: Nyika Plateau, near Chelinda Bridge, 19.iv.1986, *Philcox, Pope & Chisumpa* 10025 (BR; K; LISC; MO; NDO; SRGH). C: Kasungu, c. 1030 m., 6.vii.1970, *Brummitt* 11801 (K; MAL; SRGH). S: c. 1 km. W. of Kasupe, c. 720 m., 14.vi.1970, *Brummitt* 11389 (K; MAL; SRGH). **Mozambique**. N: c. 10 km. from Marrupa to Maua, c. 710 m., 7.viii.1981, *Jansen, de Koning & de Wilde* 127 (K).

Also in Rwanda, Burundi, Uganda, Tanzania, Zaire, Angola, Namibia, South Africa (Transvaal, Natal, Orange Free State, Cape Province) and Lesotho. Miombo woodland, wooded grassland and submontane grasslands, on rocky hillsides or grassy plains in doleritic or sandy soils.

A variable species, particularly in its habit, leaf shape and leaf size. In addition to the typical plants two variants can be recognised:
1.   a robust bushy variant occurs in the north and west of the Flora Zambesiaca area, intergrading with the typically decumbent, flexuous-stemmed plants also found there. In this variant the stems are stouter, stiff, much-branched and many-capitulate, eg. *Richards* 19057, *Lawton* 395, from Zambia N; *Mutimushi* 3444, from Zambia W; and *Pawek* 4732, from Malawi N. Similar plants occur in southern Tanzania and eastern Zaire.
2.   a wide-leaved variant occurs in the submontane and plateau grasslands and high rainfall woodlands of the Flora Zambesiaca area. In this variant the leaves are up to c. 25 mm. wide, elliptic, less membranous than in the typical plants and usually pilose-lanate on the upper surface, eg. *Rushworth* 762, from the eastern highlands of Zimbabwe and *Richards* 5494 from the Mbala area of Zambia. This variant intergrades with the more typical, sympatrically occurring, plants with linear or linear-elliptic leaves less than 5 mm. wide. Similar plants are recorded from the eastern Transvaal and Pretoria areas of South Africa, and from Ufipa in Tanzania.

11.   **Dicoma attenuata** (S. Moore) G.V. Pope in Kew Bull. **46**: 708 (1991). TAB. **8** fig. B. Type from Angola.
       Dicoma anomala var. *attenuata* S. Moore in Journ. Bot. **65**, Suppl. 2, Gamopet.: 65 (1927). —Mendonça, Contrib. Conhec. Fl. Angol., 1 Compositae: 143 (1943). Type as above.
       Dicoma anomala subsp. *cirsioides* sensu Wild in Kirkia **8**: 191, 192 (1972) pro parte quoad *Robinson* 3669.

A stiff caespitose perennial herb, 18–30 cm. tall from a tuberous rootstock; rootstock semi-woody turbinate, up to c. 3 cm. in diam., producing short perennial subterranean stems from the tuber apex. Aerial stems annual from the subterranean stems, many, stiffly erect, simple, densely uniformly leafy, closely araneose. Leaves sessile, ascending to spreading, mostly 15–45 mm. long, filiform, subobtuse mucronulate at the apex, ± revolute and obscurely sharply serrulate on the margins; gland-pitted and glabrous or puberulous on the upper surface, closely greyish araneose-lanate beneath. Capitula erect, solitary and terminal on each stem, sessile with numerous subtending leaves. Involucres mostly 22–30 × 25–35 mm., spreading to c. 45 mm. in diam., very broadly campanulate, truncate-concave below. Phyllaries numerous in many series, diverging,

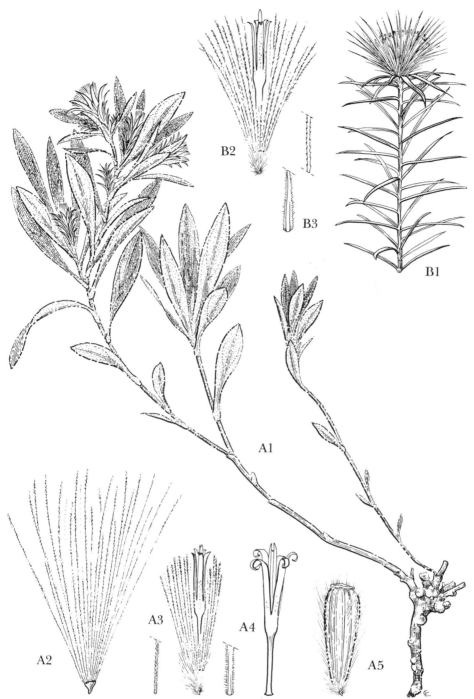

Tab. 8. A. —DICOMA NICCOLIFERA. A1, habit (× ⅔); A2, marginal floret, sterile, achene
rudimentary (× 3); A3, inner floret, pappus shorter than that of outer florets and setae wider
(× 3); A4, corolla of inner floret (× 6); A5, achene (× 8), A1–A5 from *King* 86. B. —DICOMA
ATTENUATA. B1, capitulum (× ⅔); B2, inner floret (× 3); B3, showing wider pappus seta from
inner floret, and terete seta of outer (× 6), B1–B3 from *Milne-Redhead* 4392. Drawn by Eleanor
Catherine.

from c. 5 mm. long outside to c. 30 mm. long inside, exceeding the pappus; all stiffly
subulate to narrowly triangular-subulate, tapering to fine pungent bristle-like apices,
straight or somewhat curved, minutely sharply serrulate on the margins, finely araneose-
lanate in narrow purplish grooves on either side of a broad midrib. Florets 30–45, corollas
whitish, 9–10 mm. long, infundibuliform; lobes erect, linear, glandular without near the
apex. Achenes to c. 2 mm. long, ± compressed turbinate, obscurely c. 10-ribbed, densely
strigose-hispid, glandular-viscid; pappus copious, many-seriate, setae to c. 9 mm. long,
finely barbellate, the outer filamentous, the innermost very narrowly scarious-winged in
the lower part.

Zambia. W: Mwinilunga airstrip, c. 1625 m., 16.v.1986, *Philcox, Pope, Chisumpa & Ngoma* 10316
(BR; K; LISC; M; MO; NDO; SRGH; WAG).
Also in Angola and Zaire. Watershed and dambo grassland in white sandy soil.

12. **Dicoma gerrardii** Harv. ex F.C. Wilson in Bull. Misc. Inf., Kew **1923**: 384 (1923). —Merxm., Prodr.
Fl. SW. Afr. 139: 50 (1967). Type from South Africa (Natal).
Dicoma anomala var. *microcephala* Harv. in Harv. & Sond., F.C. **3**: 517 (1865). —Oliv. & Hiern
in F.T.A. **3**: 384 (1877). Type from South Africa (Natal).
Dicoma anomala subsp. *anomala* sensu Wild in Kirkia **8**: 189 (1972) pro parte, quoad specim.
Richards 14557 et *Molife* 126; sensu Hilliard, Compos. Natal: 584 (1977) pro parte quoad specim.
Edwards 2766.

A caespitose ± bushy perennial herb, 10–30 cm. tall from a semi-woody rootstock;
rootstock narrowly obconic or turbinate, up to c. 10 × 2 cm. Stems annual many densely
leafy, branched, araneose-puberulous, somewhat glandular-viscid. Leaves subsessile,
mostly 10–45(70) × 2–3(6) mm., linear, subobtuse at the apex, sharply callose-tipped
serrulate on the margins, green drying darker and glandular otherwise glabrous on the
upper surface, densely greyish-lanate beneath. Capitula numerous, solitary and terminal
on leafy branches or short shoots, sessile subtended by 2–many leaves. Involucres mostly
10–15 mm. long, up to c. 15 mm. in diam., spreading to c. 20 mm. in diam., obconic.
Phyllaries stiffly thinly coriaceous, shortly pungent at the apex, minutely serrulate on the
margins, usually with a purplish groove on either side of a broad midrib, glabrous rarely
puberulous; the outer c. 6 series each with many phyllaries, less than half the involucre in
length, from c. 1.5 mm. long and narrowly triangular; the inner c. 3 series each with fewer
phyllaries, not overlapping above, lanceolate and ± equalling the pappus in length.
Florets 6–15; corollas purplish fading whitish, 9–11 mm. long, tubular in the lower half
widening above into a cylindric, deeply-lobed limb; lobes erect, linear, glandular without
near the apex. Achenes to c. 2.5 mm. long, turbinate, obscurely c. 10-ribbed, densely
strigose-hispid between the ribs, glandular-viscid; pappus copious, many-seriate, setae to
c. 11 mm. long, finely-barbellate, the outer filamentous, the innermost very narrowly
scarious-winged in the lower part.

Botswana. N: Border near Plumtree c. 600 m., 7.iii.1961, *Richards* 14557 (K; SRGH). SE:
Digkatlong Ranch, 25°03'E, 25°40'S, c. 1030 m., 3.ii.1977, *Hansen* 3018 (C; GAB; K; PRE; SRGH).
Zimbabwe. N: Gokwe Distr., Sengwa Res. Sta., 3.v.1974, *Guy* 2116 (K; SRGH). W: Matobo Distr.,
Hope Fountain Mission, c. 1400 m., 18.iv.1973, *Norrgrann* 340 (K; SRGH). C: Gweru Distr.,
Watershed Block, 10.iv.1968, *Biegel* 2595 (K; SRGH). E: Mutare-Umvumvumvu, v.1923, *Swynnerton*
s.n. (K). S: c. 8.6 km. S. of Mberengwa on Mnene Mission road, 4.v.1973, *Biegel, Pope & Simon* 4292
(COI; K; SRGH).
Also in Namibia and South Africa (Transvaal, Natal and N. Cape). Wooded grassland, miombo
woodland or scrubland, on Kalahari Sand, bare sandy soils or stony hillsides.

13. **Dicoma schinzii** O. Hoffm. in Engl., Bot. Jahrb. **15**: 543 (1893). —F.C. Wilson in Bull. Misc. Inf.,
Kew **1923**: 384 (1923). —Merxm., Prodr. Fl. SW. Afr. 139: 51 (1967). —Wild in Kirkia **8**: 185
(1972). Syntypes from Namibia.
Dicoma arenicola Muschl. ex Dinter in Fedde, Repert. **17**: 185 (1921) nom. nud.
Dicoma lanata Muschl. ex Dinter in Fedde, Repert. **17**: 186 (1921) nom. nud.

A prostrate or low suberect perennial herb to c. 60 cm. in diam. or to c. 25 cm. tall, from a
woody taproot. Stems annual, 1–several, branched, leafy, brownish or greyish lanate.
Leaves with a petiole up to c. 16 mm. long, broadly canaliculate, lanate; lamina mostly 1–8
× 0.5–2.4 cm., narrowly to broadly elliptic, acute at the apex, cuneate below, margins
obscured subserrate with callose teeth, greyish-lanate on the upper surface and more
densely so beneath; the greyish indumentum occasionally intermixed with brownish
hairs, margins and midrib beneath often markedly brownish-lanate. Capitula few to

many, solitary and terminal on branches or short shoots, subsessile and axillary, subtended by 1–several leaves. Involucre mostly 17–27 × 10–16 mm., spreading to c. 30 mm. wide, campanulate. Phyllaries numerous, stiffly chartaceous with narrow membranous margins, ± sharply ciliolate; the outer phyllaries from c. 5 mm. long, subulate-aristate, sparsely lanate, tapering to a long curved or straight apical bristle; the inner phyllaries to c. 26 mm. long, linear-lanceolate, tapering to a straight narrowly acute-pungent apex, glabrescent. Outer florets neuter; corollas rudimentary, c. 5 mm. long, filiform; ovary rudimentary, to c. 1.5 mm. long; pappus several-seriate of numerous barbellate setae 18–25 mm. long, not or hardly exceeding the involucre, overtopping the inner floret pappus. Inner florets c. 15, corollas whitish or reddish-tinged, 5–7 mm. long, infundibuliform, sparsely puberulous, limb 4–5 mm. long with erect linear lobes 3–4 mm. long, lobes subacute and sometimes recurved at the apex; achenes c. 4.5 mm. long, narrowly turbinate, strongly 10-ribbed, uniformly densely appressed-ascending setose between the ribs, the bristles less than half the achene in length, the lowermost tuft of hairs shorter and more numerous often spreading; pappus 2-seriate, the outer of 10–20 slender barbellate setae 7–11 mm. long, the inner of 5–6 more robust very narrowly scale-like setae, 7–12 mm. long, broadly membranous undulate-winged towards the base and ciliolate particularly in the upper part.

Caprivi Strip. E. of Kwando R., x.1945, *Carson* 1166 (PRE). **Botswana**. N: c. 32 km. SW. of Samochimo on track to Tsodilo Hllls, 2.v.1975, *Biegel, Müller & Gibbs-Russell* 5054 (K, SRGH). SW: Ghanzi Pan, 26.i.1969, *Brown* s.n. (K). SE: Thalamabele-Mosu area, near Soa Pan, 15.i.1974, *Ngoni* 337 (K; SRGH). **Zambia**. B: Siwelewele (Shangombo), c. 1036 m., 10.viii.1952, *Codd* 7499 (BM; K; PRE; SRGH). S: Mambo, c. 33 km. N. of Mapanza, c. 1036 m., 9.viii.1954, *Robinson* 871 (K). **Zimbabwe**. N: Sengwa Game Res., viii.1971, *Wild* 7841 (K; SRGH). W: Victoria Falls, c. 914 m., ix.1909, *Rogers* 5300 (BOL; K). C: c. 16 km. S. of Beatrice, Maas Plain, 18.vi.1971, *Wild* 7839 (K; SRGH). S: Masvingo, Makaholi Expt. Sta., 16.iii.1968, *Senderayi* 283 (K; SRGH).

Also in Angola, Namibia and South Africa (Transvaal, Orange Free State and Northern Cape). Lower rainfall sandy areas, in grassland or wooded grassland or in *Baikiaea, Colophospermum mopane*, miombo or mixed deciduous woodlands, often on Kalahari Sands.

14. **Dicoma arenaria** Bremekamp in Ann. Transv. Mus. **15**: 263 (1933). Syntypes from South Africa (N. Transvaal).

A decumbent, or low suberect perennial herb, to c. 50 cm. in diam. or to c. 25 cm. tall, from a woody taproot. Stems annual, several, branched, greyish-lanate. Leaves spathulate, mostly 15–40 × 10–22 mm., rotund-ovate in the upper half and abruptly narrowed to a winged midrib below, subacute to ± rounded mucronate at the apex, margins subserrate with callose-tipped teeth, often undulate, thinly greyish lanate on leaf upper surface and more densely so beneath, obscurely 3-nerved from the base. Capitula few to many, solitary and terminal on branches or very short shoots, sessile and axillary with several subtending leaves. Involucres mostly 15–20 × 15–20 mm., spreading to 30 mm. wide, campanulate. Phyllaries numerous, chartaceous with narrow membranous margins, ± sparsely lanate; the outer phyllaries from c. 5 mm. long, pungent-subulate; the inner phyllaries to c. 20 mm. long, linear-lanceolate. Outer florets neuter, corollas rudimentary; ovary rudimentary to c. 1.5 mm. long, long-setose from the base, pappus several-seriate of numerous barbellate setae 10–12 mm. long, exceeding that of the inner florets. Inner floret corollas c. 8 mm. long, funnel-shaped with long linear lobes; achenes c. 3 mm. long, narrowly turbinate, c. 10-ribbed, long appressed-ascending setose between the ribs; pappus 2-seriate, the outer of c. 10 slender barbellate setae 7–10 mm. long, the inner of 6–7 broadly membranous winged scale-like setae c. 7 mm. long.

Botswana. N: near Tsokotse (Chukutsa) Salt Pan, ii.1897, *Lugard* 222 (K).

Also in South Africa (N. Transvaal, mostly the Soutpansberg). Stony outcrops and grassland in sandy areas.

15. **Dicoma capensis** Less. in Linnaea **5**: 277 (1830). —Less., Synops. Composit.: 109 (1832). —DC., Prodr. **7**: 36 (1838). —Harv. in Harv. & Sond., F.C. **3**: 516 (1865). —O. Hoffm. in Engl., Bot. Jahrb. **15**: 543 (1893). —F.C. Wilson in Bull. Misc. Inf., Kew **1923**: 382 (1923). —Merxm., Prodr. Fl. SW. Afr. 139: 49 (1967). —Wild in Kirkia **8**: 185 (1972). Types from South Africa (Cape Province).

A prostrate or low spreading perennial herb from a woody taproot. Stems annual, many from the root crown, sometimes woody at the base, decumbent to c. 30 cm. long, branched, leafy, appressed greyish-lanate. Leaves subsessile or with the midrib petiole-like in the lower c. 10 mm., mostly 10–40 × 3–10 mm., ± narrowly elliptic to narrowly oblanceolate, obtuse mucronate at the apex, narrowly cuneate or attenuate below; margins finely undulate, minutely callose-serrulate; midrib sometimes ± recurved and the lamina conduplicate; upper surface thinly greyish araneose, paler lanate beneath. Capitula few to many, solitary and terminal on branches or short shoots, subsessile axillary, subtended by several leaves. Involucres mostly 12–20 × 6–10 mm., spreading to c. 25 mm. across the squarrose phyllaries, campanulate. Phyllaries numerous, stiffly chartaceous, usually ± squarrose in upper half, finely araneose-lanate where exposed, or glabrescent, finely serrulate-ciliate; the outer phyllaries from c. 34 mm. long, finely subulate, the inner to c. 20 mm. long and linear-lanceolate tapering to an acute or acuminate pungent apex. Outer florets neuter; corollas rudimentary, 4–5 mm. long, filiform; ovary rudimentary, finely setulose from the base; pappus several-seriate of numerous barbellate setae to (7)9–11 mm. long, shorter than the involucre. Inner floret corollas pale-mauve, to c. 7 mm. long, lobes linear erect but recurved at the apex; achenes to c. 3 mm. long, narrowly turbinate, strongly 10-ribbed, uniformly densely appressed-ascending setose-strigose between the ribs, the basal hairs shorter more numerous and ± spreading; pappus 2-seriate, the outer series of 15–20 slender subplumose-ciliolate setae c. 8 mm. long, the inner of c. 5 narrowly scale-like setae c. 8 mm. long, broadly white scarious-winged in the lower half, subplumose-ciliolate above.

Botswana. SW: Olifants Kloof, c. 16 km. NE. of Mamuno, c. 1220 m., 13.ii.1970, *Brown* 8 (K).
Also in Namibia and South Africa (Karoo, Namaqualand Northern Cape, and Orange Free State). In dry to very dry areas, grassland or scrubland, in sandy soils, sometimes at edges of pans.

16. **Dicoma niccolifera** Wild in Mitt. Bot. Staatss. Münch. **10**: 267 (1971); in Kirkia **8**: 186 (1972). TAB. 8 fig. A. Type: Zimbabwe, Bindura, Kingston Hill, *Wild* 7770 (K, holotype).
   Dicoma macrocephala sensu Wild in Kirkia **7**, Suppl.: 7, 10 etc., fig. 1, photo 6, (1970).
   Dicoma schinzii sensu F.C. Wilson in Bull. Misc. Inf., Kew **1923**: 384 (1923) pro parte quoad specim. *Walters* 2345, 2347. —Wild in Kirkia **5**: 81 (1965).

A straggling prostrate or low spreading bushy perennial herb from a woody taproot. Stems annual, several, to c. 60 cm. long, branched, leafy, greyish lanate. Leaves mostly 1.5–5 × 0.5–1.5 cm., subspathulate-oblanceolate, ± elliptic in the upper half and narrowing ± abruptly below the middle into a winged midrib 0.2–1.5 cm. long, acute at the apex, narrowly semi-amplexicaul at the base, margins obscured subserrate, uniformly silvery-grey thinly appressed-lanate on the upper surface, more densely so beneath. Capitula few–many, solitary and terminal on branches or short shoots, subsessile usually axillary, subtended by several leaves and leaf-like bracts. Involucre mostly 15–20 × 6–15 mm., spreading to c. 25 mm. widc, campanulate. Phyllaries numerous, stiffly chartaceous to subcoriacous with a narrow membranous margin, sharply ciliolate and often dull purplish towards the apex; the outer phyllaries from c. 5 mm. long, lanceolate-subulate, usually ± strongly recurved, araneose-lanate below; the inner phyllaries to c. 20 mm. long, linear-lanceolate; the innermost glabrescent, tapering to an acute or long acuminate often recurved apex. Outer floret corollas absent, ovary rudimentary 0.5–1 mm. long, pappus several-seriate of numerous barbellate setae to c. 20 mm. long, ± equalling the involucre in length. Inner florets c. 9; corollas pale pale-purple, 8–9 mm. long, narrowly funnel-shaped, lobes erect c. 4 mm. long linear, subacute and recurved at the apex; achenes c. 3 mm. long, turbinate, strongly c. 10-ribbed, densely whitish appressed-ascending setose between the ribs, most numerous towards the achene base and ± equalling the achene in length, the lowermost hairs shorter often spreading; pappus 2-seriate, the outer of c. 20 slender subplumose-ciliate setae 9–10 mm. long, the inner of c. 10 paleaceous setae scarious-winged in the lower part, c. 10 mm. long and subplumose-ciliate.

Zambia. C: Munale c. 8 km. E. of Lusaka, c. 1280 m., 5.viii.1955, *King* 86 (K). **Zimbabwe**. N: near Darwendale, 20.iv.1948, *Roden* 4345 (K; PRE). W: Bubi Distr., Inyati, Ndumba Hill, 15.vi.1947, *Keay* in FHI 21320 (K; SRGH). C: Shurugwi Distr., Umtebekwa R., 10.ix.1975, *Wild* 8013 (K; SRGH). E: Mutare, Mozambique Border Post, c. 1158 m., 4.v.1947, *Chase* 364 (COI; K; LISC; SRGH).
   Known only from the Flora Zambesiaca area. From serpentine soils with high nickel values, except for the plants recorded from Lusaka (*King* 86, K) and Hurungwe in Zimbabwe (*Jacobsen* 655, SRGH). For discussion of this species see Wild in Mitt. Bot. Staatss. Münch. **10**: 267 (1971).

17. **Dicoma macrocephala** DC., Prodr. **7**: 36 (1838). —Harv. in Harv. & Sond., F.C. **3**: 516 (1865). —Merxm., Prodr. Fl. SW. Afr. 139: 50 (1967). —Wild in Kirkia **8**: 187 (1972). —Hilliard, Compos. Natal: 584 (1977). Type from South Africa (Cape Province).

A low spreading perennial herb to c. 10 cm. tall, from a woody taproot. Stems annual, many from the root crown, slender, erect ascending or decumbent, up to c. 15 cm. long, branched, leafy, greyish lanate. Leaves mostly 2.5–6.5 × 0.4–1.5 cm., elliptic, acute mucronate at the apex, narrowing to a short winged midrib below, margins thickened subserrate with ± obscured callose teeth; upper surface thinly whitish araneose later glabrescent, lower surface densely greyish-white lanate. Capitula 1–several, solitary and terminal on branches or short shoots, subsessile, axillary, subtended by several leaves. Involucres mostly 23–33 × 7–15 mm., spreading to c. 25 mm. wide, narrowly campanulate. Phyllaries numerous, stiffly chartaceous; the outer from c. 6 mm. long, tapering-subulate, often squarrose, araneose soon glabrescent, sharply ciliolate on the margins; the inner to c. 32(35) × 3(4) mm., lorate-lanceolate, tapering acute and pungent. Receptacle shallowly alveolate. Outer floret corollas rudimentary, c. 6 mm. long, filiform; ovary rudimentary, long lanate-setose from the base; pappus of numerous barbellate setae to c. 25 mm. long, ± equalling the involucre. Inner floret corollas whitish-mauve, c. 6 mm. long, narrowly tubular; achenes c. 4 mm. long, narrowly turbinate and somewhat compressed, with c. 10 pronounced ± wing-like ribs, uniformly densely setose-strigose between the ribs, basal hairs shorter more numerous and ± spreading; pappus 2-seriate, outer of 15–20 slender subplumose-ciliolate setae to c. 12 mm. long, inner of c. 10 more robust very narrowly scale-like setae to c. 15 mm. long and broadly membranous undulate-winged towards the base, subplumose ciliolate above.

Botswana. SE: c. 6.5 km. S. of Tsessebe, 21.i.1960, *Leach & Noel* 295 (SRGH). **Zimbabwe.** W: Nkai Dam, 26.v.1970, *Wild* 7790 (K; SRGH). C: Gweru, Mlezu School, 18.xi.1965, *Biegel* 549 (SRGH). E: Nyanga Distr., St. Swithin's Tribal Area, Mika Hill, 20.iv.1972, *Wild & Pope* 7948 (K; LISC; SRGH). S: Makaholi Expt. Sta., 15.iii.1978, *Senderayi* 268A (K; SRGH). **Mozambique.** M: Maputo, vi.1914, Maputoland Exped. 133 (LISC).
Also in South Africa (Transvaal, Natal, Orange Free State, Northern Cape). Open or lightly wooded grassland, on serpentine soils, Kalahari Sand or poor stony soils.

18. **Dicoma tomentosa** Cass. in Bull. Sci. Soc. Philom. **1818**: 47 (1818). —Less., Synops. Composit.: 109 (1832). —DC., Prodr. **7**: 36 (1838). —A. Rich., Tent. Fl. Abyss. **1**: 459 (1847). —Oliv. & Hiern in F.T.A. **3**: 443 (1877). —Hook. f., Fl. Brit. Ind. **3**: 387 (1881). —O. Hoffm. in Engl., Bot. Jahrb, **15**: 513 (1893), Pflanzenw. Ost-Afr. **C**: 420 (1895). —Hiern, Cat. Afr. Pl. Welw. **1**, 3: 613 (1898). —F.C. Wilson in Bull. Misc. Inf., Kew **1923**: 381 (1923). —S. Moore in Journ. Bot. **65**, Suppl. 2: 65 (1927). —Mendonça, Contrib. Conhec. Fl. Angol., **1** Compositae: 140 (1943). —F.W. Andr., Fl. Sudan **3**: 23 (1956). —Adams in Hepper, Fl. W. Trop. Afr. ed. 2, **2**: 287 (1963). —Merxm., Prodr. Fl. SW. Afr. 139: 51 (1967). —C. Jeffrey in Kew Bull. **21**: 208 (1967). —Wild in Kirkia **8**: 188 (1972). Type from Senegal.

An erect annual herb, 16–65 cm. tall. Stems purplish, branched above, leafy, greyish-white lanate to glabrescent. Leaves discolorous, subsessile or midrib petiole-like in the lower c. 10 mm., mostly 10–75 × 3–12 mm., linear-elliptic to oblanceolate, often conduplicate, subacute to rounded mucronate or cuspidate at the apex, narrowly cuneate or attenuate below, callose-serrulate on the margins, green and thinly araneose to ± glabrescent on the upper surface, greyish-lanate beneath. Capitula numerous, subsessile or on short shoots, 1–15 arranged along each branch, singly or less often paired, subtended by 1–4 leaves. Involucres mostly 12–20 mm. long with phyllaries diverging to 15–20 mm. in diam., loosely obconic-campanulate. Phyllaries numerous, rigid dry-coriaceous, straight not reflexed, puberulous particularly the outer, usually purple-tinged with a darker purple stripe on either side of the midrib, margins sharply serrulate-ciliolate, from c. 3 mm. long and subulate outside to c. 18 mm. long and very narrowly triangular inside, apex long acuminate finely pungent. Corollas creamy-white, 5–6 mm. long, lobes linear erect straight or recurved at the apex; achenes c. 2.5 mm. long, somewhat compressed turbinate, with c. 10 pronounced ± wing-like ribs, somewhat glutinous, ± densely long hispid-setose between the ribs, the hairs ± equalling achene in length and obscuring the surface, the basal hairs shorter more numerous and ± spreading; pappus copious, dimorphic, the outer series of slender barbellate setae 5–6 mm. long, the inner of c. 10 more robust narrowly scale-like setae 5–6 mm. long and broadly scarious-winged, barbellate near the apex.

Botswana. N: Okavango Delta, Boro R., 23°31.25'E, 19°50.8'S, 7.v.1984, *P.A. Smith* 4440 (K; SRGH). SE: c. 1 km. from Francistown, c. 900 m., 7.iii.1961, *Richards* 14551 (K). SW: Ghanzi Camp, 10.ii.1969, *Brown* s.n. (K). **Zambia.** C: Katondwe, 4.iv.1966, *Fanshawe* 9655 (K; NDO; SRGH). S: Livingstone, 1.vi.1960, *Fanshawe* 5716 (K; SRGH). **Zimbabwe.** N: Binga, L. Kariba shore, 3.xii.1960, *Crozier & Mwanza* 7700 (SRGH). W: Hwange Distr., Sebungwe R. Drift, c. 80 km. NE. of

Kamativi Tin Mine, 13.v.1955, *Plowes* 1838 (K; LISC; SRGH). E: Odzi R., ii.1931, *Myres* 457 (K). S: Beitbridge Distr., Umzingwane R., Fulton's Drift, 26.ii.1961, *Wild* 5434 (K; SRGH). **Malawi**. S: Chikwawa Distr., Murukanyama Foothills by Mikombo R., c. 9 km. S. of Ngabu, c. 160 m., 22.iv.1980, *Brummitt & Osborne* 15518 (K; MAL; MO; SRGH). **Mozambique**. T: Baroma Prov., Msusa, Zambezi R., c. 213 m., 25.vii.1950, *Chase* 2792 (BM; COI; K; LISC; M; SRGH). MS: Rios de Sena, *Peters* 65 (K).

Widely spread in north tropical Africa (Senegal to the Sudan, Egypt and Ethiopia) east tropical Africa (Uganda, Kenya and Tanzania) and in Angola, Namibia and South Africa. Also in Socotra, W. Pakistan and western India. In the Flora Zambesiaca area it occurs at lower altitudes in low rainfall areas, on sandy or dry rocky river banks, rocky slopes, grassy floodplains, sand veld or *Colophospermum mopane*-grassland. Also a weed of old cultivation and disturbed areas.

7. GERBERA L.

Gerbera L., Opera Var. : 247 (1758) nom. conserv. —C. Jeffrey in Kew Bull. **21**: 211 (1967). —Wild in Kirkia **8**: 202 (1972). —Hilliard, Compos. Natal: 586 (1977). —H.V. Hansen in Opera Bot. **78**: 5–36 (1985).

Scapigerous perennial herbs from woody rootstocks; rootcrowns lanate, roots numerous spreading thong-like or fusiform. Leaves radical, entire serrulate to dentate or sinuate-pinnatifid, sometimes villose to closely tomentose on the lower surface, usually glabrescent. Scapes 1–several, ebracteate in the Flora Zambesiaca area, often appearing before or with the young leaves, sometimes elongating after seed-set, erect, densely villose-tomentose towards the apex, glabrescent in the lower part. Capitula solitary, heterogamous, bilabiate-radiate; corollas all 2-lipped, the outer lip strap-shaped or shortly elliptic (2)3-denticulate, the inner lip of 2 small linear lobes. Involucres broadly obconic; phyllaries numerous, imbricate, increasing in size to the inside, narrowly lanceolate or linear-triangular. Receptacle flat, shallowly alveolate. Outer florets female the marginal ones radiate and the submarginal ones equally bilabiate, staminodes usually present (often absent in *Gerbera piloselloides*). Inner florets hermaphrodite or functionally male. Corollas white yellow pink or red. Anthers with narrowly ovate apical appendages, and sagittate bases with tails entire or ciliate. Style branches of hermaphrodite flowers shortly and broadly lanceolate, the tips rounded or subacute with short pollen-sweeping hairs outside. Achenes narrowly flask-shaped, ± attenuate above or prolonged into an apical beak, 4–10-ribbed, sparsely puberulous. Pappus several-seriate of minutely barbellate setae.

An Old World genus of some 28 species, mostly South African with 3 species extending into central, east and west tropical Africa, of these 1 species (*Gerbera piloselloides*) also occurs in Asia. A further 8 species are confined to Madagascar and c. 6 species to Asia.

Gerbera jamesonii Adlam, the "barberton daisy", is cultivated as a garden ornamental in the Flora Zambesiaca area. It is indigenous to E. Transvaal, and easily recognised by its large (to 68 cm. long) pinnatisect leaves, its large capitula (to 4.8 cm. in diam.) and its showy red ray-florets and yellow inner florets. Many cultivars with variously coloured rays have been produced.

1. Rays only slightly exceeding the involucre; scape widened below the capitulum; achenes with a beak ± equalling the body of the achene in length - - - - - 3. *piloselloides*
   - Rays considerably longer than the involucre; scape not widened below the capitulum; achenes very shortly beaked - - - - - - - - - - - 2
2. Leaf lower surface glabrescent, or thinly pilose to somewhat villose but the lamina remaining clearly visible; rays white inside and red or purplish tinged outside - - 1. *viridifolia*
   - Leaf lower surface persistently and densely short white or lemon-yellow tomentose; rays usually golden-yellow in the Flora Zambesiaca area - - - - - - 2. *ambigua*

1. **Gerbera viridifolia** (DC.) Sch. Bip. in Flora **27**: 780 (1844). —Harv. in Harv. & Sond., F.C. **3**: 523 (1865). —Eyles in Trans. Roy. Soc. S. Afr. **5**: 522 (1916). —R.E. Fr., Wiss. Ergebn. Schwed. Rhod.-Kongo-Exped. **1**: 349 (1916). —C. Jeffrey in Kew Bull. **21**: 216 (1967). —Wild in Kirkia **8**, **2**: 202 (1972). —H.V. Hansen in Opera Bot. **78**: 24 (1985). —Lisowski, Aster. Fl. Afr. Centr. [in Fragm. Flor. Geobot. **36**, Suppl. 1]: 558 (1991). TAB. **9** fig. A. Type from South Africa.
   Lasiopus viridifolius DC., Prodr. **7**: 19 (1838). Type from South Africa.
   Gerbera abyssinica Sch. Bip. in Flora **27**: 779 (1844). —A. Rich., Tent. Fl. Abyss. **1**: 458 (1848). —Oliv. & Hiern in F.T.A. **3**: 445 (1877). —Engl., Hochgebirgsfl. Afr.: 452 (1892). —O. Hoffm. in Engl., Bot. Jahrb. **15**: 547 (1893); Pflanzenw. Ost-Afr. **C**: 420 (1895). —S. Moore in Journ. Linn. Soc., Bot. **40**: 123 (1911). —Eyles in Trans. Roy. Soc. S. Afr. **5**: 522 (1916). —Robyns, Fl. Parc Nat.

Tab. 9. A. —GERBERA VIRIDIFOLIA. A1, habit (×⅔), from *Robinson* 1277; A2, floret from capitulum margin, female, bilabiate with strap-shaped outer lip and 2-lobed inner lip (× 5); A3, outer floret, female, bilabiate (× 5); A4, floret from capitulum centre, functionally male, bilabiate, achene rudimentary (× 5), A2–A4 from *Phillips* 187. B. —GERBERA PILOSELLOIDES. B1, capitulum, note thickened woolly stalk (×⅔), from *Wiehe* N/238; B2, achene (× 3), from *Brummitt* 8577. Drawn by Eleanor Catherine.

Alb. **2**: 593 (1947). —Brenan in Mem. N.Y. Bot. Gard. **8**, 5: 488 (1954). Type from Ethiopia.

Gerbera plantaginea Harv. in Harv. & Sond., F.C. **3**: 522 (1865). —Eyles in Trans Roy. Soc. Bot. S. Afr. **5**: 522 (1916). Type from South Africa.

Gerbera lasiopus Bak. in Bull. Misc. Inf., Kew **1898**: 156 (1898). Type: Malawi, Misuku (Masuku) Plateau, *Whyte* s.n. (K, holotype).

Perdicium abyssinicum (Sch. Bip.) Hiern, Cat. Afr. Pl. Welw. **1**, 3: 615 (1898) quoad typum. Type as above.

Gerbera sp. sensu Eyles in Trans. Roy. Soc. Bot. S. Afr. **5**: 522 (1916) pro parte quoad specim. *Rogers* 4076.

Gerbera sp. cf. *G. viridifolia* sensu Merxm. in Proc. & Trans. Rhod. Sci. Ass. **43**: 65 (1951).

Piloselloides hirsuta sensu C. Jeffrey in Kew Bull. **21**: 219 (1967) pro parte, quoad syn. *Gerbera lasiopus*.

Gerbera ambigua sensu Hilliard, Compos. Natal: 591 (1977) pro parte, quoad syn. *Lasiopus viridifolius, Gerbera viridifolia, Gerbera plantaginea* et *Gerbera viridifolia* var. *woodii*.

An acaulescent perennial herb, from a thickened woody rootstock; rootcrowns densely silky-lanate; roots numerous brown thong-like, c. 2 mm. in diam. Leaves radical, ascending or spreading, very variable, usually petiolate; petiole up to c. 20 cm. long but usually shorter; lamina up to c. 22(28) × 14 cm., but usually smaller, elliptic or oblong, rounded mucronate at the apex, rounded or more usually cordate below and sometimes shortly decurrent on the petiole, sinuate-denticulate undulate or coarsely crenate-denticulate on the margins, upper surface at first pilose-hispid or sometimes sparsely long-pilose, usually glabrescent, lower surface thinly crisped pubescent to glabrescent sometimes with weak pilose hairs, the indumentum occasionally intermixed with glandular-capitate hairs. Scapes one–several, often appearing before the leaves, erect slender, up to c. 45 cm. long, densely brown-tomentose or white-lanate towards the apex, glabrescent to the base; indumentum of short crisped hairs intermixed with scattered long pilose hairs, occasionally also with few–numerous glandular-capitate hairs. Capitula solitary erect; involucres mostly 10–18 × 20–30 mm., broadly obconic; phyllaries from c. 4 mm. long outside increasing to c. 20 mm. long inside, narrowly lanceolate to narrowly elliptic, lanate-pilose to crisped-pubescent or glabrescent, often also ± densely glandular-capitate. Outer florets female, of two kinds; a marginal series of ray-florets with the outer lip erect c.(8)15(24) × 1.7 mm., oblong-elliptic, minutely 3-dentate, usually white inside and reddish or purplish on the margins and outside (in the Flora Zambesiaca area), the inner lip of 2 c. 2 mm. long linear often curled lobes, the tube c. 4(6) mm. long cylindric, pappus copius several-seriate of minutely barbellate setae 6–8 mm. long, achenes c. 6 mm. long and narrowly flask-shaped ± beaked when mature, 6–10-ribbed, minutely hispid; a submarginal series of bilabiate florets more numerous than the outer, corollas to c. 7 mm. long slender-tubular, the outer lip 2–3 mm. long and minutely 3-dentate, the inner lip of 2 erect linear lobes, the pappus and achenes as for the outer florets. Central florets numerous, bilabiate, functionally male, corollas whitish, 8–9 mm. long, the limb narrowly cylindric tapering below, 2-lipped above, the outer lip broadly elliptic and minutely 3-dentate, the inner lip with 2 linear curved lobes; achenes to c. 6 mm. long, somewhat flattened not swelling, minutely pubescent, pappus as for outer florets.

Zambia. N: N. Kalunga R., 31.xii.1907, *Kassner* 2270a (BM; K). W: Solwezi Distr., Lusala R. west of Mutanda Bridge, 17.vii.1930, *Milne-Redhead* 723 (K). C: Chakwenga Headwaters, 100–129 km. E. of Lusaka, 6.x.1963, *Robinson* 5687 (K). E: Nyika Plateau, c. 6 km. S. of Nganda Hill, c. 2133 m., 24.xi.1955, *Lees* 88 (K). S: c. 21 km. E. of Choma near Manzama R., c. 1300 m., 28.v.1955, *Robinson* 1277 (K; SRGH). **Zimbabwe**. N: Mazowe, c. 1463 m., viii.1960, *Eyles* 400 (BM). C: Harare, 6.i.1937, *Eyles* 8879 (K; SRGH). E: Mutare Distr., Imbeza Valley, c. 1100 m., 21.ix.1956, *Chase* 6206 (K; LISC; SRGH). S: Chibi Distr., Madzivire Dip, c. 6 km. N. of Runde (Lundi) R. Bridge, 3.v.1962, *Drummond* 7900 (K; SRGH). **Malawi**. N: Nyika Plateau, c. 40 km. SE. of Rest House on Nchenachena Rd., c. 2400 m., 28.x.1958, *Robson* 429 (BM; K). C: Nkhota Kota Distr., Chenga Hill, c. 1600 m., 9.ix.1946, *Brass* 17599 (K). S: Zomba Distr., Naisi, 16.viii.1967, *Salubeni* 822 (K; LISC; MAL). **Mozambique**. N: Maniamba, 24.viii.1934, *Torre* 421 (COI). MS: Chimanimani Mts., 6.vi.1949, *Munch* 168 (K; SRGH). M: Maputo Distr., Goba Fronteira, 28.vi.1961, *Balsinhas* 492 (K; LISC).

Also in Somalia, Ethiopia, Sudan, Cameroon, Zaire, Angola, Uganda, Rwanda, Burundi, Kenya, Tanzania, Lesotho, Swaziland and South Africa (Transvaal, Natal, Orange Free State and the Cape Province). A pyrophyte of submontane grasslands, less often in open *Brachystegia* woodlands and dambos.

In the Flora Zambesiaca area and further north the two species *Gerbera viridifolia* and *Gerbera ambigua* are readily separated by a clear discontinuity in features of leaf indumentum and in ray colour (see key above). South of the Limpopo River, however, these distinctions become less clear –specimens with the indumentum of the leaf underside intermediate between tomentose and pilose-glabrescent can be found. More importantly, yellow-rayed capitula are not reliably diagnostic in these plants, being found also in plants with leaves glabrous on the lower surface, ie. yellow rays are no longer only associated with leaves yellow-tomentose beneath. In the South African plants most capitula (of both taxa) have rays white inside and reddish outside.

These variations are apparently not correlated with habitat or geographical distribution, and are found in plants ranging from the Transvaal Highveld through the Natal Drakensberg, Lesotho and Swaziland to coastal Natal.

Hilliard was unable to keep *G. viridifolia* and *G. ambigua* separate and combined them, at the same time recognising the species *G. kraussii* Sch. Bip. The characters which enabled her to separate *G. kraussii* in Natal, however, break down in the Flora Zambesiaca area.

Hansen, in his revision of the Section *Lasiopus (Gerbera)*, places more emphasis on the leaf indumentum and keeps *G. viridifolia* and *G. ambigua* separate throughout their range of distribution. In this Flora treatment it is considered best to follow Hansen and keep the two separate, as indeed they are in the Flora Zambesiaca area.

2. **Gerbera ambigua** (Cass.) Sch. Bip. in Flora **27**: 780 (1844). —Harv. in Harv. & Sond., F.C. **3**: 522 (1865). —C. Jeffrey in Kew Bull. **21**: 218 (1967). —Wild in Kirkia **8**, 2: 204 (1972). —Hilliard, Compos. Natal: 591 (1977) pro parte quoad typum et syn. *Gerbera discolor.* —H.V. Hansen in Opera Bot. **78**: 29 (1985). —Lisowski, Aster. Fl. Afr. Centr. [in Fragm. Flor. Geobot. **36**, Suppl. 1]: 559, fig. 114 (1991). Type from South Africa.

Lasiopus ambiguus Cass. in Bull. Sci. Soc. Philom. Paris **1817**: 152 (1817); in Dict. Sci. Nat. **25**: 297 (1822). Type as above.

Gerbera discolor Sond. ex Harv. in Harv. & Sond., F.C. **3**: 522 (1865). —Eyles in Trans. Roy. Soc. S. Afr. **5**: 522 (1916). —Merxm. in Proc. & Trans. Rhod. Sci. Assn. **43**: 65 (1951). Syntypes from South Africa.

Gerbera welwitschii S. Moore in Journ. Bot. **54**: 284 (1916); **65**, suppl. 2: 66 (1927). —Mendonça, Contrib. Conhec. Fl. Angol., 1 Composit.: 147 (1943). Lectotype from Angola (chosen by *Hansen* loc. cit.).

Gerbera welwitschii var. *velutina* S. Moore in Journ. Bot. **54**: 285 (1916). Syntypes: Zimbabwe, Mt. Pene, *Swynnerton* 1821 (BM), 6113 (BM), Chirinda, *Swynnerton* s.n. (BM).

Gerbera flava R.E. Fr., Wiss. Ergebn. Schwed. Rhod.-Kongo-Exped. **1**: 348, t. 22, fig. 3, 4 (1916). Type: Zambia, Bwana Mkubwa, *Fries* 424 (UPS).

Gerbera randii S. Moore in Journ. Bot. **64**: 305 (1926). Type: Zimbabwe, Mwami (Miami), *Rand* 179 (BM).

An acaulescent perennial herb, from a thickened woody rootstock; rootcrowns densely silky-lanate; roots numerous, brown and thong-like, c. 2 mm. in diam. Leaves radical, ascending or spreading, very variable, usually petiolate; petiole up to c. 26 cm. long, but usually shorter; lamina up to c. 20 × 6.5(9) cm., but usually smaller, broadly oblong-elliptic or oblanceolate, rounded or sub-obtuse mucronate at the apex, cuneate to ± rounded below and shortly somewhat asymmetrically decurrent on the petiole, rarely sub-cordate, margins entire to sinuate-denticulate undulate, upper surface ± patent hispid-pilose to glabrescent, lower surface whitish to lemon-yellow appressed lanate-tomentose. Scapes one–several, often appearing before the leaves, erect slender, up to c. 45 cm. long, densely brown tomentose often becoming white lanate towards the apex, glabrescent to the base, indumentum dense of short crisped hairs intermixed with few to numerous long pilose hairs. Capitula solitary erect; involucres mostly 8–14 × 12–25 mm., broadly obconic; phyllaries from c. 4 mm. long outside to c. 14 mm. long inside, narrowly lanceolate, crisped-pubescent to tomentose or densely lanate-pilose outside, often glabrescent. Outer florets female, of two kinds; a marginal series of ray-florets with the outer lip erect 8–12 × 1–2.5 mm., oblong-elliptic or narrowly oblanceolate, minutely 3-dentate, yellow in the Flora Zambesiaca area, the inner lip of 2 c. 1.2 mm. long linear often curled lobes, pappus copius several seriate of minutely barbellate setae 4–6 mm. long, achenes to c. 6 mm. long narrowly flask-shaped and ± beaked when mature, 6–10-ribbed, sparsely to densely minutely hispid; a submarginal series of fewer bilabiate florets, corollas 6–7 mm. long, the outer lip 1.5–2 mm. long and minutely 3-dentate, the inner lip of 2 erect linear lobes, the pappus and achenes as for the outer florets. Central florets numerous bilabiate, functionally male, corollas yellow, 7–8 mm. long, the limb narrowly cylindric and tapering below, 2-lipped, the outer lip minutely 3-dentate, the inner lip with 2 linear curved lobes; achenes to c. 6 mm. long, somewhat flattened not swelling, minutely pubescent, pappus as for outer florets.

Zambia. N: Mbala Distr., Zombe Plain, Lombwe drainage, c. 457 m., 11.x.1966, *Richards* 21517 (K; M; SRGH). W: Kitwe, 17.xi.1964, *Mutimushi* 2354 (K; NDO). **Zimbabwe.** W: Matobo Distr., Besna Kobila Farm, c. 1430 m., i.1954, *Miller* 2035 (K; SRGH). C: Harare Distr., Gwebi R. near Mt. Hampden, 6.ix.1955, *Drummond* 4886 (K; LISC; S; SRGH). E: Nyanga, c. 2040 m., 18.x.1946, *Wild* 1415 (K; SRGH). **Malawi.** N: Mzimba Distr., 5 km. W. of Mzuzu, Katoto, 1310 m., 23.v.1970, *Brummitt & Pawek* 11068 (K). S: Mt. Mulanje, Litchenya Plateau, 1995 m., 8.xi.1986, *J.D. & E.G. Chapman* 8201 (K). **Mozambique.** MS: Manica Plateau, NW. of Musapa, c. 2040 m., 6.ix.1957, *Chase* 6704 (K;

SRGH). M: Libombos near Namaacha, Mt. Mpondium, 800 m., 22.ii.1955, *Exell, Mendonça & Wild* 535 (BM).

Also in Angola, Zaire, Tanzania, Lesotho, Swaziland and South Africa (Transvaal, Natal, Orange Free State and Cape Province). A pyrophyte of submontane grasslands (sometimes on termitaria) and high altitude or plateau *Brachystegia* woodlands and dambos.

See notes under *G. viridifolia.*

3. **Gerbera piloselloides** (L.) Cass. in Dict. Sci. Nat. **18**: 461 (1821). —Harv. in Harv. & Sond., F.C. **3**: 522 (1865). —Oliv. & Hiern in F.T.A. **3**: 445 (1877). —O. Hoffm. in Engl., Bot. Jahrb. **15**: 546 (1893); in Engl., Pflanzenw. Ost-Afr. **C**: 420 (1895). —S. Moore in Journ. Linn. Soc., Bot. **40**: 124 (1911); in Journ. Bot. **65**, Suppl. 2: 66 (1927). —Eyles in Trans. Roy. Soc. S. Afr. **5**: 522 (1916). —Mendonça, Contrib. Conhec. Fl. Angol., 1 Compositae: 146 (1943). —Robyns, Fl. Parc Nat. Alb. **2**: 593 (1947). —Brenan in Mem. N.Y. Bot. Gard. **8**, 5: 488 (1954). —Hilliard, Compos. Natal : 594 (1977). —H.V. Hansen in Opera Bot. **78**: 19 (1985). —Lisowski, Aster. Fl. Afr. Centr. [in Fragm. Flor. Geobot. **36**, Suppl. 1]: 556, fig. 114 (1991). TAB. **9** fig. B. Type from South Africa (Cape; not traced, a paratype at OXF fide C. Jeffrey).

Arnica piloselloides L., Pl. Afr. Rar.: 22 (1760). Type as above.

Arnica hirsuta Forssk., Fl. Aegypt.-Arab.: 151 (1775). Type from Yemen.

Perdicium piloselloides (L.) Hiern, Cat. Afr. Pl. Welw. **1**, 3: 615 (1898) nom. illegit. non Vahl (1791).

Piloselloides hirsuta (Forssk.) C. Jeffrey ex Cuf. in Bull. Jard. Bot. Nat. Belg. **37**, 3: 1180 (Sept. 1967). —C. Jeffrey in Kew Bull. **21**: 218 (Nov. 1967) pro parte excl. syn. *Gerbera lasiopus.* —Wild in Kirkia **8**, 2: 206 (1972). Type as above.

An acaulescent perennial herb, from a thickened woody rootstock; rootcrowns densely long silky-lanate; roots numerous slender thong-like. Leaves few to many, radical, ascending or spreading; petiole up to c. 10 cm. long, but usually shorter or absent; lamina usually turning dark-brown on drying, 4–20(30) × 1.5–7 cm., broadly oblanceolate, rounded or sub-obtuse mucronulate at the apex, attenuate sometimes broadly cuneate, or rounded below and ± decurrent on the petiole, margins entire, upper surface sparsely pilose-hispid to glabrescent, lower surface thinly pilose or araneose to glabrescent; midrib and margins silky-villose, more densely so towards the base and on the rootcrown. Scapes solitary sometimes several, appearing with the young leaves, erect slender, elongating after seed-set to c. 90 cm. long, widening at the apex, brown pilose to glabrescent in the lower part, villose above becoming more densely so and usually golden-brown just below the apex, the hairs one or more times obliquely septate. Capitula solitary erect; involucres mostly 16–28 × 15–35 mm., broadly obconic; phyllaries drying dark-purplish or brown, from c. 12 mm. long outside to c. 26 mm. long inside, linear-triangular tapering to an acute apex, hispid-pilose to glabrescent, the inner with hyaline often undulate margins. Outer florets female, of two kinds; a marginal series of ray-florets with the outer lip strap-shaped, usually creamy-white and tinged reddish below, or mauve to dark-red, rays erect c. 7 × 1 mm. exceeding the pappus by up to c. 6 mm., linear minutely 3-dentate, the inner lip of 2 linear often curled lobes 2–3 mm. long, the tube c. 7 mm. long cylindric, pappus copius, several seriate of minutely barbellate setae 6–11 mm. long, achenes 8–11 mm. long increasing to c. 19 mm. long including a beak to c. 14 mm. long in mature achenes, narrowly flask-shaped and long-beaked at the apex, narrowly 4–8-ribbed, sparsely puberulous; a submarginal series of fewer florets, bilabiate not markedly strap-shaped, corollas 9–10 mm. long, slender-tubular, the outer lip 1.5–2 mm. long and minutely 3-dentate, the inner lip of 2 erect linear lobes, the pappus and achenes as for the outer ray-florets. Central florets creamy-white, numerous, functionally male, corollas to c. 11 mm. long, 2-lipped, the outer lip minutely 3-dentate, the inner lip with 2 linear curved lobes; pappus and achenes as for ray-florets.

Zambia. N: Mbala Distr., Kambole Escarpment, Katanga R. above Lusango Waterfall, c. 1500 m., 14.ix.1960, *Richards* 13259 (K). W: Mwinilunga Distr., c. 0.8 km. S. of Matonchi Farm, 6.xii.1937, *Milne-Redhead* 3524 (BM; K; LISC). C: Mkushi Distr., Fiwila, c. 1220 m., 8.i.1958, *Robinson* 2687 (K; SRGH). E: Chadiza, c. 850 m., 28.xi.1958, *Robson* 769 (K; SRGH). S: Mumbwa, 1911, *Macaulay* 344 (K). **Zimbabwe.** W: Matopos Distr., American Mission, x.1905, *Gibbs* 257 (BM). C: Kadoma Distr., Chengiri N.P.A., Gadze R., 5.ix.1963, *Bingham* 822 (K; M; SRGH). E: Chimanimani, Gungunyana For. Res., c. 1066 m., x.1961, *Goldsmith* 82/61 (K; LISC; M; SRGH). **Malawi.** N: Mzimba Distr., Viphya, c. 18 km. S. of Chikangawa, c. 1615 m., 23.xii.1970, *Pawek* 4136 (K; MAL). C: Dedza, c. 1500 m., 13.ix.1944, *Brass* 17630 (K; MO; SRGH). S: Blantyre Distr., Ndirande Mt., 1450–1570 m., 12.iii.1970, *Brummitt* 9038 (K). **Mozambique.** N: Lichinga (Vila Cabral), 8.i.1935, *Torre* 293 (COI). Z: Alto Molócuè, km. 16, estrada para Alto Ligonha, 29.xi.1967, *Torre & Correia* 16273 (LISC). MS: Chimanimani Mts., 6.vi.1949, *Munch* 200 (K; SRGH). M: Maputo, 30.x.1897, *Schlechter* 11533 (BM; COI; E; K; PRE; S).

Also in Ethiopia, Somalia, Sudan, Uganda, Rwanda, Burundi, Kenya, Tanzania, Cameroons, Zaire, Angola, Lesotho, South Africa (Transvaal, Natal, Orange Free State, and Cape Province). Outside Africa it occurs in Madagascar, the Yemen, E. Himalayas, Assam, Thailand, Indo-China, South China and Java. A pyrophyte of submontane and plateau grasslands, in dambos, open plateau miombo and *Brachystegia/Uapaca* woodland.

Tribe 2. C A R D U E A E Cass.

Cardueae Cass. in Journ. Phys. **88**: 155 (1819). —C. Jeffrey in Kew Bull. **22**: 107–140 (1968). —Dittrich, *Cynareae*, Syst. Rev. in Heywood et al., Biol. & Chem. Comp.: 999–1015 (1977). "Cynareae".

Annual or perennial herbs (seldom shrubs), often spiny, usually robust. Leaves cauline and alternate, or radical. Capitula mostly homogamous and discoid or, as in *Centaurea* in the Flora Zambesiaca area, heterogamous and radiant with the marginal florets neuter and often slightly larger than the perfect inner florets. Phyllaries many-seriate, imbricate, cartilaginous or scarious, spine-tipped or with a spiny or expanded membranous appendage, margins entire to spinulose or lacerate. Receptacle densely paleaceous or setose (rarely alveolate or naked). Corollas white, purple or blue, sometimes yellow. Anther bases strongly caudate, rarely merely sagittate. Styles of hermaphrodite florets bifid, branches short flat narrow and ± obtuse, usually with a thickened often minutely hairy ring below the fork. Achenes ± terete or compressed, erostrate, attachment-scar basal and ± oblique or horizontal, or lateral and ± concave; pappus many-seriate, of narrow scales or bristles. Basic pollen type spherical, very spiny and tricolporate. Haploid chromosome numbers recorded are: 8,9,10,11,12,13,15,17,19.

A tribe of about 50 genera distributed throughout Europe and Asia, particularly in the Mediterranean region and the Near East, a few also represented in N. and S. America, Australia, North Africa and tropical east Africa.

The tribal epithet *Cardueae* has priority over the name *Cynareae* of Lessing.

Key to the genera

1. Leaves, and stem-wings if present, not spinescent at the margins; outer phyllaries with a distinct apical appendage spinose on the margins; pappus of scale-like setae free to the base; capitula heterogamous, radiant; plants not thistle-like - - - - - - **10. Centaurea**
   – Leaves and stem-wings spinescent at the margins; outer phyllaries without obvious apical appendages, apices ± spinescent, sometimes pectinate-spinescent or lacerate on the margins about the apex; pappus setae connate at the base into a ring; capitula homogamous; plants thistle-like - - - - - - - - - - - - - - - - 2
2. Pappus setae feathery; margins of outer phyllaries pectinate-spinescent **9. Cirsium**
   – Pappus setae not feathery; phyllary margins entire - - - - - **8. Carduus**

8. CARDUUS L.

Carduus L., Sp. Pl. **2**: 820 (1753); Gen. Pl., ed. 5: 358 (1754). —Kazmi in Mitt. Bot. Staatss. Münch. **5**: 139–198 (1963); **5**: 279–550 (1964). —C. Jeffrey in Kew Bull. **22**: 123 (1968).

Perennial or annual herbs. Stems erect, simple, or branched above, usually spiny-winged. Leaves usually decurrent on the stem, pinnatifid in the Flora Zambesiaca area (or undivided and serrate, or sinuate-dentate to 2-pinnatisect), spinescent at the margins. Capitula homogamous, discoid, in dense terminal clusters, sometimes solitary. Involucre globose or campanulate, glabrous to densely araneose; phyllaries imbricate in many series, ± deflexed in the upper part, coriaceous or scarious, narrowly lanceolate and spine-tipped, margins entire scabrous or ± ciliate (serrate or fimbriate). Receptacle flat or convex, pitted, densely setose. Florets hermaphrodite; corollas purple, pink or white, deeply 5-lobed (sometimes ± 2-lipped); anther bases sagittate, adjacent auricles connate and produced into an entire or lacerated tail; style shortly and obtusely 2-lobed, abruptly thickened below the fork with a ring of hairs below the thickening. Achenes glabrous, ovate-oblong, somewhat angled with a raised annular collar at the apex, attachment-scar basal and oblique or ± horizontal; pappus many-seriate of scabrid setae united at the base.

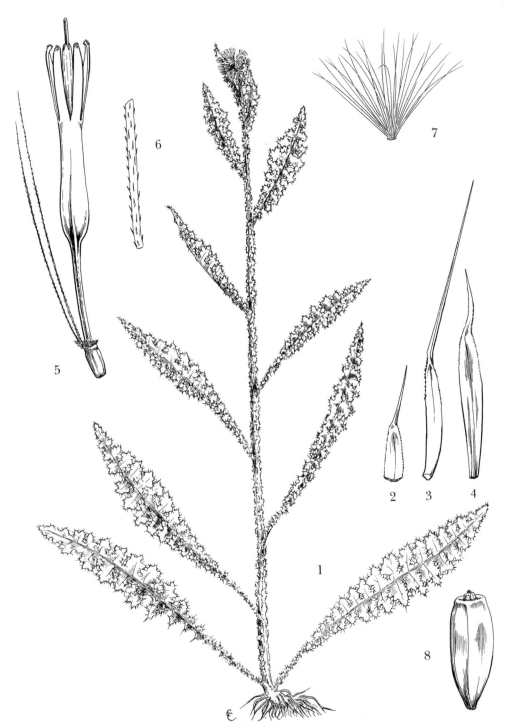

Tab. 10. CARDUUS NYASSANUS. 1, habit (×⅓); 2, outer phyllary; 3, middle phyllary; 4, inner phyllary
(all × 4); 5, floret with all except two pappus setae cut away (× 6), 1–5 from *Robson* 365; 6, detail of
pappus seta (× 16); 7, pappus showing setae united at the base (× 2); 8, achene (× 6), 6–8 from
Fanshawe 7316. Drawn by Eleanor Catherine.

A genus of about 95 species, most of which are native to Europe, central Asia, North Africa and the Canary Isles, with a few in east Africa. Several species have become widespread weeds. (Thistles).

Carduus nyassanus (S. Moore) R.E. Fr. in Acta Hort. Berg. **8**: 25, t. 3 (7a–f) (1923). —Kazmi in Mitt. Bot. Staatss. Münch. **5**: 158 (1963). —C. Jeffrey in Kew Bull. **22**: 126 (1968) pro parte excl. *C. steudneri* (Engl.) R.E. Fr. —Maquet in Fl. Rwanda, Spermat. **3**: 676, fig. 212(1) (1985). —Lisowski, Aster. Fl. Afr. Centr. [in Fragm. Flor. Geobot. **36**, Suppl. 1]: 589 (1991). TAB. **10**. Type: Malawi, Mt. Malosa, *Whyte* s.n. (K, lectotype, selected by Jeffrey); *Buchanan* 219 (K, lectoparatype), 577 (BM, lectoparatype); *Henderson* s.n. (BM, lectoparatype).

Carduus leptacanthus var. *nyassanus* S. Moore in Journ. Linn. Soc., Bot. **37**: 326 (1906). Types as above.

An erect spinose perennial herb, to c. 1.5 m. tall. Stems simple, or sparingly branched above, stout, coarsely striate, ± pubescent, ± crowded leafy below, leaves fewer and smaller above, longitudinally winged throughout with wings interrupted or sinuate; wing-segments up to c. 1.5 cm. wide and strongly spinose on the margins. Basal leaves 10–33(50) × 2–6.5 cm., narrowly oblanceolate in outline, pinnately divided, pubescent especially on the midribs, or glabrescent; leaf-segments up to c. 3.5 cm. long near the leaf apex and decreasing in size towards the leaf base, margins coarsely dentate with the teeth and apices strongly spine-tipped. Capitula many, densely aggregated in terminal clusters 3.5–7 cm. in diam.; involucres up to c. 2.5 cm. long, narrowly campanulate. Phyllaries numerous, many-seriate, narrowly lanceolate, tapering to apical spines, the spines patent and increasingly longer on the inner phyllaries, the lamina stamineous or purple-tinged, sparsely puberulent outside and ciliolate on the margins, the outer phyllaries from c. 10 mm. long, the middle to c. 20 mm. long, the innermost shorter. Corollas white, reddish-tinged or mauve, 10–13 mm. long, very narrowly infundibuliform, limb 7–8 mm. long, glabrous, lobes 2.5–4 mm. long. Achenes stramineous or pale-brown, 4–4.5 mm. long, narrowly turbinate and somewhat 3–4-ribbed, glabrous; pappus several-seriate, setae 11–12 mm. long, numerous, flattened, barbellate, united at the base in a deciduous ring.

Zambia. E: Nyika, 30.xii.1962, *Fanshawe* 7316 (K; NDO; SRGH). **Malawi**. N: Nyika Plateau, near Chelinda Chalets, 11.i.1960, *Hilliard & Burtt* 4449 (K). C: Dedza Distr., Ngoma Forest Nursery, 24.viii.1968, *Salubeni* 1133 (SRGH). S: Zomba Plateau, 28.xi.1967, *Banda* 954 (SRGH).

Also in Nigeria, Cameroon, Zaire, Rwanda, Burundi, Ethiopia, Sudan, Tanzania, Kenya and Uganda. Stream-sides and moist localities in montane grassland.

9. CIRSIUM Mill.

Cirsium Mill., Gard. Dict., abr. ed. 4 (1754). —C. Jeffrey in Kew Bull. **22**: 129 (1968).

Perennial or biennial (annual) herbs. Stems spiny-winged from decurrent leaf bases, rarely unarmed. Leaves pinnatifid in the Flora Zambesiaca area (entire) with spine-tipped teeth at the margins. Capitula homogamous, discoid, loosely aggregated (or large solitary and terminal) on the branches. Phyllaries numerous many-seriate imbricate appressed, ± patent apically and usually spine-tipped, with or without subterminal vitta, margins entire to spinulose or lacerate. Receptacle flat or convex, densely long-setose. Florets hermaphrodite (rarely unisexual); corollas white to purple (yellowish), narrowly tubular with a broad deeply 5-lobed limb. Anther bases sagittate, the adjacent auricles connate and produced into entire or lacerate tails. Style 2-branched, with an abrupt swelling below the branches and a ring of hairs at or below the swelling; the branches erect, long somewhat flat and ± connate, diverging only near the tips. Achenes ± oblong, compressed smooth, glabrous with a raised annular collar at the apex, attachment-scar sub-basal oblique or ± horizontal. Pappus many-seriate; setae plumose, connate into a ring at the base, deciduous (± persistent), longest in the innermost series.

A genus of 250–300 species, mainly in Europe, Asia and N. America with a few in North Africa. Only one species occurs naturally in the Flora Zambesiaca area. A few species are widespread weeds. (Thistles).

Cirsium buchwaldii O. Hoffm. in Engl., Bot. Jahrb. **38**: 211 (1906). —C. Jeffrey in Kew Bull. **22**: 130 (1968). —Agnew, Upland Kenya Wild Fl.: 493 (1974). —Maquet in Fl. Rwanda, Spermat. **3**: 678, fig. 210(2) (1985); in Bull. Jard. Bot. Nat. Belg. **55**: 71 (1985). —Lisowski, Aster. Fl. Afr. Centr. [in Fragm. Flor. Geobot. **36**, Suppl. 1]: 594 (1991). TAB. **11**. Types from Tanzania.

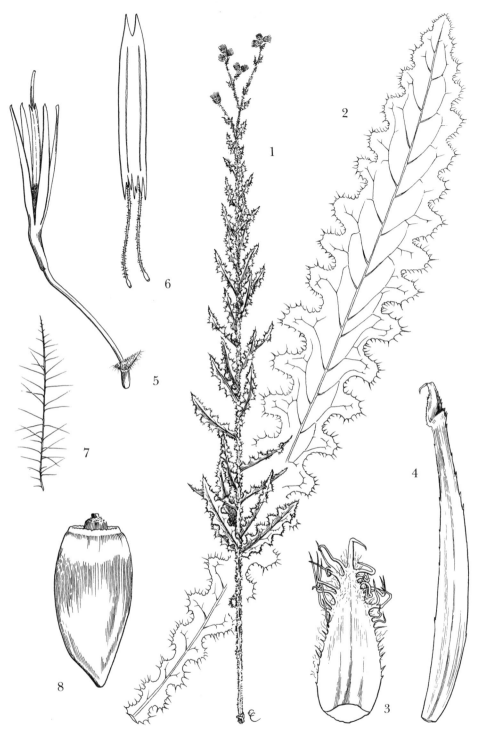

Tab. 11. CIRSIUM BUCHWALDII. 1, habit (× ½), from *Brummitt* 10491; 2, basal leaf (× ⅔); 3, outer phyllary (× 6); 4, inner phyllary (× 6); 5, floret with pappus setae cut away near the base (× 4); 6, stamens (× 8); 7, plumose pappus seta; 8, achene (× 10), 2–8 from *Richards* 4991. Drawn by Eleanor Catherine.

A robust spiny, stiffly erect, rhizomatous perennial herb. Rhizome up to c. 2.5 cm. thick with numerous white fleshy roots. Stems simple, up to c. 1.5 m. tall, exceptionally to 2 m. tall, sparingly branched above, longitudinally winged, leafy, puberulent below, white araneose above; wings interrupted, spinose-dentate. Leaves numerous, decurrent on the stem; basal leaves rosulate, spreading, up to c. 30 × 4.5 cm., very narrowly elliptic, regularly sinuately- or bluntly pinnately-lobed, less deeply cut than the cauline leaves, lobes spinose-dentate, basal leaves withering before the cauline leaves; cauline leaves discolorous, ascending-appressed, up to c. 15 × 6 cm., more usually c. 8–10 cm. long, elliptic to lanceolate in outline, pinnatilobed with the lobes usually split into 2 narrowly triangular spine-tipped segments, the lamina upper surface green and thinly araneose at first becoming scabridulous, the lower surface densely white araneose often with a brownish nervation. Capitula 3–10, ± loosely aggregated at the stem apex; involucres 13–20 mm. long, slightly longer than broad, campanulate, often subtended by 1–several very spiny foliaceous bracts. Phyllaries coriaceous, progressively longer towards the inside, from c. 4 mm. long and narrowly ovate outside to 16–20 mm. long and linear-lanceolate inside; the outer phyllaries with an indurated apical spine and strongly pectinate-ciliate margins, the spines and cilia smaller in the inner series of phyllaries; the innermost phyllaries fimbriate-ciliate, sometimes the apical cilia ± fused into an expanded hyaline margin. Receptacle densely setose, the setae c. 13 mm. long. Corollas white to pinkish-blue, glabrous, 14–16.5 mm. long; the tube narrowly cylindrical widening abruptly into a broadly cylindric deeply-lobed limb 8–9 mm. long; the lobes 4–5 mm. long, strap-shaped. Achenes stramineous or pale-brown, smooth and shiny, c. 4.5 mm. long, narrowly obovoid, obscurely 3–4-angular with a raised apical annulus; pappus several-seriate of numerous plumose setae 10–13 mm. long, the setae united at the base into a deciduous ring.

Zambia. N: Nkali (Kali) Dambo, 1530 m., 17.iii.1955, *Richards* 4991 (K; SRGH). **Malawi**. N: Rumphi Distr., Nyika Plateau below Sangule Kopje, 7 km. SW. of Chelinda Camp, 2255 m., 15.v.1970, *Brummitt* 10770 (K; SRGH).
Also in the Sudan, Uganda, Rwanda, Kenya and Tanzania. Common in marshy situations in montane grassland.
The basal leaves and rhizomes are seldom represented in herbarium material of this formidable species and hence poorly known.

10. CENTAUREA L.

Centaurea L., Sp. Pl. **2**: 909 (1753); Gen. Pl., ed. 5: 389 (1754). —Philipson in Journ. Bot. **77**: 227–233 (1939). —C. Jeffrey in Kew Bull. **22**: 134 (1968).

Perennial or annual herbs (shrubs), hirsute-scabridulous or tomentose-araneose, often with sessile glands, rarely glabrous. Leaves alternate (or radical), entire, dentate or sometimes pinnatilobed (spiny), bases sometimes decurrent on the stem. Capitula solitary (or paniculate), heterogamous, radiant. Involucre ovoid-subglobose, hemispheric or nearly cylindric, araneose or glabrous; phyllaries many-seriate, imbricate, coriaceous, with a ± patent pectinate or spinose appendage. Receptacle densely setose, the setae smooth and twisted. Florets tubular, the outer series neuter sometimes with staminodes, the inner florets hermaphrodite; corollas whitish, purple, pink, blue or yellow, slender tubular below, dilated and deeply 5(8)-lobed above; anther bases sagittate, adjacent lobes connate into a short or long tail; style 2-branched, abruptly swollen below the branches with a ring of hairs at or below the swelling; the branches erect, long somewhat flat and ± connate, diverging only near the tips. Achenes oblong or obovoid, ± laterally compressed, apex rounded to truncate, glabrous when ripe, attachment-scar lateral-oblique, ± concave. Pappus many-seriate of unequal barbellate (or plumose) scale-like setae, persistent (caducous or absent).
A genus of about 500 species, mostly natives of the Mediterranean area and SW Asia, with a few species occurring in tropical Africa and N. and S. America. Several species are widely distributed as weeds, one cultivated in gardens as an ornamental (*C. cyanus* L., "Cornflower").

Corollas whitish or purplish; leaves not decurrent on the stem; plants perennial from a woody
    rootstock; capitula often radical and precocious - - - - - - 1. *praecox*
Corollas yellow; leaves decurrent on the stem; plants annual; capitula not radical nor precocious
    2. *melitensis*.

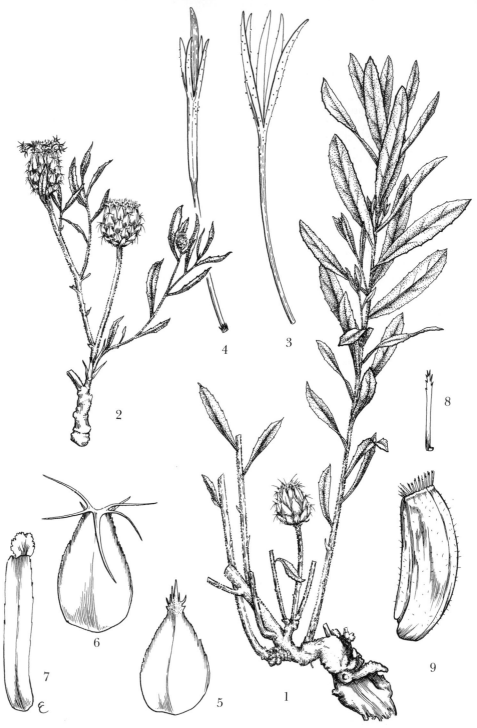

Tab. 12. CENTAUREA PRAECOX. 1, habit, capitula with mature achenes, leaves fully developed, from *Biegel* 4112; 2, habit, capitula in flower, young leaves; 3, outer floret, sterile; 4, inner floret, fertile; 5,6 & 7 outer, middle & inner phyllaries, 2–7 from *Angus* 3772; 8, pappus scale (× 20); 9, achene with lateral attachment scar (× 5), 8 & 9 from *Greenway & Brenan* 8111. (Habits × ⅔, florets × 4, phyllaries × 3). Drawn by Eleanor Catherine.

1. **Centaurea praecox** Oliv. & Hiern in F.T.A. **3**: 438 (1877). —Philipson in Journ. Bot. **77**: 231 (1939). —F.W. Andrews, Fl. Pl. Sudan **3**: 15 (1956). —Adams in F.W.T.A. 2, **2**: 291 (1963). —C. Jeffrey in Kew Bull. **22**: 134 (1968). —Lisowski, Aster. Fl. Afr. Centr. [in Fragm. Flor. Geobot. **36**, Suppl. 1]: 596 (1991). TAB. **12**. Type from Nigeria.
   Centaurea rhizocephala Oliv. & Hiern in F.T.A. **3**: 438 (1877) non Trautv. (1873) nom. illegit. —Philipson in Journ. Bot. **77**: 231 (1939). Type from the Sudan.
   Centaurea goetzeana O. Hoffm. in Engl., Bot. Jahrb. **30**: 441, t. 19, fig. H–P (1902). —Philipson in Journ. Bot. **77**: 231 (1939). Type from Tanzania.
   Centaurea tisserantii Philipson in Journ. Bot. **77**: 87, 231 (1939). Type from the Central African Republic.

A low-growing perennial herb, often caespitose, with 1–several stems from a large, branched, semi-woody rootstock. Stems up to c. 40 cm. high, erect or somewhat decumbent, sparingly branched, hispid-pubescent to glabrescent, less often granular-tomentose. Leaves subsessile, mostly 3–4 × 0.5–1.4 cm., oblong-elliptic to oblanceolate, acute mucronate at the apex, rounded to cuneate below, remotely serrate above, glandular-punctate and hispid-pubescent becoming scabridulous or glabrescent, the hairs abruptly filiform at the apices. Capitula precocious, or contemporaneous with the leaves, 1–many borne at or just above ground-level when produced before the leaves, solitary and terminal on 1–many short naked or bracteate radical stalks c. 1–5 cm. long; or the capitula solitary and terminal on foliose branches 4–18 cm. long when produced together with the leaves. Involucres up to 20 × 16 mm., hemispheric to campanulate, broadly urceolate in fruit. Phyllaries many-seriate, imbricate, coriaceous and glabrous with distinctive apical appendages; outer phyllaries c. 5 mm. long and ovate-rotund, increasing uniformly to c. 19 mm. long towards the inside becoming narrowly lanceolate; appendages of the outer 4–6 series of phyllaries usually purplish and reflexed, expanded with an apical spine and 2–3 pairs of smaller lateral spines, the spines best developed in the outermost phyllaries, up to c. 7 mm. long or very much reduced; appendages of the innermost phyllaries paler, expanded hyaline and lacerate. Receptacle densely setose, the setae 8–13 mm. long, ± flattened smooth and twisted. Florets glandular, glabrous; the marginal florets neuter, corollas whitish up to c. 25 mm. long, narrowly cylindric and deeply 5-lobed, staminodes included; inner-florets hermaphrodite, corollas white, often purple-tinged, up to c. 19 mm. long, narrowly infundibuliform, lobes 4–5 mm. long, anthers purplish. Achenes stramineous or pale-brown, 7–8 mm. long, 2.5–3 mm. in diam. at the middle, ovoid-cylindric and somewhat curved, ± obliquely truncate and shallowly cup-shaped apically, narrowly 4-grooved, sparsely finely puberulent to glabrescent, with a lateral oblique-concave attachment-scar; pappus several-seriate of persistent unequal narrow scale-like setae up to c. 1.5 mm. long, free to the base.

Zambia. N: Mbala-Kawimbe road, c. 1500 m., 16.viii.1956, *Richards* 5839 (K). W: Chingola, 2.x.1955, *Fanshawe* 2484 (K; NDO). C: 25.5 km. N. of Kapiri Mposhi, 27.ix.1963, *Angus* 3772 (FHO; K). **Zimbabwe**. N: Mazowe, 1350–1460 m., viii.1906, *Eyles* 397 (BM; SRGH). C: Harare, Greystone Park, 27.xi.1972, *Biegel* 4112 (K; LISC; PRE; SRGH). **Malawi**. N: Chisenga, foot of Mafinga Hills, 1850 m., 9.xi.1958, *Robson & Fanshawe* 525 (BM; K; LISC; SRGH). C: Dedza, Chongoni Forest, 11.ix.1967, *Salubeni* 829 (K; PRE; SRGH). **Mozambique**. N: Unango, x.1964, *Magalhães* 90 (COI). T: 20 km. from Furancungo on Angónia road, 29.ix.1942, *Mendonça* 532 (LISC).
   Also from W. Africa, Zaire, Burundi, Sudan, Ethiopia, Uganda, Kenya and Tanzania. A locally frequent, usually precocious, pyrophyte of grassland and plateau woodlands and in dambos.
   A very variable species; in W. African material the leaves are prominently 3-nerved from the base, have a white sericeous indumentum and are usually narrowly elliptic with entire margins, whereas in the Flora Zambesiaca area the leaves are at most obscurely 3-nerved from the base, hispid or glabrous, and distinctly serrate towards the apex. Leaf size is also variable; in the Central African Republic large-leaved specimens, with the lamina c. 6 cm. wide are found (described as *C. tisserantii* by Philipson). Time of flowering, that is whether or not the capitula are contemporaneous with the leaves, and the degree of development of the phyllary appendages are variable throughout the range of distribution. None of these variations appear to be sufficiently discontinuous, on their own or in combination, to allow practical subdivision of this species.

2. **Centaurea melitensis** L., Sp. Pl.: 917 (1753). —Philipson in Journ. Bot. **77**: 231 (1939). —C. Jeffrey in Kew Bull. **22**: 136 (1968). —Hilliard, Comp. Natal: 580 (1977). Type from Malta.

An erect annual herb up to c. 1 m. tall. Stems slender, stiffly branched, angular, narrowly winged, scabridulous. Basal leaves up to c. 10 cm. long, narrowly obovate to lyrate-pinnatilobed; lobes few, remote, narrow, entire or dentate. Cauline leaves sessile, up to c. 4 × 0.5 cm., oblong-lanceolate, entire, sometimes sinuate or dentate, decurrent on the stem as narrow wings, hispid-pubescent or young growth ± sericeous-araneose at first.

Capitula solitary and terminal on short lateral branches; involucres c. 12 mm. long, urceolate, araneose. Phyllaries imbricate, increasing in length towards the inside, coriaceous, glabrous, appendaged; appendages of outer phyllaries brownish or purple, each with a deflexed apical spine 5–7 mm. long and 2–3 pairs of pinnately arranged short lateral spines; appendages of innermost phyllaries with the terminal spine reduced and the lateral spines absent, often expanded and chartaceous. Receptacle densely setose; the setae c. 7 mm. long. Florets yellow, glandular otherwise glabrous; marginal florets neuter; disk-florets hermaphrodite, 8–12 mm. long, narrowly infundibuliform, limb 4.5–5.5 mm. long, lobes c. 1.5 mm. long. Achenes pale brownish-purple, c. 3 mm. long, oblong, somewhat laterally compressed, sparsely finely puberulent, attachment-scar lateral and oblique-concave; pappus many-seriate of persistent unequal paleaceous setae up to c. 3 mm. long.

Zimbabwe. E: Mutare, 1100 m., 5.x.1953, *Chase* 5106 (BM; COI; K; LISC; SRGH).

Native of S. Europe, from Portugal to the southern Balkan peninsula, also in North Africa, Madeira and the Canary Islands. Widely spread as a weed in central and northern Europe, N. and S. America, Australia and Africa. An uncommon weed, recorded from roadsides in the Flora Zambesiaca area.

Tribe 3. VERNONIEAE Cass.

Vernonieae Cass. in Journ. Phys. **88**: 203 (1819). —Wild & G.V. Pope in Kirkia **10**: 339–384, figs. 1–10 (1977). —S.B. Jones, *Vernonieae*, Syst. Rev. in Heywood et al., Biol. & Chem. Comp.; 503–521 (1977). —Wild in Kirkia **11**: 31–127 (1978). —C. Jeffrey in Kew Bull. **43**: 195–277 (1988).

Annual or perennial herbs (often suffrutescent in the Flora Zambesiaca area with annual stems from woody rootstocks) or small to large shrubs, woody scramblers or rarely small trees. Leaves alternate, sometimes opposite, rarely whorled, sometimes radical, sessile or petiolate, rarely pinnatilobed in the Flora Zambesiaca area. Capitula homogamous discoid, 1–many flowered, sometimes syncephalous; usually arranged in corymbiform cymes, sometimes scorpioidly cymose, or in various axillary or terminal clusters, spicate, paniculate, or solitary and terminal. Florets normally bisexual and fertile, rarely unisexual. Involucres usually cyathiform-campanulate or globose; phyllaries free in many series, rarely 1–2-seriate, closely or loosely imbricate, rarely decussate, herbaceous to chartaceous or coriaceous, sometimes with apical appendages. Receptacle flat or subconvex, smooth or pitted, sometimes alveolate, occasionally paleate. Corollas reddish-purple to mauve or white, rarely yellowish or orange, tubular to infundibuliform, usually regularly 5-lobed. Anthers apically appendaged, sagittate at the base with obtuse acute or rarely tailed auricles. Style branches subterete or subulate with stigmatic papillae on the inner surface towards the base, usually shortly hirsute outside. Achenes subterete to ± laterally compressed, obpyramidal and 4–5-sided, oblong-obovoid, turbinate or subfusiform, 2–20-ribbed, sometimes smooth, (rarely dimorphic), sparsely to densely glandular, variously hairy, or glabrous. Pappus of persistent or caducous setae or scales, 1- or 2-seriate, sometimes several-seriate, the outer series often reduced, sometimes pappus wanting. Pollen grains lophate, the ridges often spiny. Probable chromosome base numbers for the genera are n = 10 or 9 (mostly Old World), n = 17, 16, 15, 14, 13, 11 (New World).

A tribe of about 70 genera in both the Old and New World, of which 37 genera are monotypic. The tribe is well defined but subtribal and generic concepts within it are still uncertain. Evidence from chemistry, cytology and palynology suggests that the New-World *Vernonia* species plus satellite genera represent a major group, probably at subtribal level, which is distinct from most of the Old-World *Vernonia* species and satellite genera. Since the type species of *Vernonia, V. noveboracensis* (L.) Willd., is a New-World species it is probable that many Old-World species presently included in *Vernonia* will eventually be excluded from it. Until the taxonomic relationships are better understood, and the classification revised, *Vernonia* here is maintained in its traditional sense. For discussion on generic delimitation see C. Jeffrey in Kew Bull. **43**: 195–277 (1988).

Key to the genera

1. Capitula aggregated in many secondary capitulum-like glomerules (syncephalous); phyllaries mostly 8, decussate - - - - - - - - - - - **20. Elephantopus**
– Capitula not in bracteate glomerules though sometimes aggregated into clusters; phyllaries mostly more than 8, spirally imbricate - - - - - - - - - 2

2. Pappus 2-seriate, the inner whorl composed of setae, the outer whorl of setae or scales or reduced to a low ring of free or ± united brief scales, occasionally rim-like (pappus setae few, caducous or absent in *Vernonia lundiensis* and *Vernonia stellulifera*) - - - - 3
– Pappus 1-seriate, of free setae or scales, or ± fused and cupuliform, or pappus absent (sometimes apparently 2-seriate in *Ageratinastrum* where scales can often alternate with paleaceous setae within the same whorl) - - - - - - - - - - - - - - - 5
3. Receptacle with scales between the florets; pappus setae broadly paleaceous
19. Dewildemania
– Receptacle without scales between the florets; pappus setae various but not broadly paleaceous
4
4. Inner pappus copious, usually persistent (setae few, caducous or absent in *Vernonia lundiensis* and *Vernonia stellulifera*), mostly ± twice as long as the mature achene; outer pappus of free setae or scales, never reduced to a low rim; achenes usually variously hairy **11. Vernonia**
– Inner pappus of few to many caducous setae ± equalling the mature achene in length (up to twice as long in *Brachythrix lugarensis* but then outer pappus reduced to a low rim); outer pappus of free or ± united shorter scales, often reduced to a low rim - - - - **17. Brachythrix**
5. Leaves deeply pinnatifid; achenes narrowly 4-sulcate, densely minutely tuberculate overall, not ribbed - - - - - - - - - - - - - - **15. Rastrophyllum**
– Leaves entire, serrate or dentate; achenes not narrowly 4-sulcate and not completely covered in minute tubercules, achenes broadly to narrowly ribbed or angular - - - - 6
6. Indumentum on stems and leaf lower surfaces closely araneose-felted, hairs symmetrically T-shaped with short stalks and matted long equal arms - - - **12. Gutenbergia**
– Indumentum on stems and leaves various or absent, not felted; hairs simple or asymmetrically T-shaped with one arm much shorter than the other - - - - - - - 7
7. Pappus of persistent ± united shiny scales or scales free and alternating with narrower paleaceous setae equalling or exceeding the scales in length **18. Ageratinastrum**
– Pappus of caducous barbellate setae, not of scales, or absent - - - - - 8
8. Pappus setae subplumose or coarsely barbellate; achenes with single glandular trichomes sunk in scattered pits between the ribs - - - - - - - - - **13. Erlangea**
– Pappus setae barbellate, or pappus 0; achenes if glandular not pitted - - - 9
9. Leaves linear, less than 3 mm. wide (at least the cauline leaves); plants subaquatic, rooted in shallow water or marshy ground near water - **14. Bothriocline monocephala** and
Bothriocline emilioides
– Leaves more than 3 mm. wide; plants not subaquatic though sometimes in swampy places 10
10. Pappus absent; achene strongly 2–6-ribbed, obpyramidal or turbinate with apical tissue widely expanded and truncate or extended upward into a ± well developed cup-shaped depression - - - - - - - - - - - - - **16. Ethulia**
– Pappus sparse, of few very caducous barbellate setae, 1–3 mm. long; achenes mostly narrowly obovoid-oblong with 4–5 broad densely gland-filled grooves; (pappus sometimes absent in *Bothriocline steetziana*); achenes without gland-filled grooves in *Bothriocline monocephala, B. trifoliata* and *B. pectinata* but then involucres 12–35 mm. in diam., leaves trifoliate and phyllaries matted brown-araneose respectively - - - - - - - **14. Bothriocline**

11. VERNONIA Schreb.

Vernonia Schreb., Gen. Pl. **2**: 541 (1791) nom. conserv. —Wild in Kirkia **11**: 31–127 (1978). —C. Jeffrey in Kew Bull. **43**: 199–248 (1988).
Distephanus Cass. in Bull. Soc. Philom. Paris **1817**: 151 (1817).
Gongrothamnus Steetz in Peters, Reise Mossamb., Bot.: 336 (1864).
Triplotaxis Hutch. in Bull. Misc. Inf., Kew **1914**: 355 (1914).

Annual or perennial herbs (often suffrutescent in the Flora Zambesiaca area with annual stems from woody rootstocks) or small to large shrubs, woody scramblers or rarely small trees. Leaves alternate, rarely subopposite, sometimes radical, sessile or petiolate, penninerved but sometimes 3–5-nerved from the base, often glandular. Capitula homogamous, 1–many flowered, florets normally bisexual and fertile; capitula usually arranged in corymbiform cymes, sometimes scorpioidly cymose, or in various axillary or terminal clusters, spicate, paniculate, or solitary and terminal. Involucres cyathiform-campanulate, obconic, cylindric, or hemispheric. Phyllaries free in 2–many series, loosely or appressed imbricate; apices acute obtuse or rounded, often mucronate or aristate, sometimes with a membranous white or coloured appendage; receptacle flat or convex, plane or alveolate, the alveolae sometimes fimbriate. Corolla creamy-white, various shades of purple or blue, rarely orange, cylindric to narrowly infundibuliform or slender-tubular below and abruptly widened cylindric above, regularly 5-lobed. Anthers with a linear or lanceolate apical appendage; theca bases acute or obtuse. Style branches long tapering-subterete with stigmatic papillae on the inner surface towards the base, pilose or shortly hirsute outside, usually with sweeping hairs. Achenes subterete to ± laterally compressed, obpyramidal and 4–5-sided, oblong-obovoid, turbinate or subfusiform, 2–20-ribbed, sometimes smooth, sparsely to densely glandular, variously hairy, or glabrous, with or without a distinct basal callus. Pappus usually 2-seriate with an outer short whorl of scales or setae, sometimes caducous; inner whorl of long barbellate or subplumose, sometimes caducous, whitish purple or green setae, more rarely the outer pappus absent, sometimes inner pappus setae flattened or widened towards the apex. Pollen grains lophate, the ridges often spiny.

A very large genus of tropical origin comprising approximately 1000 species. As presently circumscribed it has two centres of distribution, one in America and one in Africa, and is found in tropical and warm temperate America, in Africa and Madagascar and in SE. Asia.

For *Vernonia* the preliminary work of H. Wild in Kirkia **11**: 31–127 (1978) is gratefully acknowledged.

Key to the species and groups

1. Plants annual, single season ephemerals; stem single from a single season tap-root
   group J page 67
 – Plants perennial, trees, or shrubs with woody stems, or suffrutices producing 1–several herbaceous scapes or leafy stems each year from a woody perennial rootstock   2
2. Stems perennial, woody; trees or shrubs, sometimes scandent, usually over 1.5 m. tall   3
 – Stems or scapes annual, herbaceous from woody rootstocks or perennial tuberous roots; suffrutices usually less than 1.5 m. tall   -   -   -   -   -   -   -   -   -   25
3. Corollas orange-yellow   -   -   -   -   -   -   -   -   -   -   1. *aurantiaca*
 – Corollas purple, mauve, bluish or whitish   -   -   -   -   -   -   -   -   4
4. Scandent shrubs; leaves 3–5-nerved from the base, ovate, deltate-ovate, elliptic or cuneate-rhomboid, subcoriaceous   -   -   -   -   -   -   -   -   **group B** page 62
 – Subshrubs, shrubs or trees; leaves not as above   -   -   -   -   -   -   5
5. Corolla tube long, slender, abruptly dilated above into a short, cylindric, shallowly-lobed limb; pappus inner setae flattened; anther-tube purple; phyllaries distinctly appendaged or appendages absent   -   -   -   -   -   -   -   -   -   -   **group I** page 65
 – Corolla funnel-shaped, tube shorter than limb, lobes long ± spreading; pappus inner setae subterete-terete; anther-tube buff, pale-brown or mauve; phyllaries usually not distinctly appendaged   -   -   -   -   -   -   -   -   -   -   -   -   6
6. Phyllaries tightly appressed-imbricate in mature capitula (excluding subulate apices of *V. muelleri* subsp. *muelleri* and *V. colorata* subsp. *oxyura*); involucre remaining subcylindric-fusiform or ovoid not ± spreading obconic, ± constricted above (cup-shaped in *V. eylesii* and *V. nepetifolia*), phyllaries persistent in fruit   -   -   -   -   -   -   -   -   -   -   -   7
 – Phyllaries loosely arranged, soon spreading; phyllaries either all persistent ± tapering and chartaceous, or inner phyllaries falling in fruit and the phyllaries short and tough   15
7. Outer phyllaries tapering into a stiff subulate or very narrow appendage-like apex 3–10 mm. long   -   -   -   -   -   -   -   -   -   -   -   -   -   -   -   8
 – Outer phyllaries various, apex not as above   -   -   -   -   -   -   -   -   9
8. Outer phyllaries with a recurved subulate apex; leaves up to 3 cm. wide, lanceolate-elliptic, sinuate-serrate   -   -   -   -   -   -   -   -   12. *muelleri* subsp. *muelleri*
 – Outer phyllaries with a narrow appendage-like apex; leaves more than 4 cm. wide, broadly elliptic, subentire to serrate   -   -   -   -   -   -   -   3. *colorata* subsp. *oxyura*
9. Pappus setae not or hardly exserted beyond phyllaries   -   -   -   -   -   10
 – Pappus setae well exserted beyond involucre, usually by more than half their length   12
10. Phyllaries glabrous or at the most sparsely puberulous, thin papery, dark-tipped; leaves more than 2 cm. wide   -   -   -   -   -   -   -   -   -   -   -   8. *pteropoda*
 – Phyllaries pubescent to tomentose, not thinly chartaceous; leaves up to 2 cm. wide   11
11. Leaves revolute, densely crowded on branches; pappus setae subplumose; involucre less than 4 mm. wide, tomentose; plants of miombo woodland   -   -   -   14. *stenocephala*
 – Leaves not revolute or densely crowded; pappus setae barbellate; involucre 5–9 mm. wide, pilose-hispid to pubescent; plants of swamp forest margins   -   -   13. *mushituensis*
12. Leaves more than 4 cm. long   -   -   -   -   -   -   -   -   **group A** page 61
 – Leaves less than 4 cm. long   -   -   -   -   -   -   -   -   -   -   13
13. Involucre narrowly cylindric and less than 4 mm. wide; leaves discolorous with a pale closely appressed indumentum on the lower surface; shrubby trees   -   -   -   5. *bellinghamii*
 – Involucre broadly cup-shaped c. 5 mm. or more wide; leaves ± densely hispid or pubescent to glabrescent on the lower surface with numerous large patent hairs on prominent main veins beneath; subshrubs   -   -   -   -   -   -   -   -   -   -   -   -   14
14. Leaves ovate, hispid, serrate-crenulate on margins   -   -   -   -   -   10. *nepetifolia*
 – Leaves lanceolate, puberulous to glabrescent, serrate on margins   -   -   11. *eylesii*
15. Involucre 10–15 mm. wide; capitula laxly arranged   -   -   -   -   -   -   16
 – Involucre 2–8(10) mm. wide; capitula numerous ± densely clustered   -   -   -   -   17

16. Leaves serrate, subspathulate, lamina ovate in upper part and abruptly narrowed below into a narrowly-winged midrib; phyllaries less than 1.5 mm. wide, finely straight bristle-tipped, inner tapering acute - - - - - - - - - - - 9. *madefacta*
   – Leaves sinuate-serrate, lanceolate-elliptic; phyllaries c. 2 mm. wide, outer recurved subulate at apex or aristate, inner not or hardly tapering to apex - - - - - 12. *muelleri*
17. Involucre c. 2.5 mm. long; phyllaries persistent, pubescent ± equalling achene in length; leaves more than 5 cm. long, ovate, acuminate at apex, pubescent on lower surface; plants large, often scandent shrubs - - - - - - - - - - - - - 6. *brachycalyx*
   – Features not combined as above, if phyllaries equalling achene then somewhat caducous, chaffy, glabrous or at the most ciliate and sparsely puberulous - - - - - 18
18. Plants loosely woolly-araneose at least on involucres, capitula stalks, twigs and young growth; leaf lower surface whitish araneose - - - - - - - - - - - 7. *holstii*
   – Indumentum variously pubescent to glabrescent not woolly or araneose; leaf lower surface pale tomentose to glabrescent - - - - - - - - - - - - - 19
19. Outer phyllaries produced into a 4–8 mm. long straight bristle, or the outermost subulate 2.5–5 mm. long - - - - - - - - - - - - - - - 20
   – Outer phyllaries not subulate or long bristle-tipped - - - - - - - 22
20. Indumentum consisting of small flagelliform hairs; upper stem and branches sparsely puberulous - - - - - - - - - - - - - 9. *madefacta*
   – Indumentum consisting of T-shaped hairs; upper stem and synflorescence branches appressed tomentellous - - - - - - - - - - - - - - 21
21. Achenes glandular otherwise glabrous; phyllaries sparsely pilose or glabrescent, middle and inner phyllaries whitish scarious often purple-tipped; corolla lobes glandular otherwise glabrous; leaves mostly somewhat truncate to cordate at base - - - 15. *syringifolia*
   – Achenes ± appressed hispid; phyllaries brownish strigulose to sericeous, greenish sometimes purple-tinged apically; corolla lobes sparsely strigose; leaves narrowly cuneate to abruptly narrowed or rounded at base, never cordate - - - - - 16. *wollastonii*
22. Leaf upper surface ± scabrid, lower surface ± densely tomentose, lanceolate to linear-elliptic, up to c. 12 × 4 cm., mostly smaller - - - - - - - - - 46. *karaguensis*
   – Leaves not scabrid above nor densely tomentose below (sometimes tomentose below in *V. myriantha* but then mostly 10–30 × 4–20 cm., and phyllaries ± caducous) - - - 23
23. Involucre 8–10 mm. long; phyllaries glabrous shiny thin-papery, c. 2 mm. wide, persistent; pappus setae white, hardly exceeding involucre - - - - - - 8. *pteropoda*
   – Involucre 3–5(8) mm. long; phyllaries puberulous to glabrous, not thin-papery, mostly to 1.5 mm. wide, caducous or persistent; pappus setae sordid or whitish, exceeding involucre in length - - - - - - - - - - - - 24
24. Erect, small or large shrubs; inner phyllaries falling in fruit; phyllaries short ± equalling or slightly longer than mature achene - - - - - - - **group C** page 62
   – Scandent or climbing shrubs; phyllaries persistent, longer than achene (if ± equalling mature achene then leaves 3-nerved from base) - - - - - - **group B** page 62
25. Phyllaries regularly shortly pectinate-ciliate all along margins (sometimes obscured by marginal tomentum); corollas bluish-mauve - - - - - - - **group D** page 63
   – Phyllaries entire to ± ciliate, not pectinate-ciliate; corollas purple, mauve or white 26
26. Corolla tube long slender, abruptly dilated above into a short, cylindric, shallowly-lobed limb; pappus inner setae flattened; anther tube purple; roots often with fusiform tuberous swellings **group I** page 65
   – Corolla ± funnel-shaped, lobes long ± spreading; pappus inner setae ± subterete; anther tube buff, pale-brown or mauve; roots not tuberous - - - - - - - 27
27. Capitula sessile or subsessile (at least the majority), few to many, agglomerated or singly; in capitate, spicate or scorpioid synflorescences - - - - - - - - 28
   – Capitula mostly stalked in various corymbiform-cymose or paniculate arrangements or capitula solitary on stem or scape - - - - - - - - - - - 33
28. Capitula agglomerated, in a large terminal cluster or in many smaller axillary clusters 29
   – Capitula solitary at nodes of stem or of synflorescence branches in a scorpioid or spicate arrangement - - - - - - - - - - - - - 32
29. Capitula numerous in a large terminal dense, subglobose cluster c. 3–4 cm. wide 30
   – Capitula in many, 2–9-capitulate, axillary clusters - - - - - - 31
30. Leaf reticulation raised on both surfaces, lamina glabrescent; phyllaries punctate 82. *adenocephala*
   – Leaves bullate, nervation impressed on upper surface, lower surface tomentose; phyllaries not punctate - - - - - - - - - - - - 49. *teucrioides*
31. Leaf lower surface tomentose, lateral nerves impressed on upper surface, lamina ± bullate; phyllaries tomentose, mucronulate - - - - - - - - 80. *bullulata*
   – Leaves glabrous on both surfaces, reticulation prominent on upper surface, lamina not bullate; phyllaries lanate-araneose, acuminate-apiculate - - - - - 79. *polysphaera*
32. Capitula scorpioidly arranged, all on one side of arching synflorescence branches; leaves 6–17 cm. long - - - - - - - - - - - 81. *catumbensis*
   – Capitula axillary at stem nodes in a spicate synflorescence; leaves up to c. 2 cm. long, often fasciculate - - - - - - - - - - - - 40. *lycioides*

31. Leaf lower surface tomentose, lateral nerves impressed on upper surface, lamina ± bullate; phyllaries tomentose, mucronulate - - - - - - - - 80. *bullulata*
 – Leaves glabrous on both surfaces, reticulation prominent on upper surface, lamina not bullate; phyllaries lanate-araneose, acuminate-apiculate - - - - - 79. *polysphaera*
32. Capitula scorpioidly arranged, all on one side of arching synflorescence branches; leaves 6–17 cm. long - - - - - - - - - - - - 81. *catumbensis*
 – Capitula axillary at stem nodes in a spicate synflorescence; leaves up to c. 2 cm. long, often fasciculate - - - - - - - - - - - - 40. *lycioides*
33. Pappus inner-setae plumose, with seta-bristles 3 or more times longer than the seta axis diam., or bristles equalling achene hairs in length - - - - - - - - - 34
 – Pappus inner-setae barbellate, the seta-barbs less than 1.5 times the seta axis diam.   35
34. Leaves more than 10 mm. wide, glabrous or if somewhat puberulous then hairs not asymmetrically T-shaped - - - - - - - - **group F** page 64
 – Leaves less than 3 mm. wide, pubescent with hairs asymmetrically T-shaped 122. *centaureoides*
35. Plants scapose or stems scapiform; leaves basal or crowded on lower stem, cauline leaves absent or few and reduced - - - - - - - - - - - - 36
 – Plants with uniformly leafy stems or leaves ± crowded below and decreasing somewhat in size up the stem - - - - - - - - - - - - - - - 44
36. Achenes obpyramidal with raised brown glands usually in ridges between well developed ribs, usually densely hispid at the base more sparsely so or glabrescent towards the apex; outer pappus whorl of long, lanceolate, overlapping scales ± equalling achene in length; receptacle alveolae walls deeply toothed; rootstock vertical - - - - **group G** page 64
 – Achenes ± narrowly cylindric-turbinate, not ridged between the ribs, uniformly densely strigose or glabrescent; outer pappus whorl of small setae, or short linear scales not overlapping (except sometimes in *V. daphnifolia*); receptacle alveolae walls not deeply toothed; rootstock various
37
37. Flowers appearing before the leaves - - - - - - **group H** page 65
 – Flowers appearing with the leaves - - - - - - - - 38
38. Phyllaries up to c. 2 mm. wide - - - - - - - - - 39
 – Phyllaries more than 3 mm. wide - - - - - - - - - 41
39. Leaves spathulate, ovate in the upper half and abruptly narrowed about the middle into a narrowly-winged midrib - - - - - - - - - 72. *praemorsa*
 – Leaves narrowly oblanceolate, ± gradually tapering into a narrowly cuneate or long-attenuate narrowly-winged midrib - - - - - - - - - - 40
40. Scapes markedly striate, puberulous above, densely brownish-lanate at the root crown; leaves up to c. 50 cm. long, long-attenuate to the base - - - - - 73. *praecox*
 – Scapes obscurely striate, appressed tomentellous above, not lanate at the base; leaves less than 15 cm. long, oblanceolate - - - - - - - - - 70. *milanjiana*
41. Capitula solitary on unbranched scapes - - - - - - - 42
 – Capitula few to many on laxly- or much-branched scapiform stems - - - 43
42. Pappus setae green; phyllaries glabrous, reddish-brown, not obviously striately-nerved
76. *mutimushii*
 – Pappus setae sordid; phyllaries hispid-pubescent to glabrescent, longitudinally striately-nerved
77. *acrocephala*
43. Phyllary apices broad, glabrous, membranous, hyaline or purple-tinged   75. *subaphylla*
 – Phyllary apices tapering, densely pubescent or tomentose - - - 71. *dewildemaniana*
44. Leaf lower surfaces appressed silvery- or brownish-sericeous; corolla lobes softly strigose, the hairs shortly-stalked long-armed T-shaped - - - - - **group E** page 64
 – Leaf lower surface tomentose to pubescent or glabrous, not silvery-sericeous; corolla lobes glabrous or glandular or with setose or acicular bristles, not with T-shaped hairs   45
45. Achenes terete - - - - - - - - - - - - 46
 – Achenes 4–10-ribbed, or angular - - - - - - - - - 47
46. Leaves up to c. 4.5 cm. long; stems purplish; hairs T-shaped with arms appressed 59. *undulata*
 – Leaves more than 4.5 cm. long; stems greenish; hairs Y-shaped, with arms spreading or ± patent - - - - - - - - - - - 53. *hirsuta*
47. Achenes distinctly 10-ribbed - - - - - - - - - 48
 – Achenes ± obscurely to clearly 4–5(8)-ribbed - - - - - - 52
48. Leaves scabrid; achene strigose-hispid, ribs usually broad - - - - - 49
 – Leaves pubescent to tomentose or glabrescent not scabrid; achenes glabrous or sparsely hispidulous, ribs narrow - - - - - - - - - - - 50
49. Involucre 12–18 mm. long and 15–30 mm. wide; subtending bracts broad, leaf-like, ovate-lanceolate, often exceeding the involucre; phyllaries mostly 2–3 mm. wide with broad pale margins, innermost series subequal - - - - - - - - 60. *purpurea*
 – Involucre less than 12 mm. long and 15 mm. wide; subtending bracts absent or if present mostly less than 5 mm. wide and not exceeding the involucre; phyllaries mostly less than 2 mm. wide with narrow margins; innermost series progressively longer - - - 61. *amblyolepis*
50. Leaf venation prominently reticulate on lower surface; leaf coriaceous, 8–14 cm. long and 1.5–5 cm. wide - - - - - - - - - - - 47. *luembensis*

– Leaf venation hardly raised on lower surface, if somewhat raised and reticulate then lamina glabrous; leaf membranous or at most thinly coriaceous, less than 7 cm. long and 1.5 cm. wide
51
51. Stems zig-zag; leaf midrib woolly-tomentellous at the base; leaf lateral nerves curving-parallel to leaf margin - - - - - - - - - - - - - - 57. *fractiflexa*
– Stems not zig-zag; leaf midrib glabrous in mature leaves; leaf lateral nerves at a ± wide angle with midrib - - - - - - - - - - - - 58. *mumpullensis*
52. Corolla lobes glabrous though often glandular - - - - - - - - 53
– Corolla lobes pubescent, pilose, setose or with acicular bristles, often also glandular 55
53. Leaves 10–20 mm. wide - - - - - - - - - 46. *karaguensis*
– Leaves mostly less than 5 mm. wide - - - - - - - - - 54
54. Involucres 10–15 mm. long; phyllaries glabrous or whitish puberulous; capitula terminal on leafy branches mostly more than 5 cm. long - - - - - - 44. *bainesii*
– Involucres mostly less than 8 mm. long; phyllaries puberulous with the hairs towards the phyllary apex reddish-brown; capitula 3–many per stem or branch, on bracteate stalks less than 5 cm. long - - - - - - - - - - - - - - - - 45. *schlechteri*
55. Achenes obpyramidal with raised brown glands usually in ridges between well developed ribs, usually densely hispid at the base and more sparsely so or glabrescent towards the apex; outer pappus whorl of long, lanceolate, overlapping scales more than half the achene in length; receptacle alveolae walls deeply toothed; rootstock vertical, narrowly turbinate
group G page 64
– Achenes ± narrowly cylindric-turbinate, not ridged between the ribs, uniformly densely strigose or glabrescent; outer pappus whorl of small setae, or short linear scales not overlapping (except sometimes in *V. daphnifolia*); alveolae walls not deeply toothed; rootstock various, if vertical then slender - - - - - - - - - - - - - - - 56
56. Leaves, at least the upper, remotely coarsely serrate - - - - - 54. *sutherlandii*
– Leaf margin entire or crispate-erose - - - - - - - - - 57
57. Stems decumbent; leaves dimorphic, with primary leaves on the stem up to c. 55 × 5 mm. subtending abbreviated axillary shoots with much smaller leaves; corollas white; achenes up to c. 2 mm. long - - - - - - - - - - - 51. *kayuniana*
– Characters not as combined above - - - - - - - - - 58
58. Involucre more than 10 mm. long; capitula solitary and terminal on long leafy branches or bracteate stalks, or capitula many in *V. musofensis* but then leaves more than 6 cm. long 59
– Involucre 4–8(10) mm. long; capitula many, clustered or lax on stalks less than 3 cm. long 61
59. Leaves 6–11 cm. long with long arching lateral veins and sometimes the tertiary veins prominent on lower surface - - - - - - - - - - - 50. *musofensis*
– Leaves less than 6 cm. long, lateral veins hardly obvious - - - - - 60
60. Phyllaries progressively longer to the inside, the outer series subulate the inner lanceolate; pappus setae 8–12 mm. long; achenes hispidulous, turbinate-cylindric 55. *daphnifolia*
– Phyllaries subequal and uniform, all subulate or linear; pappus setae c. 6 mm. long; achenes glabrous or sparsely hispidulous, narrowly cylindric - - - - - 56. *galpinii*
61. Leaves glabrous, linear less than 3 mm. wide - - - - - - 84. *helodea*
– Leaves hispid, pubescent or tomentose never glabrous, more than 10 mm. wide 62
62. Leaves petiolate, petioles more than 2 mm. long; lamina more than 3 cm. wide, broadly elliptic or ovate-elliptic; pappus several-seriate - - - - - - - 83. *turbinella*
– Leaves sessile or subsessile (if somewhat petiolate then petiole less than 2 mm. long), lamina less than 3 cm. wide, oblong lanceolate, sometimes elliptic; pappus 2-seriate - - - 63
63. Leaf lower surface densely softly uniformly pale-tomentellous, the indumentum ± obscuring a raised finely reticulate venation; lamina markedly discolorous - - - - - 64
– Leaf lower surface ± densely pubescent-hirsute or puberulous, if tomentellous then indumentum also containing long hispid-strigose hairs on main veins, if venation prominent reticulate then not obscured by tomentum on leaf lower surface; lamina usually concolorous 65
64. Leaves less than 4 cm. long, elliptic; upper surface finely bullate; inner phyllaries glabrescent
49. *teucrioides*
– Leaves mostly 5–15 cm. long, narrowly oblong; lamina hardly bullate; all phyllaries densely tomentose - - - - - - - - - - - - 48. *tanganyikensis*
65. Hairs of leaf upper surface simple, whip-like - - - - - - 52. *nestor*
– Hairs of leaf upper surface T- or Y-shaped - - - - - - - - 66
66. Phyllaries blunt, mucronate-aristate; involucre c. 6 mm. long; leaf bases cordate 53. *hirsuta*
– Phyllaries tapering to a pungent bristle-tip; involucre 7–10 mm. long; leaf bases cuneate
54. *sutherlandii*

GROUP A

Trees or shrubs; hairs multicellular, T-shaped, or apical cells excentric; phyllaries appressed imbricate; corollas mauve to whitish; achenes 2.2–5 mm. long, narrowly turbinate, 4–10-ribbed, glabrous or setose; pappus long exserted from the involucre, shaving-brush-like, outer of linear scale-like setae sometimes very caducous, inner of rigid slender setae flattened towards the apex.

1. Outer pappus of very caducous scale-like setae; involucre wider than long (length excluding the caudate phyllary apices in *V.colorata* subsp. *oxyura*); hairs each with a large excentrically placed terminal cell -   -   -   -   -   -   -   -   -   -   -   -   -   -   -   -   -   -   2
 – Outer pappus of persistent small linear scale-like setae; involucre longer than wide; hairs equal-armed, T-shaped -   -   -   -   -   -   -   -   -   -   -   -   -   -   -   -   3
2. Achenes ± glandular otherwise glabrous; leaves mostly ovate-lanceolate and widest below the middle or sometimes broadly oblong-elliptic, margins usually undulate   3. *colorata*
 – Achene setulose, glandular; leaves somewhat elliptic, tapering ± equally to apex and base or less often ovate-lanceolate, margins mostly flat -   -   -   -   -   -   -   - 2. *amygdalina*
3. Leaves mostly 8–17.5 × 4–8 cm. ± broadly lanceolate, somewhat attenuate to the apex, usually sparsely pubescent beneath; capitula numerous in a widely spreading synflorescence 8–18 cm. across -   -   -   -   -   -   -   -   -   -   -   -   -   -   -   - 4. *exsertiflora*
 – Leaves 3–6.5 × 1.5–3 cm. (seldom to 8 × 4 cm.), elliptic to oblong-ovate obtuse or rounded at the apex, not attenuate, densely greyish tomentellous beneath; capitula usually in subsessile subumbellate clusters up to c. 4 cm. across or sometimes capitula laxly corymbosely cymose   5. *bellinghamii*

GROUP B

Scramblers or climbers; leaves usually 3–5-nerved from near the base.

1. Corollas orange-yellow; leaves appressed white-tomentellous to glabrescent beneath
                                                                                       1. *aurantiaca*
 – Corollas purple or mauve fading to white; leaves pubescent to glabrescent beneath   2
2. Involucre c. 2.5 mm. long; phyllaries few, ± equalling achene in length; leaves pubescent beneath, not 3–5-nerved from the base -   -   -   -   -   - 6. *brachycalyx*
 – Involucre larger; phyllaries many, exceeding achene in length (± equalling achene in *V. anisochaetoides* but then leaves glabrous and 3-nerved from near the base) -   -   - 3
3. Upper stem and branches with T-shaped hairs; leaf margins variously serrate to crenulate or subentire; lamina ovate-lanceolate to linear-lanceolate ± acuminate apically; outer phyllaries narrowly lanceolate or subulate -   -   -   -   -   -   -   -   -   -   -   - 4
 – Upper stem and branches with 1-armed or whip-like hairs; leaf margins coarsely dentate to sublobate or entire not serrate; lamina deltate-ovate, elliptic or cuneate-rhomboid, sometimes tapering but not acuminate apically; outer phyllaries lanceolate never subulate   5
4. Achenes glandular otherwise glabrous; phyllaries sparsely pilose or glabrescent, middle and inner phyllaries whitish scarious often purple-tipped; corolla lobes glandular otherwise glabrous; leaves somewhat truncate to cordate at the base -   -   -   - 15. *syringifolia*
 – Achenes ± appressed hispid; phyllaries brownish strigulose to sericeous, greenish sometimes purple-tinged apically; corolla lobes sparsely strigose; leaves narrowly cuneate to abruptly narrowed or rounded at the base, never cordate -   -   -   -   - 16. *wollastonii*
5. Mature leaves deltate-ovate, truncate to subcordate, or abruptly narrowed then shortly tapering into the petiole, lamina membranous usually 5-nerved from near the base; achenes sharply tapered, c. 5-ribbed; corolla lobes less than half as long as tube -   -   -   - 6
 – Mature leaves cuneate below, lamina subcoriaceous strongly 3-nerved from the base; achenes obscurely 8–10-ribbed; corolla lobes more than half as long as the tube -   -   - 7
6. Leaves opposite or subopposite; achenes silky-pilose or strigose; plants of coastal forest
                                                                                       18. *angulifolia*
 – Leaves alternate; achenes shortly hispidulous or glabrescent; plants of submontane forest
                                                                                       17. *biafrae*
7. Leaves elliptic-ovate, lamina to c. 3.5 cm. long, entire, not narrowly decurrent on the petiole; florets c. 9 per capitulum; capitula in moderately dense paniculate clusters, stalks 1–3 mm. long -   -   -   -   -   -   -   -   -   -   -   -   -   - 19. *inhacensis*
 – Leaves cuneate-rhomboid, lamina mostly more than 4 cm. long coarsely toothed on upper margin or entire, strongly decurrent on the petiole; florets 15–20 per capitulum; capitula laxly paniculate on stiff divaricate branches, stalks patent 5–10 mm. long   20. *anisochaetoides*

GROUP C

Erect, small or large shrubs; stems purplish, becoming lenticellate; ± densely appressed pubescent, hairs small, T-shaped; capitula numerous, small; phyllaries ± equalling the mature achene in length, inner phyllaries falling in fruit; florets 5–15 per capitulum; achene obscurely angular and 6–8-ribbed; outer pappus of short setae, inner of fine white often long-exserted setae.

1. Leaves present at time of flowering -   -   -   -   -   -   -   -   -   -   - 2
 – Leaves absent at time of flowering -   -   -   -   -   -   -   -   -   - 7

2. Leaves sessile or petiolate, auriculate or asymmetrically cordate at the base sometimes "stipulate", grey-green pubescent to tomentose beneath, rarely glabrescent, 10–38 cm. long, margins serrate - - - - - - - - - - - 21. *myriantha*
– Leaves not as described above (appressed tomentellous beneath in *V. potamophila* but then leaves smaller, with a cuneate base and subentire to serrulate margin) - - - 3
3. Leaves markedly discolorous, appressed densely grey-tomentellous beneath, dark-green and sparsely pubescent on upper surface; inner phyllaries at least tapering to the apex; plants of Kalahari Sand - - - - - - - - - - - 23. *potamophila*
– Leaves not densely grey tomentellous beneath; inner phyllaries blunt or subacute; plants not confined to Kalahari Sand - - - - - - - - - - - 4
4. Capitula numerous, shortly stalked, in many dense subglobose clusters; clusters c. 5 cm. across, solitary or subcorymbiformly arranged on leafy branches; leaves ± pubescent beneath, the indumentum of patent long-stalked T-shaped hairs - - - - 22. *thomsoniana*
– Capitula not in dense clusters, ± laxly paniculate, often somewhat scorpioidly arranged; leaf lower surface indumentum sparse of appressed shortly-stalked T-shaped hairs or leaves glabrous - - - - - - - - - - - - - - 5
5. Achenes glandular otherwise glabrous (rarely sparsely setulose); florets c. 5 per capitulum, usually white; leaves glabrous or soon becoming so - - - 25. *glaberrima*
– Achenes hispid, glandular; florets 9–17 per capitulum, purple or mauve; leaves puberulous at least below - - - - - - - - - - - - - - 6
6. Young branches densely silky yellowish-brown to greyish felted-strigose, indumentum entirely covering the dark-purple branches; flowers appearing with leaves, December to March; leaves up to c. 4.5 × 1.2 cm.; involucre c. 3(4) mm. long - - - - - 26. *cinerascens*
– Young branches pale- to reddish-brown puberulous or tomentellous or glabrescent; flowers usually (in the Flora Zambesiaca area) appearing before leaves, mainly August to October; leaves up to c. 8 × 3 cm.; involucre (3)4–6 mm. long - - - - 24. *suprafastigiata*
7. Achenes glandular otherwise glabrous (rarely sparsely setulose); florets c. 5 per capitulum, usually white - - - - - - - - - - - 25. *glaberrima*
– Achenes hispid, glandular; florets 9–17 per capitulum; florets purple or mauve
24. *suprafastigiata*

GROUP D

Perennial herbs, usually pyrophytic; stems annual, herbaceous, leafy, from a woody rootstock; leaves serrate, sometimes scabrid; phyllaries regularly shortly pectinate-ciliate all along the margins; receptacle fimbriate-alveolate; corollas cornflower-blue or mauve, mostly 14–25 mm. long (except in *V. schweinfurthii*), lobes with acicular bristles; achenes 10–15-ribbed, 5–10 mm. long (except in *V. schweinfurthii*), strigose-hispid or glabrous; pappus setae somewhat flattened, subplumose, brown or purple-tinged, 8–17 mm. long (except in *V. schweinfurthii*).

1. Stems scapiform, simple; leaves basal or clustered below; capitula solitary and terminal
33. *gerberiformis*
– Stems leafy, usually branched; capitula 3–numerous, if solitary or if stems somewhat scapiform then leaves scabrid - - - - - - - - - - - - - 2
2. Achenes 3–4 mm. long; pappus setae barbellate, 4–7 mm. long - - 34. *schweinfurthii*
– Achenes 5–10 mm. long; pappus setae subplumose, 8–18 mm. long - - - - 3
3. Involucre and capitulum stalk usually white farinose-araneose (often also on young growth), glabrescent or hispid; leaves glabrous, margins entire rarely distantly toothed; plants of serpentine soils in Zimbabwe - - - - - - - 29. *accommodata*
– Involucre brownish pubescent or glabrescent; leaves scabrid or glabrous, margins variously serrate; plants not confined to serpentine soils - - - - - - - - 4
4. Involucre mostly less than 15 mm. long and 15 mm. wide; leaves glabrous, (if scabridulous then capitula numerous in dense clusters) - - - - - - - - - 5
– Involucre mostly more than 15 mm. long and 15 mm. wide; leaves usually coarsely scabrid; capitula mostly 1–9 laxly arranged - - - - - - - - - 6
5. Plants erect, mostly more than 40 cm. tall; stems usually robust, simple below the synflorescence; capitula numerous, corymbosely cymose, capitula stalks 0.2–3 cm. long 30. *glabra*
– Plants low-growing mostly less than 40 cm. tall; stems slender decumbent, usually branched; capitula few to many laxly arranged on stalks 3–10 cm. long - - - 31. *rosenii*
6. Basal leaves mostly more than 15 cm. long, linear-oblanceolate ± narrowly tapering below, crowded near the stem base; phyllaries mostly less than 2 mm. wide - - 32. *melleri*
– Leaves mostly less than 8 cm. long, elliptic or oblanceolate, not crowded near the stem base; at least the inner phyllaries more than 2 mm. wide - - - - - - - 7
7. Achenes densely strigose-pilose; involucre obconic, to c. 20 mm. long; phyllaries apically acute not appendaged - - - - - - - - - - 28. *kawoziensis*
– Achenes glabrous; involucre campanulate-hemispheric, 15–30 mm. long; phyllaries appendaged
27. *verrucata*

GROUP E

Perennial herbs; stems annual, herbaceous, leafy, from a woody rootstock; indumentum (at least on leaf lower surfaces) silvery-sericeous, the hairs long-armed, shortly-stalked, appressed T-shaped; corolla lobes with appressed T-shaped hairs; achenes 1.5–4 mm. long, usually densely silky-strigose; outer pappus of very short linear scales, inner of 4–6 mm. long white copious setae.

1. Capitula sessile, spicately arranged on long branches; leaves in fascicles on abbreviated side shoots - - - - - - - - - - - - - - - - 40. *lycioides*
– Capitula stalked, ± clustered in terminal corymbiform cymes; leaves alternate, not fasciculate on spur shoots - - - - - - - - - - - - - - - - 2
2. Leaves coarsely toothed at the apex; outer pappus setae very caducous; achenes not silky-strigose - - - - - - - - - - - - - - - - 39. *tigna*
– Leaves entire, ± narrowly acute at the apex; outer pappus persistent; achenes silky-strigose 3
3. Leaves linear often filiform, 6 or more times longer than wide, or if narrowly-elliptic then phyllaries less than 3 mm. long and involucre 1–1.5 times as long as the mature achenes 4
– Leaves ovate, narrowly-elliptic or lanceolate, less than 6 times as long as wide; phyllaries more than 3 mm. long and the involucre 2 or more times as long as the achenes - - 5
4. Leaf upper surface densely strigose-sericeous; phyllaries more than 3 mm. long; stems pale green-grey-sericeous; plants not slender - - - - - - - 35. *natalensis*
– Leaf upper surface thinly strigose or appressed pilose; phyllaries less than 3 mm. long; stems glabrescent and reddish-purple towards the base; plants slender - - 38. *calyculata*
5. Leaves ovate to ovate-elliptic, 1.5–3 times as long as wide, somewhat abruptly narrowed to a cuspidate apex or tapering acuminate - - - - - - - - 37. *oligocephala*
– Leaves narrowly elliptic or lanceolate, more than 3 times as long as wide, apex acute 6
6. Leaf upper surface thinly strigose or appressed pilose, green; lower surface closely silvery-sericeous often brownish-sericeous - - - - - - - - 36. *alticola*
– Leaves greenish-grey sericeous on both surfaces (at least on leaves of upper stem)
35. *natalensis*

GROUP F

Perennial herbs; rootstock woody; stems annual, herbaceous, wiry, brittle, strongly ribbed, leafy or scapiform; leaves coriaceous, glabrous or beset with minute scattered dark-purple or black glandular hairs; phyllaries densely brown tomentellous, or araneose-ciliate; corolla lobes 3–6 mm. long; achenes 3.5–8 mm. long, narrowly cylindric, 6–8-narrowly ribbed, sparsely hispid or glabrous; outer pappus of short, linear, fimbriate-ciliate scales; inner of white subplumose setae.

1. Stems scapiform; leaves subrosulate, ± crowded, prostrate, obovate or broadly oblanceolate, mostly 3–7 cm. wide - - - - - - - - - - - 43. *upembaensis*
– Stems leafy throughout; leaves ascending, narrowly oblanceolate, mostly less than 3 cm. wide
2
2. Involucre large, more than 2 cm. long and 3 cm. wide; corollas 18–22 mm. long; achenes to c. 8 mm. long; stems robust - - - - - - - - - - 42. *roseoviolacea*
– Involucre smaller, less than 1.5 cm. long and 2 cm. wide; corollas less than 15 mm. long; achenes c. 3.5 mm. long; stems wiry - - - - - - - - - 41. *subplumosa*

GROUP G

Perennial herbs; rootstock semi-woody, vertical; stems annual, herbaceous, mostly scapiform; phyllary hairs usually patent, flagelliform; receptacle alveolae often with deeply toothed margins; corolla lobes acicular-setose near the apex; achenes c. 2 mm. long, obpyramidal with raised brown glands usually in ridges between 4–5 well developed ribs, usually densely hispid at the base, more sparsely so or glabrous towards the apex; outer pappus of long lanceolate overlapping scales ± equalling the achene in length, inner pappus setae pink, grey or sordid.

1. Stems and involucres ± densely villous; lower leaves not or hardly tapering in the lower half; plants known so far only from the Mafinga Hills, Zambia - - - - 64. *rubens*
– Stems and involucres glabrous to variously hairy, not villous; lower leaves narrowed or tapering in lower half - - - - - - - - - - - - - - - - 2
2. Pappus grey; plants of mountain grassland above 1900 m. - - - 63. *griseopapposa*
– Pappus whitish, sordid, brown, reddish or purple; plants of deciduous woodland or seasonally wet grassland or dambos below 1900 m. - - - - - - - - - 3

3. Involucre less than 6 mm. long and 9 mm. in diam.; phyllaries 3–4-seriate shortly acute at the apex; stems wiry, solitary and somewhat sinuous or tufted - - - - - - 4
- Involucre 8–15 mm. long and more than 10 mm. in diam.; phyllaries numerous in many series, apically subulate or apparently appendaged; stems robust to somewhat slender, not wiry 5
4. Stems scapiform, simple or sparsely branching near the apex; cauline leaves reduced or absent; basal leaves clustered to subrosulate - - - - - - - - 69. *ianthina*
- Stems leafy, branching; leaves diminishing to the stem base, not clustered or subrosulate
68. *viatorum*
5. Basal leaves to c. 5 mm. wide, occasionally to c. 10 mm. wide, linear to linear-oblanceolate; plants somewhat glabrous - - - - - - - - - - - - 6
- Basal leaves more than 10 mm. wide, oblanceolate; plants variously hairy - - - 7
6. Radical leaves more than ¾ the stem length, much longer than the cauline leaves, upper cauline leaves or subtending bracts overtopping the capitula; outer phyllaries tapering attenuate at the apex - - - - - - - - - - - - - 66. *isoetifolia*
- Radical leaves less than ½ the stem length, ± equalling the cauline leaves in length; capitula overtopping the leaves; outer phyllaries with narrowly-lanceolate green appendages
67. *robinsonii*
7. Involucres c. 10 mm. long; phyllaries greenish with pungent purple tips; pappus sordid, occasionally reddish-purple - - - - - - - - - 62. *ugandensis*
- Involucres larger, 12–15 mm. long; phyllaries dark-purple, or at least the apex purple, broadly subulate to attenuate; pappus usually reddish-purple less often sordid
65. *violaceopapposa* subsp. *nuttii*

GROUP H

Scapose herbs, often precocious perennial pyrophytes, or stems scapiform; rootcrown brownish-lanate; aerial parts annual; leaves mostly radical, rosulate, subspathulate, the midrib rigid and narrowly-winged in the lower part; hairs mostly T-shaped although flagelliform on the phyllaries; achenes densely hispid at first, narrowly 10-ribbed.

1. Capitula numerous in a dense terminal cluster on a simple or sparsely branched scapiform stem 10–70 cm. tall, shortly stalked; stems 1–several - - - - - - 72. *praemorsa*
- Capitula solitary on scapes 1–30 cm. tall, or few to many laxly arranged on branching scapes or scapiform stems, not densely clustered; scapes 1–numerous - - - - - 2
2. Pappus setae green; phyllaries uniformly reddish brown - - 70. *mutimushii*
- Pappus setae sordid or brownish; phyllaries greenish or brownish, dark-purple or green-tipped often with hyaline or purple margins - - - - - - - - - 3
3. Capitula solitary on unbranched scapes; phyllaries acute, acuminate or obtuse 4
- Capitula few to many on ± laxly branching scapes; phyllaries acute to rounded or much expanded-membranous at the apex, (capitula sometimes solitary on unbranched scapes in *V. subaphylla* but then phyllaries expanded-membranous at the apex) - - - - 5
4. Outer phyllaries broadly ovate, mostly 4–8 mm. wide, greenish often purple-tipped, striate, phyllaries subequal; young leaves present with the capitula; scapes usually solitary, 3–30 cm. tall - - - - - - - - - - - - - 77. *acrocephala*
- Outer phyllaries lanceolate, tapering to the apex, less than 4 mm. wide, purple sometimes green, not clearly striate, phyllaries progressively longer to inside; leaves seldom present with the young capitula; scapes usually caespitose, less than 5 cm. tall - - - 78. *chthonocephala*
5. Involucres small, less than 6 × 8 mm.; capitula in a large panicle, on slender stalks less than 5 cm. long; stem scapiform sometimes with cauline leaves - - - - - 73. *praecox*
- Involucres larger, more than 10 × 12 mm.; capitula terminal on branches, or stalks usually more than 5 cm. long; cauline leaves absent or few greatly reduced - - - - - 6
6. Phyllaries with a broad membranous hyaline or purplish margin expanded and ± elongated at the apex becoming ± contracted-folded when dry; scapes usually solitary; leaves usually present with the capitula - - - - - - - - - - - - 75. *subaphylla*
- Phyllaries with a narrow hyaline or purple membranous margin, subacute to tapering-acuminate often undulate at the apex; scapes several to many, clustered; leaves seldom present with the young capitula - - - - - - - - - - - 74. *denudata*

GROUP I

Shrubs or subshrubs (one species an annual herb); indumentum of flagelliform hairs; phyllaries appendaged (at least in the outer and middle phyllaries), appendages petaloid, foliaceous or merely a membranous extension, or appendages absent; corolla tube long slender and abruptly dilated into a short, cylindric, exserted, shallowly-lobed limb; anthers purplish; achenes barrel-shaped, c. 10-ribbed with or without 10 secondary ribs; pappus multi-seriate, the setae somewhat caducous, longer to the inside, flattened towards the apex.

1. Annual herbs; outer pappus of ± persistent scales, inner pappus of longer more caducous setae; outermost phyllaries exceeding the inner series in length   -   -   -   100. *anthelmintica*
   – Perennial herbs with woody rootstocks and/or tuberous roots; outer pappus elements similar to, but smaller than, the inner setae; outermost phyllaries usually not exceeding the inner in length   -   -   -   -   -   -   -   -   -   -   -   -   -   -   -   -   2
2. Appendages of the outer phyllaries crustaceous with a prominent finely reticulate venation similar to that of the leaf lower surface, the outermost phyllaries often wholly leaf-like and ± enclosing the capitulum   -   -   -   -   -   -   -   -   -   -   93. *retifolia*
   – Appendages of the phyllaries petaloid or absent, venation if obvious not as above; outermost phyllaries never wholly leaf-like   -   -   -   -   -   -   -   -   -   -   -   3
3. Phyllaries lacking petaloid appendages, their tips not or obscurely differentiated, apically mucronate (as in *V. tuberifera* and *V. longipedunculata*), or tips differentiated being membranous in the apical portion and often subapically subaristate, the phyllary tips usually darker green or purplish often becoming withered   -   -   -   -   -   -   -   -   -   -   -   4
   – Phyllaries with distinct petaloid appendages (at least in some); appendages showy, narrow to much expanded, often longer than the phyllary base, apically mucronate and usually creamy-white, sometimes purplish   -   -   -   -   -   -   -   -   -   -   -   -   8
4. Capitula 1–3(9) per stem; leaves oblanceolate to subspathulate, tapering-attenuate to the base, concolorous   -   -   -   -   -   -   -   -   -   -   -   -   -   -   -   5
   – Capitula numerous; leaves oblong-elliptic or narrowly obovate, ± discolorous with the lower surface pale-brownish- or greyish-tomentellous to puberulous, or leaves glabrescent and concolorous with a raised finely reticulate venation beneath   -   -   -   -   -   6
5. Phyllaries subequal; capitula 1.5–2 cm. wide; stems ± leafy   -   -   -   -   99. *tuberifera*
   – Phyllaries imbricate, progressively longer to inside; capitula mostly more than 2 cm. wide; stems usually subscapose, the leaves ± crowded below or subrosulate   90. *longipedunculata*
6. Florets c. 3–4 per capitulum; involucres 4–5 mm. wide, subcylindric, hardly spreading
   96. *sunzuensis*
   – Florets c. 7–15 per capitulum; involucres 5–10(15) mm. wide, obconic-campanulate, phyllaries soon spreading or apically squarrose   -   -   -   -   -   -   -   -   -   -   7
7. Phyllary apices membranous and acute to attenuate, the outermost phyllaries ± subulate; indumentum of upper stem and synflorescence branches white woolly; capitula aggregated in a dense terminal cluster   -   -   -   -   -   -   -   -   -   -   95. *lafukensis*
   – Phyllary apices ± rounded with narrow membranous margins, outermost phyllaries ovate-lanceolate; indumentum of upper stem and synflorescence branches pale-tomentellous; capitula in ± lax corymbosely arranged clusters   -   -   -   -   -   -   97. *solweziensis*
8. Leaf venation strongly raised, finely reticulate on the lower surface, even the tertiary veins prominent, all clearly visible; leaves finely pilose or pubescent to glabrescent or glabrous on both sides, not tomentose or felted on the lower surface   -   -   -   -   -   -   9
   – Leaf venation various, often ± obscured by a pubescent to densely felted indumentum on the leaf lower surface, sometimes the venation visibly raised reticulate on the lower surface but then not finely so and the lamina scabrid above   -   -   -   -   -   -   -   -   -   11
9. Appendages of outer phyllaries crustaceous with a prominent finely reticulate venation similar to that of the leaf lower surface; outermost phyllaries often wholly leaf-like and ± enclosing the capitula   -   -   -   -   -   -   -   -   -   -   -   -   -   93. *retifolia*
   – Appendages of phyllaries ± petaloid, venation if obvious not as above; outermost phyllaries never wholly leaf-like   -   -   -   -   -   -   -   -   -   -   -   -   -   10
10. Capitula in large many-capitulate clusters terminal on stems and branches; indumentum of upper stem and branches densely greyish felted-tomentose   -   -   -   95. *lafukensis*
    – Capitula in numerous 3–9-capitulate clusters subscorpioidly arranged; indumentum of upper stem and branches crisped-pubescent to tomentose   -   -   -   -   94. *bruceana*
11. Leaves widest above the middle and tapering to a subsessile base, or leaves filiform; roots tuberous; pappus setae not readily dislodged; achene truncate below with a thin flat carpopodium; innermost phyllaries (plus appendage) exceeding others in length   12
    – Leaves widest below the middle, ± tapering towards the apex, sometimes elliptic never filiform; roots not tuberous; pappus setae easily dislodged; achene ± tapering to the base, carpopodium a well developed ± swollen smooth pale annulus; innermost phyllaries (plus appendages) usually exceeded in length by the appendages of the middle phyllaries   -   -   -   -   16
12. Leaves markedly discolorous, lower surfaces yellowish-brown tomentellous with a purple reticulate venation   -   -   -   -   -   -   -   -   -   -   -   -   98. *incompta*
    – Leaves concolorous, lower surfaces puberulous to glabrescent never felted tomentellous   13
13. Leaves filiform or linear-oblanceolate, less than 10 mm. wide; capitula mostly less than 2 cm. wide   -   -   -   -   -   -   -   -   -   -   -   -   -   -   -   14
    – Leaves oblanceolate, mostly more than 20 mm. wide; capitula mostly more than 2 cm. wide   15
14. Leaves filiform less than 5 cm. long, regularly spaced along the stem, margins revolute entire
    92. *orchidorrhiza*
    – Leaves linear-oblanceolate, the lower ones more than 5 cm. long, densely crowded on the lower stem, margins remotely serrate   -   -   -   -   -   -   -   -   -   91. *najas*

15. Outer phyllaries linear or ± subulate, loosely scattered on the lower receptacle and capitulum stalk; stems somewhat zigzag above, coarsely pubescent, leafy; leaves increasing in size and ± crowded towards the stem apex, scabrous on upper surface; plants flowering from January to July, not pyrophytic - - - - - - - - - - 89. *filipendula*
 - Outer phyllaries oblong to somewhat spathulate, appressed imbricate, absent from the lower receptacle and capitulum stalk; stems scapiform or leafy, not zigzag above, softly white pubescent to tomentose or glabrescent; leaves absent from the upper stem, or if present then decreasing in size towards the stem apex, hispid or scabridulous on upper surface; plants pyrophytic, flowering from August to November - - - - - - - - - - 90. *longipedunculata*
16. Capitula large, in lax groups of 2–5 or solitary on branches, individual capitula often with a subtending leaf; involucres 15–30 mm. long and 20–45 mm. wide (occasionally smaller); florets c. 100–300 per capitulum; corollas 15–24 mm. long; outer pappus present, consisting of a minute ± fimbriate fringe or of short linear scales - - - - - - - 85. *adoensis*
 - Capitula smaller, in many ± dense clusters of 3–12 capitula on short branches, the clusters usually with subtending leaves; involucres 8–17 mm. long and 10–25 mm. wide; florets 16–70 per capitulum; corollas 12–16 mm. long; achenes with a narrow constricted collar below the pappus but an outer pappus not differentiated - - - - - - - - - 17
17. Capitula in congested globose clusters, stalks usually 0–15 mm. long; achenes glandular otherwise glabrous; pappus setae curving outwards; phyllary appendages ± equalling the phyllary in width - - - - - - - - - - - - - 88. *tolypophora*
 - Capitula in ± open corymbiform cymes, stalks mostly more than 5 mm. long; achenes hispid to glabrescent; pappus setae straight; phyllary appendages usually much wider than the phyllary itself - - - - - - - - - - - - - - - - - 18
18. Petaloid phyllary appendages lanceolate, tapering to an acute apex, or narrowly oblong and ± obtuse at the apex; involucres obconic-campanulate; leaf usually cuneate at the base; plants of swampy or wet localities in Zambia W: - - - - - - - 87. *ringoetii*
 - Petaloid phyllary appendages broadly oblong-ovate to rotund, shortly acute to ± rounded or truncate and apiculate at the apex; involucres subspherical; leaf ± abruptly narrowed into a short narrowly cuneate base; plants of evergreen forest margins and submontane grassland of eastern Zimbabwe and central Mozambique - - - - 86. *calvoana* subsp. *meridionalis*

GROUP J

Plants annual, single season ephemerals; stems single from a single season taproot.

1. Outer phyllaries produced into long spathulate apical appendages, exceeding the inner phyllaries in length; anther tube purple - - - - - - 100. *anthelmintica*
 - Phyllaries not produced into long spathulate appendages, progressively longer to the inside; anther tube straw-coloured - - - - - - - - - - - - - 2
2. Pappus setae grey, subplumose or with well developed barbs; achene ± densely silvery-strigose, phyllaries woolly-araneose, pilose or long ciliate (except in *V. galamensis*) - - - 3
 - Pappus setae white, sordid, brownish, purplish or green; other features not as combined above 6
3. Phyllaries finely tapering to a 2–7 mm. long pungent usually squarrose awn; leaves filiform or linear mostly less than 5 mm. wide - - - - - - - - - - - - 4
 - Phyllaries mucronate or shortly aristate never long awned; leaves linear to elliptic more than 5 mm. wide - - - - - - - - - - - - - - - - 5
4. Involucre mostly more than 10 mm. long; achene 3–4 mm. long; outer pappus of short linear seta-like scales; plants of hot dry areas usually on black basaltic soils or in mopani veld 101. *fastigiata*
 - Involucre less than 10 mm. long; achene up to c. 2.5 mm. long; outer pappus of short oblong-lanceolate ± fimbriate scales; plants of granite outcrops - - - - 102. *graniticola*
5. Phyllaries glabrous or hispidulous, green, narrowly tapering apically, the outer long subulate; mature achenes firmly attached; leaves elliptic, mostly more than 2 cm. wide 104. *galamensis*
 - Phyllaries ± woolly or cobwebby, purplish and obtuse to rounded mucronate apically, outer shortly lanceolate; achenes easily dislodged; leaves up to c. 2 cm. wide 103. *kirkii*
6. Stems and branches densely leafy; leaves filiform, revolute; involucre tightly imbricate, hemispheric to cup-shaped, more than 13 mm. wide when mature; achenes c. 5 mm. long, 6–10-ribbed (*V. steetziana* has similar but smaller capitula and smaller 5-ribbed achenes) 123. *perrottetii*
 - Features not as combined above - - - - - - - - - - - - 7
7. Inner phyllaries c. 2 mm. wide; involucre 12–15 mm. long; pappus setae white, copious, easily dislodged; stem indumentum of T-shaped hairs, arms sometimes somewhat unequal 124. *amoena*
 - Features not as combined above - - - - - - - - - - - - - 8
8. Achenes terete or obscurely c. 5-ribbed, subfusiform, tapering to a somewhat abruptly narrowed base; outer pappus whorl of few to many scales, white or brownish not overlapping, sometimes fringe-like and fimbriate; inner pappus whorl of green sometimes brownish, sordid or white setae - - - - - - - - - - - - - - - - - 9

– Achenes strongly 5–12-ribbed often angular, obconic, tapering to a ± pointed base; outer pappus
   whorl of numerous scales, brownish or sordid, ± overlapping, narrowly oblong-lanceolate; inner
   pappus whorl of brown, brownish-purple or sordid setae, sometimes wanting   17
9. Capitula numerous small, involucres 4–6 mm. long; pappus setae white, copious, sometimes
    much reduced or absent; stems with short-stalked long-armed T-shaped hairs   10
– Capitula few to many, involucres more than 6 mm. long; pappus setae brown, brownish-purple
   or green; stem indumentum lacking T-shaped hairs   -   -   -   -   -   -   12
10. Inner pappus whorl absent or of a few bristle-like scales up to c. 1.4 mm. long   111. *stellulifera*
– Inner pappus whorl of setae 2.5–5.0 mm. long   -   -   -   -   -   -   11
11. Outer pappus whorl of narrow, distinct, lanceolate scales 0.2–0.6 mm. long; plants of lower
    altitude mostly from sea level to 800 m.   -   -   -   -   -   -   112. *cinerea*
– Outer pappus whorl of short distinct or ± confluent, oblong-rounded or subtruncate scales
   0.1–0.2 mm. long; plants of higher altitudes, mostly above 800 m.   113. *meiostephana*
12. Pappus setae brown, brownish-purple or sordid; corollas purple   -   -   -   13
– Pappus setae bright- or dull-green; corollas creamy white, mauve or purple   -   -   14
13. Phyllaries subacute-obtuse, often subapically mucronate, brownish strigose-pubescent especially
    towards the apex; involucres cylindric-campanulate, the phyllaries hardly diverging
                                                                                  114. *sylvicola*
– Phyllaries gently tapering to a bristle-tipped apex; sparsely puberulous or scabridulous;
   involucres obconic, the phyllaries diverging   -   -   -   115. *jelfiae* var. *jelfiae*
14. Phyllaries narrowly subobtuse submucronate; involucres ± rounded at the base, broad,
    spreading-campanulate, silvery-white pubescent; capitula stalks stout, often widely diverging;
    pappus setae bright green   -   -   -   -   -   -   -   -   117. *chloropappa*
– Phyllaries gently tapering to a sharp bristle-tipped apex; involucres obconic or narrowly
   cyindric-campanulate, brownish or white pubescent; capitula stalks slender; pappus setae green
   or brownish-green   -   -   -   -   -   -   -   -   -   -   -   15
15. Inner pappus whorl of subplumose setae, 9–17 per achene; outer pappus whorl of narrowly
    triangular scales, fimbriate-ciliate, brownish; corollas cream-coloured or sometimes mauve
                                                                           115. *jelfiae* var. *albida*
– Inner pappus whorl of barbellate setae, 6–9 sometimes 12 per achene; outer pappus whorl of
   broad scales, fimbriate-toothed on the upper margin, white or brownish; corollas purple or
   cream-coloured   -   -   -   -   -   -   -   -   -   -   -   -   16
16. Corollas purple; inner pappus setae 8–12, usually 9, per achene; involucres obconic, sparsely
    brownish-pubescent   -   -   -   -   -   -   -   -   115. *jelfiae* var. *jelfiae*
– Corollas cream, white or mauve; inner pappus setae c. 6 per achene; involucres narrowly
   cylindric-campanulate, white pubescent   -   -   -   -   -   -   116. *mbalensis*
17. Inner pappus whorl consisting of few (2–3) setae or wanting, outer pappus whorl of persistent
    scales; capitula small   -   -   -   -   -   -   -   -   -   108. *lundiensis*
– Inner pappus whorl consisting of many or copious setae, outer pappus whorl of scales;
   involucres mostly more than 6 mm. long and 8–15(20) mm. wide   -   -   -   -   18
18. Leaves usually more than 10 mm. wide, hardly revolute (c. 5 mm. wide in *V. zambiana*); achenes ±
    rugulose between ribs; indumentum containing large erect flagelliform hairs (at least on lower
    stems); receptacle usually alveolate; phyllaries longitudinally ribbed   -   -   -   19
– Leaves mostly less than 5 mm. wide, + linear, often filiform, revolute; achenes not rugulose
   between ribs; erect flagelliform hairs absent; receptacle not alveolate; phyllaries not
   longitudinally ribbed   -   -   -   -   -   -   -   -   -   -   -   24
19. Corollas creamy white to pale-mauve; pappus setae subplumose; leaves mostly c. 5 mm. wide
                                                                                  105. *zambiana*
– Corollas purple; pappus setae barbellate; leaves mostly more than 10 mm. wide   20
20. Capitula overtopped by subtending leaves   -   -   -   -   -   -   -   21
– Capitula exceeding the subtending leaves, or subtending leaves absent   -   -   -   22
21. Plants stunted, less than 30 cm. tall; involucre ± obconic, tapering to a narrow base; phyllaries
    mostly straight, tapering to a greenish or purplish tip   -   -   -   -   106. *ambigua*
– Plants 10–120 cm. tall; involucre broadly cup-shaped ± flattened at the base and clearly wider
   than long (when mature); phyllaries mostly purple-tipped and recurved   107. *petersii*
22. Pappus scales of outer whorl whitish, broad (width more than half the length), pappus setae of
    inner whorl relatively few (5–10), straw-coloured or purple-tinged; phyllaries with distinct purple
    pungent apices; involucre broadly obconic-campanulate   -   -   -   109. *miombicola*
– Pappus scales of outer whorl brownish or sordid, narrowly oblong (width less than half the
   length); pappus setae of inner whorl more than 10, brownish or sordid; phyllaries acute or finely
   tapering, sometimes purple at the apex; involucre broadly cup-shaped or obconic   23
23. Involucre becoming broadly cup-shaped and wider than long (when mature), densely white
    sericeous-lanate below; phyllaries narrowly lanceolate, at least some ± recurved   107. *petersii*
– Involucre obconic becoming hemispheric, sparsely brownish pilose-hispid; phyllaries subulate
   (at least the outer) all straight and finely tapering apically   -   -   -   110. *acuminatissima*
24. Pappus setae of inner whorl plumose, the seta bristles spreading, equalling or exceeding the
    achene hairs in length   -   -   -   -   -   -   -   -   -   122. *centaureoides*
– Pappus setae of inner whorl barbellate, the seta barbs ascending-appressed shorter than the
   achene hairs   -   -   -   -   -   -   -   -   -   -   -   -   25

25. Achenes 8–12-ribbed - - - - - - - 118. *poskeana* subsp. *botswanica*
   - Achenes 5–7-ribbed - - - - - - - - - - - - 26
26. Phyllary midrib continued as a 1 mm. or more long, bristle-like mucro, purple and squarrose or somewhat spiralled, seldom straight - - - - - - - 120. *erinacea*
   - Phyllaries ± mucronate, the mucro short straight (up to c. 0.5 mm. long in *V. steetziana* but then bristle not purple and involucre rounded at the base and more than 5 mm. wide) 27
27. Phyllaries all, or at least the innermost 2–3 series, widening towards the apex with the margins broadly hyaline or purple-tinged, abruptly narrowed to a rounded or truncate-emarginate apex, mucronate or shortly aristate; involucre campanulate, rounded at the base with phyllaries appressed imbricate - - - - - - - - - - - 121. *steetziana*
   - Phyllaries all gently tapered or somewhat rounded at the apex, not expanded apically, ± mucronate; involucre obconic, subfusiform or campanulate, usually ± tapered at the base 28
28. At least some leaves deeply 3-lobed - - - - - - - 119. *rhodesiana*
   - Leaves never 3-lobed - - - - - - - - - - - - - 29
29. Involucre 10 mm. or more long, more than 6 mm. wide, ± campanulate or turbinate; capitula stalks not slender, the majority more than 2 cm. long; achenes 3–4 mm. long
      118. *poskeana* subsp. *samfyana*
   - Involucre less than 10 mm. long (sometimes to c. 10 mm. long but then less than 6 mm. wide), obconic, subfusiform to hemispheric; capitula stalks slender, the majority less than 2 cm. long; achenes 1.5–3 mm. long - - - - - - - 118. *poskeana* subsp. *poskeana*

1. **Vernonia aurantiaca** (O. Hoffm.) N.E. Br. in Bull. Misc. Inf., Kew **1909**: 116 (1909). —Brenan in Mem. N.Y. Bot. Gard. **8**, 5: 461 (1954). —White in F.F.N.R.: 429 (1962). —Hilliard in Ross, Fl. Natal: 356 (1972); Compos. Natal: 34 (1977). —Wild in Kirkia **11**: 123 (1978). —Maquet in Fl. Rwanda, Spermat. **3**: 560 (1985). —C. Jeffrey in Kew Bull. **43**: 211 (1988). Type from Tanzania.
   Gongrothamnus divaricatus Steetz in Peters, Reise Mossamb., Bot.: 342 (1864) non *Vernonia divaricata* Sw. (1806). —Bentham in Hook., Icon. Pl. **12**: 36, t. 1140 (1872). —Oliv. & Hiern in F.T.A. **3**: 401 (1877). —Merxm., Prodr. Fl. SW. Afr. 139: 85 (1967). Type: Mozambique, Inhambane, *Peters* (B†).
   Gongrothamnus aurantiacus O. Hoffm. in Engl., Bot. Jahrb. **30**: 433 (1901). Type as above.
   Vernonia vitellina N.E. Br. in Bull. Misc. Inf., Kew **1909**: 116 (1909). Type as for *Gongrothamnus divaricatus*.
   Distephanus divaricatus (Steetz) H. Robinson & B. Kahn in Proc. Biol. Soc. Wash. **99**: 499 (1986). Type as for *Gongrothamnus divaricatus*.

A divaricately branched scandent shrub or scrambler, to c. 8 m. tall, often scrambling over other vegetation; branches whitish tomentellous or puberulous to glabrescent, ribbed, older branches lenticellate purplish. Leaves with a petiole up to c. 2 cm. long; lamina 2.5–12 × 1.5–10 cm., broadly ovate to lanceolate, apex ± long acuminate, base abruptly cuneate or cuneate and shortly decurrent on upper petiole, margins entire to distantly serrate, upper surface green and sparsely puberulous, the lower surface hoary or whitish-tomentellous, sometimes thinly so or glabrescent, strongly 3-nerved from near the base. Capitula numerous, in ± dense subglobose clusters at the ends of the branches. Involucres 4–7 × 3–6 mm., obconic or obconic-campanulate. Phyllaries ± numerous, tapering to an acute mucronate apex, farinose-puberulous outside and ± ciliate on the margins or glabrescent; the outer phyllaries c. 3 mm. long, lanceolate-subulate, extending onto the capitulum stalk; the middle phyllaries 4–5 mm. long, oblong-lanceolate; the inner linear-oblong. Corolla orange-yellow, sweetly scented, 7–10 mm. long, tubular widening towards the apex. Achenes 1.2–2.5 mm. long, very narrowly obovoid-subcylindric, tapering to the base, somewhat angular or faintly 6–10-ribbed, densely strigose; outer pappus of short linear scale-like setae, inner of stramineous barbellate setae 4–6.5 mm. long.

Botswana. N: c. 29 km. SW. of Maun, 6.v.1967, *Lambrecht* 180 (COI; K; PRE; SRGH). SE: Mahalapye, 28.iv.1961, *Yalala* 133 (BM; COI; SRGH). **Zambia**. B: Nangweshi, Zambezi R., 22.vii.1952, *Codd* 7152 (BM; K; PRE; SRGH). N: Chienge, 19.viii.1958, *Fanshawe* 4745 (BR; K). W: Mwinilunga, Musombosombo R., 10.vi.1963, *Edwards* 715 (K; M; PRE; SRGH). C: Lusaka, 11.vii.1963, *Best* 384 (M; SRGH). S: c. 27 km. NE. of Choma, 11.vii.1930, *Hutchinson & Gillett* 3540 (BM; K; LISC; SRGH). **Zimbabwe**. N: Gokwe, Sessami Tsetse Sta., 28.v.1963, *Bingham* 774 (K; M; SRGH). W: Shangani, Gwampa For. Res., v.1955, *Goldsmith* 155/55 (K; LISC; SRGH). C: between Kadoma and Kwekwe, 2.v.1958, *West* 3593A (K; SRGH). E: Mutare, 4.vii.1952, *Chase* 4565 (BM; BR; K; LISC; MO; SRGH). S: Gwanda, Doddieburn Ranch, c. 800 m., 11.v.1972, *Pope* 741 (K; SRGH). **Malawi**. N: Rumphi, c. 920 m., 25.vi.1972, *Pawek* 5483 (K; SRGH). C: Salima–Balaka road, c. 12 km. S. of Chipoka, 4.v.1980, *Blackmore, Brummitt & Banda* 1440 (K). S: L. Malawi, Palm Beach, 28.v.1961, *Leach & Rutherford-Smith* 11054 (LISC; SRGH). **Mozambique**. N: Cabo Delgado, between Nangororo and Rio Tati, 30.v.1959, *Gomes e Sousa* 4475 (COI; K; PRE; SRGH). T: Tete, Chissua, 2.x.1972, *Macedo* 5248 (LISC; PRE; SRGH). MS: R. Zangué, 3.vii.1946, *Simão* 705 (PRE;

SRGH). GI: entre Moambe e Magude, 2.v.1944, *Torre* 6546 (LISC; PRE). M: Santaca, 18.viii.1947, *Gomes e Sousa* 3601 (SRGH).

Also in Kenya, Tanzania, Rwanda, Zaire, Angola and South Africa (Natal and Transvaal). Usually at lower altitudes, frequently on termitaria, in deciduous woodland, riverine fringe vegetation and also on granite outcrops.

2. **Vernonia amygdalina** Delile, Cent. Pl. Afr., Voy. Méroé, Cailliaud: 41 (1826). —Oliv. & Hiern in F.T.A. **3**: 284 (1877). —Mendonça, Contrib. Conhec. Fl. Angol., **1** Compositae: 22 (1943). —Brenan, Check-list For. Trees Shrubs Tang. Terr.: 165 (1949); Mem. N.Y. Bot. Gard. **8**, 5: 460 (1954). —White, F.F.N.R.: 429 (1962). —Adams in F.W.T.A. ed. 2, **2**: 277 (1963). —Hilliard in Ross., Fl. Natal: 356 (1972); Compos. Natal: 29 (1977). —K. Coates Palgrave, Trees Southern Afr.: 902 (1977). —Wild in Kirkia **11**: 95 (1978). —Maquet in Fl. Rwanda, Spermat. **3**: 560, fig. 169, 1 (1985). —C. Jeffrey in Kew Bull. **43**: 212 (1988). TAB. **13** fig. C. Type from Sudan.

Vernonia randii S. Moore in Journ. Bot. **37**: 369 (1899). Syntypes: Zimbabwe, Harare, *Rand* 495 (BM); 1372 (BM).

A many-stemmed shrub or small tree up to c. 8 m. tall. Stems slender with a pale-brown closely longitudinally furrowed bark; branches and twigs greyish pubescent, the indumentum of 1-armed T-shaped hairs. Leaves with petioles 0.2–1.5 cm. long; lamina mostly 7–14 × 3–4 cm., sometimes smaller or occasionally up to c. 28 × 9 cm., c. 4 times as long as wide, narrowly elliptic or lanceolate-elliptic, apex tapering acute to sub-acuminate, base ± narrowly cuneate, margins subentire to serrulate, upper surface puberulous to glabrescent, lower surface paler and ± tomentose or glabrescent with venation ± reticulate and somewhat raised. Capitula very numerous in terminal, divaricately branching, rounded panicles; stalks 3–12 mm. long, stiff, tomentellous. Involucres 3–5 mm. long and 3–6 mm. wide, mostly wider than long, subglobose becoming hemispheric or broadly campanulate. Phyllaries many, tightly appressed-imbricate at first, thinly cartilaginous, sometimes puberulous and glandular at the apex and ciliate on the margins, otherwise glabrous; the outer phyllaries from c. 1 mm. long, ovate-elliptic; the middle phyllaries broadly oblong-elliptic, rounded to obtuse at the apex; the innermost up to c. 5 mm. long, narrowly oblong-oblanceolate with a rounded to acute apex. Florets 10–24 per capitulum; corollas exserted above the involucre, whitish, sweet-scented, 6–9 mm. long, narrowly funnel-shaped. Achenes 2.2–3.5 mm. long, narrowly turbinate to subcylindric tapering to the base, c. 10-ribbed, setulose particularly on the ribs, glandular; pappus exserted beyond the involucre, outer pappus of very caducous linear scale-like setae 1–3 mm. long, inner pappus of more persistent sordid to brownish, rigid, slender setae widening and somewhat flattened towards the apex 4–6 mm. long and barbellate.

Botswana. N: c. 42 km. NE. of Maun, 29.vi.1937, *Pole Evans* 4129 (K; PRE; SRGH). **Zambia**. B: Kabompo R., 21.vi.1954, *Gilges* 391 (K; SRGH). N: Mbala, Kawimbe, c. 1680 m., 11.ix.1956, *Richards* 6157 (K; SRGH). W: Ndola, Kafue R., Hippo Pools, c. 1350 m., 10.vii.1961, *Linley* 165 (K; SRGH). C: Lusaka, 28.vii.1963, *Robinson* 5591 (K; M; SRGH). E: Luangwa Valley, c. 19 km. W. of Jumbe, c. 750 m., 13.x.1958, *Robson* 89 (BM; K; LISC; SRGH). S: Victoria Falls, Palm Grove, 27.viii.1911, *Rogers* 7423 (K; MO; SRGH). **Zimbabwe**. N: Hurungwe (Urungwe), c. 4.8 km. E. of Kariba, 19.vi.1956, *Goodier* 84 (K; SRGH). W: Victoria Falls, 19.vii.1947, *Keay* in FHI 21397 (K; SRGH). C: Harare, Cleveland Dam, 17.viii.1965, *Biegel* 383 (MO; SRGH). E: Birchenough Bridge-Chimanimani road, 800 m., 26.viii.1960, *Rutherford-Smith* 5 (K; LISC; SRGH). S: Great Zimbabwe, 19.viii.1965, *West* 6737 (SRGH). **Malawi**. N: Mzimba Distr., Mzuzu, c. 1370 m., 11.x.1969, *Pawek* 2900 (K). C: Kasungu, c. 1000 m., 24.viii.1946, *Brass* 17407 (K; MO; PRE; SRGH). S: Chikwawa, Lower Mwanza R., 3.x.1946, *Brass* 17930 (K; MO; SRGH). **Mozambique**. N: Cabo Delgado, Montepuez, 29.viii.1948, *Barbosa* 1917 (LISC). T: between Tete and Zóbuè, c. 450 m., 25.vi.1947, *Hornby* 2772 (K; SRGH). MS: Maronga Forest, 15.viii.1945, *Simão* 473 (SRGH). GI: Guijá, Machino-Estivane, 2.vi.1947, *Pedro & Pedrógão* 2012 (PRE).

Throughout tropical Africa, the islands of the Gulf of Guinea and South Africa (Natal). Riverine forest, woodland or thicket and grassland bordering rivers and streams or on the margins of seasonally waterlogged grassland (vleis), usually on alluvium, occasionally on termitaria.

3. **Vernonia colorata** (Willd.) Drake in Grandidier, Hist. Madag. Pl. 6, Atlas 4, t. 466 (1897); Bull. Soc. Bot. Fr. **46**: 230 (1899). —Mendonça, Contrib. Conhec. Fl. Angol., **1** Compositae: 22 (1943). —Brenan Check-list For. Trees Shrubs Tang. Terr.: 165 (1949). —White, F.F.N.R.: 430 (1962). —Adams in F.W.T.A. ed. 2, **2**: 277 (1963). —Hilliard in Ross, Fl. Natal: 356 (1972); Compos. Natal: 29 (1977). —K. Coates Palgrave, Trees Southern Afr.: 904 (1977). —Wild in Kirkia **11**: 93 (1978). —C. Jeffrey in Kew Bull. **43**: 212 (1988). Type from Senegal.

Eupatorium coloratum Willd. in L., Sp. Pl. ed. 4, **3**: 1768 (1803). Type as above.

Baccharis senegalensis Pers., Syn. Pl. **2**: 424 (1807). Type from Senegal.

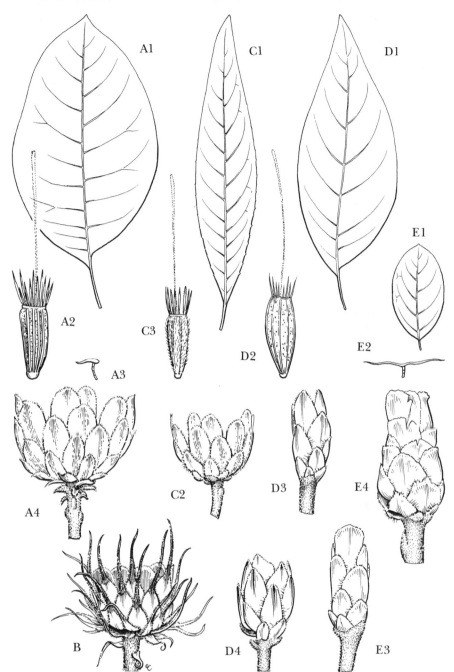

Tab. 13. A. —VERNONIA COLORATA subsp. COLORATA. A1, leat; A2, achene, glandular otherwise glabrous, outer pappus caducous; A3, trichome from twig; A4, involucre, A1–A4 from *Richards* 12903. B. —VERNONIA COLORATA subsp. OXYURA, involucre, from *Pawek* 2249. C. —VERNONIA AMYGDALINA. C1, leaf; C2, involucre, C1 & C2 from *Fanshawe* 134; C3, achene, glandular and pubescent, outer pappus caducous, from *Angus* 108. D. — VERNONIA EXSERTIFLORA var. TENUICALYX. D1, leaf; D2, achene, outer pappus persistent; D3, involucre, D1–D3 from *Fanshawe* 8640; D4, var. EXSERTIFLORA, involucre, from *Brummitt* 10645. E. —VERNONIA BELLINGHAMII. E1, leaf; E2, trichome from twig; E3, involucre, E1–E3 from *Brummitt* 9276; E4, involucre, distinctly stalked, from *Simon* 266. (Leaves ×½, trichomes ×20, involucres ×4, achenes ×6 showing only one of the numerous inner pappus setae). Drawn by Eleanor Catherine.

Gymnanthemum quercifolium Steetz in Peters, Reise Mossamb., Bot.: 334 (1864). Types: Mozambique, Tete, *Peters* (B†); Sena, *Peters* (B†).

Vernonia senegalensis (Pers.) Less. in Linnaea, **4**: 265 (1829). —Harv. in Harv. & Sond., F.C. **3**: 50 (1865). —Oliv. & Hiern in F.T.A. **3**: 283 (1877). —Sim, For. Fl. Port. E. Afr.: 77 (1909). —Eyles in Trans. Roy. Soc. S. Afr. **5**: 505 (1916). Type as above.

A shrub or small tree up to c. 6 m. tall (c. 10 m. fide *White* 2476); stems up to c. 7 cm. diam.; branches and twigs fulvous-pubescent or tomentose; indumentum of 1-armed T-shaped hairs. Leaves with a petiole 0.5–3 cm. long; lamina 6–21 × 3–10 cm., elliptic to broadly oblong-elliptic or ovate, apex obtuse to acute, base cuneate less often somewhat rounded, margins subentire sometimes serrulate, undulate, upper surface thinly pilose becoming scabridulous or glabrescent, lower surface ± thinly pubescent to tawny tomentose, occasionally glabrescent with venation reticulate and somewhat raised. Capitula numerous in terminal divaricate paniculate corymbs up to c. 28 cm. across; stalks 3–25 mm. long, stiff, tomentellous. Involucres 3–14 mm. long (including caudate phyllary apices in subsp. *oxyura*) and 4–14 mm. wide, wider than long, subglobose-cyathiform. Phyllaries many, cymbiform, tightly appressed imbricate, ecaudate or with long caudate ± curved or reflexed appendages; outer phyllaries from c. 1 mm. long and ovate-acuminate or long subulate, the caudate tip if present up to c. 6 mm. long, the outermost sometimes extending briefly down the capitulum stalk; middle phyllaries c. 4 mm. long and narrowly obovate rounded and darkly mucronate-apiculate or ovate-elliptic with an attenuate-caudate tip up to c. 10 mm. long; the innermost to c. 6 mm. long and oblong-oblanceolate, rounded and darkly mucronate or sometimes caudate at apex; lamina cartilaginous, puberulous or glandular at apex, ciliate on margins otherwise glabrous; caudate apices withering early and drying brownish or purplish. Florets 10–26 per capitulum, sweetly scented; corollas mauve fading to whitish, 8–11 mm. long, narrowly funnel-shaped. Achenes 3–4 mm. long tapering-subcylindric, c. 10-ribbed, glabrous, ± glandular between ribs; pappus exserted beyond involucre, outer pappus of relatively few very caducous linear scale-like setae 1–4 mm. long, inner pappus of more persistent copious, reddish, rigid slender setae widening and somewhat flattened towards the apex, 7–8 mm. long and barbellate.

O. Hoffmann's species *V. oxyura*, which has larger capitula and subulate or long-caudate phyllaries, cannot be kept separate at the species level. Discontinuity between it and typical *V. colorata* is incomplete with forms, particularly to the north of the Flora Zambesiaca area, combining features of both.

Phyllaries ecaudate, at most apiculate at apex, the outermost ovate-acuminate rarely extending onto capitulum stalk; involucre mostly less than 7 mm. wide; corolla 7–9(11) mm. long; inner pappus setae c. 6(8) mm. long   -   -   -   -   -   -   -   -   -   -   -   subsp. *colorata*

Phyllaries long caudate, the outermost usually subulate, curved or ± reflexed and extending briefly down the capitulum stalk; involucre mostly more than 7 mm. wide; corolla tubes 9–12 mm. long; inner pappus setae 7–8 mm. long   -   -   -   -   -   -   -   -   subsp. *oxyura*

Subsp. **colorata**, TAB. **13** fig. A.

Botswana. N: Chobe, v.1952, *Miller* B1321 (K; PRE). **Zambia.** N: Kalambo Falls, 16.vii.1950, *Bullock* 2997 (BR; K; SRGH). C: Chilanga, Mt. Makulu Res. Sta., 26.v.1957, *Angus* 1611 (BR; K; MO). S: Choma, Siamambo For. Res., 9.vi.1952, *White* 2936 (BR; FHO; K; MO). **Zimbabwe.** N: Hurungwe, near Harare-Chirundu road, 10.v.1951, *Lovemore* 42 (K; SRGH). W: Nkayi (Nkai), Gwampa For. Res., x.1958, *Goldsmith* 29/59 (BR; M; MO; SRGH). C: Kwekwe, Hunters Road, 15.iv.1967, *Biegel* 2072 (K; SRGH). E: Chipinge, Dakata R., 15.v.1956, *Mowbray* 108 (K; SRGH). S: Chibi, Nyoni Mts., c. 700 m., 6.vii.1968, *Müller* 794 (K; SRGH). **Malawi.** S: Chikwawa Distr., Lengwe Game Res., 21.vii.1970, *Hall-Martin* 852 (K; SRGH). **Mozambique.** N: c. 26 km. E. of Malema, c. 700 m., 24.v.1961, *Leach & Rutherford-Smith* 10992 (K; LISC; SRGH). Z: Mocuba, Namagoa, 6.vii.1943, *Faulkner* 133 (K; PRE; SRGH). T: between Fingoè and Vila Vasco da Gama, 27.vi.1949, *Barbosa & Carvalho* 3337 (K; SRGH). MS: c. 48 km. E. of Vila de Manica, 14.vi.1959, *Leach* 9103 (K; SRGH). GI: Macia, 11.vii.1947, *Pedro & Pedrógão* 1450 (K; SRGH). M: Goba, Umveluzi R., c. 120 m., 4.vi.1947, *Codd & Dyer* 2935 (K; PRE).

Throughout west, central and south tropical Africa, East Africa and South Africa (Natal, Transvaal) and Swaziland. Found in a variety of habitats including miombo, mixed deciduous and escarpment woodlands, riverine vegetation, grassland and sometimes on termitaria.

Subsp. **oxyura** (O. Hoffm.) C. Jeffrey in Kew Bull. **43**: 213 (1988). TAB. **13** fig. B. Type: Malawi, *Buchanan* 41 (BM, lectotype fide Smith loc. cit.; K).

Vernonia oxyura O. Hoffm. in Engl. Pflanzenw. Ost-Afr. **C**: 403 (1895). —Brenan, Check-list For. Trees Shrubs Tang. Terr.: 162 (1949). —Earle Sm. in U.S. Dept. Agric. Handbook **396**: 43 (1971). —Wild in Kirkia **11**: 102 (1978). Type as above.

Vernonia polyura O. Hoffm. in Engl., Bot. Jahrb. **30**: 422 (1901). —Brenan, Check-list For. Trees Shrubs Tang. Terr.: 162 (1949); Mem. N.Y. Bot. Gard. **8**, 5: 461 (1954). —White, F.F.N.R.: 429 (1962). Type from Tanzania.

Vernonia cirrifera S. Moore in Journ. Linn. Soc., Bot. **35**: 320 (1902). Type: Malawi, *Buchanan* 370 (BM, holotype).

Vernonia longipetiolata Muschl. in Engl., Bot. Jahrb. **46**: 74 (1911). Type from Zaire.

Zambia. N: Mpika-Isoka road, c. 37 km. north of Chinsali turnoff, 14.iv.1986, *Philcox, Pope & Chisumpa* 9924 (K; MO; NDO; SRGH). E: Chipata, c. 48 km. on Lundazi Road, 26.iv.1952, *White* 2476 (BR; FHO; K; MO). **Malawi**. N: Rumphi, c. 80 km. from Chisungu on Njakwa-Chisungu road, 22.vi.1962, *Richards* 16791 (K; M). C: Kasungu Hill, 28.viii.1946, *Brass* 17452 (BR; K; MO; SRGH). S: Zomba, 14.viii.1962, *Adlard* 475 (K; SRGH). **Mozambique**. N: Massangulo, iv.1955, *Gomes e Sousa* 1397 (COI; K). Z: c. 77 km. from Mocuba between Mocuba and Milange, 9.vi.1949, *Barbosa & Carvalho* 2999 (K; SRGH).

Also in Kenya, Tanzania and Zaire. Rocky slopes and hillsides in miombo or mixed deciduous woodland.

4. **Vernonia exsertiflora** Bak. in Bull. Misc. Inf., Kew **1898**: 147 (July, 1898). —Wild in Kirkia **11**: 89 (1978). —C. Jeffrey in Kew Bull. **43**: 213 (1988). Type: Malawi, between Kondowe and Karonga, *Whyte* 95 (K, holotype).

Vernonia kreismannii Welw. ex Hiern, Cat. Afr. Pl. Welw. **1**, 3: 517 (Dec., 1898). —Mendonça, Contrib. Conhec. Fl. Angol., 1 Compositae: 7 (1943). —Wild in Kirkia **11**: 90 (1978). Syntypes from Angola.

Vernonia sphaerocalyx O. Hoffm. in Engl., Bot. Jahrb. **30**: 423 (1902). Type from Tanzania.

A bushy shrub or small, branched tree up to c. 6 m. tall. Stems with grey, rough, longitudinally fissured bark. Branches densely appressed-pubescent, the indumentum of T-shaped hairs. Leaves petiolate; petiole up to c. 4 cm. long, indumentum similar to that of branches; lamina up to c. 23 × 9 cm., ovate to lanceolate, tapering to apex, base rounded sometimes unequally, or broadly cuneate, margins entire sometimes obscurely serrate, upper surface glabrous and somewhat shiny, lower sparsely to densely pubescent or appressed tomentellous, indumentum of fine ± matted T-shaped hairs, nervation somewhat prominent below. Capitula very numerous in ± dense clusters up to c. 18 cm. across, or much fewer in lax terminal clusters, capitula stalks 3–15 mm. long densely pubescent. Involucres mostly 4.5–6 × 2–3.5 mm., subspherical to narrowly ellipsoid-cylindrical, ± contracted at mouth, less often 8–16 × 6–11 mm. and obovoid-oblong; phyllaries tightly appressed imbricate increasing in size towards the inside, outermost from c. 1 mm. long and ovate-acute, inner mostly to c. 6 mm. long, less often to c. 11 × 2–3 mm. and oblanceolate to very narrowly elliptic, apices acute to rounded submucronate, lamina somewhat cartilaginous, soon glabrescent, margins subscarious and ciliate. Florets 4–20 per capitulum; corollas purplish-mauve or fading white, 7–11 mm. long narrowly funnel-shaped glabrous. Achenes 3–5 mm. long, narrowly 2–3-angular turbinate, obscurely 5–10-ribbed or smooth, glabrous and ± glandular between ribs; pappus exserted beyond involucre and shaving-brush-like, outer pappus of short linear-lanceolate scales; inner of white or sordid flattened setae 6–9 mm. long closely serrulate-barbellate.

Distinguished from *V. bellinghamii* S. Moore on leaf size and shape *V. exsertiflora* exhibits a similar pattern of variation in the inflorescence and involucre size and shape, namely a reduction in the number of capitula with a corresponding increase in size.

Involucre oblong-ovoid to subspherical, less than twice as long as wide; florets 9–11 per capitulum rarely as few as 7; usually found on rocky hillsides in dry deciduous woodland or scrubland
var. *exsertiflora*
Involucre narrowly ellipsoid, more than twice as long as wide; florets usually 5 per capitulum, less often 4 or 6; usually found in riverine vegetation or dry evergreen thickets (mateshi)
var. *tenuicalyx*

Var. **exsertiflora**, TAB. **13** fig. D4

Zambia. N: c. 95 km. from Mbala on Nakonde Road, c. 1650 m., 23.iv.1986, *Philcox, Pope & Chisumpa* 10091 (K; NDO; SRGH). W: Solwezi Distr., Mutanda Bridge, 14.v.1986, *Philcox, Pope & Chisumpa* 10257 (K; NDO; SRGH). **Zimbabwe**. C: Harare Distr., Spelonken, Bot. Gard. ext., 20.vii.1974, *Pope* 1360 (K; SRGH). **Malawi**. N: c. 30 km. W. of Rumphi on road to Katumbi, c. 1250 m., 12.v.1970, *Brummitt* 10646 (K).

Also in Tanzania. On rocky hillsides in deciduous woodland often with *Brachystegia glaucescens*, or scrubland.

Var. **tenuicalyx** G.V. Pope in Kew Bull. **43**: 279 (1988). TAB. **13** fig. D. Type: Zambia, Mbala Distr., Inono Stream, *Richards* 1529 (K, holotype).

Zambia. N: Mweru-Wantipa, near Selemani, 17.iv.1961, *Phipps & Vesey-FitzGerald* 3280 (K; LISC; SRGH). W: Chingola, Luano For. Res., 9.v.1968, *Mutimushi* 2608 (K; NDO; SRGH).
Also in Tanzania and Zaire. Riverine vegetation and dry evergreen thicket (mateshi).

5. **Vernonia bellinghamii** S. Moore in Journ. Bot. **38**: 155 (May, 1900). —Brenan in Mem. N.Y. Bot. Gard. **8**, 5: 460 (1954). —White, F.F.N.R.: 430 (1962). —K. Coates-Palgrave, Trees Southern Afr.: 903 (1977). —Wild in Kirkia **11**: 89 (1978). —C. Jeffrey in Kew Bull. **43**: 213 (1988). TAB. **13** fig. E. Syntypes: Malawi, Likoma, Lake Malawi, *Bellingham* s.n. (BM); without precise locality, *Buchanan* 1124 (K); 1136 (BM; E; K); Mozambique, Unango to Lake Shire, *Johnson* 48 (K).
        Vernonia goetzeana O. Hoffm. in Engl., Bot. Jahrb. **28**: 503 (July, 1900), non *goetzei* Muschl. (1911). Type from Tanzania.
        Vernonia elisabethvilleana De Wild. in Fedde, Repert. **13**: 208 (1914). Type from Zaire.
        Vernonia fraterna N.E. Br. in Bull. Misc. Inf., Kew **1921**: 295 (1921). Syntypes: Zambia, Bwana Mkubwa, *Rogers* 8310 (K); Mukwela, *Rogers* 26027 (K).

A shrub or small tree 1–4 m. tall; branches leafy, densely greyish appressed tomentellous at first, becoming less densely so and exposing lenticels on older branches, the indumentum of matted T-shaped hairs. Leaves petiolate; petiole 2–8 mm. long with an indumentum similar to that of branches; lamina mostly 3–6.5 × 1.5–3 cm., sometimes smaller, less often up to c. 8 × 4 cm., elliptic to oblong-ovate or obovate, apex subacute to rounded, base cuneate sometimes unequal, margins subentire to obscurely serrulate, subcoriaceous and discolorous, upper surface green, glandular otherwise glabrous, lower surface densely greyish tomentellous. Capitula massed in 1–many small dense subumbellate clusters 1–4 cm. across, the clusters 4–15-capitulate and arranged at ends of twigs or sometimes the capitula becoming fewer with subumbels reduced to 2–3 distinctly stalked, laxly arranged capitula in abbreviated panicles. Involucres usually 6–8 × 2.5 mm., subcylindric gradually narrowing to apex, sometimes up to c. 10 × 4.5 mm. and narrowly ovoid. Phyllaries tightly appressed-imbricate, outermost c. 1 mm. long and broadly ovate-acute, longer towards the inside becoming oblong-oblanceolate, rounded and scarious at apex; inner up to 8–10 mm. long, lamina glabrous with margins ± araneose-ciliate. Florets 4–12 per capitulum; corollas white or pale-mauve, up to c. 9 mm. long, narrowly funnel-shaped. Achenes 3–4 mm. long, narrowly turbinate to subcylindric and tapering towards the base, somewhat 3–4-angular and 4–6-ribbed, glabrous seldom sparsely hispidulous; pappus long-exserted beyond involucre, outer pappus of short linear-lanceolate scales, inner of whitish or straw-coloured flattened setae 5–6 mm. long and serrulate-barbellate on edges.

Zambia. N: Kasama, 24.iii.1960, *Angus* 2158 (FHO; K). W: Ndola, 18.vi.1954, *Fanshawe* 1284 (BR; K). C: Serenje, Kundalila Falls, c. 13 km. SE. of Kanono, 17.ii.1970, *Drummond & Williamson* 9671 (BR; K; SRGH). E: Chipata, Township, 23.iv.1952, *White* 2451 (FHO; K). **Zimbabwe**. N: Gokwe, Charama Plateau, 11.iv.1963, *Bingham* 618 (K; M; SRGH). **Malawi**. N: c. 16 km. N. of Mzimba, c. 1380 m., 9.viii.1946, *Brass* 17144 (BM; BR; K; MO; PRE; SRGH). C: Lilongwe Distr., Dzalanyama For. Res., c. 6 km. SE. of Chaulongwe Falls, 1230–1290 m., 22.iii.1970, *Brummitt* 9276 (BR; K). S: 1891, *Buchanan* 1136 (BM; E; K). **Mozambique**. N: Unango to Lake Shire, 1899, Johnson 48 (K).
Also in Tanzania and Zaire. Rocky outcrops, hillsides and escarpments in miombo and deciduous woodland.
        The subumbellate masses of narrowly cylindric c. 4-flowered capitula, so characteristic for the Malawian and eastern Zambian plants, undergo a gradual reduction in the number of capitula per inflorescence in plants further to the west and south west. As the capitula become fewer they increase in size with a corresponding increase in the number of florets — up to 12 per capitulum, and they also become more laxly arranged with each borne on a distinct stalk. The leaves too exhibit some variation changing from the typically tough elliptic-obovate leaves in the east to larger slightly thinner-textured leaves in the west, eg. *Fanshawe* 8647 (K) from Mansa in northern Zambia. This distinctive form, described as *V. fraterna* by N.E. Brown, cannot satisfactorily be separated from *V. bellinghamii*.

6. **Vernonia brachycalyx** O. Hoffm. in Engl., Pflanzenw. Ost-Afr. **C**: 405 (1895). —Brenan, Check-list For. Trees Shrubs Tang. Terr.: 163 (1949). —Wild in Kirkia **11**: 46 (1978). —Maquet in Fl. Rwanda, Spermat. **3**: 559 (1985). —C. Jeffrey in Kew Bull. **43**: 218 (1988). Syntypes from Kenya, Tanzania and Zaire.

A shrub, often scandent, up to c. 4 m. tall; stems densely pale-brown tomentellous, glabrescent. Leaves petiolate; petiole 5–15 mm. long, tomentellous; lamina up to 10 × 6 cm., ovate, or lanceolate when young, apex acuminate to acute, base ± rounded and then abruptly narrowly and shortly attenuate into the petiole, margin serrate, upper surface hispid-pubescent, lower surface ± sparsely tomentose to pubescent, venation reticulate. Capitula numerous in a terminal paniculate arrangement, c. 10 cm. in diam.; capitula stalks 2–3 mm. long pale-brown tomentose. Involucre c. 2.5 mm. long, spreading-campanulate; phyllaries c. 2-seriate, the inner ± equalling achene in length, ovate to lanceolate, apex acute or mucronate, densely pubescent, inner with membranous margins. Corolla lavender to violet, c. 5 mm. long, narrowly funnel-shaped, glandular, lobes narrowly triangular acute, 1.3 mm. long. Achenes c. 2 mm. long, obconic, 4-sided, minutely glandular, setulose on the angles; outer pappus of short, narrow, scale-like setae, inner of whitish barbellate setae to 4 mm. long.

Zambia. N: Mbala Distr., Lumi R. Marsh, c. 1680 m., 30.iii.1957, *Richards* 8942 (BR; K). **Malawi**. N: Rumphi Distr., South Rukuru R., Njakwa Gorge, c. 1066 m., 4.iv.1969, *Pawek* 1913 (K).
Also in Ethiopia, Uganda, Rwanda, Kenya, Tanzania, Zaire and Angola. Riverine thicket vegetation and river gorges.

7. **Vernonia holstii** O. Hoffm. in Engl., Bot. Jahrb. **20**: 220 (1894); in Engl., Pflanzenw. Ost-Afr. **C**: 402 (1895). —Brenan, Check-list For. Trees Shrubs Tang. Terr.: 163 (1949); in Mem. N.Y. Bot. Gard. **8**, 5: 459 (1954). —Wild in Kirkia **11**: 48 (1978). —Maquet in Fl. Rwanda, Spermat. **3**: 559 (1985). —C. Jeffrey in Kew Bull. **43**: 214 (1988). Type from Tanzania.
Vernonia assimilis S. Moore in Journ. Bot. **56**: 211 (1918). Type from Zaire.

A lax, branching shrub up to c. 4 m. tall; stems villous at first. Leaves shortly petiolate; lamina up to 17 × 8 cm., ovate to elliptic, apex acuminate, base cuneate, margins sinuate-serrate or dentate, upper surface pubescent, lower surface whitish-grey-tomentose or araneose. Synflorescence a wide panicle; capitula stalks slender, villous, to 1 cm. long. Involucre c. 8 mm. long, obconic; phyllaries spreading, increasing in length to the inside, lanceolate-acuminate, margins whitish-membranous, outermost phyllaries villous-araneose, the remainder glabrescent outside. Corolla whitish, 6–7 mm. long, narrowly funnel-shaped, lobes 1.3 mm. long linear acute, glandular. Achenes c. 3 mm. long, oblong-cylindric, 10-ribbed, pilose between the ribs; outer pappus of short narrow scales, inner of 7 mm. long slender exserted barbellate setae.

Zambia. N: Mafinga Hills, 24.v.1973, *Chisumpa* 77 (K; NDO). **Zimbabwe**. C: Wedza Mt., 14.v.1964, *Wild* 6560 (K; M; SRGH). E: Chimanimani, Tarka, vii.1971, c. 1300 m., *Goldsmith* 24/71 (K; M; SRGH). **Malawi**. N: Rumphi Distr., Manchewe Falls, 1160 m., 24.ii.1977, *Phillips* 2632 (K; MAL; MO). S: Mulanje Distr., Likabula Gorge, c. 840 m., 20.vi.1946, *Brass* 16382 (K; MO; PRE; SRGH). **Mozambique**. M: Chimoio, Garuso Mt., vii.1964, *Wild* 6573 (K; M; SRGH).
Also in Cameroons, Rwanda, Kenya, Tanzania and Zaire. Forest edges in high rainfall submontane areas.

8. **Vernonia pteropoda** Oliv. & Hiern in F.T.A. **3**: 283 (1877). —Brenan in Mem. N.Y. Bot. Gard. **8**, 5: 460 (1954). —Wild in Kirkia **11**: 93 (1978). —C. Jeffrey in Kew Bull. **43**: 215 (1988). Type: Malawi, Manganja Hills, Mt. Chiradzulu, *Meller* s.n. (K, holotype).
Vernonia urophylla Muschl. in Engl., Bot. Jahrb. **46**: 86 (1911). Type from Tanzania.

A subshrub up to c. 2.5 m. tall. Stems simple below, becoming somewhat woody at the base, upper stem and branches finely striate, coarsely pilose to glabrescent; indumentum a mixture of few to very numerous, purplish, multicellular somewhat fleshy patent-pilose hairs and smaller whitish appressed hairs; synflorescence branches bracteate, bracts foliaceous up to 7 × 2 cm. Leaves petiolate; petiole up to c. 10 mm. long; lamina up to c. 25 × 8.5 cm. elliptic, tapering to a long acuminate apex, ± abruptly contracted in the lower third to a narrowly winged midrib gradually tapering into the petiole, margins incised-dentate or serrate with short setiform teeth, upper surface thinly pilose or glabrescent, lower surface ± sparsely pilose on the main nerves and minutely gland-pitted. Capitula numerous in many terminal and lateral, bracteate clusters of 6–30 capitula; capitulum stalk filiform, 1–15 mm. long. Involucres 8–11 × 5–8 mm., campanulate-obconic. Phyllaries chartaceous; the outer broadly ovate acute, the middle ovate-lanceolate, the inner up to c. 10 mm. long and linear-oblong, the apex subacute to rounded, sometimes mucronulate, lamina thinly pilose-puberulous becoming hispidulous or glabrescent, somewhat striate, greenish to stramineous, darker-tipped. Corolla white or lilac, 5–9 mm. long, narrowly funnel-shaped, shortly tubular below widening above. Achenes 2–2.5 mm.

long, subcylindric narrowing to the base, narrowly 5–8-ribbed, glabrous or hispidulous on the ribs; outer pappus a fringe of very short setae; inner of copious silky-white ± caducous barbellate setae 5–7 mm. long.

Zambia. E: Makutus Mts., 27.x.1972, *Fanshawe* 11590 (K; NDO). **Zimbabwe**. E: Chirinda Forest, 22.viii.1965, *Chase* 8304 (BR; K; LISC; MO; SRGH). **Malawi**: N: Karonga, Misuku Hills, Wilindi For. Res., 16.ix.1970, *Müller* 1678 (SRGH). S: Thyolo Mt., 20.ix.1946, *Brass* 17679 (BR; K; MO; PRE; SRGH). **Mozambique**. Z: Massingire (at top of) Serra da Morrumbala, 6.viii.1942, *Torre* 4529 (LISC). MS: Garuso Mts., viii.1948, *Munch* 130 (K).
Also in Tanzania and Kenya. Evergreen rainforest understorey.

9. **Vernonia madefacta** Wild in Kirkia **11**: 9; 48 (1978). Type: Zambia, Mbala Distr., Chilongowelo, *Richards* 5815 (K, holotype; BR; LISC).

A perennial herb or subshrub c. 1 m. tall; stems slender, pubescent. Leaves up to c. 15 × 7 cm., broadly ovate, membranous, long acuminate at the apex, rounded and abruptly contracted below into a long ± tapering-winged midrib, wing 4–6 mm. wide tapering into a short petiole, upper surface glabrous, lower surface very thinly pilose; margins serrate with apiculate teeth. Capitula laxly paniculately arranged on bracteate stalks; lower bracts leaf-like with subulate awns c. 3 mm. long; stalks slender, to c. 25 mm. long, thinly pubescent and glandular. Involucre c. 9 mm. long, obconic-campanulate; phyllaries scarious-membranous, not recurved, spreading, increasing in length towards the inside, outermost subulate becoming shortly lanceolate each with a long straight aristate awn at apex, inner narrowly lanceolate and acute-pungent at apex, purplish and pubescent where exposed. Corolla pale-mauve, c. 10 mm. long, widening very slightly upwards, lobes 3 mm. long, linear. Achenes (immature) 1.5 mm. long, obconic-cylindric, narrowly 8–10-ribbed, strigose pilose; outer pappus short, narrowly scale-like and barbellate, inner of whitish barbellate setae c. 7 mm. long.

Zambia. N: Mbala Distr., Chilongowelo, top of Tasker's Deviation, c. 1493 m., 22.v.1955, *Richards* 5815 (BR; K; LISC).
Not known elsewhere. On damp rocks by waterfall.

10. **Vernonia nepetifolia** Wild in Kirkia **11**: 1; 51 (1978). Type: Zimbabwe, Chimanimani Mts., *Swynnerton* 1909 (BM, holotype; K).
  Vernonia gracilipes var. *minor* S. Moore in Journ. Linn. Soc., Bot. **40**: 106 (1911). Type as above.

A woody subshrub to 1.3 m. tall; branches fulvo-tomentose and glandular to glabrescent. Leaves shortly petiolate; lamina usually up to 4 × 3 cm., ovate to broadly ovate, apex obtuse or rounded mucronate, base truncate and abruptly cuneate to shallowly cordate, margins crenate, upper surface hispid, lower surface densely hispid, glandular, with nerves prominent. Capitula stalked, few to many in paniculate clusters, terminal on branches; stalks 6–20 mm. long. Involucres ± campanulate, c. 5 mm. long; phyllaries 4-seriate, outer very shortly ovate-lanceolate, the innermost up to c. 5 mm. long and oblong to narrowly oblong, apex obtuse mucronate, densely pubescent. Corolla mauve, c. 9 mm. long, narrowly funnel-shaped, lobes up to 4 mm. long, linear acute, minutely glandular. Achenes c. 3 mm. long, narrowly oblong-cylindric, faintly 8–10-ribbed, shortly strigose, glandular; outer pappus of short narrow scales; inner of whitish barbellate setae 5–6 mm. long.

Zimbabwe. E: Chimanimani Mts., 5.ix.1961, *Loveridge* A46 (K; M; SRGH).
Known so far only from the Chimanimani Mts. of Zimbabwe. Rocky slopes, on quartzite and on Umkondo sandstones on Mt. Peza and Mt. Pene.

11. **Vernonia eylesii** S. Moore in Journ. Linn. Soc., Bot. **47**: 261 (1925). —Wild in Kirkia **11**: 50 (1978). Type: Zimbabwe, Makoni Distr., Forest Hill, *Eyles* 737 (BM, holotype; K; SRGH).

A branching subshrub c. 1.5 m. tall; stems brownish-purple, branches stiffly spreading-pubescent. Leaves subsessile, up to 4 × 1.5 cm., elliptic to ovate-lanceolate, apex acute minutely mucronate, base broadly cuneate, margin entire to serrate-dentate, upper surface hispid-pubescent, lower surface pubescent, minutely glandular. Capitula in ± dense 2–15-capitulate clusters at the ends of branches; stalks up to c. 5 mm. long, densely pubescent. Involucre c. 5 mm. long, campanulate; phyllaries imbricate, increasing in length to inside, shortly lanceolate to oblong-lanceolate, obtuse, pubescent, apex dark-tipped. Corollas purple, c. 9 mm. long, narrowly funnel-shaped. Achenes obconic-cylindric, obscurely 4–5-ribbed, appressed-setulose; outer pappus of narrow short scales; inner of whitish barbellate setae 5–6 mm. long.

Zimbabwe. C: Rusape, Valhalla, vii.1952, *Dehn* R15 (M; MO; SRGH). E: Mutare, Zimunya Communal Land, 3.vi.1956, *Chase* 6139 (K; LISC; M; SRGH). S: Masvingo, Kyle Dam, Popoteke R., 23.v.1971, *Grosvenor* 538 (SRGH).
Known only from eastern Zimbabwe. Granite hills and associated sand-veld.

12. **Vernonia muelleri** Wild in Kirkia **11**: 3; 47 (1978). —C. Jeffrey in Kew Bull. **43**: 226 (1988). Type: Mozambique, Manica e Sofala, Chimanimani Mts., *Müller & Kelly* 1098 (SRGH, holotype; K).

A lax bushy perennial herb or subshrub to c. 2.6 m. tall; stems and branches ± densely appressed brownish-pubescent, indumentum of T-shaped hairs. Leaves petiolate, up to 16 × 5 cm., oblong-elliptic or broadly lanceolate, apex acute to acuminate, often mucronate, base attenuate narrowly tapering into petiole or cuneate, margin ± coarsely sinuate- or serrate-dentate, upper surface somewhat harshly pubescent to glabrescent, lower surface appressed tomentellous to pubescent or glabrescent, the hairs T-shaped. Capitula laxly arranged in many 3–9-capitulate clusters at the ends of branches, stalks up to c. 5 cm. long, densely appressed pubescent. Involucre 10–15 mm. long, broadly campanulate; phyllaries imbricate, outermost shortly ovate to lanceolate and recurved subulate or bristle-tipped, middle phyllaries lanceolate with or without a curved apical bristle up to 8 mm. long, innermost to c. 13 mm. long, linear-oblong, ± expanded membranous and aristate or mucronate apically, purplish towards the apex, pubescent or puberulent outside, margins ciliate to strongly pectinate. Corolla pale-mauve to whitish, c. 10 mm. long, narrowly funnel-shaped, lobes to c. 3 mm. long linear acute, glandular-puberulent outside. Achene c. 3 mm. long, subcylindric, narrowly 8–10-ribbed, sparsely hispid and glandular between the ribs; outer pappus of short, narrowly oblong scales c. 0.6 mm. long, inner of caducous stramineous barbellate setae 6–8 mm. long.

Outer phyllaries with a recurved, subulate-setaceous apex 3–8 mm. long; leaves narrowly attenuate at
 base, puberulent beneath; plants of the Chimanimani Mts. on the Mozambique-Zimbabwe
 border - - - - - - - - - - - - - - - subsp. *muelleri*
Outer phyllaries pungent-aristate or cuspidate, apicle bristles less than 2 mm. long; leaves cuneate at
 base, sparsely puberulous to glabrescent beneath; plants from N. Zambia, adjacent Tanzania,
 Zaire and Malawi - - - - - - - - - - - - subsp. *brevicuspis*

Subsp. **muelleri**

Zimbabwe. E: Chimanimani Mts., Outward Bound School, 23.ix.1966, *Simon* 901 (K; SRGH).
Mozambique. MS: Southern tip of Chimanimani Mts., above Haroni-Makurupini Forest, 28.v.1969, *Müller & Kelly* 1098 (K; SRGH).
Not known elsewhere. Evergreen forest margins and adjacent woodlands.

Subsp. **brevicuspis** G.V. Pope in Kew Bull. **43**: 279 (1988). Type: Zambia, Mbala, Lunzaw Falls [?Lunzua], *Richards* 22262 (K, holotype).
   Vernonia jugalis var. *jugalis* sensu Wild in Kirkia **11**, 1: 46 (1978).
   Vernonia jugalis var. *dekindtii* sensu Wild in Kirkia **11**, 1: 46 (1978) non (O.Hoffm.) Hiern (1898).

Zambia. N: Mbala, 9.v.1952, *Richards* 1161 (K; SRGH). **Malawi**. N: Nyika Plateau, 1200–1800 m., Jan. to July 1896, *Whyte* s.n. (K). S: Shire Highlands, x.1880, *Buchanan* 111 (E; K).
Also in Tanzania and Zaire. High-rainfall woodland, often beside rivers.

13. **Vernonia mushituensis** Wild in Kirkia **11**: 4; 78 (1978). Type: Zambia, Chingola, *Fanshawe* 10901 (SRGH, holotype; K; NDO).

A weak subshrub to 2.6 m. tall. Stems slender, leafy, branching, ribbed, ± appressed pubescent; hairs long unequal-armed T-shaped, or hairs with the terminal cell ± excentric on a short base; branches slender, arising in axils of largest leaves. Leaves subsessile or with slender petioles to c. 4 mm. long; lamina to c. 9 × 1.8 cm., smaller on branches, narrowly lanceolate to narrowly elliptic, apex subacute to obtuse mucronate, base cuneate, margins subentire to remotely subserrate with callose-tipped teeth; membranous, sparsely puberulous and gland-pitted, indumentum of small multicellular, uniseriate hairs usually densest on petiole and midrib below. Capitula numerous in lax subpanicles on the main branches, stalks to c. 4.5 cm. long, slender, pubescent; bracts few or absent, to c. 5 mm. long. Involucres 9–11 × 5–9 mm., broadly subfusiform to narrowly

campanulate; phyllaries many-seriate, appressed imbricate, increasing regularly in size
from short, ovate-lanceolate outer ones to linear-lanceolate inner ones up to c. 10 mm.
long, apices acute and ± densely fuscus-hairy, margins pale membranous. Corollas
mauve, to c. 9 mm. long, tubular and gradually slightly wider above, lobes acute c. 1.5 mm.
long. Achenes 3–4 mm. long, subfusiform-subcylindric with 8–10 narrow ribs, hispidulous
mainly on the ribs; outer pappus of short oblong-lanceolate scales 0.5–1 mm. long, inner
pappus of brownish barbellate setae to c. 6 mm. long.

Zambia. N: Kasama, Munguri, 2.x.1960, *Robinson* 3894 (K; M; SRGH). W: Chingola, 25.viii.1954,
Fanshawe 1486 (K).
Not known from elsewhere. Mushitu (swamp forest) margins.

14. **Vernonia stenocephala** Oliv. in Hook., Ic. Pl. 14, **2**: 35, t. 1349A (1881). —Wild in Kirkia **11**: 77
(1978). —C. Jeffrey in Kew Bull. **43**: 242 (1988). Type from Tanzania.
   Vernonia oocephala Bak. in Bull. Misc. Inf., Kew **1895**: 68 (1895). Type: Zambia, Fwambo,
Carson 74 (K, holotype).
   Vernonia oocephala var. *angustifolia* S. Moore in Journ. Bot. **52**: 334 (1914). —Mendonça,
Contrib. Conhec. Fl. Angol., **1** Compositae: 12 (1943). Type from Angola.
   Vernonia luteoalbida De Wild. in Fedde, Repert. **13**: 207 (1913). Syntypes from Zaire.

An erect leafy perennial herb to c. 1.3 m. tall. Stems 1–several from a woody rootstock,
strict, soon becoming leafless below, usually much-branched above, subterete or ribbed,
tomentellous-pubescent, glabrescent below; branches stiffly ascending, c. 5–35 cm. long,
densely leafy. Leaves ascending, subsessile, variable; cauline leaves up to c. 7 × 0.6 cm. and
linear-elliptic, or up to c. 7 × 2.2 cm. and lanceolate to oblong-elliptic, apices acute to
rounded mucronate, margins revolute entire, midribs prominent below, lamina upper
surface puberulous, lower surface finely pubescent, becoming scabridous; branch leaves
mostly ericaceous, 1–3 × 0.1–0.3 cm., linear, sometimes up to c. 4 × 0.4 cm. and lorate-
elliptic, scabridous or puberulous, revolute and entire on the margins, midrib prominent
below. Capitula very numerous, ± densely paniculate at the stem apex. Involucres up to c.
12 mm. long, ellipsoid-campanulate, softly tomentellous; phyllaries numerous,
appressed-imbricate, the outermost very shortly ovate and extending onto the capitulum
stalk, middle phyllaries oblong, becoming linear-lanceolate and c. 7 mm. long inside,
tomentellous to lorate where exposed. Receptacle flat, c. 1.3 mm. in diam. Corollas white
or creamy-white, up to c. 9 mm. long, tubular below, widening above. Achenes up to c. 4
mm. long, subcylindric, tapering to the base, obscurely 6–8-narrowly-ribbed, ± densely
strigose; outer pappus of short lanceolate, ciliate scales; inner pappus of copious, pale-
brown, subplumose setae c. 6 mm. long.

Zambia. N: Mbala, c. 1680 m., 6.v.1951, *Bullock* 3850 (BR; K; SRGH). W: Mufulira, 21.v.1934, *Eyles*
8306 (K; SRGH). C: c. 45 km. SW. of Kabwe, c. 1160 m., 13.vii.1930, *Hutchinson & Gillett* 3611 (BM; K;
LISC; SRGH). E: Chipata, Lunkwakwa, 28.ix.1966, *Mutimushi* 1497 (K; M; NDO; SRGH). S: c. 40
km. from Mumbwa on Kafue Hoek road, 7.xi.1959, *Drummond & Cookson* 6187 (K; SRGH). **Malawi.** N: Nyika Plateau, c. 2270 m., ix.1902, *McClounie* (K). **Mozambique.** N: Maniamba, x.1964,
Magalhães 66 (COI). T: between Casula and Furancungo, 70 km. from Casula, 9.vii.1949, *Barbosa &
Carvalho* 3532 (K; SRGH). MS: Manica, Serra de Vumba, 25.iii.1948, *Garcia* 718A (LISC).
Also in Nigeria, Burundi, Tanzania, Zaire and Angola. Miombo woodland.

15. **Vernonia syringifolia** O. Hoffm. in Engl., Pflanzenw. Ost-Afr. **C**: 405 (1895). —Wild in Kirkia **11**:
49 (1978). —Maquet in Fl. Rwanda, Spermat. **3**: 559 (1985). —C. Jeffrey in Kew Bull. **43**: 217
(1988). TAB. **14** fig. B. Type from Tanzania.
   Vernonia calongensis Muschl. in Engl., Bot. Jahrb. **46**: 89 (1911). Type from Zaire.

A diffuse or scandent shrub with stems to c. 6 m. long. Branches alternate, striate,
pubescent, glandular; indumentum of short-stalked T-shaped hairs. Leaves membranous,
petiolate; petiole 1–3 cm. long; lamina to 13 × 6 cm., ovate-lanceolate, apex tapering-
acuminate, base broadly cuneate to truncate in upper leaves, cordate often unequally so
in lower leaves, sometimes narrowly decurrent on petiole, margins serrate or serrulate,
lamina upper surface sparsely pilose to scabridulous, lower surface thinly pilose, the hairs
one-armed or unequal-armed T-shaped on upper surface and T-shaped beneath.
Capitula numerous, grouped in moderately dense clusters 3–6 cm. across, borne in the
upper leaf axils, or capitula in large terminal compound panicles; synflorescence
branches sparsely linear-bracteate; capitula stalks mostly 1–10 mm. long, sometimes to c.
40 mm. long, linear-bracteate, tomentellous to appressed-pubescent. Involucres to c. 6 × 6
mm., spreading to c. 10 mm. wide, obconic-campanulate; phyllaries few-seriate, outer 2–4

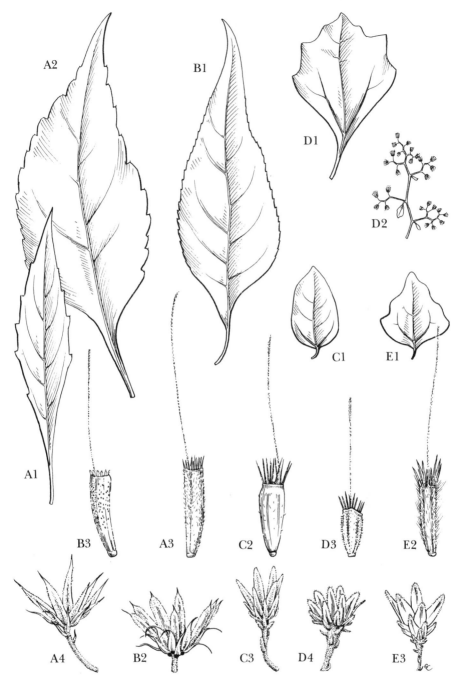

Tab. 14. A. —VERNONIA WOLLASTONII. A1, leaf, from *Pope* 114; A2, leaf; A3, achene; A4, involucre, A2–A4 from *Whellan* 1145. B. —VERNONIA SYRINGIFOLIA. B1, leaf; B2, involucre, B1 & B2 from *Brummitt & Synge* WC 79; B3, achene, glandular, from *Brummitt* 11955. C. —VERNONIA INHACENSIS. C1, leaf; C2, achene, C1 & C2 from *Mogg* 28430; C3, involucre, from *Mogg* 31840. D. —VERNONIA ANISOCHAETOIDES. D1, leaf; D2, part of divaricately branching synflorescence; D3, achene; D4, involucre, D1–D4 from *Bayliss* 5800 (Cape Province). E. —VERNONIA ANGULIFOLIA. E1, leaf; E2, achene; E3, involucre, E1–E3 from *Nuvunga* 229. (Leaves × ⅔, involucres × 3, achenes × 8 showing only one of the numerous inner pappus setae). Drawn by Eleanor Catherine.

mm. long and subulate to linear-lanceolate with a subulate apex, inner to 6 mm. long and narrowly oblong-elliptic, subacute mucronate or aristate, scarious sometimes with a hyaline margin, thinly pilose outside or glabrescent. Florets 15–20 per capitulum. Corollas pale-mauve to whitish, sweetly-scented, c. 6 mm. long, narrowly funnel-shaped, widening slightly to apex. Achenes pale-brownish, 2.5–3 mm. long, narrowly subcylindric, tapering slightly to base, sharply narrowly 5-ribbed, glandular otherwise glabrous, at the most sparsely hispidulous; outer pappus fringe-like, of narrow minute lacerate scales, inner of white barbellate setae 5–6 mm. long.

Zambia. E: Nyika, 9.vii.1962, *Lawton* 923 (K; NDO). Malawi. N: Nyika Plateau, c. 10 km. from Chelinda Camp, c. 2200 m., 10.vii.1970, *Brummitt* 11955 (K; LISC; MAL; PRE; SRGH; UPS). C: Dedza Mountain, 4.v.1968, *Jeke* 186 (K; MAL; SRGH).
Also in Sudan, Uganda, Rwanda, Kenya, Tanzania and Zaire. Submontane evergreen forest margins.

16. **Vernonia wollastonii** S. Moore in Journ. Linn. Soc., Bot. **38**: 257 (1908); in Journ. Bot. **52**: 334 (1914). —Wild in Kirkia **11**: 88 (1978). —C. Jeffrey in Kew Bull. **43**: 224 (1988). TAB. **14** fig. A. Type from Uganda.
    Vernonia gracilipes S. Moore in Journ. Linn. Soc., Bot. **40**: 105 (1911). Syntypes: Zimbabwe, Chimanimani Mts., *Swynnerton* 1830 (BM), 1830a (BM; K), Mt. Pene, *Swynnerton* 6126 (BM), 6127 (BM).
    Vernonia umbratica Oberm. in Journ. S. Afr. Bot. **2**: 164 (1936). —Hilliard in Ross, Fl. Natal: 357 (1972); Compos. Natal: 33 (1977). Type from South Africa (Transvaal).
    Vernonia transvaalensis Hutch., Bot. S. Afr.: 374 (1946). Type from South Africa (Transvaal).

A lax somewhat delicate scrambling shrub, or diffuse subshrub, to c. 3 m. tall. Stems to c. 5 m. long; branches striate, appressed pubescent, indumentum of short-stalked T-shaped hairs. Leaves membranous, petiolate; petiole mostly 3–5 mm. long, sometimes to c. 45 mm. long; lamina mostly to 5–8(12) × 1–4 cm., ± narrowly lanceolate or elliptic-lanceolate, apex tapering acuminate, base cuneate and decurrent on petiole, or lamina to c. 12 × 6.5 cm. and ovate-lanceolate to ovate with apex acute and base abruptly contracted and narrowly decurrent on petiole; margins crenate-serrate with callose-tipped teeth or coarsely serrate; sparsely pilose or puberulent to glabrescent on both surfaces, minutely gland-pitted beneath; indumentum of one-armed or unequal-armed T-shaped hairs. Capitula numerous, in lax clusters 4–10 cm. across, terminal on branches, or capitula in a large terminal panicle; synflorescence branches sparsely linear-bracteate; capitula stalks mostly 4–22 mm. long, appressed-pubescent. Involucres 3–6 × 4–7 mm., obconic-campanulate; phyllaries few-seriate, the outer 2–4 mm. long and narrowly lanceolate with an aristate or subulate apex, the inner to c. 6 mm. long and narrowly oblong-elliptic with an acute mucronate-aristate to tapering-acuminate and sharply-tipped apex, greenish, sometimes purple-tinged above, brownish strigulose or sericeous. Florets 14–20 per capitulum. Corollas mauve, fading to white, c. 7 mm. long, narrowly funnel-shaped. Achenes 2–3 mm. long, subcylindric to subfusiform, obscurely narrowly c. 5-ribbed, ± appressed-hispid glandular; outer pappus of short, narrow, scale-like setae, inner of barbellate setae 5–7 mm. long.

Zambia. E: Mukutus, 28.x.1972, *Fanshawe* 11618 (K). Zimbabwe. E: Mutare, 12.ix.1956, *Whellan* 1145 (K; SRGH). S: Bikita, Turgwe R., above gorge, c. 940 m., 8.v.1969, *Pope* 114 (K; SRGH). Malawi. N: Misuku Hills, c. 1650 m., 10.i.1959, *Richards* 10594 (BR; K; SRGH). S: Chambe Plateau, Mulanje Mt., 20.v.1971, *Salubeni* 1614 (K; SRGH). Mozambique. Z: Namuli Peaks, c. 1380 m., 26.vii.1962, *Leach & Schelpe* 11471A (K; LISC; SRGH). MS: Tsetserra, c. 1925 m., 6.vi.1971, *Biegel* 3571 (K; LISC; SRGH).
Also in Sudan, Ethiopia, Uganda, Rwanda, Kenya, Tanzania, Zaire and South Africa (Transvaal and Natal). Evergreen submontane forest margins usually amongst rocks beside rivers and streams, also in montane grassland.

17. **Vernonia biafrae** Oliv. & Hiern in F.T.A. **3**: 270 (1877). —C. Jeffrey in Kew Bull. **43**: 216 (1988). Type from Cameroons.
    Vernonia tufnelliae S. Moore in Journ. Bot. **46**: 292 (1908). Type from Uganda.
    Vernonia leptolepis O. Hoffm. in Engl., Pflanzenw. Ost-Afr. **C**: 405 (1895) non Bak. —Wild in Kirkia **11**: 49 (1978). Type from Tanzania.
    Vernonia verschuerenii De Wild. in Bull. Jard. Bot. Brux. **5**: 94 (1915). Syntypes from Zaire.

A diffuse ± scandent shrub to c. 3 m. high. Branches alternate, striate, pubescent, glandular. Leaves membranous, petiolate; petiole to c. 1.5 cm. long; lamina to c. 7 × 4.5 cm., deltate-ovate, apex acute, base truncate to cordate and very narrowly decurrent on the petiole, repand-lobulate or coarsely toothed about the lower margins, entire towards the apex, rarely all entire; upper leaves smaller, ovate-lanceolate and rounded at the base, often entire; sparsely pilose on both sides, hairs densest on nerves, or glabrescent; lower 2 pairs of lateral nerves arising close together near the base, not pronounced. Capitula numerous, grouped in moderately dense 5–30-capitulate clusters 3–6 cm. across, or capitula in large terminal panicles; capitula stalks slender to c. 10 mm. long, bracteate. Involucres to c. 7 × 4–8 mm., obconic campanulate. Phyllaries several-seriate, sparsely puberulent and glandular outside, sometimes weakly ciliolate, scarious subhyaline on the margins, sometimes ± expanded towards the apex; the outer 1–2 mm. long, ovate-lanceolate, acute, extending onto the capitulum stalk; the inner phyllaries progressively longer, to c. 6 mm. long, lanceolate to narrowly lanceolate-elliptic, obtuse. Florets 8–10 per capitulum. Corollas pale-mauve to creamy-white, c. 6 mm. long, narrowly tubular but widening slightly to apex. Achenes pale-brownish, 2.5–3 mm. long, narrowly cylindric, tapering somewhat to the base, sharply narrowly 5-ribbed, sparsely hispidulous or glabrescent; outer pappus of short linear seta-like scales, inner of white barbellate setae 5–6 mm. long.

Zambia. N: Isoka Distr., Mafinga Hills, 21.viii.1965, *Lawton* 1281 (K; SRGH). **Malawi**. N: Mzimba Distr., Mzuzu, Marymount, c. 1380 m., 8.xi.1972, *Pawek* 5935 (K). S: Zomba Plateau, below road to summit opposite Malosa Saddle, c. 1900 m., 2.viii.1970, *Brummitt & Banda* 12375 (EA; K; LISC; MAL; PRE; SRGH; UPS; WAG).
Widespread in east, west and central tropical Africa, Sudan and Ethiopia. Submontane forest margins.

18. **Vernonia angulifolia** DC., Prodr. **5**: 29 (1836). —Harv. in Harv. & Sond., F.C. **3**: 49 (1865). —Hilliard in Ross, Fl. Natal: 356 (1972); Compos. Natal: 33 (1977). —Wild in Kirkia **11**: 125 (1978). TAB. **14** fig. E. Type from South Africa (Natal).
    Distephanus angulifolius (DC.) H. Robinson & B. Kahn in Proc. Biol. Soc. Wash. **99**: 499 (1986). Type as above.

A scandent shrub or climber to c. 6 m. tall. Branches usually opposite or subopposite, striate, puberulous, glandular. Leaves mostly opposite, membranous, drying dark, petiolate; petiole to c. 15 mm. long; lamina to c. 5.5 × 5 cm., mostly smaller, deltate-ovate, apex subacute to obtuse, base truncate or abruptly cuneate and decurrent on the petiole, margins coarsely 3–9-toothed to sublobate, leaves smaller, entire and ± ovate on younger growth; sparsely pilose to glabrescent on upper surface, puberulous beneath, gland-pitted, strongly 3–5-nerved from near base. Capitula very numerous in moderately dense clusters; clusters 3–6 cm. across, borne in the upper leaf axils and terminal on the branches; capitulum stalks slender, to c. 6 mm. long, bracteate. Involucres 4–6 × 4 mm., spreading to c. 10 mm. wide, obconic-campanulate. Phyllaries several-seriate, midrib slightly fleshy towards the apex, margins narrowly membranous, lamina puberulent outside; the outer phyllaries short, narrowly ovate-triangular, extending briefly onto the capitulum stalk, the inner phyllaries progressively longer, to c. 6 mm. long, ± narrowly lanceolate-elliptic, subacute. Florets c. 11 per capitulum, sweetly-scented; corollas pale-mauve to creamy-white, c. 6 mm. long, tubular but widening slightly to the apex. Achenes 1.5–2.5 mm. long, narrowly turbinate, sharply narrowly 5-ribbed, ± densely appressed silky-pilose or strigose; outer pappus of short, linear, seta-like scales, inner pappus of white barbellate setae 4–5 mm. long.

Mozambique. M: Maputo, Marracuene (Bobole) For. Res., 11.vii.1980, *Nuvunga* 229 (K; LMU). Also in South Africa (Natal). Dune and coast forest margins.

19. **Vernonia inhacensis** G.V. Pope in Kew Bull. **43**: 280 (1988). TAB. **14** fig. C. Type: Mozambique, Xai-Xai (João Belo), Praia Sepulveda, *Barbosa & Lemos* 7843 (K, holotype; COI; LISC).

A scrambling or climbing shrub. Branches and twigs alternate, striate, puberulous and glandular to glabrescent. Leaves thinly coriaceous, petiolate; petiole to c. 8 mm. long; lamina 2–5 × 0.8–3 cm., elliptic-ovate, apices subacute to obtuse, bases ± broadly cuneate or somewhat rounded in mature leaves, margins entire often ± revolute, glabrous or at first minutely puberulous, obscurely gland-pitted on lower surface, 3-nerved from the base. Capitula numerous, in moderately dense clusters 3–8 cm. across and terminal on short leafy twigs, or capitula in large terminal panicles; capitulum stalks 1–5 mm. long, bracteate. Involucres to c. 4 × 4 mm., spreading to c. 6 mm. wide, obconic-campanulate. Phyllaries several-seriate, progressively longer to inside, margins very narrowly hyaline, the lamina puberulous outside where exposed, setulose and ciliolate apically; the outer narrowly triangular, extending briefly onto the capitulum stalk; the inner to c. 3.5 mm. long, linear-lanceolate. Florets c. 9 per capitulum. Corollas pale-mauve, c. 5 mm. long, shortly tubular below with a deeply-lobed funnel-shaped limb; lobes c. 2 mm. long, more than half as long as limb. Achenes 2–2.5 mm. long, subcylindric to very narrowly turbinate, obscurely 8–10-ribbed sparsely hispidulous; outer pappus of short linear scales, inner of barbellate setae to c. 6 mm. long.

Mozambique. GI: Inhambane, Inharrime, Ponta Zavora, 16.x.1957, *Barbosa & Lemos* 8069 (COI; LISC). M: Inhaca Island, 11.ix.1964, *Mogg* 31840 (K).
Not known outside the Flora Zambesiaca area. Littoral scrub forest.

20. **Vernonia anisochaetoides** Sond. in Linnaea **23**: 61 (1850). —Harv. in Harv. & Sond., F.C. **3**: 49 (1865). —Hilliard in Ross, Fl. Natal: 356 (1972); Compos. Natal: 34 (1977). —Wild in Kirkia **11**: 125 (1978). TAB. **14** fig. D. Syntypes from South Africa (Natal).
    Distephanus anisochaetoides (Sond.) H. Robinson & B. Kahn in Proc. Biol. Soc. Wash. **99**: 499 (1986). Type as above.

A vigorous scandent or climbing shrub. Branches alternate, striate, puberulent and glandular to glabrescent. Leaves alternate, petiolate; petiole to c. 1.5 cm. long; lamina of lower leaves to c. 8.5 × c. 7 cm., very broadly ovate to cuneate-rhomboid, usually coarsely 3–9-toothed in upper half otherwise entire, bases broadly cuneate to abruptly narrowed before narrowly tapering onto the petiole, puberulous to glabrescent, gland-pitted beneath, strongly 3–5-nerved from near the base; upper leaves smaller, ovate-elliptic, rounded to acute at the apex, cuneate at the base. Capitula numerous in large stiffly divaricately-branched panicles; stalks to c. 10 mm. long, strongly ribbed, bracteate. Involucres 3–4 × 3–4 mm., broadly campanulate. Phyllaries several-seriate, midribs somewhat prominent and fleshy, margins narrowly membranous, lamina densely pilose-lanate to glabrescent outside, ciliate; the outer phyllaries short, narrowly triangular, extending briefly onto the capitulum stalk; the inner phyllaries progressively longer, to c. 4 mm. long, lanceolate to linear-oblong, subacute. Florets c. 15–20 per capitulum, sweetly-scented; corollas pale-mauve fading to creamy-white, 4–5 mm. long, tubular below, narrowly funnel-shaped above. Achenes 1.5–2 mm. long, narrowly subcylindric, 8–10-ribbed, sparsely hispidulous, glandular; outer pappus of short linear scales, inner of barbellate setae 3–5 mm. long.

Mozambique. GI: Gaza, Masieni, ix.1924, *van Dam* in Herb. Trans. Mus. 25349 (PRE). M: ? Maputo, 1905, *Alexander* 1907 (E).
Also in South Africa (Natal). Coast forest and riverine vegetation of coastal areas.

21. **Vernonia myriantha** Hook.f. in Journ. Linn. Soc., Bot. **7**: 198 (1864). —C. Jeffrey in Kew Bull. **43**: 219 (1988). TAB. **15** fig. C. Types from the Cameroons.
    Vernonia podocoma Sch. Bip. ex Vatke in Linnaea **39**: 476 (1875). —Oliv. & Hiern in F.T.A. **3**: 296 (1877). —S. Moore in Journ. Linn. Soc., Bot. **40**: 107 (1911). —Eyles in Trans. Roy. Soc. S. Afr. **5**: 505 (1916). —Mendonça, Contrib. Conhec. Fl. Angol., **1** Compositae: 32 (1943). Type from Ethiopia.
    Vernonia cylindrica sensu A. Rich., Tent. Fl. Abyss. **1**: 374 (1848) pro parte, non Sch. Bip. ex Walp. (1843).
    Vernonia subuligera O. Hoffm. in Engl., Pflanzenw. Ost-Afr. **C**: 403 (1895). —Maquet in Fl. Rwanda, Spermat. **3**: 552 (1985). Type from Tanzania.
    Vernonia stipulacea Klatt in Bull. Herb. Boiss. **4**: 457 (1896). —Hilliard & Burtt in Notes Roy. Bot. Gard. Edinb. 32, **3**: 386 (1973). —Hilliard in Compos. Natal: 30 (1977). —Wild in Kirkia **11**: 121 (1978). Type from South Africa (Transvaal).
    Vernonia ampla O. Hoffm. in Engl., Bot. Jahrb. **30**: 423 (1901). —Brenan in Mem. N.Y. Bot. Gard. **8**, 5: 462 (1954). —Adams in F.W.T.A. ed. 2, **2**: 277 (1963). Type from Tanzania.
    Vernonia lujae De Wild., Pl. Nov. Herb. Hort. Then. **2**: 119, t. 96 (1910). Type: Mozambique, Morrumbala, *Luja* 342 (holotype BR).
    Vernonia myrianthoides Muschl. in Engl., Bot. Jahrb. **46**: 79 (1911). Type from Tanzania.
    Vernonia uhligii Muschl., in Engl., Bot. Jahrb. **46**: 79 (1911). Type from Tanzania.
    Vernonia oliveriana Pichi Sermolli in Webbia **7**: 345 (1950) nom. illegit. superfl. Type as for *V. podocoma* Sch. Bip. ex Vatke.

A spreading shrub 1.5–6 m. tall. Stems few to many, stout; branches lax, ± angular when young, often sparsely tuberculate-lenticellate, pubescent to tawny-tomentose, occasionally glabrescent, granulose-glandular; indumentum of flagelliform hairs. Leaves petiolate, occasionally sessile; petiole up to c. 7 cm. long, sometimes very narrowly-winged, sometimes with suborbicular stipule-like structures at the base, pubescent to tomentose; lamina up to 40 × 15 cm. and oblong-elliptic to broadly lanceolate, or up to c. 30 × 20 cm. and ovate, apex subacute to obtuse, base ± asymmetrically cordate or auriculate, sometimes broadly cuneate to attenuate in sessile leaves, with or without basal auricles, margins irregularly serrate to coarsely serrate-dentate with callose-tipped teeth, upper surface sparsely pubescent, lower surface densely grey-green pubescent to sericeous-tomentose, sometimes glabrous on both surfaces. Capitula small and very numerous, in many subglobose clusters paniculately arranged; capitula subsessile. Involucres 5–7 × 2.5(5) mm., subcylindric becoming narrowly obconic. Phyllaries many-seriate, closely imbricate at first, scarious, becoming caducous, inner ones falling in fruit, mostly glabrous, finely ciliate and narrowly hyaline on the margins; outer phyllaries up to c. 1 mm. long, triangular-ovate; the inner up to c. 6 mm. long, narrowly oblong-elliptic, rounded to subacute at apex. Florets 2–6 per capitulum, corollas whitish or pale-mauve to purple, c. 7 mm. long, funnel-shaped. Achenes 3–4 mm. long, subcylindric, tapering to base, narrowly c. 8-ribbed, thinly hispid; outer pappus of short flattened barbellate setae, inner of white barbellate setae c. 6 mm. long.

Zambia. N: Mbala, Old Kasama Road, 1500 m., 25.viii.1960, *Richards* 13157 (K; M; SRGH). W: Ndola, 7.viii.1958, *Fanshawe* 4680 (K; LISC; M; NDO; SRGH). C: Chilanga, c. 920 m., 10.x.1909, *Rogers* 8534 (K). E: Lunkwakwa/Chipata, 20.ix.1966, *Mutimushi* 1441 (K; NDO). S: Mazabuka, c. 1000 m., 14.viii.1931, *Trapnell* in C.R.S. 417 (K; PRE). **Zimbabwe**. W: Matopos Hills, near the View, ix.1905, *Gibbs* 66 (BM). C: Harare, 3.ix.1932, *Gundy* in GHS 6109 (K; SRGH). E: Mutare, 13.ix.1965, *Plowes* 2706 (K; LISC; SRGH). S: Masvingo, iv.1936, *Eyles* 8804 (K; SRGH). **Malawi**. N: Mzimba Distr., Viphya Plateau, c. 18 km. SW. of Mzuzu, c. 1460 m., 25.ix.1975, *Pawek* 10171 (K; MAL; MO). C: Dedza Plateau, c. 50 km. S. of Dedza, c. 1375 m., 15.ix.1972, *Pawek* 5709 (K; SRGH). S: Thyolo Mt., c. 1200 m., 25.ix.1946, *Brass* 17805 (BR; K; MO; PRE; SRGH). **Mozambique**. N: Nampula, rio Monapo, 27.ix.1936, *Torre* 640 (COI). Z: Tumbini Mt., 1.ix.1942, *A.J.W. Hornby* 2781 (K; LISC; PRE). MS: Mt. Maruma, c. 1070 m., 12.ix.1906, *Swynnerton* 113a (BM; K). M: Maputo, Namaacha, 5.vii.1973, *Marques* 2490 (K).

Widespread in tropical Africa; also in South Africa (the Transvaal and Natal). In moist localities, including high rainfall woodland, riverine vegetation, wet or marshy grassland and evergreen forest margins.

This is a variable species exhibiting a gradation in the number of florets per capitulum from c. 3 in the south of the Flora Zambesiaca area to c. 12 per capitulum in the north. In addition, leaf shape and degree of pubescence is rather variable, this however, does not follow any pattern. For the most part the leaves are petiolate with the lamina oblong-elliptic and narrowly cordate or auriculate at the base. Occasionally the leaves are sessile and attenuate or cuneate at the base, with or without auricles. In some cases stipule-like structures occur at the petiole base. The leaves are usually tomentose or pubescent below but can sometimes be almost glabrous.

It is not considered desirable or practical to subdivide this taxon on present evidence.

Vernonia auriculifera (Welw.) Hiern, from Angola, Zaire and E. Africa, is closely related to *V. podocoma* but is readily distinguished by its uniformly 1-flowered capitula.

22. **Vernonia thomsoniana** Oliv. & Hiern ex Oliv. in Trans. Linn. Soc. **29**: 91 (1873); in F.T.A. **3**: 295 (1877). —O. Hoffm. in Bol. Soc. Brot. **13**: 16 (1897). —Hiern, Cat. Afr. Pl. Welw. **1**, 3: 538 (1898). —S. Moore in Journ. Bot. **65**, Suppl. 2, Gamopet.: 49 (1927). —Mendonça, Contrib. Conhec. Fl. Angol., **1** Compositae: 31 (1943). —Adams in F.W.T.A. ed. 2, **2**: 277 (1963). —Wild in Kirkia **11**: 120 (1978). —Maquet in Fl. Rwanda, Spermat. **3**: 552 (1985). —C. Jeffrey in Kew Bull. **43**: 217 (1988). Type from Uganda.

Vernonia livingstoniana Oliv. & Hiern in F.T.A. **3**: 295 (1877). —S. Moore in Journ. Linn. Soc., Bot. **40**: 107 (1911). —Eyles in Trans. Roy. Soc. S. Afr. **5**: 504 (1916). Syntypes: Malawi, Manganja Hills, *Meller* (K); Shire R., *Stewart* (K).

Vernonia cruda Klatt in Bull. Herb. Boiss. **4**: 456 (1898). Type from Zaire.

Vernonia densicapitulata De Wild. in Bull. Jard. Bot. Brux. **5**: 92 (1915). Type from Zaire.

Vernonia thomsoniana var. *livingstoniana* (Oliv. & Hiern) Pichi Sermolli in Webbia **7**: 340 (1950). —Brenan in Mem. N.Y. Bot. Gard. **8**, 5: 462 (1954). Syntypes as above.

An erect slender, sparingly-branched shrub up to c. 3(5) m. tall. Stems 1–several, pale or reddish-brown, up to c. 5 cm. in diam., upper stem and branches finely ribbed, granular-glandular and densely appressed-puberulous or tomentellous; indumentum of brownish small T-shaped hairs. Leaves petiolate; petiole up to c. 18 mm. long; lamina mostly to c. 10 × 5 cm., but up to c. 17 × 10 cm. in lower leaves, lanceolate to ovate, apex acute, base narrowly to shortly cuneate and somewhat decurrent on the petiole, margin subentire to serrate, upper surface ± sparsely hispid-pubescent or glabrescent, lower surface pubescent to tomentose; hairs patent, T-shaped; venation reticulate and ± prominent beneath. Capitula small, numerous in ± contracted globose-paniculate clusters c. 5 cm. across; clusters few to very many, terminal on leafy branches; capitula stalks 1–8 mm.

long. Involucres up to c. 5 mm. long, narrowly campanulate-obconic. Phyllaries few-seriate, somewhat loosely arranged soon spreading, becoming caducous with the inner ones being shed first, scarious, thinly puberulous where exposed otherwise glabrous, ± densely glandular at the apex, the margins ± membranous and finely-ciliate, the midrib becoming ± prominent and dark-green or purple towards the apex and usually continued into a stiff point; the outer phyllaries from c. 1 mm. long, ovate-lanceolate; the inner phyllaries up to c. 4 mm. long and ± narrowly lanceolate or elliptic, apices obtuse to rounded and mucronulate. Florets 4–9 per capitulum, honey-scented; corollas white or mauve, 6–7 mm. long, gradually widening from the base. Achenes c. 2 mm. long, narrowly obovoid-turbinate, tapering to the base, faintly 4–6-ribbed or somewhat angular, minutely glandular, glabrous; outer pappus of small, caducous, linear, scale-like setae, inner pappus of white barbellate setae 4–5 mm. long.

Zambia. N: Mbala, Lunzua Hydro-electric station, c. 900 m., 11.vii.1960, *Richards* 12865 (K; SRGH). W: Kitwe, 27.vii.1957, *Fanshawe* 3373 (BR; K; LISC). C: near Mumbwa (?) 15°S 28°E, 1911, *Macaulay* 812 (K). E: Lundazi, c. 1100 m., 1.vi.1954, *Robinson* 807 (K). S: Mumbwa, 1.vi.1961, *Fanshawe* 6640 (K; M). **Zimbabwe**. C: Chegutu (Hartley), Umfuli R., viii.1933, *Eyles* 7508 (K; SRGH). E: Mutare, Ishakwe Mt., Burma Farm, c. 1310 m., 21.viii.1955, *Chase* 5723 (BM;BR; K; LISC; SRGH). **Malawi**. N: Chitipa Distr., Lufira Bridge, c. 1220 m., 31.vii.1977, *Phillips* 2676 (K; MO). C: Ntchisi Mt., 9.ix.1946, *Brass* 17612 (K; MO; SRGH). S: Thyolo (Cholo) Mt., 22.ix.1946, *Brass* 17734 (K; MO; PRE; SRGH). **Mozambique**. N: Lichinga (Lixinga), 13°20'S, 35°30'E, viii.1931, *Gomes e Sousa* 732 (K). Z: c. 5.8 km. from crossroads to Morrumbala and Mopeia, 31.vii.1949, *Barbosa & Carvalho* 3789 (K; SRGH). T: c. 37.9 km. from Vila Gamito on road to Furancungo, 11.vii.1949, *Barbosa & Carvalho* 3570 (K; SRGH). MS: between Catandica (Vila Gouveia) and Chimoio (Vila Pery), 18.vii.1969, *Leach & Cannell* 14337 (K; LISC; SRGH).

Throughout tropical Africa from Guinea and the Sudan southwards to Angola and Mozambique. Submontane forest margins and grassland to high rainfall mixed deciduous woodland and grassland at medium altitudes.

23. **Vernonia potamophila** Klatt in Ann. K.K. Naturhist. Hofmus. Wien **7**: 100 (1892). —Mendonça, Contrib. Conhec. Fl. Angol., 1 Compositae: 33 (1943). —Wild in Kirkia **11**: 117 (1978). Type from Angola.

A lax subshrub 1–1.5 m. tall, or a ± diffuse shrub to 2.6 m. high. Branches appressed grey-tomentose to glabrescent; hairs shortly-stalked, long-armed, T-shaped. Leaves discolorous, petiolate; petiole 2–10 mm. long, tomentose; lamina becoming somewhat bullate, up to c. 10 × 3 cm., oblanceolate to narrowly oblong-elliptic, apex rounded to obtuse and mucronate, base cuneate, margins ± remotely repand to serrulate, upper surface green and sparsely pubescent, lower surface densely appressed grey-tomentose with a prominent reticulate venation; hairs shortly-stalked, long-armed, T-shaped. Capitula small, numerous in ± contracted paniculate clusters up to c. 15 cm. across, terminal on the branches; capitula stalks 2–10 mm. long, grey-tomentose. Involucres 3–5 mm. long, obconic, soon spreading. Phyllaries few-seriate, somewhat loosely arranged, becoming caducous with the inner ones being shed first, scarious, thinly puberulous where exposed or glabrescent, finely ciliate and ± glandular apically, midrib expanding and becoming raised and dark-green to purple towards the apex, usually continued into a stiff point; the outer phyllaries from c. 1 mm. long, ovate-lanceolate; the inner phyllaries up to c. 4 mm. long, ± narrowly lanceolate, apices tapering-acute mucronate. Florets 2–6 per capitulum; corollas mauve, up to c. 7 mm. long, gradually widening from near the base. Achenes almost equalling the involucre in length when mature, 2.5–3 mm. long, subcylindric ± tapering to the base, faintly 4–6-ribbed, or somewhat angular, sparsely hispidulous; outer pappus of short barbellate setae, inner pappus of long-exserted, whitish, barbellate setae 6–7 mm. long.

Caprivi Strip. c. 69 km. from Singalamwe on road to Katima Mulilo, c. 1000 m., 3.i.1959, *Killick & Leistner* 3277 (K; PRE; SRGH). **Zambia**. B: Zambezi (Balovale), 26.ii.1964, *Fanshawe* 8346 (K; NDO). Also in Angola and Zaire. Dry evergreen woodland dominated by *Cryptosepalum-Isoberlinia* sp. (mavunda) on Kalahari Sand.

24. **Vernonia suprafastigiata** Klatt in Bull. Herb. Boiss. **4**: 458 (1896). —R.E. Fr., Wiss. Ergebn. Schwed. Rhod.-Kongo-Exped. 1911–1912 **1**: 325 (1916). —Wild in Kirkia **11**: 119 (1978). —C. Jeffrey in Kew Bull. **43**: 220 (1988). TAB. **15** fig. A. Types from Zaire.
  Vernonia brachylaenoides S. Moore in Journ. Bot. **51**: 184 (1913). Type from Zaire.
  Vernonia lescrauwaetii De Wild. in Bull. Jard. Bot. Brux. **4**: 228 (1914). Type from Zaire.
  Vernonia multiflora De Wild. in Fedde, Repert. **13**: 208 (1914). Syntypes from Zaire.

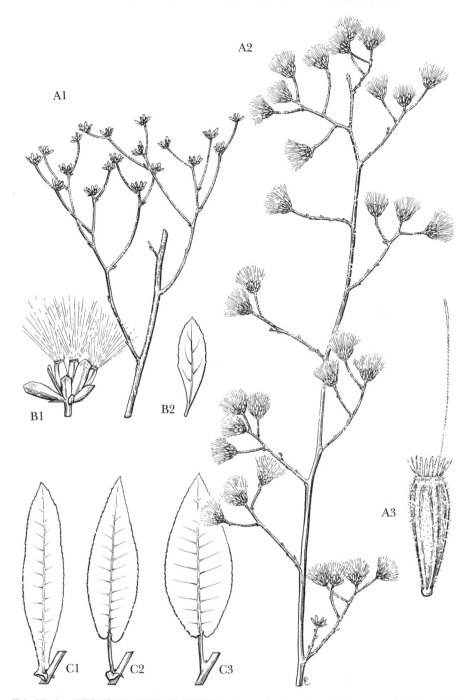

Tab. 15. A. —VERNONIA SUPRAFASTIGIATA. A1, twig with old capitula, some phyllaries shed (×⅔), from *Chisumpa* 113; A2, flowering twig (×⅔); A3, achene, showing outer pappus and only one of the numerous inner pappus setae (× 8), A2 & A3 from *Robson* 19. B. —VERNONIA GLABERRIMA. B1, involucre, mature with some of the inner phyllaries already fallen (× 3), from *Richards* 15360; B2, leaf (×⅔), from *Fanshawe* 5838. C. —VERNONIA MYRIANTHA. C1, leaf, narrowed to a sessile auriculate base (× ½), from *Greenway & Trapnell* 5556; C2, leaf with stipule-like base (× ½), from *Loveridge* 1085; C3, leaf, petiolate (× ½), from *Pawek* 2620. Drawn by Eleanor Catherine.

An erect spreading laxly branched subshrub or shrub, 0.15–4 m. tall, from a woody rootstock, usually leafless at flowering time. Stems and branches green and angular when young, becoming purplish-grey and subterete with pale prominent lenticels; branches ascending, densely appressed-pubescent especially on young shoots, soon glabrescent, sometimes tomentellous; hairs short-stalked T-shaped. Leaves usually appearing after the flowers, petiolate; petiole 2–10 mm. long, pubescent; lamina eventually up to c. 8.5 × 3 cm., narrowly elliptic to obovate-oblanceolate, apex acute to obtuse, base narrowly cuneate, margins subentire to subserrate towards the apex, sometimes ± revolute, thinly coriaceous, upper surface glabrescent, lower surface ± sparsely appressed-pubescent, or both surfaces glabrous, hairs short-stalked T-shaped; nervation finely reticulate and ± raised on both surfaces. Capitula borne on numerous ± abbreviated lateral shoots; shoots 1–5 cm. long, bracteate, with 1–7 capitula racemosely to subpaniculately arranged, or capitula more laxly arranged in large diffuse panicles with capitula often scorpioidly cymose on the synflorescence branches, stalks 0.3–3 cm. long. Involucres 3–6 mm. long, obconic to broadly cup-shaped, later spreading. Phyllaries few-seriate, becoming caducous, scarious, thinly pubescent where exposed, or glabrescent, finely ciliate and densely glandular apically, midrib ± fleshy raised and purple towards the apex, margins narrowly hyaline; the outer phyllaries from c. 1 mm. long, the inner to c. 5 mm. long, narrowly oblong-elliptic, ± rounded and mucronate at the apex. Florets 9–12(17) per capitulum; corollas purple or mauve, paler below, 6–8 mm. long, widening towards the apex. Achenes 3–4 mm. long, almost equalling involucre in length when mature, tapering to base, 6–8-ribbed, uniformly puberulous and glandular; outer pappus of short, barbellate, scale-like setae, inner of long-exserted, whitish, barbellate setae 5–7 mm. long.

Zambia. B: Lukolwe, 9.viii.1952, *Gilges* 159 (K; PRE; SRGH). N: Mbala, 14.ix.1950, *Bullock* 3321 (BR; K; SRGH). W: Ndola, x.1906, *Allen* 303 (K). C: Chakwenga Headwaters, 100–129 km. E. of Lusaka, 29.ix.1963, *Robinson* 5675 (K). E: c. 11 km. E. of Katete, c. 1100 m., 8.x.1958, *Robson & Angus* 19 (BM; K; LISC; PRE; SRGH). S: Kalomo, 8.x.1955, *Gilges* 466 (K; SRGH). **Zimbabwe**. N: Sebungwe, Gokwe side of Vulanduli, ix.1955, *Davies* 1535 (BR; K; LISC; SRGH). **Malawi**. N: Nyika Plateau, c. 2290 m., ix.1902, *McClounie* 65 (K).

Also in Tanzania, Zaire and Angola. In various types of mixed deciduous woodland, including miombo, *Brachystegia/Uapaca, Colophospermum mopane* and *Kirkia* woodlands, often on rocky hillsides and escarpments.

The capitula are usually borne on numerous, abbreviated, lateral shoots arising in leaf axils of the previous seasons growth. They also occur more laxly arranged in large ± diffuse panicles, where the branches and capitula are contemporaneous.

The specimen *Angus* 566 (K) from Mwinilunga in Zambia is remarkable for the size of the plant (described as a "12 foot tree") the fissured corky bark and the brown densely tomentose indumentum.

25. **Vernonia glaberrima** Welw. ex O. Hoffm. in Bol. Soc. Brot. **13**: 15 (Sept., 1896). —Hiern, Cat. Afr. Pl. Welw. **1**, 3: 537 (1898). —Mendonça, Contrib. Conhec. Fl. Angol., **1** Compositae: 33 (1943). —Wild in Kirkia **11**: 118 (1978). —C. Jeffrey in Kew Bull. **43**: 220 (1988). TAB. **15** fig. B. Syntypes from Angola.

   Vernonia hensii Klatt in Bull. Herb. Boiss. **4**: 828 (Dec., 1896). Type from Zaire.

   Vernonia mashonica N.E. Br. in Bull. Misc. Inf., Kew **1906**: 108 (1906). Syntypes: Zimbabwe, Harare, *Evelyn Cecil* 70 (K); Mutare, *Cecil* 229 (K).

A much-branched subshrub to c. 1 m. tall, occasionally up to c. 2 m. tall, from a woody rootstock, usually with young leaves or sometimes leafless at flowering. Stems and branches purplish, at first strongly angular, later ± ribbed or subterete with pale prominent lenticels, glabrous, or ± densely granulose-glandular particularly on young growth, sometimes also sparsely appressed puberulous; hairs T-shaped. Leaves usually appearing at the time of flowering or occasionally later, shortly petiolate; petiole up to c. 3 mm. long; lamina eventually up to ?c. 9 × 4.5 cm., elliptic to oblanceolate or obovate, apex acute to obtuse-rounded, base cuneate, margins subentire to coarsely dentate-serrate in mature leaves, coriaceous, glabrous or sometimes sparsely crisped-pubescent, hairs T-shaped with short crispate arms, nervation finely reticulate and somewhat raised on both surfaces. Capitula very numerous, paniculate, or sometimes subumbellate on numerous short lateral branches; lateral branches 3–11 cm. long, green, appressed-pubescent; capitula stalks 2–30 mm. long. Involucres 4–5 mm. long, obconic, spreading later. Phyllaries few-seriate, becoming caducous with the inner ones being shed first, strap-shaped, ± abruptly narrowed to an obtuse-acuminate often recurved apex, scarious, thinly puberulous where exposed, finely ciliate and upper surface densely glandular apically, the midrib becoming somewhat fleshy ± raised and purple towards the apex, the margins narrowly hyaline; the outer phyllaries from c. 1 mm. long, ovate-lanceolate; the inner phyllaries up to c. 5 mm. long, oblong elliptic to oblanceolate. Florets 5–6 per capitulum, sweetly scented; corollas white, 6–9 mm. long, widening towards the apex. Achenes 3–4 mm. long, narrowly turbinate, tapering to the base, 6–8-ribbed, sparsely to very densely glandular otherwise glabrous, rarely with a few hairs; outer pappus of short barbellate setae, inner pappus of long-exserted whitish barbellate setae 6–7 mm. long.

Zambia. B: road from Kaoma (Mankoya) to Mongu, 1.x.1957, *West* 3482 (K; SRGH). N: Mporokoso-Sanga Hill road, c. 1500 m., 25.xi.1961, *Richards* 15360 (K). W: c. 13 km. W. of Mwinilunga on Matonchi road, 24.x.1969, *Drummond & Williamson* 9360 (K; SRGH). C: Serenje, 18.ii.1955, *Fanshawe* 2074 (BR; K). **Zimbabwe.** C: Goromonzi Distr., Domboshawa, 25.xi.1960, *Rutherford-Smith* 387 (K; SRGH). E: Mutare Distr., Tsonzo Div., Kukwanisa, 21.xii.1967, *Biegel* 2399 (K; SRGH). **Malawi.** N: Mzimba Distr., Viphya Plateau, 4.xi.1968, *Salubeni* 1190 (K; SRGH). S: Mt. Mulanje, Chambe cableway path, 25.i.1967, *Hilliard & Burtt* 4617 (E; K). **Mozambique.** N: Negomano, 2.xi.1960, *Gomes e Sousa* 4590 (COI; K; M; PRE). Z: Alto Molócuè, c. 60 km. on road to Ribáuè, c. 600 m., 4.xii.1967, *Torre & Correia* 16347 (LISC). MS: Mozambique/Zimbabwe border opposite Gairesi Ranch, c. 1000 m., 19.xi.1956, *Robinson* 1953 (K; LISC; NDO; SRGH).

Also in west, central and south tropical Africa and in Zaire and Angola. In high rainfall miombo woodland, *Baikiaea* woodland on Kalahari Sand, low altitude evergreen forest and submontane grassland.

26. **Vernonia cinerascens** Sch. Bip. in Schweinf., Beitr. Fl. Aethiop.: 162 (1867). —Mendonça, Contrib. Conhec. Fl. Angol., 1 Compositae: 15 (1943). —Adams in F.W.T.A. ed. 2, **2**: 283 (1963). —Merxm., Prodr. Fl. SW. Afr. 139: 162 (1967). —Wild in Kirkia **11**: 79 (1978). —C. Jeffrey in Kew Bull. **43**: 222 (1988). Type from Ethiopia.

A laxly-branched twiggy subshrub to 1.3(3) m. tall. Stems several, stiff, woody, appressed greyish-tomentellous, glabrescent; indumentum dense, hairs very shortly stalked T-shaped; branches and twigs leafy, brittle, greyish- to yellowish-brown appressed-tomentellous. Leaves singly, or many in fascicles, from numerous tomentose cushions, sessile or shortly petiolate, up to 4.5 × 1.2 cm., usually smaller, narrowly obovate-spathulate to oblanceolate, apex truncate or rounded, base tapering-attenuate, margin crenulate-dentate about the apex otherwise entire, pubescent to glabrescent, gland-pitted, hairs T-shaped usually with one arm longer than the other. Capitula very numerous, paniculate or scorpioidly cymose, capitula stalks 0–16 mm. long. Involucres up to c. 4 mm. long, campanulate, soon loosely spreading-hemispheric. Phyllaries few, not firmly attached, soon diffusely arranged, apices purple ± truncate mucronate and somewhat serrulate, bases distinctly narrowed, margins ciliate, pubescent outside, hairs unequal-armed T-shaped; the outer phyllaries short, elliptic-ovate; the inner up to 3.5 mm. long, oblong-oblanceolate. Corolla purple, up to c. 6 mm. long. Achenes c. 2.5 mm. long, somewhat cylindric, tapering to a sharp base, narrowly 6–8-ribbed, strigose-hispidulous; outer pappus very short of linear-acuminate scales; inner pappus of white barbellate setae c. 4 mm. long.

Botswana. SE: Lothlekane, 20.iii.1924, *Allen* 55 (PRE). **Zimbabwe.** S: Beitbridge Distr., near Shashi R., 6.i.1961, *Wild* 5312 (K; M; MO; SRGH).

From Angola, Namibia, South Africa (Transvaal), E. Africa, Somalia, Ethiopia, Sudan, and north-western India. Low altitude, low rainfall tree or bush savanna with *Acacia*, *Terminalia* or *Colophospermum mopane*.

27. **Vernonia verrucata** Klatt in Ann. K.K. Naturhist. Hofmus. Wien, **7**: 99 (1892). —O. Hoffm. in Bol. Soc. Brot. **10**: 172 (1893). —Mendonça, Contrib. Conhec. Fl. Angol., 1 Compositae: 24 (1943). —Wild in Kirkia **11**: 99 (1978). TAB. **16** fig. C. Type from Angola.
Vernonia chlorolepis S. Moore in Journ. Bot. **52**: 92 (1914). —Mendonça, Contrib. Conhec. Fl. Angol., 1 Compositae: 23 (1943). Syntypes from Angola.

A stiff erect, sparingly branched, perennial herb up to c. 75 cm. tall, from a woody rootstock. Stems annual, usually purple, simple below, uniformly leafy, glabrous or pilose-scabrid; hairs becoming brittle with bases persisting as rigid points; branches up to c. 22 cm. long, leafy. Lower leaves with petioles up to c. 5 mm. long; leaves 5–15 × 1.2–4.5 cm., oblanceolate-elliptic to narrowly elliptic, apices obtuse ± apiculate, bases cuneate, margins coarsely serrate to subentire, ± coriaceous, glabrous, or more usually with white hispid hairs on both surfaces becoming scabrous, nerves ± prominent beneath. Capitula terminal, solitary, or up to c. 4 subcorymbiformly cymose; capitula stalks arising in upper leaf axils, bracteate, glabrous or scabrous; bracts 1–several, up to c. 10(15) mm. long, linear. Involucre 15–30 × 25–50 mm., broadly campanulate to hemispheric. Phyllaries numerous, many-seriate, appendaged, all ± similar but progressively longer towards the inside, 10–25 × 2.5(3) mm., stramineous and linear-oblong below the appendage, coriaceous, glabrous sometimes hispidulous, margins with fine comb-like teeth; appendages green, thinly coriaceous with slightly prominent nerves, ± equalling the phyllary base in length and usually exceeding it in width. Receptacle minutely fimbriate-alveolate. Corolla 17–22 mm. long, the lower two-thirds narrowly cylindric, slightly widened and deeply lobed at the apex, limb cornflower-blue. Achenes pale-brown when mature, up to c. 10 mm. long, subcylindric to subfusiform, c. 10-ribbed, glabrous with dark-brown gland spots; outer pappus of narrow scale-like setae up to c. 3 mm. long; inner pappus of brown or purple-tinged, 14–15(18) mm. long, ± flattened setae, bristled on margins.

Zambia. W: Mwinilunga, between Ikelenge and Kalene, c. 1625 m., 15.v.1986, *Philcox, Pope & Chisumpa* 10305 (BR; K; MO; NDO; SRGH).
Also in Angola. *Brachystegia* woodland, often on dry rocky hillsides.

28. **Vernonia kawoziensis** F.G. Davies in Kew Bull. **31**: 170 (1976). —Wild in Kirkia **11**: 97 (1978).
    Type: Malawi, Mt. Kawozya, *Brummitt & Synge* WC178 (K, holotype; SRGH).

A tufted harsh perennial herb, up to c. 75 cm. tall from a woody rootstock. Stems annual, many, simple, erect, ribbed and sparsely hispid. Leaves subsessile, up to 6 × 2.3 cm., oblanceolate to elliptic, apex acute mucronate, base cuneate, margins subentire to sparsely serrate, lamina leathery, scabrous on both sides; lowermost leaves smaller, upper leaves becoming bract-like. Capitula 3–13, corymbiformly cymose at the stem apex on stiff bracteate stalks up to c. 7 cm. long; bracts subulate and crowded below the capitula. Involucres up to c. 20 mm. long, obconic. Phyllaries many-seriate, increasing markedly in size towards the inside, cartilaginous, hispid-scabridulous outside and hispid-muricate inside the apex, the margins pectinate-denticulate; the outer phyllaries c. 4 mm. long, lanceolate-subulate and extending onto the capitulum stalk; the innermost up to c. 18 mm. long, strap-shaped tapering to an acuminate purple-tipped aristate apex. Receptacle shallowly alveolate, alveolae walls fimbriate. Corolla deep blue, 17–19 mm. long, slender-tubular below widening slightly above into a short limb, lobes 4–5 mm. long. Achenes 7–9 mm. long, subcylindric, tapering to the base, c. 10-ribbed, densely strigose; outer pappus of short, linear, acuminate, scale-like setae; inner pappus of copious, flattened, subplumose setae c. 13 mm. long.

Malawi. N: Mt. Kawozya, 1890 m., 10.viii.1972, *Brummitt & Synge* WC178 (K; MAL; SRGH).
Known only from the type gathering. Miombo woodland.

29. **Vernonia accommodata** Wild in Kirkia **5**: 82 (1965); **11**: 123 (1978). TAB. **16** fig. D. Type: Zimbabwe, Mpingi Pass through Great Dyke, *Wild* 5773 (COI; K; SRGH, holotype).

A bushy perennial herb to c. 1 m. tall. Stems branched leafy striate, white araneose-lanate on young growth, or glabrescent, glabrous below; branches stiffly ascending. Leaves subcoriaceous, subsessile, to c. 8.5 × 2 cm., narrowly oblanceolate, apices acute to obtuse, base narrowly cuneate, margins entire or distantly toothed, white araneose-lanate on young leaves soon glabrescent, gland-pitted on both sides, venation usually obscure. Capitula numerous with 3–10 heads laxly subpaniculately arranged at the ends of branches; capitula stalks 1–6 cm. long, lanate at first, bracteate; bracts 1–6 mm. long, linear-lanceolate or subulate, appressed and more numerous towards the stalk-apex and grading into the phyllaries. Involucres 8–15 × 8–10 mm., spreading to c. 15 mm. wide, obconic-campanulate. Phyllaries many-seriate, progressively longer towards the inside, greenish-stramineous, stiffly coriaceous, araneose-lanate or glabrous, occasionally hispid, margins minutely pectinate; the outer phyllaries from c. 2 mm. long, triangular or linear-lanceolate and extending onto the capitulum stalk; middle phyllaries lanceolate, green-tipped and stramineous below; the inner to c. 15 mm. long, linear-oblong. Receptacle fimbriate-alveolate. Corollas blue to purplish sometimes white, to c. 14 mm. long, narrowly tubular but expanding slightly into the deeply-lobed limb. Achenes up to c. 5 mm. long, narrowly subcylindrical to subfusiform, c. 10-ribbed, strigose-hispid; outer pappus of linear, scale-like setae to c. 3 mm. long, inner pappus setae brownish to purplish-tinged or sordid, somewhat flattened and subplumose, up to c. 11 mm. long.

Zimbabwe. N: Makonde Dist., W. side of Great Dyke, N. of Mpinga, 14.iii.1961, *Drummond & Rutherford-Smith* 6890 (BR; K; LISC; M; SRGH). C: Kwekwe Distr., Sebakwe, 8.iv.1975, *Wild* 7998 (K;

SRGH). E: Mutare, S. of Odzi Road, 13.v.1969, *Wild* 7771 (COI; K; M; SRGH). S: Mberengwa, S. end of Great Dyke near Otto Mine, 17.iii.1964, *Wild* 6394 (K; SRGH).
Known only from serpentine soils in Zimbabwe. Spontaneous on chrysotyle asbestos mine dumps.

30. **Vernonia glabra** (Steetz) Vatke in Oesterr. Bot. Zeit. **27**: 194 (1877). —Oliv. & Hiern in F.T.A. **3**: 286 (1877). —S. Moore in Journ. Linn. Soc., Bot. **40**: 106 (1911). —Eyles in Trans. Roy. Soc. S. Afr. **5**: 503 (1916). —Adams in F.W.T.A. ed. 2, **2**: 280 (1963). —Merxm., Prodr. Fl. SW. Afr. 139: 182 (1967). —Hilliard in Ross, Fl. Natal: 356 (1972); Compos. Natal: 40 (1977). —Wild in Kirkia **11**: 95 (1978). —C. Jeffrey in Kew Bull. **43**: 232 (1988). TAB. **16** fig. B. Syntypes: Mozambique, Boror, *Peters* (B†); Rios de Sena, *Peters* (B†).

Linzia glabra Steetz in Peters, Reise Mossamb., Bot.: 353 (1864). Syntypes as for *V. glabra*.

Linzia glabra var. *confertissima* Steetz in Peters, Reise Mossamb., Bot.: 354 (1864). nom. non rite public. Syntypes as for *V. glabra*.

Vernonia obconica Oliv. & Hiern in F.T.A. **3**: 286 (1877). Type: ? Zambia or Tanzania, Lake Tanganyika, *Cameron* s.n. (K, holotype).

Vernonia pogosperma Klatt in Ann. K.K. Naturhist. Hofmus. Wien **7**: 99 (1892). Type from Angola.

Vernonia ondongensis Klatt in Bull. Herb. Boiss. **3**: 430 (1895). Type from Namibia.

Vernonia glabra var. *ondongensis* (Klatt) Merxm. in Mitt. Bot. Staatss. Münch. **2**: 37 (1954); Prodr. Fl. SW. Afr. 139: 183 (1967). Type as for *V. ondongensis*.

Vernonia roseopapposa Gilli in Ann. K.K. Naturhist. Hofmus. Wien **78**: 165 (1974). Type from Tanzania.

An erect, often robust, sparingly branched perennial herb to, c. 3 m. tall from a large woody rootstock. Stems annual, solitary or several clustered, simple below, uniformly leafy, striately ribbed, puberulous to pilose-tomentose, glabrescent; synflorescence branches stiffly ascending or somewhat laxly spreading, up to c. 30 cm. long. Leaves subsessile on the upper stem, or with a petiole up to c. 10 mm. long on the lower stem; lamina 3–18 × 0.6–5 cm., oblong-elliptic to oblanceolate, lanceolate in uppermost leaves, apices rounded-obtuse to acute, bases rounded to subcordate or broadly cuneate, margins ± coarsely serrate, puberulous to glabrescent, sometimes hispid on the main veins beneath and then scabridulous. Capitula numerous, in few to many dense clusters terminal on branches, or capitula in a large ± lax terminal paniculate arrangement; capitula stalks mostly 0.2–3 cm. long, often bracteate; bracts up to c. 4 mm. long, scale-like and merging into the phyllaries; synflorescence branches with narrowly-oblong leaves up to c. 5 cm. long. Involucres 7–15 × 5–15 mm., narrowly cylindric-campanulate to broadly obconic-campanulate, glabrous or hispidulous. Phyllaries usually somewhat differentiated apically, progressively longer to the inside, greenish-stramineous, stiffly coriaceous, hispidulous to pubescent or glabrescent; apices puberulous, glandular, dark-green or purplish, sometimes somewhat expanded; margins pectinate-denticulate or pungent-ciliolate; the outer phyllaries from c. 2 mm. long, ovate to linear-lanceolate; the inner phyllaries to c. 14 mm. long, linear-lanceolate to narrowly oblong-lanceolate, acute or obtuse-mucronate. Receptacle fimbriate-alveolate. Corollas blue, up to c. 14 mm. long, narrowly tubular with a deeply lobed limb. Achenes dark-brown when mature, 5–7 mm. long, narrowly subcylindrical to subfusiform, c. 10-ribbed, strigose-hispid; outer pappus of short, irregular, 0.5–2 mm. long, narrow, scale-like setae, inner pappus of brown-purplish, somewhat flattened, subplumose setae 8–10 mm. long.

Involucre usually 10 mm. or more long; capitula numerous in large ± dense clusters; outer phyllaries usually many, ± subulate, often extending onto capitulum stalk; erect robust plants of woodlands, dambos and river fringing vegetation, roadsides, less often on alluvium   var. *glabra*
Involucre mostly less than 10 mm. long; capitula laxly arranged usually on long stalks, not densely clustered; or capitula in small clusters laxly paniculately arranged; outer phyllaries relatively fewer, ovate-lanceolate, seldom extending onto capitulum stalk; lax low-growing less robust plants of flood plain alluvium, on heavy clay or sandy soil   -   -   -   -   - var. *laxa*

Var. **glabra**

Botswana. SE: Mbati, 21.iv.1965, *Child* in GHS 200939 (K; SRGH). **Zambia**. N: Mporokoso Distr., Mweru-Wantipa, Kangiri, c. 1050 m., 7.iv.1957, *Richards* 9090 (K). W: Kitwe, Ichimpi, 18.v.1963, *Mutimushi* 306 (K; NDO). C: Mkushi, 2.v.1957, *Fanshawe* 3252 (K). E: Lundazi, c. 1096 m., 31.v.1954, *Robinson* 792 (K). S: Gwembe Valley near Nangombe R., E. of Sinazeze, 17.vi.1961, *Angus* 2914 (FHO; K). **Zimbabwe**. N: Bumi R., Nebiri Camp, c. 610 m., ix.1955, *Davies* 1543 (BR; K; SRGH). W: Matobo Distr., Matopos Res. Sta., c. 1380 m., iii.1954, *Miller* 2238 (K; LISC; SRGH). C: Harare, Makabusi R., c. 1460 m., 23.xi.1962, *Lewis* 6256 (K). E: Lusitu R. Valley, c. 600 m., vii.1973, *Goldsmith* 28/73 (K; SRGH). S: Runde R., 3.vi.1930, *Hutchinson & Gillett* 3311 (BM; BR; K). **Malawi**. N: Mzimba

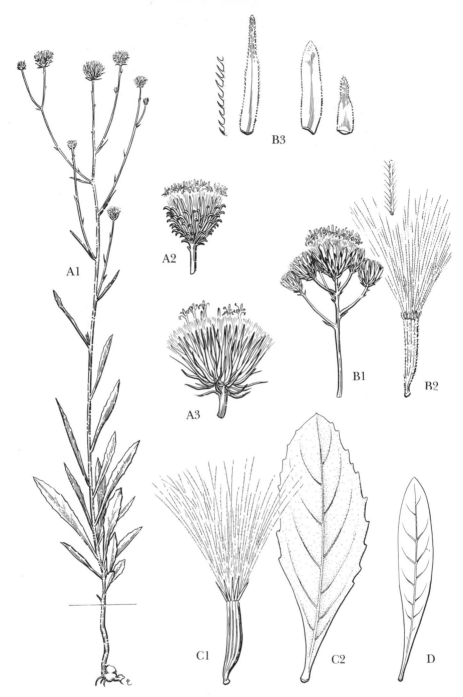

Tab. 16. A. —VERNONIA MELLERI var. MELLERI. A1, habit ($\times \frac{2}{9}$); A2, capitulum, A1 & A2 from *Pawek* 6577; A3, var. SUPERBA, capitulum, from *Pawek* 1803. B. —VERNONIA GLABRA. B1, cluster of capitula; B2, achene, with detail of subplumose pappus seta ($\times$ 20); B3, phyllaries, inner middle and outer, detail showing pectinate phyllary margin ($\times$ 20), B1–B3 from *Miller* 2769. C. —VERNONIA VERRUCATA. C1, achene; C2, leaf, C1 & C2 from *Milne-Redhead* 2551. D. —VERNONIA ACCOMMODATA, leaf, from *Wild* 6502. (Capitula and leaves $\times \frac{2}{3}$, phyllaries and achenes $\times$ 3). Drawn by Eleanor Catherine.

Distr., Mzuzu, Marymount, c. 1370 m., 25.vii.1970, *Pawek* 3633 (K). C: Lilongwe, c. 1100 m., 29.iii.1970, *Brummitt & Little* 9497 (K). S: c. 29 km. NW. of Blantyre, c. 790 m., 13.v.1961, *Leach & Rutherford-Smith* 10832 (K; LISC; SRGH). **Mozambique**. N: Erati, 3.v.1962, *Lemos & Marrime* 340 (BM; COI; K; LISC; MO; SRGH). Z: c. 8 km. from Nampevo to Mugeba, 31.v.1948, *Barbosa & Carvalho* 2948 (K). T: Changara Distr., between Tete and Mutoko, c. 762 m., 13.v.1949, *Gerstner* 7051 (K; PRE). MS: c. 48 km. E. of Vila de Manica, c. 600 m., 14.vi.1950, *Leach* 9105 (K; SRGH).

Also in Kenya, Tanzania, Zaire, Namibia and South Africa (Transvaal). High rainfall woodlands, dambos and river fringing vegetation, often on roadsides.

Var. **laxa** (Steetz) Brenan in Mem. N.Y. Bot. Gard. **8**, 5: 460 (1954). —C. Jeffrey in Kew Bull. **43**: 232 (1988). Syntypes: Mozambique, Sena, *Peters* (B†); Zambezi R., *Peters* (B†).
Linzia glabra var. *laxa* Steetz in Peters, Reise Mossamb., Bot.: 354 (1864).

Caprivi Strip. Katima Mulilo, c. 920 m., 24.xii.1958, *Killick & Leistner* 3065 (K; PRE). **Botswana**. N: Nata R., c. 6 km. upstream from Nata R. Delta, 21.iv.1976, *Ngoni* 527 (K; SRGH). **Zambia**. B: Sesheke Distr., *Gairdner* 576 (K). C: Chingombe, 26.ix.1957, *Fanshawe* 3732 (K). E: Luangwa River bridge, 5.ix.1947, *Greenway & Brenan* 8047 (K). S: Mazabuka Distr., Kafue Pilot Polder, 30.viii.1962, *Angus* 3312 (FHO; K). **Zimbabwe**. S: Gona-re-Zhou, Tambaharta Pan, between Chitsa's Store and Save/Runde R. junction, 31.v.1971, *Grosvenor* 594 (K; SRGH). **Malawi**. S: Chikwawa Distr., lower Mwanza River, 180 m., *Brass* 18018 (K). **Mozambique**. Z: Namacurra, Rio Licungo, 28.viii.1949, *Barbosa & Carvalho* 3865 (K; LISC). T: Mazowe R., near Dique, c. 300 m., 21.ix.1948, *Wild* 2568 (BR; K; SRGH). MS: Beira Distr., Vundudzi delta area, N. of Urema Lake, ix.1971, *Tinley* 2167 (K; SRGH). GI: Guijá para Mabalane, proximo da Aldeia da Barragem, 3.vi.1959, *Barbosa & Lemos* 8574 (COI; K; LISC). M: Manhica, estrada Nacional Velha na area de Taninga, 29.viii.1980, *Nuvunga & Boane* 297 (BM; K; LMU).

Also in Kenya, Tanzania, Namibia and South Africa (Natal, Transvaal) and Swaziland. Flood plain alluvium, usually in heavy black clay or as a pioneer on flood plain sandbanks and disturbed ground.

A western variant of this variety may be distinguished, recognisable by shortened ± blunt, purple-tipped phyllaries and relatively wider more campanulate involucres. In typical var. *laxa* the phyllaries are more tapered apically and the involucre narrowly obconic, being longer than wide. This western variant occurs on the flood plain systems of the Kafue and Zambezi Rivers in S. Zambia, the Okovango Swamp and River in N. Botswana and adjacent Namibia, and the Limpopo River in the Transvaal and Mozambique.

31. **Vernonia rosenii** R.E. Fr., Wiss. Ergebn. Schwed. Rhod.-Kongo-Exped. 1911–1912, **1**: 324 (1916). Type from Zaire.
Vernonia glabra sensu Wild in Kirkia **11**: 96 (1978) pro parte.

A low-growing perennial herb, up to c. 43 cm. tall from a woody rootstock. Stems annual, 1–several, somewhat decumbent, branched, leafy, white-woolly or pubescent at first soon ± glabrescent. Lower leaves with petioles up to c. 6 mm. long, other leaves subsessile; lamina 1.5–8 × 0.5–2.5 cm., narrowly oblong-elliptic to oblong-oblanceolate, apices acute to rounded, bases cuneate, margins strongly serrate to subentire, pubescent soon glabrescent, minutely glandular-punctate. Capitula many, large, laxly arranged, terminal and solitary on the branches, or 2–4 corymbiformly cymose on long stalks at the ends of the branches; stalks (1)3–10 cm. long, ascending, usually bracteate; bracts up to c. 5 mm. long linear, scale-like. Involucres 12–15 × 12–18 mm., ± broadly turbinate-campanulate. Phyllaries many-seriate, progressively longer towards the inside, greenish-stramineous, the tips dark-green or purplish, hispidulous to glabrescent, margins pectinate-denticulate or pungent-ciliate; the outer phyllaries from c. 2 mm. long, linear-lanceolate, acute mucronulate; the inner phyllaries up to c. 15 mm. long, narrowly oblong and somewhat tapered or abruptly narrowed to a mucronulate apex. Receptacle fimbriate-alveolate. Corollas cornflower-blue or mauve, up to c. 18 mm. long, narrowly tubular, gradually widening to a deeply lobed limb. Achenes purplish-brown when mature, 6–8 mm. long, narrowly subcylindrical to subfusiform, c. 10-ribbed, strigose-hispid; outer pappus of irregular, narrow, scale-like setae 1–3 mm. long; inner pappus of brown, purplish-tinged or sordid ± flattened, subplumose setae 8–11 mm. long.

Botswana. N: Chobe Nat. Park, Kasane, 29.viii.1970, *Mavi* 1142 (K; LISC; SRGH). **Zambia**. B: Nangweshi, c. 1040 m., 26.vii.1952, *Codd* 7197 (BM; K; PRE). C: Munali, c. 6.4 km. E. of Lusaka, c. 1280 m., 14.viii.1955, *King* 91 (K). S: Mazabuka, 4.viii.1959, *West* 4028 (K; SRGH). **Zimbabwe**. N: Mwami (Miami), c. 1370 m., 4.x.1946, *Wild* 1264 (K). C: Harare, c. 1900 m., 1.xi.1968, *Biegel* 2662 (K; SRGH).

Also in Zaire. A pyrophyte of floodplain and dambo grassland and sometimes deciduous woodland.

32. **Vernonia melleri** Oliv. & Hiern in F.T.A. **3**: 282 (1877). —Eyles in Trans. Roy. Soc. S. Afr. **5**: 504 (1916). —Mendonça, Contrib. Conhec. Fl. Angol., **1** Compositae: 20 (1943). —Brenan in Mem.

N.Y. Bot. Gard. **8**, 5:60 (1954). —Wild in Kirkia **11**: 72 (1978). —C. Jeffrey in Kew Bull. **43**: 233 (1988). Type: Malawi, Shire Highlands, Manganja Hills, *Meller* s.n. (K, holotype).

An erect scabrid perennial herb, 0.3–1.5 m. tall from a woody rootstock; root-crown sometimes with lanate tufts of hairs. Stems annual, simple or laxly branched in the upper half, uniformly leafy or scapiform, striately ribbed, brittle, scabrous-pubescent; branches up to c. 30 cm. long. Leaves 4–19 × 2–6 cm., obovate to narrowly oblanceolate, acute to obtuse mucronate at the apex, tapering to attenuate to the base, margins subentire to remotely coarsely serrate, the teeth callose-tipped, thinly coriaceous and scabrous-pubescent on both sides, or lower surface hispidulous-scabridulous; midrib and lateral nerves prominent beneath. Capitula 1–many, solitary and terminal, or several laxly corymbiform cymose on branches or stem apex. Involucres 1.5–3 × 1.5–4 cm., very broadly campanulate to spreading obconic. Phyllaries numerous, often extending onto the capitulum stalk, progressively longer to the inside, from 4–10 mm. long outside to c. 25 mm. long inside, dark-green or purplish towards the apex, cartilaginous, scabridulous-hispid outside and densely muricate-scabridulous inside towards the apex, sharply pectinate-denticulate on the margin, narrowly triangular to linear-lanceolate, ± strongly recurved, or straight and stiffly tapering-attenuate apically. Receptacle alveolate, alveolae walls brown, short, fimbriate. Corolla 18–25 mm. long; tube slender, yellowish-brown, widening slightly into a cornflower-blue or purplish limb; lobes 4–8.5 mm. long, ± equalling limb in length; limb exserted beyond pappus at anthesis. Achenes 6–11 mm. long, narrowly cylindric-fusiform, c. 10-ribbed, glandular and shortly setulose or glabrous; outer pappus of 2.5–5 mm. long, narrow, scale-like setae, barbellate on the margins; inner pappus copious, composed of ± flattened barbellate setae, 10–14 mm. long, brownish-purple in upper part and often sordid below.

Achenes setulose-hispid sometimes sparsely so; involucre mostly c. 15 mm. long; capitula 4–21 seldom fewer; phyllaries shortly tapering, often ± recurved apically, at least the outer

var. *melleri*

Achenes glabrous or sparsely setulose; involucre mostly c. 20 mm. or more long; capitula 1–4, seldom more; phyllaries narrowly tapering-attenuate, straight or somewhat flexuous but not recurved

var. *superba*

Typical *V. melleri*, readily recognised by its smaller more numerous capitula, extends from southern Tanzania westwards to Angola and southwards to central Mozambique and Zimbabwe. In Malawi and in N. and W. Zambia it overlaps with the northern variety, var. *superba*.

Var. **melleri**, TAB. **16** fig. A.

    Vernonia scabrifolia var. *amplifolia* O. Hoffm. in Warb., Kunene-Samb.-Exped. Baum: 403 (1903). Type from Angola.

Botswana. N: Chobe, Panda Forest, v.1966, *Mutakela* 67 (SRGH). **Zambia**. N: Mpika, Danger Hill, 31°30'E, 11°40'S, c. 1525 m., 27.v.1959, *Stewart* 182 (BR; K). W: Mufulira, 3.vi.1934, *Eyles* 8231 (BR; K; SRGH). C: c. 21 km. F. of Lusaka, 24.v.1953, *Best* 40 (BR; K; SRGH). E: Chipata, 2.vi.1958, *Fanshawe* 4508 (BR; K). S: c. 35 km. NE. of Choma, 7.iv.1958, *Robinson* 2832 (BR; K; M; PRE; SRGH). **Zimbabwe**. N: Guruve (Sipolilo), Chenanga, 27.v.1965, *Bingham* 1530 (MO; SRGH). W: Shangani Distr., Gwaai Valley, 30.vi.1949, *West* 2926 (MO; SRGH). C: Chegutu (Hartley), Poole, 3.iv.1946, *Wild* 1003 (BR; SRGH). E: Mutare, Commonage, 6.vii.1952, *Chase* 4569 (COI; LISC; SRGH). S: Mushandike, 30.iv.1971, *Wright* 7320 (M; SRGH). **Malawi**. N: Rumphi-Nyika road, c. 1375 m., 26.iv.1973, *Pawek* 6577 (K; MO; MAL). C: Ntchisi, 1.viii.1946, *Brass* 17088 (BR; K; MO; PRE; SRGH). S: c. 1 km. W. of Machinga (Kasupe), c. 720 m., 14.vi.1970, *Brummitt* 11394 (K). **Mozambique**. N: Unango to Lake Shire (?L. Malawi) 1899, *Johnson* 31 (K). Z: Chibisa to Tshinmuze, ix.1859, *Kirk* s.n. (K). T: between Furancungo and Vila Coutinho, 15.vii.1949, *Barbosa & Carvalho* 3618 (K; SRGH).

Also in Angola, Zaire and Tanzania. Miombo and other deciduous woodlands or grasslands.

Var. **superba** (O. Hoffm.) C. Jeffrey in Kew Bull. **43**: 233 (1988). TAB. **16** fig. A. Type from Tanzania.

    Vernonia superba O. Hoffm. in Engl., Pflanzenw. Ost-Afr. C: 406 (1895). —Wild in Kirkia **11**: 73 (1978). Type as above.

    Vernonia scabrifolia O. Hoffm. in Engl., Bot. Jahrb. **30**: 424 (1901) non Hieron (1897). Type from Tanzania.

    Vernonia paludigena S. Moore in Journ. Bot. **52**: 91 (1914). Type from Zaire.

    Vernonia vanmeelii Lawalrée, Expl. Hydrobiol. Lac Tanganyika, Res. Sc. 6, **2**: 59 (1955). Type as for *V. scabrifolia* O. Hoffm.

Zambia. N: c. 30 km. from Makumbe School on Nakonde road, c. 1500 m., 22.iv.1986, *Philcox, Pope & Chisumpa* 10066 (BR; K; MO; SRGH). W: Mufulira, Njiri Forest, 12.iii.1964, *Mutimushi* 695

(K). C: c. 45 km. SW. of Kabwe, 13.vii.1930, *Hutchinson & Gillett* 3632 (BM; K; LISC; SRGH). E: Katete, St. Francis' Hospital, c. 1065 m., 19.ii.1956, *Wright* 77 (K). S: Mapanza Mission, 14.ii.1954, *Robinson* 531 (BR; K; SRGH). **Malawi**. N: Viphya, Kalapya Dome, 15.ii.1968, *Simon, Williamson & Ball* 1809 (K; SRGH). C: Mchinji (Ft. Manning), c. 1250 m., 7.i.1959, *Robson* 1069 (K; LISC; PRE; SRGH). S: Dedza-Golomoti road, c. 1250 m., 19.iii.1955, *Exell, Mendonça & Wild* 1035a (BM). **Mozambique**. N: Amaramba, 20 km. from Nova Freixo (Cuamba), E. of serra Mitucué, c. 800 m., 15.ii.1964, *Torre & Paiva* 10603 (LISC).

Also in Tanzania and Zaire. Miombo and other deciduous woodland.

33. **Vernonia gerberiformis** Oliv. & Hiern in F.T.A. **3**: 285 (1877) *"gerberaeformis"*. —S. Moore in Journ. Linn. Soc., Bot. **40**: 106 (1911). —Eyles in Trans. Roy. Soc. S. Afr. **5**: 503 (1916). —Mendonça, Contrib. Conhec. Fl. Angol., 1 Compositae: 24 (1943). —Adams in F.W.T.A. ed. 2, **2**: 279 (1963). —Earle Sm. in U.S. Dept. Agric. Handb. **396**: 61, figs. 42, 43, 44 (1971). —Wild in Kirkia **11**: 103 (1978). —C. Jeffrey in Kew Bull. **43**: 234 (1988). Type from Sudan.

A tufted perennial herb, 4–45 cm. tall from a stout vertical rootstock; scapose with a rosette of basal leaves, or stems scapiform. Scapes or stems 1–several, simple or 1–2-branched from near the base, striately ribbed, bracteate, ± densely whitish pubescent or woolly, especially below the capitulum and near the base, to glabrescent, or glabrous, brownish lanate at the rootcrown; bracts few, up to c. 15 mm. long, linear. Leaves numerous, basal, or clustered on lower stem, sessile, often appearing after the capitula, eventually up to c. 15(36) × 5(12) cm., oblanceolate to ± spathulate, linear-oblanceolate, or elliptic-obovate, obtuse to rounded at apex or tapering acute, tapering-attenuate from about the middle to a ± clasping base, margins serrulate-denticulate to subentire, membranous or thinly coriaceous, occasionally glaucous, thinly villous at first, soon puberulous to glabrescent, or sometimes hispid and scabridulous, or glabrous, midrib prominent beneath. Capitula solitary, terminal; involucre 12–25(30) × 15–45(50) mm., broadly obconic-campanulate, later widely spreading. Phyllaries many-seriate, often purple-tinged towards the apex, coriaceous, finely woolly and soon puberulous to hispidulous, or glabrescent and glandular outside, the margins pectinate-denticulate or stiffly ciliate; the outer phyllaries from c. 4 mm. long, shortly lanceolate with an acute or long attenuate apex, or linear and gradually tapering to the apex; the middle phyllaries up to c. 20 mm. long, ± lanceolate, or up to 8 mm. wide and oblong-ovate tapering shortly to an acute apex; the inner phyllaries narrower, up to c. 25 mm. long. Receptacle fimbriate-alveolate. Corollas cornflower-blue to mauve in upper portion, 15–30 mm. long, narrowly tubular in the lower two-thirds widening to a campanulate limb with lobes 4–6 mm. long, bristled outside at apex. Achenes up to c. 11 mm. long, subcylindric, tapering slightly to the base, c. 10-ribbed, densely hispid to sparsely hispidulous; outer pappus of relatively few irregular, scale-like bristles 1–4 mm. long, inner pappus of sordid-brown, somewhat flattened barbellate setae 10–22 mm. long.

Plants slender scapose, flowers somewhat precocious; leaves subrosulate, up to c. 20 mm. wide, eventually some to c. 40 mm. wide, puberulous or glabrous; involucres up to c. 30 mm. wide rarely more; phyllaries mostly less than 3 mm. wide - - - - - subsp. *gerberiformis*
Plants robust, flowers and leaves contemporaneous; stems scapiform, cauline leaves 3–7 sometimes 0, larger leaves usually more than 40 mm. wide, glabrous or glabrescent; involucres 30–40 mm. wide; phyllaries mostly 4 mm. or more wide - - - - - subsp. *macrocyanus*

Subsp. **gerberiformis**

Phyllaries, at least the inner, lanceolate-lorate, subobtuse; scapes often white-woolly pubescent; leaves oblanceolate to subspathulate, rounded or obtuse apically, drying dark greenish-brown - - - - var. *gerberiformis*
Phyllaries tapering-acuminate; scapes mostly glabrescent; leaves linear to narrowly oblanceolate tapering attenuate or acute - - - - - - - - - - - - var. *hockii*

Var. **gerberiformis**

Vernonia macrocyanus var. *ambacensis* Hiern, Cat. Afr. Pl. Welw. **1**, 3: 532 (1898). —Mendonça, Contrib. Conhec. Fl. Angol., 1 Compositae: 23 (1943). Type from Angola.

Low-growing, scapose herbs; scapes usually white-woolly pubescent; root crown lanate. Leaves to c. 2 cm. wide, eventually some to 4 cm. wide, oblanceolate, rounded to subobtuse at apex, usually appearing after the capitula, usually drying dark greenish-brown. Involucres up to c. 3 cm. wide, occasionally more; phyllaries, at least the inner, lanceolate-lorate, subobtuse.

Zambia. N: Mbala, Nkali Dambo, c. 1740 m., 1.ix.1960, *Richards* 13181 (K; M; SRGH). W: Solwezi Distr., Mbulungu Dambo, W. of Mutanda bridge, 17.vii.1930, *Milne-Redhead* 720 (BR; K). **Zimbabwe**. C: Harare Distr., c. 3 km. along Norton Road from Harare-Beatrice road, c. 1800 m., 30.viii.1960, *Rutherford-Smith* 20 (BR; K; LISC; M; SRGH). E: Chimanimani Mts., c. 1525 m., 9.vi.1949, *Wild* 2964 (K; SRGH). **Malawi**. C: Dedza, Chongoni For. Res. boundary, 11.ix.1967, *Salubeni* 828 (COI; K; SRGH).

Also in the Sudan, Uganda, Kenya, Tanzania, Zaire and Angola. A pyrophyte of dambos, seasonally wet grassland and submontane grassland.

Var. **hockii** (De Wild. & Muschl.) G.V. Pope in Kew Bull. **43**: 280 (1988). Type from Zaire.
  Vernonia hockii De Wild. & Muschl. in Bull. Soc. Roy. Bot. Belg. **49**: 240 (1913). —Wild in Kirkia **11**: 104 (1978). Type as above.
  Vernonia pristis Hutch. & B.L. Burtt in Rev. Zool. Bot. Afr. **23**: 38 (1932). Type from Zaire.

Slender scapose herbs; scapes pubescent soon glabrescent, seldom lanate at the base. Leaves mostly less than 1 cm. wide, linear to narrowly oblanceolate, tapering-attenuate or acute at the apex, usually drying greenish-brown. Involucres up to c. 3 cm. wide; phyllaries ± narrowly tapering-acuminate at apex.

Zambia. W: Solwezi, c. 65 km. W. of Chingola, 21.x.1969, *Drummond & Williamson* 9216 (BR; K; SRGH). C: c. 45 km. N. of Kabwe on Great North Road, 23.ix.1947, *Brenan & Greenway* 7921 (FHO; K).

Also in Zaire and Angola. Pyrophyte of dambos.

Subsp. **macrocyanus** (O. Hoffm.) C. Jeffrey in Kew Bull. **43**: 234 (1988). Type from Angola.
  Vernonia macrocyanus O. Hoffm. in Bol. Soc. Brot. **13**: 20 (1896). —Hiern, Cat. Afr. Pl. Welw. **1**, 3: 531 (1898). —S. Moore in Journ. Bot. **65**, Suppl. 2, Gamopet.: 48 (1927). —Mendonça, Contrib. Conhec. Fl. Angol., **1** Compositae: 24 (1943). Type as above.
  Vernonia primulina O. Hoffm. in Warb., Kunene-Sambesi Exped. Baum: 402 (1903). Type from Angola.
  Vernonia towaensis De Wild. in Bull. Jard. Bot. Brux. **5**: 96 (1915). Type from Zaire.

Robust herbs; stems scapiform, often with 3–7 cauline leaves, puberulous or glabrescent. Largest leaves more than 4 cm. wide, elliptic-obovate to oblanceolate, rounded to subobtuse at apex, contemporaneous with capitula, usually drying pale-green, sometimes brownish. Involucres 3–5 cm. wide; phyllaries mostly more than 4 mm. wide, ovate-lanceolate, tapering-acute apically.

Zambia. W: Ndola, 24.ix.1953, *Fanshawe* 308 (BR; K; NDO). C: Kabwe, vi.1909, *Rogers* 8263 (K).
Also in Tanzania, Burundi, Zaire, Angola, Cameroon and Nigeria. A pyrophyte of dambo margins and wooded grasslands.

34. **Vernonia schweinfurthii** Oliv. & Hiern in F.T.A. **3**: 285 (1877). —C. Jeffrey in Kew Bull. **43**: 228 (1988). Syntypes from the Sudan.
  Vernonia asterifolia Bak. in Bull. Misc. Inf., Kew **1898**: 146 (1898). Type: Malawi, Zomba, *Whyte* s.n. (K, holotype).
  Vernonia katangensis O. Hoffm. in Ann. Mus. Congo Ser. IV, **1**: IX (1902). Type from Zaire.
  Vernonia glabra sensu Wild in Kirkia **11**: 96 (1978) pro parte quoad *V. asterifolia* Bak.

A low-growing tufted, slender to ± bushy perennial herb, 6–30(45) cm. tall from a small woody rootstock; roots numerous, stout, thong-like. Stems annual, few to many, erect or ascending, frequently much branched, stem and branches leafy, faintly ribbed, pubescent. Leaves subsessile, 0.5–3.5(4.5) × 0.8 cm., linear-elliptic to narrowly oblong-elliptic, apices acute to obtuse often mucronate, bases ± narrowly cuneate, margins serrate to subentire, puberulent or glabrescent, lower surface minutely glandular-punctate. Capitula numerous, borne above the foliage, either solitary on branches up to c. 18 cm. long, or heads 2–7 on each branch, corymbiformly cymose on bracteate stalks up to c. 6.5 cm. long. Involucres 5–12 × c. 12 mm., turbinate-hemispheric. Phyllaries greenish-stramineous, the tips dark-green or purplish, hispidulous, margins pectinate-denticulate, or minutely pungent-ciliolate; the outer phyllaries from c. 1 mm. long, lanceolate, tapering to a mucronate apex; the inner 5–7 mm. long, narrowly oblong-oblanceolate and ± abruptly narrowed to a mucronate apex, or up to c. 10 mm. long and linear-lanceolate, tapering to an aristate apex. Receptacle shallowly alveolate, alveolae walls slightly produced at the corners, not fimbriate. Corollas mauve or purple, 5–11 mm. long, tubular, widening slightly above. Achenes brown when mature, 2.5–4 mm. long, narrowly subcylindric-turbinate, narrowly c. 10-ribbed, strigose-hispid; outer pappus of short, linear, scale-like setae; inner of brownish purple-tinged or sordid barbellate setae 4–7 mm. long.

Zambia. W: Kitwe, 6.xi.1966, *Fanshawe* 9828 (K; NDO). C: Lusaka Distr., c. 11 km. S. of Makeni Police Post, 8.ix.1957, *Simwanda* 106 (K; SRGH). **Malawi.** S: Zomba and vicinity, c. 760–1060 m., xii.1896, *Whyte* s.n. (K).

Also from Ivory Coast to the Sudan, Uganda, Kenya, Tanzania, Rwanda, Burundi and Zaire. A pyrophyte of seasonally wet grassland or wooded grassland.

35. **Vernonia natalensis** Sch. Bip. ex Walp., Repert. Bot. Syst. **2**, Suppl. 1: 947 (1843). —Harv. in Harv. & Sond., F.C. **3**: 51 (1865). —Oliv. & Hiern in F.T.A. **3**: 277 (1877). —S. Moore in Journ. Linn. Soc., Bot. **40**: 106 (1911). —Eyles in Trans. Roy. Soc. S. Afr. **5**: 504 (1916). —Mendonça, Contrib. Conhec. Fl. Angol., 1 Compositae: 16 (1943). —Brenan in Mem. N.Y. Bot. Gard. **8**, 5: 459 (1954). —Hilliard in Ross, Fl. Natal: 356 (1972); Compos. Natal: 42 (1977). —Wild in Kirkia **11**: 82 (1978). —Hilliard & Burtt, Bot. S. Natal Drakensb.: 218 (1987). —C. Jeffrey in Kew Bull. **43**: 223 (1988). TAB. **17** figs. A. & D. Type from South Africa (Cape Province).

> *Webbia aristata* DC., Prodr. **5**: 73 (1836). Type as above.
> *Vernonia aristata* (DC.) Sch. Bip. in Flora **27**: 667 (1844) in sched. non Less. (1829). Syntypes as for *Webbia aristata*.
> *Vernonia pseudonatalensis* Wild in Kirkia **11**: 11; 83 (1978). Type from South Africa (Transvaal).
> *Vernonia smithiana* sensu Wild in Kirkia **11**: 82 (1978) pro parte.

An erect ± tufted perennial herb, to c. 120 cm. tall from a woody rootstock; indumentum silvery-grey-silky, hairs closely appressed T-shaped. Stems annual, many, simple or rarely branched, strict, leafy, ribbed. Leaves numerous appressed-ascending, variable, usually 3.0–10.5 × 0.3–2.0 cm., linear-lanceolate, lanceolate or narrowly elliptic, apex acute or obtuse mucronate, base cuneate, often with a petiole 1–2 mm. long, margins entire, lamina silvery or greenish-grey sericeous, the indumentum usually obscuring the lamina surface, nervation prominent beneath; or (in central and southern Mozambique) cauline leaves to c. 10 × 3 cm., broadly oblanceolate, sericeous, concolorous, ± coriaceous; or (in northern Malawi and Zambia) cauline leaves c. 5 × 1.6 cm., narrowly elliptic to elliptic, ± sparsely strigose on upper surface, ± discolorous, ± membranous. Capitula numerous in a dense terminal corymbiform cymose arrangement up to c. 24 cm. across. Involucres 5–11 mm. long, campanulate to somewhat spreading. Phyllaries few-seriate, greenish or stramineous to reddish-brown, variable, up to c. 11 mm. long (including setaceous tip) narrowly lanceolate, apex purple and curved-subulate or gently tapered to a flexuous bristle up to c. 4 mm. long, or phyllaries shorter and narrowly ± oblong, acute to abruptly narrowed at apex and bristle-tipped; the inner phyllaries equal to or very much shorter than the pappus in length, the outer phyllaries shorter and narrower than the inner, densely silky-silvery pubescent or sparsely pubescent to glabrescent. Corolla purple, 5–7 mm. long, funnel-shaped; lobes 1.5–2 mm. long, linear-oblong, strigose-puberulent. Achenes 2–2.5 mm. long, turbinate to subcylindric, ± tapered below, densely covered with silky-strigose hairs; outer pappus of very short, linear, acuminate scales; inner pappus 4.5–6 mm. long, of white copious, barbellate setae.

Zambia. N: Mpika, Lake Chibakabaka, 15.x.1963, *Robinson* 5753 (K; M; SRGH). C: Serenje, 16.xi.1971, *Richards* 27611 (M). E: Nyika Plateau, 28.xii.1962, *Fanshawe* 7276 (K). **Zimbabwe**. N: Tatagura Valley, c. 4.8 km. W. of Mazowe, 9.xi.1962, *Angus* 3419 (FHO; K). C: Harare, Teviotdale Road, 17.xii.1975, *Pope & Biegel* 1543 (K; SRGH). E: Mutare, Tsonzo Div., Kukwanisa, c. 1440 m., 12.xii.1967, *Biegel* 2388 (K; SRGH). S: Great Zimbabwe, 18.xii.1962, *Grosvenor* 4 (M; SRGH). **Malawi.** N: Viphya, c. 61 km. SW. of Mzuzu, 1680 m., 9.ix.1969, *Pawek* 2980 (K). C: Dedza Mt., 23.x.1956, *Banda* 292 (BM; K; SRGH). S: Machinga Distr., Chikala Hills, 800 m., 13.xi.1977 *Brummitt & Patel* 15094 (K). **Mozambique.** Z: Namagoa Estate, Mocuba, 60–120 m., vii-viii.1943, *Faulkner* Pretoria 134 (K; PRE). MS: Serra da Gorongosa, monte Nhandore, c. 1840 m., 19.x.1965, *Torre & Pereira* 12418 (LISC). GI: Linedela, Inhambane/Maxixe junction, c. 60 m., 4.x.1963, *Leach & Bayliss* 11813 (K; LISC; SRGH). M: Namaacha, ix.1930, *Sousa* 149 (K).

Also in Kenya, Tanzania, Angola and South Africa. A pyrophyte of dambos, grassland, miombo and *Brachystegia-Uapaca* woodlands.

Three distinctive, but intergrading, variants of *V. natalensis* can be recognised in the Flora Zambesiaca area. The most widespread of these is recorded from Zambia, Tanzania, Malawi, Zimbabwe, N. Transvaal and Angola. It is characterised by linear-lanceolate leaves, greenish-grey-sericeous on both surfaces, and by softly subulate or flexuously bristle-tipped phyllaries.

The second is a low altitude variant from central and southern Mozambique and N. Natal. This is distinguished by its leaves, which are thick and broadly oblanceolate, whilst possessing the same characteristic indumentum of the variant above. The phyllaries, however, vary from being long and softly subulate apically (eg., *Sousa* 149 (K), from Namaacha) to abbreviated and abruptly contracted apically (eg. *Sousa* 443 (K) also from Namaacha).

The third is the northern, submontane variant from the grasslands of the Viphya and Nyika Plateaux in Malawi, from Mbala in Zambia and from the adjacent Tanzanian districts of Ufipa and

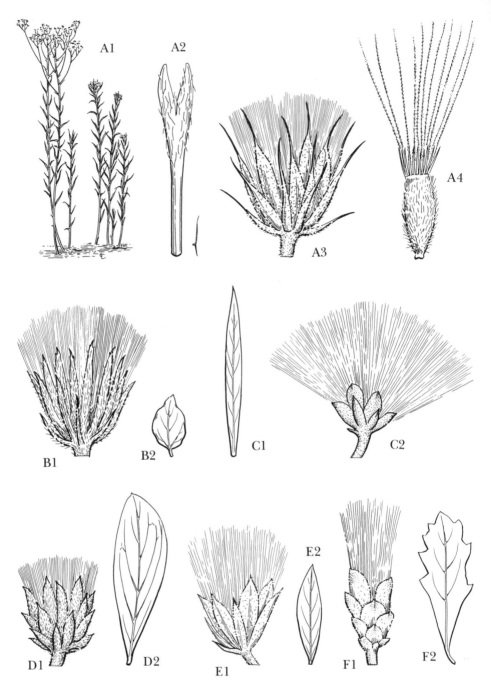

Tab. 17. A. —VERNONIA NATALENSIS. A1, habit (× 1/15), from *Brummitt & Patel* 15094; A2, corolla (× 8) and trichome (× 20); A3, involucre, phyllaries flexuously bristle-tipped; A4, achene (× 8), A2–A4 from *Pope & Biegel* 1543. B. —VERNONIA OLIGOCEPHALA. B1, involucre; B2, leaf, B1 & B2 from *Richards* 5918. C. —VERNONIA CALYCULATA. C1, leaf; C2, involucre, C1 & C2 from *Robson* 1128. D. —VERNONIA NATALENSIS. D1, involucre, phyllaries acute; D2, leaf, broadest above the middle, D1 & D2 from *Faulkner* 134. E. —VERNONIA ALTICOLA. E1, involucre; E2, leaf, E1 & E2 from *Richards* 10508. F. —VERNONIA TIGNA. F1, involucre; F2, leaf, F1 & F2 from *Marques* 2452. (Leaves × 2/3, involucres × 4). Drawn by Eleanor Catherine.

Songea. In this variant the leaves are more elliptic, shorter more discolorous and thinner in texture than in the variants above. The phyllaries are less frequently softly subulate, being instead ± abruptly contracted apically and sparsely pubescent to glabrescent exposing a stramineous to reddish-brown lamina.

V. smithiana (DC.) Less. from Uganda, Kenya, Tanzania and West tropical Africa, may be distinguished from *V. natalensis* by its leaves, especially the lower cauline, which have a prominent reticulate venation on both the upper and lower surfaces.

36. **Vernonia alticola** G.V. Pope in Kew Bull. **43**: 281 (1988). —C. Jeffrey in Kew Bull. **43**: 223 (1988). TAB. **17** fig. E. Type: Malawi, *Robson & Angus* 223 (BM; LISC; K, holotype).

An erect ± tufted perennial herb, 9–40(100) cm. tall from a woody rootstock; indumentum brownish-sericeous, hairs closely appressed T-shaped. Stems annual, many, simple or sometimes branched, strict, leafy, ribbed, sericeous; branches when present arising in leaf axils throughout the length of the stem, up to c. 20 cm. long. Leaves numerous, appressed-ascending to spreading; cauline leaves mostly 1.5–5 × 0.4–1.1 cm., elliptic, less often to 3.6 × 1.8 cm. and broadly elliptic-ovate, tapering to an acute ± apiculate apex and to a cuneate or a somewhat rounded base, shortly petiolate, margins entire or sometimes subserrate; lamina markedly discolorous, upper surface green sometimes drying brown, thinly silky pubescent or glabrescent, lower surface densely brownish-silvery-sericeous; nervation ± prominent beneath. Capitula few to numerous in a terminal corymbiform cymose arrangement. Involucres 4–6 mm. long, campanulate. Phyllaries purplish or purple-tipped, few-seriate, longer towards the inside, the outer linear-lanceolate with a subulate-aristate tip, the inner up to c. 6 mm. long and oblong-lanceolate tapering to an acute, acuminate or aristate apex, sometimes ± abruptly contracted to a mucronate tip, appressed puberulous to glabrescent. Corollas mauve or purple, 7–8 mm. long, funnel-shaped, deeply lobed, strigose-puberulent. Achenes 2.5–3 mm. long, subcylindric, densely covered with silky-strigose hairs; outer pappus of very short linear-lanceolate scales; inner pappus of 4–5.5 mm. long, sordid to stramineous, barbellate setae.

Zambia. N: Nkali (Kali) Dambo, c. 1525 m., 5.i.1955, *Richards* 3923 (K). E: Nyika Plateau, Chowo For., c. 2230 m., 14.xi.1967, *Richards* 22533 (K). **Malawi**. N: Nyika Plateau, Chelinda Dam, c. 2340 m., 29.ix.1969, *Pawek* 2790 (K).

Also in Tanzania. Sub-montane grassland, often in moist localities.

37. **Vernonia oligocephala** (DC.) Sch. Bip. ex Walp., Repert. Bot. Syst. **2**, Suppl. 1: 947 (1843). —Merxm., Prodr. Fl. SW. Afr. 139: 183 (1967). —Hilliard in Ross, Fl. Natal: 356 (1972); Compos. Natal: 43 (1977). —Wild in Kirkia **11**: 81 (1978). —Hilliard & Burtt, Bot. S. Natal Drakensb.: 218 (1987). —C. Jeffrey in Kew Bull. **43**: 223 (1988). TAB. **17** fig. B. Type from South Africa (Natal).
      Webbia oligocephala DC., Prodr. **5**: 73 (1836). Type as above.
      Webbia elaeagnoides DC., Prodr. **5**: 73 (1836). Type from South Africa (Natal).
      Vernonia elaeagnoides (DC.) Sch. Bip. in Flora **27**: 668 (1844) non H.B. Kunth (1820). Type as for *Webbia elaeagnoides*.
      Vernonia kraussii Sch. Bip. ex Walp., Repert. Bot. Syst. **2**, Suppl. 1: 947 (1843). —Harv. in Harv. & Sond., F.C. **3**: 51 (1865). —Oliv. & Hiern in F.T.A. **3**: 276 (1877). —S. Moore in Journ. Linn. Soc., Bot. **40**: 106 (1911). —Eyles in Trans. Roy. Soc. S. Afr. **5**: 504 (1916). Type as for *Webbia elaeagnoides*.

An erect usually tufted perennial herb, 30–80 cm. tall from a woody rootstock; indumentum silvery-grey-sericeous, hairs closely appressed T-shaped. Stems annual, many, simple or rarely branched, leafy, ribbed. Leaves numerous, appressed-ascending, up to c. 4 × 2.2 cm., ovate to lanceolate or elliptic, apex somewhat abruptly narrowed and cuspidate or tapering-acuminate mucronate, base cuneate to rounded, petiole 1–4 mm. long, margins entire often ± undulate especially towards the apex; lamina markedly discolorous, upper surface green, thinly sericeous or glabrescent, lower surface densely grey-silvery- or brownish-silvery-sericeous with the nervation prominent. Capitula numerous, in a dense terminal corymbose arrangement up to c. 15 cm. across. Involucres 3–6 mm. long, campanulate. Phyllaries few-seriate, silky-silvery pubescent or somewhat glabrescent, variable; the outer from 1.5 mm. long, lanceolate; the inner up to c. 6 mm. long, oblong to ± lanceolate, ± abruptly contracted to a mucronate tip or more gradually acute. Corolla purple, 5–7 mm. long, funnel-shaped; lobes 1.5–2 mm. long, linear-oblong, strigose. Achenes 2–2.5 mm. long, subcylindric, densely covered with silky-strigose hairs; outer pappus of very short linear-lanceolate scales; inner of white, copious, barbellate setae, 4–6 mm. long.

Botswana. SE: 3 km. E. of Kanye, 18.i.1960, *Leach & Noel* 198 (K; SRGH). **Zambia**. N: Mbala, Ndundu, c. 1740 m., 29.xi.1964, *Richards* 19282 (K; M). C: near Chilanga, 12.xi.1963, *van Rensburg* 2618 (K; M; SRGH). **Zimbabwe**. N: Gokwe, Sengwa Res. Sta., 14.i.1969, *Jacobsen* 453 (SRGH). W: Matopos, xi.1922, *Eyles* 3766 (K; SRGH). C: Harare, Alpes Road, xii.1975, *Pope & Biegel* 1514 (K; SRGH). E: Odzani, 1915, *Teague* 449 (K; BOL). S: Masvingo Distr., Mushandike Nat. Park, 25.xi.1974, *Bezuidenhout* 132 (K; SRGH). **Mozambique**. GI: Bilene Distr., planicies de Magul, 14.xii.1940, *Torre* 2412 (LISC). M: Maputo, Goba, near Maiuana R., 7.xi.1960, *Balsinhas* 217 (K). M: Ressano Garcia, 17.xii.1952, *Myre & Carvalho* 1356 (LISC).

Also in Tanzania, Zaire, Namibia and South Africa. A pyrophyte of grassland, usually on dambo margins, and also of wooded grassland with *Acacia* species or mixed deciduous shrubs.

Material of this species from around Mbala (Zambia), at the northern end of its geographical range, differs somewhat from that to the south in having generally fewer and larger capitula. The leaves however, are those of *V. oligocephala* and this material is taken to represent a variation of that species.

38. **Vernonia calyculata** S. Moore in Journ. Linn. Soc., Bot. **35**: 316 (1902). —Wild in Kirkia **11**: 84 (1978). —C. Jeffrey in Kew Bull. **43**: 223 (1988). TAB. **17** fig. C. Syntypes: Malawi, Mt. Mulanje, *Whyte* (BM); Shire Highlands, *Scott-Elliot* 8471 (BM; K).

   Vernonia natalensis var. sensu Britten in Trans. Linn. Soc., ser. 2, Bot. **4**: 18 (1894).

An erect slender perennial herb, up to c. 75 cm. tall from a woody rootstock. Stems annual, 1–several, simple or rarely branched, strict, faintly ribbed, purple at least below, appressed-pubescent, glabrescent below; hairs short-stalked long-armed T-shaped. Leaves few to numerous, ascending-spreading, up to c. 4 cm. long and linear, or up to 9.5 × 1.2 cm. and very narrowly elliptic, gradually tapered to the apex and narrowly cuneate base; margins entire, flat to strongly revolute; lamina discolorous, upper surface green, thinly and sparsely appressed pubescent or glabrescent, lower surface closely grey-silvery-silky or brownish-silvery-silky, the hairs T-shaped, nervation prominent beneath. Capitula in 1–many dense terminal clusters, corymbiform cymosely arranged in a larger synflorescence. Involucres 3–4 mm. long, hemispheric to campanulate. Phyllaries 2–3 seriate, appressed-pubescent; the inner oblong, mostly 1–2 times as long as the achene, ± abruptly contracted to a mucronate tip, sometimes acute; the outer shorter and narrower. Corolla purple, 4–7 mm. long, broadly funnel-shaped, deeply lobed, strigose-puberulous. Achenes 1.5–2 mm. long, subcylindric, ± tapered below, densely silky-strigose; outer pappus of very short, linear-lanceolate scales; inner pappus of 4–5 mm. long, sordid, copious, barbellate setae.

   Zambia. N: Kasama Distr., Chilubula, 24.x.1960, *Robinson* 4007 (K; M; SRGH). W: Solwezi-Kasempa, 30.ix.1947, *Greenway & Brenan* 8133 (K; SRGH). **Malawi**. N: Rumphi Distr., Kaziwiziwe R., 8.i.1959, *Richards* 10564 (BR; K; LISC; SRGH). C: Kasungu-Bua road, c. 1000 m., 13.i.1959, *Robson* 1128 (K; LISC; SRGH). S: Mt. Mulanje, Chambe cableway, 25.i.1967, *Hilliard & Burtt* 4620 (K). **Mozambique**. N: near "L. Nyasa", 1902, *Johnson* 458 (K). Z: Alto Molócuè, 29.xi.1967, *Torre & Correia* 16271 (LISC).

Also in Tanzania, Zaire and Angola. A pyrophyte of dambos, or high rainfall miombo and mixed deciduous woodland.

V. capensis (Houttuyn) Druce, from South Africa, also has linear leaves but is distinguished from *V. calyculata* by its stouter, more clustered stems which are much more leafy than in the latter species. Its involucres, too, are usually larger than in *V. calyculata*.

Although the leaves are typically linear in *V. calyculata* variation does occur and unusually broad leaves are seen in the Malawi material from Mt. Mulanje, of *Hilliard & Burtt* 4620 (K) and *Brummitt* 11792 (K) and in a *Milne-Redhead & Taylor* specimen, number 8425 (K) from Songea in Tanzania.

39. **Vernonia tigna** Klatt in Bull. Herb. Boiss. **4**: 829 (1896). TAB. **17** fig. F. Type: Mozambique, Baia de Maputo (Delagoa Bay), *Junod* 367 (isotype, BR).

   Staehelina corymbosa L.f., Suppl.: 359 (1781). Type from South Africa (Cape Province).

   Vernonia corymbosa (L.f.) Less. in Linnaea **6**: 647 (1831) non Schweinitz (1824). —Harv. in Harv. & Sond., F.C. **3**: 50 (1865). —Hilliard in Ross, Fl. Natal: 356 (1972). Type as for *Staehelina corymbosa*.

   Cacalia corymbosa (L.f.) Kuntze, Rev. Gen. **3**: 138 (1898). Type as for *Staehelina corymbosa*.

   Vernonia neocorymbosa Hilliard in Notes Roy. Bot. Gard. Edinb. **32**: 385 (1973); Compos. Natal: 31 (1977). —Wild in Kirkia **11**: 126 (1978). Type as for *Staehelina corymbosa*.

An erect subshrub to c. 1.8 m. tall, from a ?woody rootstock; indumentum silvery-grey sericeous, hairs appressed T-shaped. Stems several, simple or sparingly branched above, strict, leafy, striate, tomentose. Leaves discolorous, numerous, ± appressed-ascending; petiole to c. 5 mm. long, narrowly winged; lamina variable, up to 6 × 4 cm., obovate to cuneate-oblong or deltoid, coarsely toothed about the apex, narrowly cuneate to the base, upper surface green drying olive-green, glabrescent, lower surface densely silvery-grey sericeous with a prominent venation. Capitula numerous in moderately dense paniculate clusters; clusters 2–4 cm. across, terminal on branches; capitula stalks mostly 1–3 mm. long. Involucres 4–6 × 3–4.5 mm., oblong-cylindric. Phyllaries 4–5-seriate, appressed imbricate, silky-silvery pubescent or glabrescent; the outer phyllaries c. 1 mm. long, triangular-ovate; the inner to c. 5.5 mm. long, narrowly oblong-oblanceolate, rounded mucronulate. Florets 4–6 per capitulum. Corolla mauve, fading to whitish, 6–7 mm. long, tubular below with a campanulate limb. Achenes 2.5–3 mm. long, narrowly turbinate-cylindric, somewhat 3–4 angular, c. 10-ribbed, hispid on ribs, glandular between; outer pappus of copious readily caducous setae, 1–2 mm. long; inner of copious sordid more persistent barbellate setae, to c. 6 mm. long.

Mozambique. M: Lebombo Mts., near Namaacha, Mt. Mpondium, c. 800 m., 22.ii.1955, *Exell, Mendonça & Wild* 513 (BM; LISC; SRGH).
Also in South Africa (Transvaal, Natal, Transkei and Eastern Cape Province). Wooded grassland and forest margins.

40. **Vernonia lycioides** Wild in Kirkia **11**: 18, fig. 10; 119 (1978). Type: Zambia, Mwinilunga, *Drummond & Williamson* 9559 (BR; K; LISC; SRGH, holotype).

A lax subshrub to c. 1 m. tall. Branches purplish, simple, flagelliform, appressed-pubescent, ± glabrescent; indumentum of short-stalked long-armed T-shaped hairs. Leaves on very abbreviated shoots, or in fascicles of 5–9, subsessile or with a short, woolly petiole to c. 2 mm. long; lamina to c. 2 × 0.3 cm., narrowly oblanceolate, apices subacute to rounded, bases cuneate, margins entire, subrevolute, discolorous, upper surface green and thinly appressed pubescent, lower surface densely appressed grey-tomentose; hairs shortly-stalked, long-armed, T-shaped. Capitula numerous small solitary, subsessile on abbreviated tomentose bracteate shoots, in spike-like synflorescences c. 30 cm. long; bracts leaf-like, to c. 4 mm. long. Involucres c. 8 mm. long, obconic, soon spreading. Phyllaries few-seriate, somewhat loosely arranged, becoming caducous, the inner being shed first, ± lanceolate, scarious, densely appressed sericeous-lanate where exposed outside, midrib becoming somewhat pronounced and purplish towards the apex and usually continued into a stiff point; the outer phyllaries from c. 2 mm. long; the inner phyllaries to c. 7 mm. long, acute mucronate or acuminate. Florets c. 10 per capitulum; corollas purple, to c. 7 mm. long, gradually widening from near the base. Achenes 3–4 mm. long, subcylindric, ± tapering to the base, c. 8-ribbed, sparsely hispidulous; outer pappus of short, barbellate setae, inner of sordid, barbellate setae c. 6 mm. long.

Zambia. W: c. 32 km. S. of Mwinilunga on road to Kabompo, 28.x.1969, *Drummond & Williamson* 9559 (K; LISC; SRGH).
Known only from this gathering. Woodland.

41. **Vernonia subplumosa** O. Hoffm. in Warb., Kunene-Samb.-Exped. Baum: 404 (1903). — Mendonça, Contrib. Conhec. Fl. Angol., 1 Compositae: 10 (1943). —Wild in Kirkia **11**: 67 (1978). TAB. **18** fig. B. Type from Angola.

A tough perennial herb up to c. 65 cm. tall from a woody, fibrous-rooted rootstock. Stems annual solitary erect wiry, leafy, branching above, somewhat angled and strongly ribbed with numerous scattered, minute, often black, glandular hairs, otherwise glabrous. Leaves many, ascending, subsessile, 5–14 × 0.6–2.8 cm., the largest midcauline, linear-elliptic to elliptic, apex acute mucronate, base cuneate, margins entire to remotely denticulate-crenulate, lamina coriaceous, glabrous, nervation reticulate and prominent on both sides. Capitula few to many, 1–2 per branch, occasionally solitary, capitula stalks stiff, up to c. 10 cm. long. Involucres 10–14 × 10–15 mm., broadly campanulate, later spreading. Phyllaries coriaceous, numerous, many-seriate, tightly imbricate, shortly brownish-woolly to densely pubescent outside, or glabrescent, ciliate or cobwebby on the margins; the outer phyllaries from 1.5 mm. long, ovate-lanceolate; the inner phyllaries to c. 12 mm. long, narrowly oblong and gently tapered to a blunt mucronate apex. Receptacle shallowly alveolate. Corollas purplish, up to c. 11 mm. long, tubular, widening into a limb with lobes to c. 4 mm. long. Achenes c. 3.5 mm. long, subcylindric, somewhat angular with 5–8 narrow, raised ribs, sparsely strigose to glabrous and gland-dotted; outer pappus of linear, tapering, fimbriate-ciliate, seta-like scales c. 2 mm. long, inner of copious, sordid, subplumose setae 6–7 mm. long.

Zambia. B: Kataba, 14.vii.1960, *Fanshawe* 5787 (K; SRGH). W: Mwinilunga Distr., Zambezi

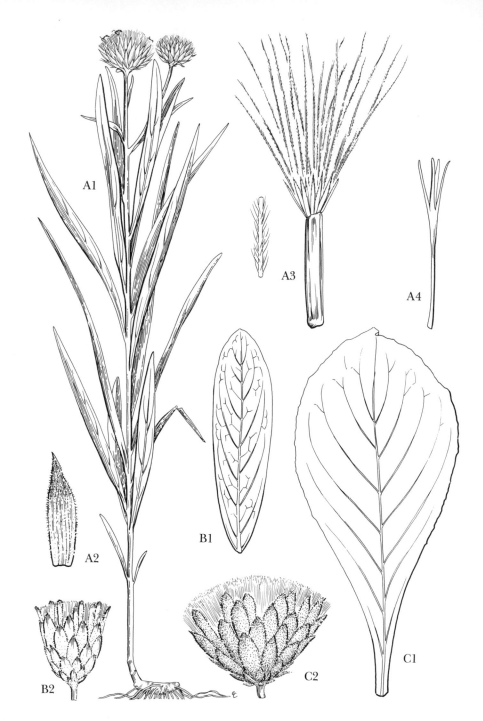

Tab. 18. A. —VERNONIA ROSEOVIOLACEA. A1, habit (× ⅓); A2, phyllary (× 1½); A3, achene (× 4) and detail of subplumose pappus seta (× 12); A4, corolla (× 2), A1–A4 from *Philcox, Pope & Chisumpa* 10079. B. —VERNONIA SUBPLUMOSA. B1, leaf, note prominent reticulate venation (× ⅔); B2, capitulum (× 1½), B1 & B2 from *Philcox, Pope, Chisumpa & Ngoma* 10296. C. —VERNONIA UPEMBAENSIS. C1, leaf (× ⅔); C2, capitulum (× 1½), C1 & C2 from *Philcox, Pope, Chisumpa & Ngoma* 10297. Drawn by Eleanor Catherine.

Source Memorial, c. 1700 m., 15.v.1986, *Philcox, Pope, Chisumpa & Ngoma* 10296 (BR; CAL; GA; K; LISC; MO; NDO; SRGH).

Also in Angola. *Cryptosepalum* forest fringes, or in chipya or miombo woodland on Kalahari Sand.

42. **Vernonia roseoviolacea** De Wild. in Bull. Jard. Bot. Brux. **5**: 98 (1915). —C. Jeffrey in Kew Bull. **43**: 227 (1988). TAB. **18** fig. A. Type from Zaire.

A robust perennial herb to c. 60 cm. tall from a woody, fibrous-rooted rootstock. Stems annual erect simple leafy ± angular, very strongly ribbed and becoming deeply sulcate above, sparsely puberulous, soon glabrescent. Leaves many, ascending, sessile, 8–25 × 1–2(5) cm., linear-oblanceolate to oblanceolate, ± narrowly tapering-acute at apex, attenuate to a narrowly cuneate base, margins entire to remotely repand-crenulate, lamina coriaceous, glabrous or minutely sparsely puberulous, nervation laxly reticulate, slightly raised. Capitula 1–4, corymbiform cymose, sometimes solitary; stalks to c. 9.5 cm. long, bracteate; involucres 2–2.5 × 3–4.5 cm., broadly campanulate to subhemispheric. Phyllaries numerous, many-seriate, coriaceous, purple-tinged towards the apex, brownish-lanate to densely puberulous outside, or somewhat glabrescent, margins finely araneose-ciliate; the outer phyllaries from c. 5 mm. long, narrowly triangular to ovate-lanceolate, tapering to an acute often acuminate apex; the middle phyllaries to c. 5 mm. wide, oblong-lanceolate and acuminate, mucronate or pungent-aristate; the inner phyllaries narrower, to c. 22 mm. long, linear, tapering to a fine point. Corollas mauve, 18–22 mm. long, narrowly tubular in lower c. two-thirds, widening to a campanulate limb with lobes 5–6 mm. long. Achenes to c. 8 mm. long, subcylindric with c. 8 narrow, raised ribs, glabrous, glandular; outer pappus of linear, fimbriate scales c. 4 mm. long, inner of white, plumose, slightly flattened setae 10–13 mm. long.

Zambia. N: Nakonde, near Old Fife Coffee Res. Sta., c. 1850 m., 22.iv.1986, *Philcox, Pope & Chisumpa* 10079 (K; NDO; BR).

Also in Tanzania and Zaire. Miombo woodland and wooded grassland.

43. **Vernonia upembaensis** Kalanda in Bull. Jard. Bot. Nat. Belg. **52**: 127 (1982). TAB. **18** fig. C. Type from Zaire.

A scapiform perennial herb to c. 90 cm. tall, from a small fibrous-rooted woody rootstock. Stems annual, erect, simple or sparingly branched above, brittle, ± angular, sharply strongly ribbed, becoming sulcate, sparsely appressed pilose, or glabrescent. Leaves 5–10, subrosulate prostrate sessile, 12–18 × 1.5–7 cm., oblanceolate to obovate or broadly elliptic, rounded above with a shortly acuminate apex in broad leaves, tapering-subobtuse in narrow leaves, base attenuate, margins irregularly repand-crenulate; lamina softly coriaceous, minutely, sparsely puberulous on both surfaces; nervation laxly reticulate and somewhat raised below. Capitula solitary, or 2–3 on bracteate stalks to c. 14 cm. long; bracts up to c. 10 mm. long, subulate. Involucres c. 15 × 15–18 mm., broadly campanulate, later spreading-hemispheric to c. 30 mm. wide. Phyllaries stiffly coriaceous, numerous, many-seriate, imbricate, shortly whitish- or brownish-woolly outside, finely-ciliate or cobwebby on the margins; outer phyllaries from c. 2 mm. long, triangular to ovate mucronate; middle phyllaries c. 10 × 3 mm., oblong, tapering or ± abruptly narrowed to a mucronate apex; inner phyllaries narrower to c. 12 mm. long, linear. Receptacle shallowly alveolate. Corollas ?purple, c. 10 mm. long, narrowly tubular, widening into a limb with lobes 2–3 mm. long. Achenes c. 6 mm. long, subfusiform-cylindric, stramineous, with c. 7 narrow raised ribs, glabrous, glandular between ribs; outer pappus of fimbriate seta-like scales, 1.5–2 mm. long, inner of white plumose setae c. 8 mm. long.

Zambia. W: Mwinilunga Distr., Zambezi River Source Memorial, c. 1700 m., 15.v.1986, *Philcox, Pope & Chisumpa* 10297 (BR; K; LISC; MO; NDO; SRGH).

Also in Zaire. Deciduous woodland in sandy soil.

44. **Vernonia bainesii** Oliv. & Hiern in F.T.A. **3**: 272 (1877). —S. Moore in Journ. Linn. Soc., Bot. **40**: 104 (1911). —Eyles in Trans. Roy. Soc. S.Afr. **5**: 503 (1916). —Wild in Kirkia **11**: 53 (1978). Type: Zimbabwe, "South African Goldfields", *Baines* s.n. (K, holotype).

A diffuse, or somewhat spreading, perennial herb to 60(120) cm. tall from a woody rootstock; stems annual, decumbent or ascending, 1–many, slender, often branched above, glabrescent, glandular. Leaves ± crowded, ± overlapping ascending, up to c. 3.5 cm. long and usually less than 2 mm. wide, linear with margins ± revolute, sometimes up to c. 5 mm.

wide and narrowly elliptic-lorate, glabrous to sparsely scabridulous. Capitula solitary at the ends of the ± numerous ascending branches. Involucres 10–15 mm. long, obconic-campanulate. Phyllaries purple-tipped, 6–8-seriate, closely appressed and overlapping, not or hardly spreading with age, obtuse or ± rounded mucronate, or (in subsp. *brassii*) tapering to an acuminate apex, pubescent or glabrescent and ± glandular outside, ciliate; the outer phyllaries c. 2 mm. long, ovate-lanceolate; the inner phyllaries up to 15 mm. long, linear. Corolla purple, c. 10 mm. long, narrowly funnel-shaped; lobes c. 2 mm. long, linear, tapering to an acute apex, minutely glandular. Achenes 3–4 mm. long, subcylindric, tapering to the base, 5–6-angled with hispid ribs on the angles; outer pappus of short narrow scales, inner of whitish, or sordid, barbellate setae 6–8 mm. long.

1. Leaves filiform or linear, 6–30 mm. long and up to 2 mm. wide, markedly revolute
                                                                                    subsp. *bainesii*
– Leaves narrowly elliptic, 10–25 × 2–7 mm., margins if revolute not markedly so   2
2. Phyllaries coriaceous, rounded to obtuse mucronate at the apex   -   -   subsp. *wildii*
– Phyllaries chartaceous, acuminate at the apex   -   -   -   -   -   -   subsp. *brassii*

Subsp. **bainesii**

> *Vernonia buchananii* Bak. in Bull. Misc. Inf., Kew **1898**: 146 (1898). Syntypes: Malawi, Nyika Plateau, *Whyte* 160 (E; K); between Mpata and Nyasa-Tanganyika Plateau, *Whyte* s.n. (K); Shire Highlands, *Buchanan* 139 (E; K).

Leaves filiform or linear, 6–30 mm. long and up to 2 mm. wide, markedly revolute. Phyllaries coriaceous, rounded or obtuse mucronate to acute at the apex.

Zambia. N: Mbala Distr., Kawimbe, 20.iii.1957, *Richards* 8829 (K; SRGH). E: Nyika Nat. Park, c. 0.5 km. SW. of Zambian Govt. Rest House, 17.iv.1986, *Philcox, Pope & Chisumpa* 9961 (BR; K; MO; NDO; SRGH). **Zimbabwe**. W: Matopos, iii.1902, *Eyles* 1045 (K; SRGH). C: Shurugwi (Selukwe) Peak, 19.iii.1964, *Wild* 6432 (K; M; SRGH). E: Nyanga (Inyanga), Juliasdale, 1.iv.1961, *Leach* 10763 (K; LISC; M; SRGH). S: Mberengwa Distr., Mt. Buhwa, 3.v.1973, *Pope, Biegel & Simon* 1093 (K; LISC; PRE; SRGH). **Malawi**. N: Viphya, opposite Rumphi Gorge, *Chapman* 172 (BM; SRGH). C: Dedza, 6.vi.1938, *Pole Evans & Erens* 593 (BR; E; K; PRE; S; SRGH). S: Kirk Range, *Young* 226 (BM; K; SRGH). **Mozambique**. T: Angónia, Posto Zootécnico, 14.v.1948, *Mendonça* 4229 (LISC; LMU). MS: Chimoio, Barùè, Serra de Choa, 25.v.1971, *Torre & Correia* 18640 (LISC; LMA; LMU; C).
Also in Tanzania and South Africa (Transvaal). Usually in submontane grassland.

Subsp. **wildii** (Merxm.) Wild in Kirkia **11**: 1; 54 (1978). Type: Zimbabwe, Marondera (Marandellas), *Dehn* 248 (BR; M, holotype; SRGH).

> *Vernonia wildii* Merxm. in Trans. Rhod. Sci. Ass. **43**: 72 (1951). Type as above.

Leaves narrowly elliptic, 10–25 × 2–7 mm., margins not markedly revolute. Phyllaries coriaceous, rounded to obtuse mucronate. Otherwise similar to subsp. *bainesii*.

Zimbabwe. N: Gt. Dyke, E. end of Vanad Pass, 6.viii.1974, *Wild* 7969 (K; SRGH). C: Rusape, 9.ii.1931, *Norlindh & Weimarck* 4947 (LD). E: Nyanga (Inyanga), 23.iii.1966, *Simon* 760 (K; SRGH).
Not known outside Zimbabwe. Grassland, on granite sands or serpentine soils. Also in *Brachystegia speciformis* woodland. Usually occurring in somewhat lower rainfall areas than subsp. *bainesii*.

Subsp. **brassii** Wild in Kirkia **11**: 1; 54 (1978). Type: Malawi, Ntchisi Mt., *Brass* 16961 (SRGH, holotype; BR; K; MO; PRE).

Leaves narrowly lanceolate, ± attenuate at the apex, 15–35 × 2–6 mm., revolute. Phyllaries chartaceous, acuminate at the apex. Otherwise similar to subsp. *wildii*.

Malawi. C: Ntchisi For. Res., 9.vi.1970, *Brummitt* 11555 (K; LISC; MAL; SRGH).
Endemic as a subspecies to the Ntchisi District. Submontane miombo woodland.

45. **Vernonia schlechteri** O. Hoffm. in Engl., Bot. Jahrb. **24**: 466 (1898). Type from South Africa, Transvaal, Lydenburg, *Wilms* 660 (K, lectotype chosen here).

A bushy perennial herb up to c. 50(75) cm. tall from a small woody rootstock. Stems annual, 1–many, stiffly branched, puberulous. Leaves ± subsessile, ± crowded, up to 55 × 2.8 mm., linear, much smaller on branches and abbreviated shoots, apex obtuse mucronate, margins entire and revolute, lamina glandular-punctate, hispidulous to scabridulous. Capitula few to numerous, racemose or laxly paniculate; involucres up to

c. 10 mm. long, hemispheric-campanulate, ± wider than long. Phyllaries appressed-imbricate, equalling the pappus in length, araneose-pubescent, the hairs on the apical part of the phyllaries often brownish, margins ciliate-pilose above; outermost phyllaries ovate to oblong and acute-mucronate; the inner phyllaries longer, lanceolate-oblong; the innermost phyllaries ± appendaged with the apex expanded purplish-hyaline and minutely aristate. Corollas purplish, 5–9 mm. long. Achenes pale-brown, c. 2.5–2.75 mm. long, obpyramidal, 4–5-ribbed, hispidulous on the ribs and glandular between; outer pappus of short, narrow, brownish scales, the inner of brownish, barbellate setae 4–5 mm. long.

Zimbabwe. E: Nyanga Distr., Juliasdale, c. 51 km. from Rusape, c. 1828 m., 1.iv.1961, *Leach* 10762 (K; SRGH). **Mozambique.** MS: Beira coast, iv.1895, *Schlechter* s.n. (K).
Also recorded from South Africa (Transvaal) and Swaziland. Submontane grassland and rocky *Brachystegia/Protea* hillsides.

46. **Vernonia karaguensis** Oliv. & Hiern in Trans. Linn. Soc., Lond. **29**: 91 (1873); in F.T.A. **3**: 280 (1877). —Maquet in Fl. Rwanda, Spermat. **3**: 558 (1985). —C. Jeffrey in Kew Bull. **43**: 221 (1988). Type from Tanzania.
   Vernonia cistifolia O. Hoffm. in Engl., Pflanzenw. Ost-Afr. **C**: 404 (1895). —Brenan in Mem. N.Y. Bot. Gard. **8**, 5: 459 (1954). —Wild in Kirkia **11**: 50 (1978). Syntypes from Tanzania.
   Vernonia cistifolia var. *rosea* O. Hoffm. in Engl., Pflanzenw. Ost-Afr. **C**: 404 (1895). —S. Moore in Journ. Linn. Soc., Bot. **40**: 105 (1911). —Eyles in Trans. Roy. Soc. S. Afr. **5**: 503 (1916). Syntypes as for *V. cistifolia*.
   Vernonia elliotii S. Moore in Journ. Linn. Soc., Bot. **35**: 315 (1902). Type from Kenya.
   Vernonia bothrioclinoides C.H. Wright in Bull. Misc. Inf., Kew **1906**: 108 (1906). Syntypes: Malawi, Namasi, *Cameron* 40 (K); Mt. Chiradzulu, *Whyte* s.n. (K).
   Vernonia porphyrolepis S. Moore in Journ. Bot. **46**: 39 (1908). Type: Zimbabwe, Mazowe, *Eyles* 371 (BM, holotype; BOL; SRGH).
   Vernonia campanea S. Moore in Journ. Bot. **54**: 251 (1916). Type from Uganda.
   Vernonia cistifolia var. *bothrioclinoides* (C.H. Wright) Brenan in Mem. N.Y. Bot. Gard. **8**, 5: 460 (1954). Type as above.

A slender erect perennial herb, to 2 m. tall from a small woody rootstock, often with short stoloniferous rhizomes. Stems annual, 1–several, branched above, ± densely pubescent, glabrescent below. Leaves subsessile, up to c. 12 × 4 cm., lanceolate-oblong or oblong, apex acute mucronate, base obtuse to broadly cuneate, margin entire or irregularly crenulate and ± revolute, upper surface scabrous, lower surface ± densely tomentose. Capitula very numerous, in clusters arranged in a large terminal panicle-like synflorescence, or capitula scorpioidly cymose being subsessile and secund on synflorescence branches with short arching internodes; branches densely pubescent to tomentellous. Involucres 4–7 mm. long, campanulate to obconic. Phyllaries purple-tinged, 3–5-seriate, pubescent or glabrous, glandular; the outer ± reflexed, shortly lanceolate; the inner longer, linear-lanceolate, acuminate. Corolla purple, 7–8 mm. long, narrowly funnel-shaped. Achenes oblong-cylindric, c. 2 mm. long, narrowly 4-ribbed, glandular and minutely pubescent; outer pappus of minute, narrow scales, inner of c. 5 mm. long, whitish, barbellate setae.

Zambia. N: Mbala Distr., Chianga to Kambole road, 2.vi.1957, *Richards* 22295 (K; M; SRGH). W: Mwinilunga Distr., L. River (Lunga R.?), 17.v.1969, *Mutimushi* 3460 (K; NDO). **Zimbabwe.** N: Mazowe Distr., below Iron Mask Range, 1.vi.1965, *Dale* 117 (SRGH). C: Shurugwe (Selukwe) Peak, 6.vii.1967, *Biegel* 2191 (K; SRGH). E: Mutare Distr., Dungari Mt., 20.vii.1952, *Chase* 4668 (BR; MO; SRGH). S: Mberengwa Distr., Buhwa Mt., 5.vii.1968, *Müller* 781 (SRGH). **Malawi.** N: Mzimba Distr., Mzuzu, c. 1390 m., 2.vi.1974, *Pawek* 8789 (K; MO; MAL). C: Dedza Mt., 20.vii.1970, *Salubeni* 1479 (LISC; SRGH). S: Mulanje, Chambe Basin, i.1959, *Richards* 16677 (K; M). **Mozambique.** N: 8 km. S. of Massangulo, 26.v.1961, *Leach & Rutherford-Smith* 11027 (LISC; M; SRGH). Z: Mugema-Gurué, 17.xi.1949, *Barbosa & Carvalho* 4470 (SRGH). T: Zóbuè, 28.vi.1947, *Hornby* 2768 (K; PRE; SRGH). MS: Mossurize, prox. Missão Católica de Espungabera, 8.vi.1942, *Torre* 4257 (LISC).
Also in Nigeria, Sudan, Ethiopia, Uganda, Kenya, Tanzania, Rwanda, Zaire and Angola. High rainfall miombo woodland often near streams.

47. **Vernonia luembensis** De Wild. & Muschl. in Bull. Soc. Bot. Belg. **49**: 244 (1913). —Wild in Kirkia **11**: 100 (1978). —C. Jeffrey in Kew Bull. **43**: 228 (1988). Type from Zaire.
   Vernonia sciaphila S. Moore in Journ. Bot. **56**: 208 (1918). —Earle Sm. in U.S. Handb. **396**: 59, figs. 39, 40 (1971). Type: Zambia, Sangolo Stream, *Kassner* 2106 (BM, holotype; K).

A stout ± tufted perennial herb, 45–100(150) cm. tall from a woody rootstock; roots numerous, thong-like. Stems annual, 1–several, erect, unbranched below the synflorescence, leafy in the upper c. two-thirds with leaves much reduced or absent below, densely sordid- or greyish-tomentose. Leaves sessile, up to c. 14 × 5 cm., oblong-elliptic to oblanceolate, apex obtuse to subacute, base obtuse, margins entire or crenulate, softly coriaceous; the upper surface sparsely puberulous, sometimes scabridulous with the hairs becoming rigid towards the base; the lower surfaces greyish-green tomentose with nervation reticulate and prominent. Capitula 1–12, solitary and terminal on the stem and branches, capitula stalks or branches 0.5–20 cm. long, stout, tomentose, bracteate; bracts few, intergrading with leaves. Involucres 1.5–2.3 × 2–3 cm., hemispheric. Phyllaries numerous, imbricate, 4–5-seriate, purple tipped with hyaline, usually glabrous margins, ± densely white-tomentose outside, at least where exposed; outer phyllaries from c. 4 mm. long and ± broadly triangular-ovate; inner phyllaries increasing uniformly in length to c. 22 mm. long inside, changing from ovate to lorate. Receptacle shallowly alveolate. Corollas 14–23 mm. long, narrowly tubular with a slightly wider limb, glabrous; limbs purple or mauve, well exserted at flowering time, lobes c. 5 mm. long, bristled at the apex. Achenes 6–8 mm. long, narrowly cylindric, c. 10-ribbed, glabrous; outer pappus of short, linear, scale-like setae 2–3 mm. long; inner pappus of stramineous, slightly flattened barbellate setae 9–11 mm. long.

Zambia. N: Mbala Distr., Sand Pits, c. 1500 m., 21.i.1965, *Richards* 19574 (K; M; SRGH). W: between Kitwe and Ndola, c. 12°53'S, 28°26'E, 11.ii.1967, *Earle Smith & Fanshawe* 4660 (BM; EA; K; SRGH; US). C: Serenje Distr., road to Kundalila Falls, c. 13°08'S, 30°40'E, c. 1600 m., 5.ii.1973, *Kornaś* 3216 (K).
Also in Tanzania and Zaire. Miombo woodland and wooded grassland often in sandy soil.

48. **Vernonia tanganyikensis** R.E. Fr., Wiss. Ergebn. Schwed. Rhod.-Kongo-Exped. 1911–1912, **1**: 325 (1916). —Wild in Kirkia **11**: 85 (1978). Type: Zambia, Lake Tanganyika, *R.E. Fries* 1267 (UPS, holotype).
    Vernonia sculptifolia sensu Hutch., Bot. S. Afr.: 508 (1946) non Hiern (1898).

An erect slender perennial herb, to c. 1 m. tall from a woody rootstock. Stems annual, 1–many, branched and leafy above, terete, yellowish-brown tomentose. Leaves subsessile, up to c. 13.5 × 3 cm., narrowly oblong to narrowly oblong-lanceolate, rounded or obtuse at the apex, cuneate to obtuse at the base, margins irregularly crenate or crenulate to remotely dentate; lamina discolorous, upper surface hispid, greenish, becoming bullate and scabridous, lower surface softly felted and greyish-tomentose with venation reticulate and prominent, the midrib and sometimes the veins brownish-tomentose. Capitula numerous, in terminal clusters on branches, or arranged in a large panicle-like synflorescence. Involucres 7–10 mm. long, broadly obconic. Phyllaries numerous, spreading, linear-lanceolate with sharply acute purplish aristate apices, very densely strigose or woolly, increasing from c. 2 mm. long outside to c. 8 mm. long inside. Corollas purple, 5–8 mm. long, very narrowly funnel-shaped, strigose-pubescent on the lobes. Achenes 2–2.5 mm. long, subcylindric to somewhat 4–5-angular, hispidulous; outer pappus of short, linear scales or paleaceous setae; inner pappus of white barbellate setae 7–8 mm. long.

Zambia. N: 48 km. N. of Kasama, 18.vii.1930, *Hutchinson & Gillett* 3805 (BM; K; LISC; SRGH).
Known only from Zambia. Miombo and escarpment miombo woodland.

49. **Vernonia teucrioides** Welw. ex O. Hoffm. in Bol. Soc. Brot. **10**: 171 (1893). —Hiern, Cat. Afr. Pl. Welw. **1**, 3: 522 (1898). —S. Moore in Journ. Bot. **64**, Suppl. 2, Gamopet.: 46 (1927). —Mendonça, Contrib. Conhec. Fl. Angol., **1** Compositae: 17 (1943). —Wild in Kirkia **11**: 86 (1978). Type from Angola.

An erect perennial herb up to ?1 m. tall, from a woody rootstock. Stems annual, 1–many, branched, leafy, yellowish-brown tomentose; branches more densely leafy than the stem. Leaves subsessile, 3–8(?) × 0.8–2(?) cm., elliptic to oblanceolate, widest about or above the middle, apex acute or obtuse, base cuneate, margins repand or crenulate, lamina finely bullate, the upper surface green and thinly pilose, the lower surface densely yellowish- or grey-tomentose with a prominent reticulate venation. Capitula many, in a condensed terminal cluster, or ± laxly subpaniculate. Involucres up to c. 6(9) × 6(10) mm., campanulate to obconic. Phyllaries many, several-seriate, appressed imbricate, later spreading, thinly pilose to woolly where exposed, otherwise glabrous; outer phyllaries lanceolate, the inner longer and oblong-lanceolate, abruptly contracted apically and shortly acuminate-apiculate. Florets c. 11; corollas mauve or purple, up to c. 7 mm. long, narrowly funnel-shaped. Achenes 2–2.5(3) mm. long, subcylindric to turbinate, ± obscurely 6–8-ribbed, hispidulous; outer pappus of very short setae, inner pappus of barbellate setae c. 7 mm. long.

Zambia. B: Kaoma (Mankoya) Distr., c. 43 km. on Mongu Road, 8.xi.1959, *Drummond & Cookson* 6242 (K; LISC; M; PRE; SRGH).
Also in Angola. Open grassland with *Parinari capensis* on Kalahari Sand.
V. sculptifolia Hiern, from Angola, is probably conspecific with this species.

50. **Vernonia musofensis** S. Moore in Journ. Bot. **56**: 206 (1918). —C. Jeffrey in Kew Bull. **43**: 244 (1988). Type from Zaire.

An erect slender perennial herb, 20–50 cm. tall from a slender ± contorted woody taproot. Stems annual, usually 1–2, branched above, leafy, ribbed, sparsely roughly pubescent to tomentose or strigose-hispid; indumentum a variable mixture of appressed hairs interspersed with few–many long patent hairs, one or other hair type predominating especially on the lower stem. Leaves sessile, up to c. 11 × 1.5 cm. and narrowly oblong-lanceolate to oblanceolate, or to c. 8 × 2.5 cm. and elliptic, apex acute, base broadly cuneate to rounded, margins subentire, hispid-strigose to pubescent or sparsely patent-pilose on both surfaces, often more densely so beneath, becoming scabrous; nervation prominent on the lower surface with the lateral nerves at an acute angle to the midrib, the basal nerves long and running almost parallel to the margin. Capitula few to numerous, corymbiformly cymose on the stem or branches, sometimes solitary. Involucres up to c. 20 mm. long, obconic, or broadly campanulate to spreading. Phyllaries numerous, increasing from c. 5 mm. long outside to c. 20 mm. long inside, stiffly linear-lanceolate, tapering to a straight finely subulate often purplish apex, or somewhat recurved apically, densely strigose-sericeous to tomentose. Receptacle shallowly alveolate. Corollas purple or creamy-white, 9–12 mm. long, cylindric, widening slightly in upper half. Achenes 3–4 mm. long, subterete to 4–5-angular and somewhat ribbed on the angles, uniformly setulose; outer pappus of 1–2 mm. long, subulate-lanceolate setae, inner pappus of white, 8–11 mm. long, somewhat flattened setae, barbed mainly on the margins.

Stems hispid-tomentose or densely strigose, patent multicellular hairs absent or few on the lower
stem; corollas usually creamy-white; phyllaries somewhat recurved at the apex; involucres
mostly less than 15 mm. long - - - - - - - - - var. *musofensis*
Stems sparsely patent-pilose or glabrescent, the hairs large, multicellular; corollas usually purple;
phyllaries pungent-subulate usually straight; involucres mostly more than 15 mm. long
var. *miamensis*

Var. **musofensis**

Zambia. N: Chinsali Distr., between Mpika and Shiwa Ngandu turnoff, 52 km. N. of Mpika, 12.iv.1986, *Philcox, Pope & Chisumpa* 9889 (BR; K; MO; NDO; SRGH). W: Mufulira, 8.v.1934, *Eyles* 8126 (K; SRGH). C: c. 48 km. E. of Lusaka on Reitfontein road, 17.ix.1958, *Angus* 2029 (K). **Malawi.** S: W. slopes of Mulanje Mt., c. 920 m., 13.iv.1984, *Christenson & Patel* 1459 (K).
Also in Zaire. Deciduous woodland.

Var. **miamensis** (S. Moore) G.V. Pope in Kew Bull. **43**: 281 (1988). Type: Zimbabwe, Mwami (Miami), *Rand* 100 (BM, holotype).
    Vernonia miamensis S. Moore in Journ. Bot. **64**: 304 (1926). Type as above.
    Vernonia lappoides O. Hoffm. in Bol. Soc. Brot. **13**: 19 (1896) nom. illegit., non Bak. (1873).
    —Mendonça, Contrib. Conhec. Fl. Angol., 1 Compositae: 11 (1943). Types from Angola.
    Dicoma ringoetii De Wild. in Fedde, Repert. **13**: 210 (1914). Type from Zaire.
    Vernonia hoffmanniana Hutch. & Dalz., F.W.T.A. **2**: 167 (1931) non S. Moore (1900). Types as
for *V. lappoides* O. Hoffm.
    Vernonia philipsoniana Lawalrée, Expl. Hydrobiol. Lac Tanganyika (1946–47) Rs. Sc. 4, **2**: 59 (1955) nom. superfl. illegit. —Wild in Kirkia **11**: 75 (1978). Types as for *V. lappoides* O. Hoffm.

Zambia. N: Chisimba Falls, c. 30 km. WNW. of Kasama, 31.iii.1984, *Brummitt & Chisumpa* 17052 (K). W: c. 88 km. W. of Chingola on Solwezi road, 16.iii.1961, *Drummond & Rutherford-Smith* 6944 (BR; K; LISC; M; MO; SRGH). C: c. 10 km. E. of Lusaka, c. 1280 m., 22.ii.1956, *King* 328 (K). S: Choma, Mapanza, c. 1060 m., 4.v.1958, *Robinson* 2860 (K; M; SRGH). **Zimbabwe.** N: Mazowe Distr., viii.1932, *Eyles* 7176 (K; SRGH). C: Harare, c. 1500 m., iii.1918, *Walters* in GHS 2482 (K; SRGH). E: Mutare, Gimbokke Farm, c. 920 m., 10.iii.1957, *Chase* 6358 (K; LISC; M; SRGH). **Malawi.** N: Mzimba Distr., Champira forest, c. 1680 m., 20.iv.1974, *Pawek* 8452 (K; MO). C: Ntchisi

For. Res., c. 1590 m., 26.iii.1970, *Brummitt* 9420 (K). S: Shire Highlands, 1887, *Last* s.n. (K).
Mozambique. N: Lichinga (Vila Cabral), v.1934, *Torre* 48 (COI). MS: Manica, between Mavita and Vila de Manica (Macequece), 7.iv.1948, *Barbosa* 1413 (LISC).
Also in Nigeria, Cameroons, Zaire, Angola and Tanzania. Miombo woodland or grassland.

51. **Vernonia kayuniana** G.V. Pope in Kew Bull. **45**: 697 (1990). Type: Malawi, Karonga Distr., Kayuni Hill, *Brummitt* 18496 (BR; K, holotype; LISC; MAL).

A low spreading perennial herb from a woody rootstock. Stems annual, several, decumbent, to c. 45 cm. long, branching towards the apex, densely leafy and mostly with a short leafy shoot in each axil, pubescent; indumentum, particularly on the lower stem, a mixture of long ± patent hairs and smaller appressed hairs. Leaves sessile, somewhat secund, 3–5.5 × 2–5 mm., linear-elliptic, margins entire, ± revolute, puberulous on both surfaces, the hairs flagelliform with an erect ± rigid base and a long erect or ± transverse terminal cell; leaves of axillary shoots less than half as long as the primary leaves and linear-oblanceolate. Capitula 1–6 on each stem, corymbiformly cymose. Involucres 7–9 mm. long, broadly campanulate. Phyllaries numerous, densely strigose-sericeous, increasing from c. 3 mm. long and flattened-subulate outside to c. 8 mm. long and lanceolate inside. Receptacle shallowly alveolate. Corollas creamy-white, 6–8 mm. long, very narrowly funnel-shaped, glandular and puberulous outside with short acicular hairs on the upper half of the lobes. Achenes c. 2 mm. long, subterete to narrowly turbinate, obscurely 4–5-angular or ribbed, uniformly setulose; outer pappus of persistent, 1–1.5 mm. long, fimbriate, linear-lanceolate scales; inner pappus of brownish, c. 6 mm. long, barbellate setae.

Malawi. N: Karonga Distr., Kayuni Hill, 33°39'E, 10°01'S, 1676 m., 11.vi.1989, *Brummitt* 18496 (BR; K; LISC; MAL).
As yet known only from this locality. Rocky ridges and hill tops above the tree line; 1676 m.

52. **Vernonia nestor** S. Moore in Journ. Linn. Soc., Bot. **35**: 317 (1902). —Brenan in Mem. N.Y. Bot. Gard. **8**, 5: 459 (1954). —Adams in F.W.T.A. ed. 2, **2**: 282 (1963). —Wild in Kirkia **11**: 76 (1978). —C. Jeffrey in Kew Bull. **43**: 244 (1988). Syntypes: Malawi, *Buchanan* 44 (BM; K), 129 (BM; MO).

A strict erect hispid-villous perennial herb, up to c. 90 cm. tall from a slender ± contorted woody rootstock. Stems annual, 1–several, branched at the apex, densely leafy; indumentum a mixture of ± appressed matted hairs, with a long contorted terminal cell on a short stalk, interspersed with large flagelliform patent-pilose hairs, with a bristle-like terminal cell on a simple many-celled stalk. Leaves sessile, stiff, scabrous on upper surface, sericeous-tomentose on lower surface; mid- and lower-cauline leaves up to 9.5 × 2 cm., narrowly oblong to oblanceolate-oblong, apex acute, base subcordate to semi-amplexicaul, margins subentire to remotely serrulate; lowermost leaves smaller and ovate; upper leaves progressively smaller becoming lanceolate and tapering to a narrowly acute apex. Capitula numerous, shortly-stalked, in clusters at the ends of the branches; stalks 0–10 mm. long, branches up to c. 20 cm. long, brownish hispid-tomentose. Involucres 7–10 mm. long, obconic to narrowly campanulate. Phyllaries extending briefly onto the capitulum stalk, progressively longer towards the innermost, up to c. 10 mm. long, oblong-oblanceolate, acuminate-apiculate, densely brownish- to silvery-sericeous. Corolla purple, up to c. 8 mm. long, widening gradually above. Achenes 1.5–2.5 mm. long, oblong-turbinate, narrowly 5–6-ribbed, uniformly strigulose-pubescent; outer pappus of short linear, acuminate scales, 1–1.5 mm. long; inner pappus of white, somewhat flattened, barbellate setae, 6.5–7 mm. long.

Zambia. N: Mbala Distr., Manyesi R., 1500 m., 12.v.1962, *Richards* 16443 (K; SRGH). W: Kitwe, 7.iv.1969, *Mutimushi* 3031 (K; NDO; SRGH). C: Kabwe Distr., c. 3 km. S. of Kapiri Mposhi, 12.vi.1960, *Leach & Brunton* 10008 (K; SRGH). E: Chipata, 21.ix.1966, *Mutimushi* 1450 (K). **Malawi**. N: Mzimba Distr., Mzuzu, 25.vi.1973, *Pawek* 6947 (K). C: Ntchisi Distr., Ntchisi Mt., 24.vii.1946, *Brass* 16894 (K; MO; SRGH). S: Ntcheu, Chilobwe, 19.vii.1971, *Salubeni* 1671 (K; MO; SRGH). **Mozambique**. N: Lichinga (Vila Cabral), 12.ix.1958, *Monteiro* 56 (LISC). T: Casula to Furancungo, 9.vii.1949, *Barbosa & Carvalho* 3534 (K).
Also in W. Africa and Tanzania. Miombo woodland. Often in sandy soils.

53. **Vernonia hirsuta** (DC.) Sch. Bip. ex Walp., Repert. Bot. Syst. **2**, Suppl. 1: 947 (1843). —Harv. in Harv. & Sond., F.C. **3**: 51 (1865). —Eyles in Trans. Roy. Soc. S. Afr. **5**: 503 (1916). —Mendonça, Contrib. Conhec. Fl. Angol., **1** Compositae: 14 (1943). —Hilliard in Ross, Fl. Natal: 356 (1972); Compos. Natal: 44 (1977). —Wild in Kirkia **11**: 85 (1978). —Hilliard & Burtt, Bot. S. Natal

Drakensb.: 218 (1987). Syntypes from South Africa.
  Webbia hirsuta DC., Prodr. **5**: 73 (1836). Syntypes as above.

An erect, sometimes tufted, ± densely hispid-tomentose perennial herb, up to c. 1 m. tall from a woody rootstock; indumentum containing Y-shaped hairs with widely spreading straight or ± curled arms on short simple stalks. Stems annual, 1–many, stout, simple, ribbed, leafy throughout, tomentose or pubescent, with few to many Y-shaped hairs particularly on the lower stem. Leaves sessile, up to c. 7.5(8) × 3.5(5) cm., oblong to lanceolate-oblong, the lower cauline often broadly oblong-elliptic, apex acute or acuminate, base cordate or subcordate, margin subentire to sparsely serrate, upper surface thinly pilose and scabridulous becoming somewhat bullate above, lower surface more densely pilose-tomentose and drying paler, nervation prominent beneath. Capitula numerous, in dense terminal clusters. Involucres up to c. 6 mm. long, campanulate. Phyllaries several-seriate, ± densely pilose to tomentellous, becoming purple at the apex; the outer subulate to linear-lanceolate; the inner longer and broader, oblong or usually wider above, ± abruptly contracted into a mucronate-aristate tip. Corollas purple, sometimes pinkish, 5–7 mm. long, narrowly funnel-shaped. Achenes 2–2.5 mm. long, subcylindric, obscurely c. 4-ribbed, densely long sericeous; outer pappus of very short, linear, scale-like setae; inner of copious, whitish, barbellate setae, 4–5 mm. long.

Zimbabwe. E: Chimanimani, Tarka For. Res., c. 1040 m., x.1968, *Goldsmith* 131/68 (COI; K; SRGH). **Mozambique.** MS: Chimanimani Mts., 6.vi.1949, *Munch* 171 (K; SRGH). M: Namaacha, 9.xii.1948, *Myre* 295 (K; SRGH).

Also in Zaïre, Angola and South Africa (Cape Province, Transvaal and Natal). Submontane grassland.

54. **Vernonia sutherlandii** Harv. in Harv. & Sond., F.C. **3**: 52 (1865). —Hilliard in Ross, Fl. Natal: 356 (1972); Compos. Natal: 45 (1977). —Wild in Kirkia **11**: 74 (1978). Type from South Africa (Natal).

An erect perennial herb, up to 40(60) cm. tall from a woody rootstock; indumentum containing Y-shaped hairs composed of a simple multicellular stalk bearing a terminal cell with widely spreading arms. Stems annual, 1–several, simple, ± densely leafy in the lower portion and almost leafless above, ribbed, coarsely pubescent to tomentose especially towards the base, indumentum interspersed with few many longer purplish hairs. Leaves up to c. 7 × 3 cm., elliptic-oblong, apex acute sometimes obtuse, base cuneate and ± petioled, margin coarsely remotely serrate to subentire, both surfaces densely to sparsely hispid, scabridulous, gland-pitted; lateral nerves prominent beneath; upper cauline leaves few and smaller. Capitula few to many, corymbiformly cymose or ± paniculate at the stem apex, capitula stalks up to c. 10 cm. long with 0–4 linear bracts. Involucres up to c. 10 mm. long, broadly obconic-campanulate. Phyllaries few-seriate, appressed pilose with longer Y-shaped hairs at the apex on the midrib; outer phyllaries few, up to c. 5 mm. long, linear-subulate; the inner phyllaries c. 10 mm. long, narrowly lanceolate, acuminate. Receptacle shallowly alveolate, ± toothed. Corollas purplish or violet, 8–10 mm. long, funnel-shaped above. Achenes c. 3 mm. long, subterete, densely white strigose, obscurely narrowly ribbed; outer pappus of short, setaceous scales; inner pappus of white, copious, slender, barbellate setae c. 6 mm. long.

Mozambique. M: Namaacha, Matianini, 7.ix.1955, *Lemos* 81 (LISC).

Also in South Africa (Natal, Transvaal) and Swaziland. In grassland and thornscrub usually in warm dry river valleys.

55. **Vernonia daphnifolia** O. Hoffm. in Bol. Soc. Brot. **13**: 18 (1896); in Warb., Kunene-Samb.-Exped. Baum: 403 (1903). —Hiern, Cat. Afr. Pl. Welw. **1**, 3: 519 (1898). —Mendonça, Contrib. Conhec. Fl. Angol., **1** Compositae: 13 (1943). —Wild in Kirkia **11**: 70 (1978). Syntypes from Angola.

An erect tufted perennial herb, up to 20(30) cm. tall from a woody rootstock. Stems annual, 1–many, simple, ± densely leafy in the lower two-thirds, ribbed, coarsely pilose-pubescent to tomentose; indumentum brownish strigose interspersed with long pilose hairs. Leaves sessile, 1–2(3) × 0.3(0.8) cm., narrowly oblong-lanceolate to linear, apex acute, base ± cuneate, densely pilose on both surfaces, gland-pitted; upper cauline leaves absent or 1–3, bract-like. Capitula mostly solitary and terminal on the stem, sometimes 2–3 in leaf axils below, these usually bud-like. Involucres up to c. 15 mm. long, broadly campanulate to widely spreading and hemispheric. Phyllaries many, strigose to glabrescent below and patent-pilose towards the apex; the outer c. 5–7 mm. long, subulate or narrowly lanceolate with an attenuate apex; the innermost up to c. 15 mm. long, lanceolate with an acuminate apex. Corollas pale-mauve, up to c. 12 mm. long, funnel-shaped. Achenes up to c. 4 mm. long, narrowly turbinate-cylindric, 4–5 ribbed, ± densely appressed-setulose, glandular; outer pappus of short, narrowly lanceolate scales; inner pappus of copious, brownish, barbellate setae, 8–10(12) mm. long.

Zambia. N: Mpika, Shiwa Ngandu, 1375 m., 31.viii.1938, *Greenway* 5645 (BR; EA; K; PRE).
Also in Angola and Zaire. High rainfall mixed deciduous woodland.

56. **Vernonia galpinii** Klatt in Bull. Herb. Boiss. **4**: 827 (1896). —Hilliard in Ross, Fl. Natal.: 356 (1972); Compos. Natal: 35 (1977). —Wild in Kirkia **11**: 100 (1978). Type from South Africa (Transvaal).
    Vernonia monocephala Harv. in Harv. & Sond., F.C. **3**: 53 (1865) nom illegit., non Gardn. (1847). —Wood & Evans, Natal Pl. **4**, 2: t. 331 (1904). —Eyles in Trans. Roy. Soc. S. Afr. **5**: 504 (1916). Syntypes from South Africa (Transvaal and Zululand).

An erect perennial herb, 8–45(60) cm. tall from a woody rootstock; roots long thong-like. Stems annual, 1–several, simple, densely appressed leafy, the leaves diminishing or absent towards the stem apex and base, coarsely pubescent to tomentose, the hairs curved-appressed flagelliform. Mid-cauline leaves subsessile, up to c. 5.5 × 2 cm., oblong-lanceolate, apex rounded to obtuse or acute mucronate, base broadly cuneate to sub-cordate, margins entire and scabrid-ciliate, lamina sparsely hispid-scabridulous on both surfaces, or glabrescent. Capitula solitary and terminal on the stem, or occasionally up to 3–4 clustered. Involucres up to c. 12 mm. long, broadly campanulate to widely spreading and hemispheric. Phyllaries few-seriate, subequal, c. 12 mm. long, linear-subulate to lorate-lanceolate, attenuate and purple-tipped, strigose or patent pilose especially on the midrib. Receptacle shallowly alveolate. Corollas purple, c. 9 mm. long. Achenes up to c. 4.5 mm. long, narrowly cylindric, narrowly 4–8-ribbed, glabrous or sparsely setulose, glandular; outer pappus of short setaceous scales; inner pappus of copious, white, barbellate setae, c. 6 mm. long.

Botswana. SE: Otse Mountain, c. 1485 m., 1.i.1979, *Woollard* 493 (K; SRGH). **Zimbabwe**. E: Gairesi Ranch, c. 10 km. N. of Troutbeck, c. 1740 m., 18.xi.1956, *Robinson* 1940 (K; LISC; SRGH). **Mozambique**. MS: Inyamatshira Mts. near Machipanda, 29.i.1950, *Chase* 1951 (BM; SRGH). M: Namaacha, 9.i.1947, *Barbosa* 80 (SRGH).
Also in South Africa (the Transvaal and Natal), Swaziland and the Transkei. Submontane and high plateau grassland.

57. **Vernonia fractiflexa** Wild in Kirkia **11**: 8; 71 (1978). Type: Malawi, Nyika Plateau, *Richards* 22687 (K, holotype).

An erect perennial herb to c. 28 cm. tall, from a woody rootstock. Stems annual, several, simple, ± zigzag, leafy, finely striate, thinly whitish-tomentose or pubescent, glandular, ± lanate at the rootcrown. Leaves sessile, to c. 5.5 × 1.4 cm., narrowly oblong to oblong-lanceolate or somewhat elliptic, apices obtuse or subacute, bases rounded to cuneate, margins entire to obscurely serrulate, gland-pitted and sparsely finely pilose on both surfaces, lateral veins almost parallel to midrib. Capitula solitary and terminal on stems; involucres to c. 15 × 25 mm., campanulate to hemispheric. Phyllaries many-seriate, acute, thinly sparsely pilose to pubescent and glandular outside, margins ciliate; the outer phyllaries from c. 5 mm. long and narrowly lanceolate, the inner to c. 15 mm. long and narrowly-lanceolate to strap-shaped. Corollas purple, 12–14 mm. long, narrowly funnel-shaped. Achenes stramineous, 4–5 mm. long, narrowly cylindric with c. 10 pronounced narrow ribs, sparsely puberulous and glandular between the ribs, or somewhat glabrescent; outer pappus of linear, scale-like, barbellate setae to c. 2 mm. long, inner of brownish, barbellate setae to c. 10 mm. long.

Malawi. N: Nyika Plateau, towards Nganda Peak, c. 2440 m., 9.i.1974, *Pawek* 7911 (K; MO; MAL; SRGH).
Not known elsewhere. Montane grassland.

58. **Vernonia mumpullensis** Hiern, Cat. Afr. Pl. Welw. **1**, 3: 530 (1898). Type from Angola.
    Vernonia milne-redheadii Wild in Kirkia **11**: 8; 70 (1978). Type: Zambia, Mwinilunga, *Milne-Redhead* 2783 (BR; K, holotype).

An erect perennial herb to c. 20 cm. tall, from a thick succulent brittle rootstock with tuberous roots. Stems annual, several, simple, sulcate, leafy, softly white-pilose to glabrescent, at first lanate at the base. Leaves sessile, to c. 7 × 1.2 cm., narrowly oblanceolate to narrowly oblong-oblanceolate, apices obtuse, bases tapering or narrowly cuneate, margins subentire to sparsely subserrate, finely white pilose-pubescent to glabrescent on both surfaces, minutely glandular-punctate; lateral veins arising at an acute angle and almost parallel to the midrib. Capitula solitary, rarely 2, terminal on the stem; involucres to c. 18 × 20–30 mm., widely obconic. Phyllaries many-seriate, chartaceous or thinly coriaceous, acute-acuminate, appressed-pubescent and glandular outside with ciliate margins, or glabrescent; outer phyllaries from c. 5 mm. long and narrowly lanceolate, middle phyllaries to c. 4 mm. wide and oblong-lanceolate; the inner phyllaries narrower, to c. 18 mm. long and strap-shaped. Corollas ?purple, to c. 11 mm. long, funnel-shaped. Achenes stramineous, to c. 9 mm. long, narrowly cylindric to subfusiform with c. 10 pronounced, narrow, wing-like ribs, glandular otherwise glabrous; outer pappus of scale-like barbellate setae to c. 1.5 mm. long, inner pappus of greenish-stramineous or sordid, barbellate setae to c. 11 mm. long.

Zambia. W: Mwinilunga Distr., Chibara's Plain, 14.x.1937, *Milne-Redhead* 2783 (K). Also in Angola. A pyrophyte of wooded grassland.

59. **Vernonia undulata** Oliv. & Hiern in F.T.A. **3**: 276 (1877). —Mendonça, Contrib. Conhec. Fl. Angol., **1** Compositae: 13 (1943). —Adams in F.W.T.A. ed. 2, **2**: 283 (1963). —Wild in Kirkia **11**: 81 (1978). —Maquet in Fl. Rwanda, Spermat. **3**: 558, fig. 168, 3 (1985). —C. Jeffrey in Kew Bull. **43**: 223 (1988). Type from Sudan.
   Vernonia dupuisii Klatt in Bull. Herb. Boiss. **4**: 825 (1896). Type from Zaire.

An erect, somewhat diffusely branched, perennial herb 30–100 cm. tall, from a woody rootstock; indumentum of ± unequal-armed, short- and long-stalked T-shaped hairs. Stems annual, purplish, 1–several, strict, branching above, leafy, ribbed, pubescent. Leaves ± appressed-ascending, shortly petiolate, up to 4.5(8) × 2(3.5) cm., ovate to lanceolate, somewhat acuminate and obtuse to acute mucronate at the apex, rounded to subcordate or cuneate at the base, margins undulate-repand or crenate, pubescent on both surfaces with T-shaped hairs; upper leaves diminishing in size. Capitula many, in corymbiform cymose clusters at the stem apex; stem branches up to c. 30 cm. long; capitula stalks 1–15 mm. long. Involucres 4–5 mm. long, campanulate, later spreading. Phyllaries c. 3-seriate, acuminate-cuspidate at the apex and narrowed to the base, pubescent, somewhat ciliate; the outer phyllaries from 1.5 mm. long, narrowly lanceolate or subulate; inner phyllaries up to c. 5 mm. long, oblong-oblanceolate. Corollas pinkish or pale-mauve, 5–8 mm. long, cylindric, tapering to the base. Achenes 2–2.5 mm. long, terete, not ribbed, narrowed below, somewhat curved, uniformly strigose; outer pappus of very short setiform scales; inner pappus of white, 4–5(6) mm. long, barbellate setae.

Zambia. W: Mwinilunga Distr., near Matonchi, 1200 m., 18.xi.1962, *Richards* 17280 (K; M). Also in the Sudan, W. and E. Africa, Rwanda, Zaire and Angola. Miombo woodland.

60. **Vernonia purpurea** Sch. Bip. ex Walp., Repert. Bot. Syst. **2**, Suppl. 1: 946 (1843). —Oliv. & Hiern in F.T.A. **3**: 281 (1877). —Adams in F.W.T.A. ed. 2, **2**: 280 (1963). —Wild in Kirkia **11**: 91 (1978). —Maquet in Fl. Rwanda, Spermat. **3**: 552 (1985). —C. Jeffrey in Kew Bull. **43**: 227 (1988). Type from Ethiopia.
   Vernonia inulifolia Steud. ex Walp., Repert. Bot. Syst. **2**: 946 (1843), "*inulaefolia*". —Oliv. & Hiern in F.T.A. **3**: 286 (1877). Type from Ethiopia.
   Vernonia rigorata S. Moore in Journ. Bot. **41**: 155 (1903). Type from Kenya.
   Vernonia scabrida C.H. Wright in Bull. Misc. Inf., Kew **1906**: 21 (1906). —Wild in Kirkia **11**: 91 (1978). Type: Malawi, Namasi, *Cameron* 41 (K, holotype).
   Vernonia duemmeri S. Moore in Journ. Bot. **52**: 91 (1914). Types from Uganda.
   Vernonia pascuosa S. Moore in Journ. Linn. Soc., Bot. **47**: 263 (1925). Type from Angola.
   Vernonia keniensis R.E. Fr. in Acta Hort. Berg. **9**: 114 (1929). Type from Kenya.

An erect slender scabrid perennial herb, 30–170 cm. tall, from a small woody rootstock with fibrous roots; indumentum of simple many-celled patent hairs, ± rigid when dry. Stems annual, usually solitary, branched above, leafy, strongly ribbed becoming ± angular above, pilose-scabrid, or glabrescent; branches short, stiffly ascending. Leaves subsessile, 4–15 × 0.3–4.5 cm., oblong-elliptic to narrowly ovate, or narrowly oblong to linear-lanceolate, apex acute-apiculate to tapering acuminate-aristate, base cuneate to narrowly rounded or subcordate, margins subentire to serrulate and somewhat revolute, upper surface scabrous, lower surface densely hispid-pubescent to sparsely pilose, or glabrescent, spreading hispid on prominent veins, paler and minutely gland-pitted beneath. Capitula few to numerous in a leafy terminal corymbiform cyme up to c. 40 cm.

long, capitula stalks mostly more than 5 cm. long, or capitula solitary and terminal on short synflorescence branches. Capitula subtended by few to many, foliaceous bracts which equal or exceed the involucre in length ± obscuring the phyllaries, bracts up to c. 3 × 1.5 cm. Involucre 12–18 × 15–30 mm., broadly campanulate to spreading. Phyllaries numerous, ± loosely arranged, the innermost 2–3 series subequal, pilose-araneose where exposed outside, hispidulous or often glabrescent below, ciliolate or cobwebby on upper margins; outer phyllaries from c. 3 mm. long, ovate-lanceolate, ± apiculate; middle phyllaries to c. 10 × 2–3 mm., oblong with pale broad scarious margins, pungent-aristate at the apex with aristae mostly 1–2 mm. long, occasionally to c. 7 mm. long; innermost phyllaries to c. 13 mm. long, narrower, subequal. Florets c. 80 per capitulum. Corollas purple or mauve, 9–12 mm. long, narrowly tubular, slightly expanded upwards, lobes setulose, or glandular only, at the apex. Achenes 2–3.2 mm. long, subcylindric with c. 10 narrow, raised ribs, strigose-hispid; outer pappus of short linear scale-like setae; inner pappus of sordid or brownish, barbellate, subterete setae 5–7 mm. long.

Zambia. N: Isoka, 14.viii.1965, *Fanshawe* 9269 (K; NDO; SRGH). W: Kitwe, 19.vii.1967, *Mutimushi* 1997 (K; M; NDO; SRGH). C: near Mumbwa, 1911, *Macaulay* 661 (K). E: Lundazi Distr., Lukusuzi Nat. Park, 12.iv.1971, *Sayer* 1145 (SRGH). S: Mapanza, 28.iii.1954, *Robinson* 639 (K; SRGH). Malawi. N: Chimaliro For. Res., c. 1400 m., 5.vii.1976, *Pawek* 11464 (K; M; MO; SRGH). C: Lilongwe Agric. Res. Sta., 1.iv.1955, *Jackson* 1553 (K; LISC; SRGH). S: Namwera Escarpment, Jalasi, c. 1120 m., 15.iii.1955, *Exell, Mendonça & Wild* 903 (BM; LISC; SRGH). Mozambique. N: Belém (Amaramba), Cuamba (Nova Freixo)-Mecanhelas, 17.ii.1964, *Torre & Paiva* 10631 (K; LISC; LMU; C). Z: between Lugela (Muobede) and Tacuane, 24.v.1949, *Barbosa & Carvalho* 2841 (K; SRGH). T: Angónia, monte Dómuè, c. 1450 m., 9.iii.1964, *Torre & Paiva* 11080 (LISC).

Also in Zaire and tropical W. Africa, Sudan, Ethiopia, Rwanda and tropical E. Africa. Dambo margins, miombo woodlands, riversides and termitaria.

A particularly harsh and narrow-leaved variant of this species from central Malawi was described as *V. scabrida* by C.H. Wright.

61. **Vernonia amblyolepis** Bak. in Bull. Misc. Inf., Kew **1898**: 146 (1898). —Wild in Kirkia **11**: 98 (1978). —C. Jeffrey in Kew Bull. **43**: 228 (1988). Syntypes: Malawi, Nyika Plateau, *Whyte* 204 (K); between Mpata and commencement of Tanganyika Plateau, *Whyte* s.n. (K).

An erect slender stiff scabrid perennial herb, up to 1.7 m. tall, from a small woody rootstock with long fibrous roots; indumentum of simple many celled hairs, rigid when dry. Stems annual, 1–several, branched above, leafy, ribbed becoming sulcate to somewhat angular above, scabrous-pubescent, or glabrescent; branches up to c. 20 cm. long, stiffly ascending. Leaves ± petiolate; petiole 0–3 mm. long; lamina 3–12 × 0.3–3 cm., ± oblong-lanceolate, sometimes linear-lanceolate, apex acute-apiculate to acuminate-aristate, base narrowly rounded to subcordate, margins subentire, somewhat revolute, upper surface ± scabrous, lower surface densely hispidulous to sparsely pilose, or glabrescent, spreading hispid on prominent veins, paler beneath. Capitula numerous, ± shortly-stalked, in ± dense clusters at the ends of bracteate synflorescence branches 2–15 cm. long, or sometimes fewer more laxly arranged; bracts similar to leaves but smaller; capitula occasionally subtended by short narrow bracts. Involucre 5–10 × 4–15 mm., campanulate or narrowly obovoid to spreading obconic, subtending bracts absent, or if present narrow and mostly less than 5 mm. wide. Phyllaries appressed-imbricate at first, progressively longer to the inside, mostly less than 2 mm. wide, all araneose-woolly where exposed outside, hispidulous or glabrescent below, long ciliolate or cobwebby on upper margins, apices purple squarrose sometimes ± expanded (at least in the inner series); outer phyllaries subulate to lanceolate-ovate, pungent-aristate; inner phyllaries up to c. 10 mm. long, oblong-lanceolate or linear, abruptly narrowed to a pungent-aristate showy purplish apex. Florets up to c. 50 per capitulum. Corollas purple, 6–9 mm. long, tubular, slightly expanded upwards, lobes sparsely shortly setulose or glandular only. Achenes 2–3.2 mm. long, sybcylindric with c. 10 narrow raised, somewhat obscured ribs, densely strigose-hispid; outer pappus of very short linear scale-like setae; inner of sordid, or brownish barbellate subterete setae c. 6 mm. long.

Zambia. N: Nakonde, Old Fife Coffee Res. Sta., c. 1850 m., 22.iv.1986, *Philcox, Pope & Chisumpa* 10073 (K; MO; NDO; SRGH). Malawi. N: Nyika Plateau, c. 8 km. NW. of Lake Kaulime, c. 2020 m., 16.v.1970, *Brummitt* 10826 (K; LISC; MAL; SRGH). S: Zomba Plateau near Kuchawe Inn, 1530 m., 14.iii.1970, *Brummitt* 9086 (K; LISC; MAL; SRGH).

Also in Tanzania. Submontane grassland and moist places beside streams and at lower altitudes in dambos in high rainfall miombo woodland.

Three distinctive variants of this species can be recognised in the Flora Zambesiaca area; the "NE.

Zambian variant", remarkable for its bristle-tipped phyllaries in which the midribs are produced into 3–7 mm. long, curved apical bristles (*Chisumpa* 76 (K); *Philcox, Pope & Chisumpa* 10073 (K; MO; NDO; SRGH)). The "Nyika variant" (*amblyolepis* sens. str.) with numerous small, very shortly-stalked capitula ± densely clustered at the ends of the branches (*Brummitt* 10826 (K)). The "Zomba variant" with few to many heads on longer stalks and more loosely clustered than in the Nyika variant (*Brummitt* 9086 (K)).

62. **Vernonia ugandensis** S. Moore in Journ. Linn. Soc., Bot. **35**: 314 (1902). —C. Jeffrey in Kew Bull. **43**: 245 (1988). TAB. **19** fig. A. Type from Uganda.

    Vernonia caput-medusae S. Moore in Journ. Linn. Soc., Bot. **37**: 166 (1905). Type from Uganda.
    Vernonia fontinalis S. Moore in Journ. Bot. **52**: 90 (1914). —Mendonça, Contrib. Conhec. Fl. Angol., 1 Compositae: 12 (1943). —Wild in Kirkia **11**: 60 (1978). Type from Angola.
    Vernonia punctulata De Wild. in Bull. Jard. Bot. Brux. **5**: 100 (1915). Type from Zaire.
    Vernonia proclivicola S. Moore in Journ. Linn. Soc., Bot. **47**: 262 (1925). Type from Zaire.
    Vernonia mgetae Gilli in Ann. Naturhist. Mus. Wien **78**: 164 (1974). Type from Tanzania.
    Vernonia barbosae Wild in Kirkia **11**: 6, figs. 3, 4; 62 (1978). Type: Mozambique, Zambezia, between Mocuba and Regulo Namabida, *Barbosa & Carvalho* 2963 (K; SRGH, holotype).

A robust perennial herb to c. 80 cm. tall, from a thickened vertical rootstock with numerous fibrous roots. Stems erect, branching above, leafy, ribbed, ± lanate or softly appressed-pubescent, glabrescent below; indumentum a mixture of appressed straight hairs and few to numerous long patent many-celled hairs. Basal leaves usually subrosulate, up to c. 16 × 4 cm., oblanceolate, apex obtuse to rounded, base cuneate or attenuate, margins subentire or serrulate, lamina sparsely pilose to puberulous or glabrescent; cauline leaves subsessile, oblanceolate becoming oblong-lanceolate, diminishing in size upwards; upper leaves broadly lanceolate to ovate-acuminate, sometimes subtending and overtopping the capitula. Capitula 3–numerous rarely solitary, cymosely arranged, central capitulum shortly-stalked and overtopped by lateral capitula, sometimes capitula densely clustered; capitula stalks to 0–40 mm. long, lanate or appressed pilose-sericeous. Involucres to 12 mm. long, mostly broader than long when mature, broadly campanulate or spreading obconic. Phyllaries numerous, appressed to somewhat spreading, often pungent purple-tipped, densely sericeous below to coarsely hirsute-lanate and ciliate towards the apex; the outer from c. 4 mm. long, narrowly lanceolate-attenuate; the inner increasing to c. 12 mm. long, narrowly lanceolate, tapering to an acuminate apex, sometimes sinuate-subulate. Receptacle deeply alveolate, alveolae walls produced upwards at the corners into teeth often equalling achene in length. Corolla purple to whitish, c. 9 mm. long, narrowly funnel-shaped; lobes narrow acute, 1.5 mm. long, strigillose at tips outside, minutely glandular. Achenes c. 2 mm. long, turbinate, 4–5-ribbed, densely hispid at the base, setulose on the ribs, horizontally wrinkled and glandular between ribs; outer pappus of numerous, overlapping, hyaline-stramineous, lanceolate-acuminate scales ± equalling achene in length, inner pappus of pale-brownish or white, rarely purplish, barbellate setae, c. 5.5 mm. long.

Zambia. B: c. 77 km. from Zambezi (Balovale) on Kabompo Road, 26.iii.1961, *Drummond & Rutherford-Smith* 7368 (K; M; SRGH). N: Mafinga Mts., 24.v.1973, *Fanshawe* 11975 (NDO; SRGH). W: Mufulira, Njiri, 12.iii.1964, *Mutimushi* 707 (K; NDO; SRGH). C: Serenje Distr., Kaombi, c. 1370 m., iv.1930, *Lloyd* s.n. (BM). E: Katete, c. 1070 m., 29.i.1957, *Wright* 136 (K). **Malawi**. N: Chitipa Distr., Mafinga Mts., 24.ii.1970, *Drummond & Williamson* 9874 (SRGH). C: Dzalanyama For. Res., near Chionjeza (Chiunjiza), 9.ii.1959, *Robson* 1535 (BM; K). S: Zomba Plateau, 6.vi.1946, *Brass* 16274 (K). **Mozambique**. N: Unango to L. Shire, 1899, *Johnson* 176 (K). Z: Mocuba, c. 30 km. on Maganja da Costa road, 8.xi.1966, *Torre & Correia* 14487 (K; LISC; LMU).

Also in Uganda, Burundi, Tanzania, Zaire and Angola. Dambos, grassland, miombo and mixed woodland.

A variable species in which the capitula are usually laxly arranged, with or without small subtending bracts. Variation seen throughout its range of distribution is towards a plant in which the capitula are shortly-stalked and clustered, usually with large foliaceous subtending bracts. In extreme cases these bracts partly envelop the capitula. Such plants were described as *V. barbosae* by Wild. In addition the phyllaries exhibit a gradation from the appressed straight bracts characteristic for the Flora Zambesiaca area to the longer more sinuate-subulate phyllaries of typical *V. ugandensis* further north. A plant of this shaggy-headed variant in which the capitula are tightly clustered was described by S. Moore as *V. caput-medusae*.

63. **Vernonia griseopapposa** G.V. Pope in Kew Bull. **43**: 282 (1988). Type: Zambia, Nyika Plateau, c. 0.5 km. SW. of Govt. Rest House, *Philcox, Pope & Chisumpa* 9967 (K, holotype; BR; GA; LISC; M; MO; NDO; SRGH; UPS; WAG).

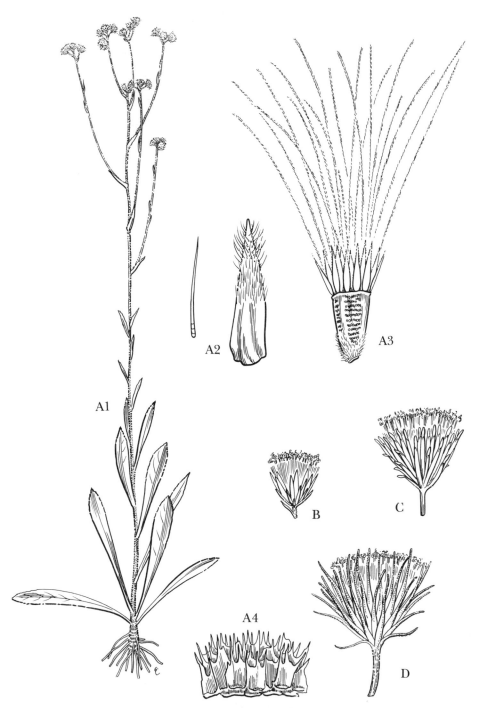

Tab. 19. A. —VERNONIA UGANDENSIS. A1, habit (× $\frac{1}{8}$); A2, phyllary (× 5) and phyllary seta (× 32); A3, achene (× 10); A4, part of receptacle showing toothed, alveolae walls (× 10), A1–A4 from *Brass* 16274. B. —VERNONIA IANTHINA, capitulum (× 1$\frac{1}{2}$), from *Fanshawe* 1610. C. — VERNONIA ROBINSONII, capitulum, phyllaries ± appendaged (× 1$\frac{1}{2}$), from *Milne-Redhead* 4495. D. —VERNONIA VIOLACEOPAPPOSA subsp. NUTTII, capitulum (× 1$\frac{1}{2}$), from *Richards* 8207. Drawn by Eleanor Catherine.

A perennial herb up to c. 40 cm. tall, from a thickened vertical rootstock with numerous fibrous roots. Stems erect, densely strigose or tomentose, sometimes interspersed with a few brownish patent multicellular shaggy hairs on the capitula stalks. Basal leaves crowded or subrosulate, up to c. 12.5 × 1.8 (14 × 2.5) cm., linear-oblanceolate, apex acute, widest above the middle and tapering gradually to a sessile base, margins subentire to serrulate, lamina membranous, sparsely hispid or glabrous, gland-pitted; cauline leaves sometimes larger than the basal leaves but decrease in size towards the stem apex becoming lorate or lanceolate; the uppermost leaves attenuate at the apex and somewhat conduplicate below. Capitula 1–2 at the ends of branches, or ± cymose at the stem apex, the central capitulum overtopped by lateral capitula. Involucres up to c. 10 mm. long, broadly campanulate to widely-spreading. Phyllaries numerous, pungent-tipped and spreading-pilose to sericeous, ciliate on margins where exposed; the outer phyllaries c. 4 mm. long, linear-lanceolate; the inner 7–10 mm. long, lanceolate. Receptacle deeply alveolate, alveolae walls produced into teeth often equalling achene in length. Corollas pale-mauve or whitish, c. 8 mm. long, narrowly infundibuliform, hispidulous on the lobes, glandular. Achenes 1.5–2 mm. long, turbinate, 4–5-ribbed, hispid, densely so at the base and often on the ribs, horizontally wrinkled between the ribs; outer pappus of numerous grey overlapping narrowly lanceolate scales as long as or shorter than the achene body; inner pappus of grey barbellate setae up to c. 6 mm. long.

Malawi. N: Rumphi Distr., Nyika Plateau, c. 2200 m., 11.iv.1969, *Pawek* 2105 (K).
Also in Burundi. Montane grassland usually over 2000 m.

64. **Vernonia rubens** Wild in Kirkia **11**: 5; 61 (1978). Type: Malawi, Mafinga Hills, *Tyrer* 627 (BM; BR; SRGH, holotype).

A stout erect perennial herb up to c. 30 cm. tall, from a thickened rootstock with spreading fibrous roots. Stems densely pale-villous. Leaves sessile, the lower cauline leaves crowded, the lowermost radical, c. 6 × 0.8 cm., narrowly lorate-oblanceolate or linear, apex acute, base widening and semi-amplexicaul, pilose on both surfaces, becoming scabridulous; upper cauline leaves diminishing in size up the stem, lanceolate to very narrowly triangular, the apices tapering acute, bases ± stem-clasping, densely pilose. Capitula 1–9, crowded at the stem apex, ± cymosely arranged, the central capitulum overtopped by the lateral capitula; stalks stout and shortly villous. Involucre up to c. 15 mm. long, broadly campanulate to spreading. Phyllaries very numerous, increasing in length towards the inside, linear-lanceolate with dark subulate tips, densely pale grey-villous. Receptacle alveolate, alveolae walls long-toothed. Corollas mauve, c. 10 mm. long, very narrowly funnel-shaped. Achene pale-brown, c. 2 mm. long, oblong-turbinate, 4–5-ribbed, densely setose at the base and along the ribs; outer pappus of narrow scale-like setae, c. 2.5 mm. long; the inner pappus of barbellate setae c. 9 mm. long, rose-coloured in the upper part.

Malawi. N: Mafinga Hills, c. 4.8 km. W. of Chisenga, 26.viii.1962, *Tyrer* 627 (BM; BR; SRGH).
Known only from this specimen. Exposed mountain ridge in skeletal soil at c. 2100 m.

65. **Vernonia violaceopapposa** De Wild. in Bull. Jard. Bot. Brux. **5**: 89 (1915). —C. Jeffrey in Kew Bull. **43**: 245 (1988). Type from Zaire.

An erect perennial herb up to 55 cm. tall, from a thickened vertical rootstock with numerous spreading fibrous roots. Stems solitary, branched above; indumentum a mixture of appressed-sericeous hairs and ± numerous patent many-celled hairs, or glabrescent. Lower leaves ± crowded, radical or basal, sometimes withering early, up to c. 18 × 3 cm., narrowly obovate to oblanceolate, obtuse to acute at the apex, subentire to remotely serrate, narrowly cuneate or attenuate to the base, petiolate, sparsely pilose or glabrous; cauline leaves decreasing in size upwards, oblong-lorate to narrowly lanceolate, acute, ± abruptly narrowed to a sessile base; the uppermost leaves narrowly ovate-lanceolate, subtending and conduplicate about the synflorescence branches. Capitula few to many in a terminal stiff ± cymose arrangement, the central capitulum short-stalked and overtopped by the lateral capitula. Involucres 12–15 mm. long, campanulate, later spreading. Phyllaries dark-purple or at least the apex purple, rarely green, extending onto the capitulum stalk and becoming somewhat foliaceous, narrowly lanceolate to linear, tapering to an attenuate-subulate apex, obscurely longitudinally ribbed, ± appressed sericeous or glabrescent below becoming spreading pilose above; the outer phyllaries from c. 6 mm. long; the inner phyllaries increasing to 10–13 mm. long and overtopping the pappus at anthesis. Receptacle alveolate, alveolae walls shallowly dentate. Corollas purple or brownish, 9–11 mm. long, tapering to base, lobes puberulent. Achenes 1.5–2.5 mm. long, turbinate, c. 5-ribbed, horizontally ridged and glandular between the ribs, villous at the base, ± setulose on the ribs; outer pappus of lanceolate-acuminate scales ± equalling the achene in length; inner pappus of barbellate setae 6–9 mm. long, purple, brownish-purple or sordid.

Subsp. **nuttii** G.V. Pope in Kew Bull. **43**: 283 (1988). —C. Jeffrey in Kew Bull. **43**: 245 (1988). TAB. **19** fig. D. Type: Zambia, Mbala, *Bullock* 2635 (K, holotype; LISC).

Upper cauline leaves narrowly ovate-lanceolate, subtending and conduplicate about the synflorescence branches; phyllaries dark-purple, subulate to tapering acuminate in the upper part, ± sparsely patent-pilose.

Zambia. N: Mbala Distr., Ndundu, 1500 m., 16.ii.1957, *Richards* 8207 (K).
Also in Tanzania. In sandy soil in open deciduous woodland, tall grassland and dambos.
Subsp. *violaceopapposa* is confined to Zaire.

66. **Vernonia isoetifolia** Wild in Kirkia **11**: 7; 71 (1978). Type: Zambia, Kambole-Mbala road, *Richards* 18919 (K, holotype; M).

An erect perennial herb up to c. 20 cm. tall, from a stout vertical rootstock with fibrous lateral roots. Stems purple, at least below, solitary, unbranched, with radical leaves exceeding the stems in length, pilose with white multiseptate hairs, or glabrescent. Leaves both basal and cauline; basal leaves up to 23 × 0.8 cm., linear-oblanceolate, subacute or obtuse at the apex, cuneate-attenuate towards the base there widening briefly and stem-clasping, margins entire or obscurely remotely serrulate, lamina glabrous; cauline leaves sessile, narrowly strap-shaped to linear-lanceolate, those subtending the capitula stalks conduplicate and usually overtopping the capitula. Capitula up to c. 10, ± cymose at the stem apex, less often solitary, stalks 1–6.5 cm. long. Involucre up to c. 14 mm. long, broadly obconic to hemispheric. Phyllaries numerous, ± narrowly lanceolate and attenuate apically, sparsely pilose-ciliate at the apex otherwise glabrous, increasing in length towards the inside; the inner phyllaries up to c. 12 mm. long and purple at the apex. Receptacle alveolate, alveolae walls toothed. Corolla purple, c. 10 mm. long, funnel-shaped with lobes c. 2 mm. long and sparsely bristled at the apex. Achenes (immature) c. 2 mm. long, 4–5-ribbed, rugose between the ribs, densely strigose at the base and often along the ribs; outer pappus of brownish overlapping lanceolate scales 1.5–2 mm. long, inner pappus of purple barbellate setae c. 7 mm. long.

Zambia. N: Mbala, Old Katwe road, c. 1520 m., 28.ii.1955, *Richards* 1719 (K).
Known so far only from around Mbala. Moist sandy grassland.

67. **Vernonia robinsonii** Wild in Kirkia **11**: 5, fig. 2; 68 (1978). TAB. **19** fig. C. Type: Zambia, 100 km. W. of Solwezi, *Robinson* 3547 (K; M; SRGH, holotype).
　　　Vernonia lawalreeana Kalanda in Bull. Jard. Bot. Nat. Belg. **52**: 122 (1982). Type from Zaire.
　　　Vernonia malaissei Kalanda in Bull. Jard. Bot. Nat. Belg. **52**: 124 (1982). Type from Zaire.

An erect slender perennial herb up to 40 cm. tall from a thickened rootstock, lanate at the rootcrown. Stems purplish, 1–many, simple, ribbed, swelling somewhat at the base, appressed-pubescent or glabrescent. Leaves often purple-tinged, ± coriaceous when dry, midrib prominent beneath; basal leaves subrosulate or crowded, up to c. 8.5 × 0.5 cm., linear-oblanceolate, apex acute, base long attenuate, margins entire and ± revolute, midrib glabrous or pubescent; upper cauline leaves decreasing in size towards the stem apex, narrowly strap-shaped. Capitula 1, or up to c. 8 corymbiformly cymose on stalks 1–6 cm. long; involucre c. 13 mm. long, broadly obconic. Phyllaries purplish, numerous, increasing in length towards the inside, coriaceous, pilose or araneose above or glabrescent; outer phyllaries recurved, subulate; inner phyllaries up to c. 13 mm. long, linear-lanceolate, ± apiculate. Receptacle deeply alveolate. Corolla purple or violet, c. 10 mm. long, pubescent outside; lobes 1–2 mm. long, linear, acute, strigillous near the tip. Achenes 1.5–2.3 mm. long, oblong-turbinate, 4–5-angular, ribbed on the angles, densely strigose on the ribs, rugulose on the faces; outer pappus of broad lanceolate-acuminate brownish scales c. 2 mm. long; the inner of subplumose, sordid or purple-tinged setae c. 7 mm. long.

Zambia. W: Mwinilunga Distr., lake 2 km. on Kanyama Road, 25.ii.1975, *Hooper & Townsend* 367 (K; SRGH).

Also in Zaire. Moist sandy grassland and dambos in Kalahari Sand.

68. **Vernonia viatorum** S. Moore in Journ. Linn. Soc., Bot. **35**: 315 (1902). —Wild in Kirkia **11**: 63 (1978). Type from Stevenson Road between L. Malawi and L. Tanganyika, *Scott-Elliot* 8271 (BM, holotype; K).

A low-growing perennial herb, 10–30 cm. tall from a slender woody rootstock. Stems annual 1–many wiry, usually branching above, leafy, pubescent. Leaves sessile, 1.5–4 × 0.3–0.7 cm., oblanceolate, acute or obtuse at the apex, tapering to the base, entire or remotely serrulate, sparsely puberulous and gland-pitted; upper cauline leaves decreasing in size and becoming strap-shaped. Capitula few to numerous, corymbiformly cymose, stalks slender up to c. 8 cm. long. Involucres up to c. 6 mm. long, obovoid to broadly campanulate when older. Phyllaries ± lanceolate, the outer from c. 2 mm. long, the inner up to c. 6 mm. long, purple-tipped, pubescent. Receptacle shallowly alveolate. Corolla purple, c. 6 mm. long, narrowly funnel-shaped. Achenes c. 1.5 mm. long, turbinate, 4–5-angled or -ribbed, hispidulous, especially on the ribs, rugulose between the ribs; outer pappus of purple-tinged or sordid, overlapping, lanceolate-acuminate scales usually shorter than the achene; inner setae usually purple-tinged, 3.5–4 mm. long, barbellate.

Zambia. N: Mbala, 10.vii.1960, *Robinson* 3775 (K; SRGH).
Not known outside the Flora Zambesiaca area. A pyrophyte of short dry grassland often with scattered trees.

69. **Vernonia ianthina** Muschl. in Engl., Bot. Jahrb. **46**: 93 (1911). —Mendonça, Contrib. Conhec. Fl. Angol., **1** Compositae: 13 (1943). —Wild in Kirkia **11**: 68 (1978). TAB. **19** fig. B. Type from Angola.
   Vernonia leptoblasta Wild in Kirkia **11**: 6; 69 (1978). Type: Zambia, Mwinilunga, *Drummond & Williamson* 9341 (SRGH, holotype).

A slender perennial herb up to c. 30 cm. tall, from a vertical woody rootstock with numerous thong-like roots; rootcrowns lanate. Stems annual, 1–several, erect wiry and somewhat flexuous, simple or sparingly branched, more densely leafy at the base, pilose; branches up to c. 12 cm. long, ascending, straight, slender. Basal leaves few to numerous, densely clustered, up to c. 8 × 1 cm., obovate-oblanceolate to linear-oblanceolate, apices rounded to obtuse, tapering-attenuate below before widening abruptly into overlapping leaf bases, margins entire or somewhat crispate, upper surface puberulous to glabrescent, lower surface thinly pilose especially on midrib; cauline leaves few, widely spaced, sessile, c. 3.5 cm. long, narrowly oblanceolate, decreasing in size to the stem apex. Capitula 3–many, corymbiformly cymose, sometimes solitary, stalks up to c. 5 cm. long. Involucres up to 5 mm. long, broadly campanulate-spreading; phyllaries ± lanceolate, progressively longer towards the inside, mucronulate at apex, somewhat obscurely longitudinally 3-ribbed, puberulous to glabrescent and glandular outside. Receptacle convex, shallowly alveolate. Corolla purple, 7–9 mm. long, narrowly funnel-shaped, lobes 1.5 mm. long. Achenes c. 1.5 mm. long, turbinate, narrowly c. 5-ribbed, setulose particularly at the base and on the ribs, rugulose between the ribs; outer pappus of membranous, oblong-lanceolate, ± acuminate scales equalling or exceeding the achene in length; inner pappus of purple or whitish, barbellate setae up to c. 7 mm. long.

Zambia. W: Mwinilunga Distr., c. 26 km. W. of Kabompo R., 11.ix.1930, *Milne-Redhead* 1111 (K).
Also in Zaire and Angola. A pyrophyte of dambo margins and moist sandy localities.

70. **Vernonia milanjiana** S. Moore in Journ. Linn. Soc., Bot. **35**: 318 (1902). —Wild in Kirkia **11**: 87 (1978). Type: Malawi, Mt. Mulanje, *Whyte* 194 (BM, holotype).

An erect subscapose perennial herb, up to c. 30 cm. tall from a slender vertical woody rootstock with many lateral roots. Stems scapiform, annual, very densely golden-strigose above, less densely so below, the hairs T-shaped, leaves mostly subrosulate or crowded near the base, with 1–3 leaves cauline. Basal leaves 3–12 × 0.8–3.5 cm., obovate to oblanceolate, apex subacute to obtuse, sometimes rounded, base narrowly cuneate, gently tapering into a short petiole, margins subentire to serrate-crenate, lamina thinly pubescent above, less densely so below; upper leaves smaller and narrower. Capitula 3–11 in a ± congested terminal cluster; stalks 1–8 mm. long, densely strigose. Involucres 5–6 × 5 mm., campanulate; phyllaries several-seriate, remaining ± appressed-imbricate, purplish-tipped and thinly puberulent, the outer c. 2 mm. long and subulate to linear-lanceolate, the inner up to c. 5 mm. long and narrowly oblong-lanceolate, acuminate. Corollas purple, c. 7 mm. long, narrowly funnel-shaped. Achenes up to c. 3 mm. long, narrowly subcylindric, densely hispidulous; outer pappus reduced to a very short fringe of linear scales, inner pappus of copious setae 4–5 mm. long.

Malawi. S: Mt. Mulanje, Ruo Basin, c. 2130 m., 7.iv.1970, *Brummitt* 9704 (K).
Endemic to Mt. Mulanje. Montane grassland.

71. **Vernonia dewildemaniana** Muschl. in Fedde, Repert. **9**: 384 (1911). Type from Zaire.
    Vernonia verdickii O. Hoffm. & Muschl. in Engl. Bot. Jahrb. **46**: 93 (1911) non O. Hoffm. (1902). Type as above.
    Vernonia entohylea Wild in Kirkia **11**: 12, fig. 7; 102 (1978). Type: Zambia, Mwinilunga, *Drummond* 8363 (BR; K; LISC; SRGH, holotype).

A robust erect perennial herb to c. 85 cm. tall from a woody rootstock, often flowering before, or with the young leaves. Stems annual scapiform, usually solitary, simple below the much branched synflorescence, with large leaves arising at or near the base and a few much smaller leaves on the stem above these, obscurely few-ribbed, densely brownish-tomentose. Leaves sessile; basal leaves ascending, up to c. 30 × 8 cm., oblanceolate to elliptic-oblanceolate, apex subacute to obtuse or rounded, gradually tapering in the lower half to a very narrowly winged midrib before widening to a ± stem-clasping base, margins serrate-dentate to subentire, upper surface thinly puberulous or glabrescent, lower surface pubescent to glabrescent and glandular-punctate, midrib large prominent and ± tomentose beneath; upper cauline leaves c. 10 × 2 cm., narrowly oblong-elliptic, decreasing in size upwards. Capitula numerous, in terminal clusters in a paniculate arrangement up to c. 25 cm. across, synflorescence branches up to c. 25 cm. long, tomentose, capitula stalks 0.1–7 cm. long. Involucres 10–20 × 12–20 mm., obconic-campanulate, spreading. Phyllaries ± loosely arranged, c. 4-seriate, extending briefly onto the capitulum stalk, longitudinally ribbed, tomentose where exposed, purple-tipped; outer phyllaries from c. 2 mm. long and subulate; middle phyllaries ovate-elliptic and ± recurved acuminate-attenuate at the apex; inner phyllaries narrower, up to c. 14 mm. long, elliptic-linear, tapering to a finely pointed, ± curved, 1–5 mm. long tip. Receptacle shallowly alveolate. Corollas purple, 7–11 mm. long, narrowly cylindric, widening slightly above, lobes with few to many sericeous hairs c. 1 mm. long. Achenes pale-brown or stramineous, 3–4 mm. long, narrowly subcylindric-turbinate, somewhat 4–5-angular, sparsely glandular puberulous; outer pappus short, consisting of few 1–2 mm. long linear scales or paleaceous setae, inner pappus of whitish or sordid, barbellate to subplumose bristles 6–9 mm. long.

Zambia. N: Mbala Distr., Kalambo Falls, c. 1200 m., 22.vii.1960, *Richards* 12901 (K). W: Mwinilunga Distr. near Zambezi River, above rapids, c. 9 km. NE. of Kalene Hill Mission, 16.vi.1963, *Drummond* 8363 (BR; K; LISC; SRGH).
Also in Zaire. In dry deciduous miombo or *Uapaca-Protea* woodland.

72. **Vernonia praemorsa** Muschl. in Engl., Bot. Jahrb. **46**: 68 (1911). —Earle Sm. in U.S. Dept. Agric. Handb. **396**: 69 (1971). —Wild in Kirkia **11**: 108 (1978). —C. Jeffrey in Kew Bull. **43**: 229 (1988). TAB. **20** fig. C. Type from Tanzania.
    Vernonia castellana S. Moore in Journ. Bot. **52**: 95 (1914). —Mendonça, Contrib. Conhec. Fl. Angol., **1** Compositae: 29 (1943). Type from Angola.
    Vernonia kuluina S. Moore in Journ. Bot. **56**: 209 (1918). —Mendonça, Contrib. Conhec. Fl. Angol., **1** Compositae: 29 (1943). Type from Angola.

An erect perennial herb, 10–70 cm. tall from a woody rootstock; rootcrowns densely brown-lanate, roots thong-like, spreading. Stems annual, ± scapiform, 1–several, simple, occasionally sparingly branched at the apex, leafy at the base and sparsely leafy above, ribbed, densely appressed tomentose or pubescent; indumentum of ± unequal-armed T-shaped hairs. Leaves appearing before, or with the flowers, mostly subrosulate, ascending, 2–8 basal or ± radical, up to c. 27 × 10 cm., ± spathulate, being ovate to ± broadly elliptic in the upper half and abruptly narrowed about the middle into a narrowly-winged petiole-like midrib to c. 10 mm. long, apex ± obtuse, margin undulate-dentate to serrate, upper surface ± pubescent, sometimes glabrous, lower surface more densely pubescent particularly on the main nerves, the hairs long-stalked T-shaped; nervation ± prominently reticulate beneath; midribs becoming rigid towards the base; cauline leaves 0–4 similar to, but smaller than, the basal leaves. Capitula numerous in a dense terminal cluster 2.5–6 cm. across; stalks 2–15 mm. long, tomentose, often bracteate, bracts up to

c. 10 mm. long, linear; involucres 8–10 × 10–14 mm., campanulate. Phyllaries green with a narrow, membranous, hyaline or purple margin, not expanded at the apex, imbricate, progressively longer to the inside; the outer from c. 3 mm. long and oblong-elliptic to oblong-obovate, obtuse-rounded at the apex; the inner phyllaries up to c. 10 mm. long, linear-oblong and tapering purplish, undulate-membranous above, sparsely hispidulous-pubescent and glandular outside, the hairs whip-like often purple. Corollas purple to mauve, 6–10 mm. long, narrowly tubular, widening above, lobes bristled at the apex. Achenes 3–3.5 mm. long, subcylindric, ± obscurely 8–10-ribbed, hispidulous, glandular; outer pappus of short, linear, acuminate scales; inner pappus of sordid or brownish barbellate setae 5.5–7.5 mm. long.

Zambia. N: Mbala, Uningi Pans, c. 1500 m., 19.xii.1959, *Richards* 11993 (K; M; SRGH). W: Kitwe, Baluba R., 2.xii.1963, *Fanshawe* 8166 (K; M; NDO; SRGH). E: Makutus Mts., 28.x.1972, *Fanshawe* 11619 (K; NDO). Malawi. N: Nyika Plateau, Buma area, road to Rukuru Bridge, c. 2250 m., 20.xi.1967, *Richards* 22638 (K; M; SRGH).
Also in Tanzania, Zaire and Angola. In dambo and submontane grassland and in high rainfall miombo woodland.

73. **Vernonia praecox** Welw. ex O. Hoffm. in Bol. Soc. Brot. **13**: 16 (1896). —Hiern, Cat. Afr. Pl. Welw. 1, 3: 532 (1898). —S. Moore in Journ. Bot. **65**, Suppl. 2, Gamopet.: 50 (1927). —Mendonça, Contrib. Conhec. Fl. Angol., **1** Compositae: 34 (1943). —Earle Sm. in U.S. Dept. Agric. Handb. **396**: 74, figs. 52, 53 (1971). —Wild in Kirkia **11**: 107 (1978). TAB. **20** fig. D. Syntypes from Angola.
Vernonia violacea sensu Klatt in K.K. Naturhist. Hofmus. Wien, **7**: 99 (1892) quoad *Teucz* 174 in *Mechow* solum.

An erect perennial herb, 40–100 cm. tall from an enlarged woody rootstock; rootcrowns densely brown-lanate, roots thong-like, spreading; flowers produced before the leaves. Stems annual, solitary, branched, ribbed, puberulous to glabrescent. Leaves radical and cauline; radical leaves 1–5 ascending, up to c. 50 × 10(15) cm., narrowly oblanceolate to narrowly obovate, apex subacute to rounded, base ± gradually tapering into a long attenuate, very narrowly winged midrib up to c. 20 cm. long, midrib becoming ± rigid in the lower portion and often persisting from previous season, margin irregularly undulate-crenate, the teeth callose-tipped, sparsely pilose to glabrescent on both surfaces, or glabrous on upper surface and sparsely puberulous to glabrescent on lower surface, nervation ± prominently reticulate; cauline leaves few, grading into synflorescence bracts, similar to radical leaves but smaller, up to c. 18 × 2 cm. Capitula many in a lax terminal panicle, or sometimes in clusters at the ends of branches; capitula stalks slender, 0.5–5 cm. long. Involucres 4–6 × 6–8 mm., ± shallowly campanulate to widely spreading. Phyllaries few-seriate, often purple-tipped, hispidulous outside; the outer from c. 1 mm. long, linear, tapering to the apex, the inner up to c. 6 mm. long, linear-elliptic, tapering to an acute ± mucronate or shortly apiculate apex, narrowed to the base. Corollas purple, 5–7.5 mm. long, narrowly tubular, widening above into a shortly lobed limb, glabrous, minutely glandular. Achenes c. 3.5 mm. long, subcylindric, ± 4–5-sided to 8–10-ribbed, ± densely hispidulous, glandular; outer pappus of short linear scales or setae; inner pappus of sordid barbellate setae 4–5 mm. long.

Zambia. W: Mwinilunga Distr., Lisombo R., c. 15 km. SW. of Kalene Hill Mission, 26.x.1969, *Drummond & Williamson* 9456 (BR; K; SRGH).
Also in Zaire and Angola. A pyrophyte of grassland and high rainfall deciduous woodland.

V. divulgata S. Moore may be distinguished from this species by its distinctive phyllaries which are much expanded and membranous at the apex and not acute mucronate.

74. **Vernonia denudata** Hutch. & B.L. Burtt in Rev. Zool. Bot. Afr. **23**: 37 (1932). —Earle Sm. in U.S. Dept. Agric. Handb. **396**: 74, figs. 48, 51 (1971). —Wild in Kirkia **11**: 106 (1978). —C. Jeffrey in Kew Bull. **43**: 228 (1988). TAB. **20** fig. A. Type from Zaire.

A strongly branching scapose tufted perennial herb, c. 10–40 cm. tall from an enlarged woody rootstock; rootcrowns densely brown-lanate, roots thong-like, spreading; flowers produced before the leaves on branching stalks. Leaves 1–several, radical, ascending, eventually up to ?45 × 13 cm., apparently spathulate, rounded at the apex, abruptly narrowed from, or above, the middle into a long, very narrowly winged midrib, margins irregularly dentate, usually fringed with purple hairs; upper surface pubescent with long-stalked T-shaped hairs; lower surface paler, tomentose to glabrescent with short-stalked long-armed ± matted T-shaped hairs; midrib rigid towards the base, the lower portion often persisting from previous season, tomentose. Scapes annual, green to

Tab. 20. A. —VERNONIA DENUDATA. A1, habit, note leaf base from previous season, from *Richards* 2040; A2, leaf, from *Richards* 11426; A3, corolla; A4, achene, A3 & A4 from *Milne-Redhead* 1126. B. —VERNONIA CHTHONOCEPHALA. B1, habit, note leaf base from previous season; B2, achene, B1 & B2 from *Wiehe* N/264. C. —VERNONIA PRAEMORSA, capitula, from *Pawek* 10369B. D. —VERNONIA PRAECOX, capitula, from *Drummond & Williamson* 9456. E. —VERNONIA ACROCEPHALA, capitulum, from *Phillips* 4049. F. — VERNONIA SUBAPHYLLA, capitulum, from *Pawek* 7477. (Habit, leaf and capitula × $\frac{2}{3}$, corolla and achenes × 6). Drawn by Eleanor Catherine.

purplish, 1–many tufted, 10–40 cm. tall, sparingly to much branched, ribbed, bracteate, ± appressed pubescent, the hairs short-stalked T-shaped with ± unequal arms, densely lanate at the base; bracts 0.3–3.6 × up to c. 0.2 cm.; branches 3–22 cm. long, mostly simple, ± ascending. Capitula ± numerous, solitary and terminal on the branches; involucres 10–16 × 12–20 mm., obconic-campanulate, becoming widely spreading. Phyllaries green with narrow, membranous, purple margins and apices, loosely imbricate, sparsely hispid and glandular outside, ciliate on margins; the hairs patent, usually purple-stalked, whip-like; the outer from c. 2 mm. long, linear-lanceolate, the middle narrowly oblong and narrowly ovate-obtuse at the apex, the inner up to c. 15 mm. long, linear-lanceolate and subacute to tapering-acuminate at the apex, sometimes somewhat membranous and undulate. Corollas purple to pale-mauve, 7–12 mm. long, narrowly tubular, widening above into a shortly-lobed limb, glabrous, minutely glandular. Achenes 3–5 mm. long, subcylindric, narrowly 8–10-ribbed, densely hispid at first, glandular; outer pappus of short linear acuminate scales or scale-like setae 1–3 mm. long, inner pappus of whitish barbellate setae 8–11 mm. long.

Zambia. N: Mbala, Kalambo Road, c. 1500 m., 4.ix.1959, *Richards* 11426 (K; M; SRGH). W: Mwinilunga, c. 12.8 km. W. of Kakomo, 28.ix.1952, *Angus* 567 (BR; BM; K; MO).
Also in Tanzania and Zaire. A pyrophyte of dambos and high rainfall woodland.

75. **Vernonia subaphylla** Bak. in Bull. Misc. Inf., Kew **1895**: 290 (1895). —Earle Sm. in U.S. Dept. Agric. Handb. **396**: 69, figs. 48, 49, 50 (1971). —Wild in Kirkia **11**: 107 (1978). —C. Jeffrey in Kew Bull. **43**: 229 (1988). TAB. **20** fig. F. Type: Zambia, Mweru, Kalungwishi, *Carson* 10 (K, holotype; SRGH, photo).
   Vernonia armerioides O. Hoffm. in Engl., Bot. Jahrb. **24**: 462 (1897). Type from Zaire.
   Vernonia agricola S. Moore in Journ. Bot. **56**: 208 (1918). Type: Zambia, Lukanda R., *Kassner* 2136 (BM, holotype; K).

A scapose perennial herb, 25–60 cm. tall from an enlarged woody rootstock; rootcrowns densely brown-lanate, roots thong-like, spreading. Leaves appearing with, or sometimes after the flowers, mostly 4–8 basal, rarely a few cauline; lower leaves eventually up to c. 25 × 2.5 cm., linear-oblanceolate to narrowly oblong-oblanceolate, subacute at the apex and tapering to an attenuate very narrowly winged midrib; upper leaves 4–10 × 1.7–3.5 cm., elliptic to narrowly obovate, ± rounded at the apex and narrowly cuneate at the base; margins slightly revolute, subentire to serrate with callose-tipped teeth, upper surface ± densely pubescent to glabrous, sometimes shiny and minutely gland-pitted, lower surface tomentose to puberulous, glabrescent, gland-pitted; hairs long-armed T-shaped; nervation reticulate, ± prominent beneath; midribs ± rigid towards the base. Scapes annual, greenish, usually solitary 25–60 cm. long, simple or branched above, ribbed, bracteate, appressed pubescent, the hairs shortly-stalked T-shaped; bracts 0.5–6.5 cm. long, linear, seldom foliaceous; branches 2–20 cm. long, ascending. Capitula solitary and terminal on the branches, involucres 10–15 × 13–18 mm., subglobose-campanulate. Phyllaries ± appressed imbricate, ± sparsely pubescent or glabrous, glandular outside, green with broad membranous, hyaline or purple margins, tapered-elongate at the apices, apical margins becoming contracted-folded when dry; the outer phyllaries from c. 2 mm. long, narrowly obovate, the middle phyllaries up to c. 6 mm. wide, obovate-lanceolate, the inner phyllaries narrower up to c. 13 mm. long, narrowly oblong-obovate to oblong-lanceolate. Corollas purple to mauve, 9–11 mm. long, narrowly tubular, widening above, glabrous, lobes up to c. 3 mm. long. Achenes 3.5–4 mm. long, subcylindric to subfusiform, narrowly 8–10-ribbed, hispidulous, glandular; outer pappus of short linear acuminate scales; inner pappus of sordid or brownish barbellate setae 5–8 mm. long.

Zambia. N: Kasama, Mungwi, 4.xii.1960, *Robinson* 4149 (K; M; SRGH). W: Chingola-Solwezi, 29.ix.1947, *Greenway & Brenan* 8127 (K; SRGH). C: Serenje Distr., Mkushi R., 27.xi.1962, *Richards* 17489 (K; M). S: Mazabuka, 15.i.1960, *White* 6280 (K; FHO; SRGH). **Malawi**. N: Mzimba Distr., Mzuzu, Marymount, 7.xi.1973, *Pawek* 7477 (K; MO). C: Dzalanyama For. Res., Chaulongwe Falls, 14.i.1967, *Hilliard & Burtt* 4488 (K).
Also in Nigeria, Cameroon, Zaire, Angola, and Tanzania. A pyrophyte of dambo and submontane grasslands, also in miombo and deciduous woodland often in rocky places.

76. **Vernonia mutimushii** Wild in Kirkia **11**: 14; 105 (1975). Type: Zambia, Kawambwa, Manchele, *Mutimushi* 389 (K; NDO, isotypes; SRGH, holotype).

An erect perennial herb, 7–26 cm. tall from a woody rootstock; roots thong-like, spreading. Stems scapiform, annual, usually reddish, solitary with leaves subrosulate at or just above the base, otherwise leafless, striate, lanate, becoming sparsely pilose; bracts 0–2, linear-elliptic. Leaves 4–8, basal subsessile, to c. 4 × 2.5 cm. (?fully expanded), elliptic to broadly obovate, apices rounded, bases cuneate or ± tapering to a winged midrib, margins entire, lamina membranous, glabrous on upper surface, lanate or ± pilose on lower surface; hairs patent flagelliform. Capitulum solitary, involucre 12–15 × c. 12(25) mm., campanulate to widely spreading. Phyllaries several-seriate, purple-brown chartaceous, outer from c. 5 mm. long, lanceolate-acuminate, sparsely puberulous to glabrous outside; the inner to c. 14 mm. long, lanceolate to linear-lanceolate, cuspidate to tapering-acuminate at apex. Corollas purple, c. ?7 mm. long. Achenes (immature) to c. 5 mm. long, narrowly subcylindric with 8–10 narrow wing-like ribs, glabrous; outer pappus of short barbellate setae, inner of clear green barbellate setae c. 7 mm. long.

Zambia. N: Kawambwa, Ntenke, 2.ix.1963, *Mutimushi* 472 (K; NDO; SRGH).
Known only from Zambia. Slightly moist dambos.

77. **Vernonia acrocephala** Klatt in Ann. K.K. Naturhist. Hofmus. Wien **7**: 100 (1892). —Mendonça, Contrib. Conhec. Fl. Angol., **1** Compositae: 24 (1943). —Earle Sm. in U.S. Dept. Agric. Handb. **396**: 67 (1971). —Wild in Kirkia **11**: 104 (1978). —C. Jeffrey in Kew Bull. **43**: 229 (1988). TAB. **20** fig. E. Type from Angola.

A scapose perennial herb, 3–30 cm. tall from a thickened woody rootstock; rootcrowns densely brown-lanate, roots thong-like, spreading. Leaves appearing with or after the flowers, 3–8 radical, at first ascending, later spreading; petiole up to c. 4 cm. long, grey- or brownish-seriaceous; lamina membranous, up to c. 8 × 3 cm. in flowering plants, eventually up to ?11 × 4 cm., narrowly oblanceolate becoming elliptic or ovate, apices rounded or subacute, bases cuneate and tapering to a narrowly winged petiole or obtuse, margins entire and ± densely fringed with purple-stalked Y-shaped hairs; upper surface ± sparsely hispid with long-stalked unequally-forked Y-shaped hairs, or glabrescent, lower surface ± densely appressed-sericeous to tomentose with shortly-stalked long-armed T-shaped or Y-shaped hairs. Scapes annual, purplish, 2–28 cm. long, unbranched, widening somewhat towards the apex, with 0–3 linear bracts up to c. 3 cm. long, grey- or brownish-tomentose above and ± appressed-sericeous on the lower portion; hairs short-stalked long-armed T-shaped or Y-shaped. Capitulum solitary; involucre 11–17 × 15–20 mm., broadly campanulate to hemispheric. Phyllaries 2–3-seriate, subequal or the outer at least three quarters the length of the inner, ± appressed-sericeous outside and ciliate on the margins at least above, the hairs often purple-stalked flagelliform; the outer phyllaries 4–8 mm. wide, ovate or oblong-ovate, acute to cuspidate, the inner phyllaries 10–15 × 3–4 mm., narrowly oblong, tapering to a purple acuminate apex. Corollas purple or mauve, 10–15 mm. long, narrowly tubular, widening gradually upwards, lobes c. 3 mm. long, linear. Achenes 2–3.5 mm. long, subcylindric, c. 10-ribbed, hispid; outer pappus of short linear acuminate scales, inner pappus of sordid to brownish barbellate setae 9–11 mm. long.

Zambia. N: Kasama Distr., path to Chambeshi Flats via Mbwayalalo Village, 10.xii.1964, *Richards* 19346 (K; M). **Malawi**. N: Nyika Plateau, Nganda Peak, c. 2470 m., 9.i.1974, *Pawek* 7915 (K; MO; SRGH; MAL).
Also in Tanzania, Zaire, Angola and Cameroon. A pyrophyte of dambos, submontane grassland and moist deciduous woodlands.

78. **Vernonia chthonocephala** O. Hoffm. in Bol. Soc. Brot. **13**: 17 (1896). —Hiern, Cat. Afr. Pl. Welw. **1**, 3: 532 (1898). —S. Moore in Journ. Linn. Soc., Bot. **37**: 312 (1906); in Journ. Bot. **65**, Suppl. 2, Gamopet.: 50 (1927). —Mendonça, Contrib. Conhec. Fl. Angol., **1** Compositae: 34 (1943). —Earle Sm. in U.S. Dept. Agric. Handb. **396**: 78, figs. 53, 54 (1971). —Wild in Kirkia **11**: 106 (1978). —Maquet in Fl. Rwanda, Spermat. **3**: 552 (1985). —C. Jeffrey in Kew Bull. **43**: 228 (1988). TAB. **20** fig. B. Types from Angola.
Vernonia perparva S. Moore in Journ. Linn. Soc., Bot. **35**: 324 (1902). Type from Tanzania.

A dwarf scapose, usually tufted, perennial herb with a large woody rootstock, flowering before the leaves appear; rootcrowns densely brown-lanate, roots thong-like, spreading. Leaves 1–5, radical, rosulate, ascending, eventually up to ?26(50) × 8 cm., oblong-oblanceolate to elliptic, apex obtuse, base tapering-attenuate into a long narrowly winged midrib, margins subentire to irregularly dentate or undulate; upper surface coarsely hispid and somewhat scabridulous or glabrescent, the hairs long-stalked T-shaped; lower surface paler tomentose to glabrous, the hairs short-stalked, long-armed, matted T-shaped; midrib rigid towards the base, the lower portion often persisting into next season. Scapes 1–several, 1–4 cm. long, very densely brownish-lanate in the lower portion, appressed sericeous to tomentose above with short-stalked T-shaped hairs. Capitula solitary on the scapes; involucre 13–20 × 15–25 mm., broadly campanulate. Phyllaries dark-purple or purple-tipped, loosely imbricate, progressively longer towards the inside, sparsely pilose outside and ciliate on the margins towards the apex, the hairs patent often purple-stalked whip-like; the outer phyllaries from c. 4 mm. long, tapering to a subacute apex; the inner phyllaries up to c. 18 mm. long, narrowly oblong-lanceolate, apices obtuse sometimes expanded above the tip into a purple undulate ovate-lanceolate appendage. Corollas blue or purplish, 8–14 mm. long, narrowly tubular, widening gradually upwards, lobes 2–3 mm. long. Achenes 3–4.5 mm. long, subcylindric, narrowly 8–12-ribbed, densely hispid at first, later glabrescent, minutely glandular; outer pappus irregular, of barbellate setae 2–4 mm. long, inner pappus of yellowish-brown somewhat flattened barbellate setae 5–10 mm. long.

Zambia. N: Top of Kambole escarpment, 1500 m., 11.ix.1960, *Richards* 13222 (K; M). C: Mumbwa, *Macaulay* 827 (K). E: Makutus Mt., 28.x.1972, *Fanshawe* 11625 (K; NDO). **Malawi**. C: Dedza Distr., near Milonde School, 2.x.1967, *Salubeni* 845 (K; SRGH). S: Nzama Road, 7.x.1949, *Wiehe* N/264 (K). Also in Sudan, Kenya, Uganda, Tanzania, Rwanda, Burundi, Zaire, Angola and W. tropical Africa. A pyrophyte of submontane grassland and high rainfall woodlands.

79. **Vernonia polysphaera** Bak. in Bull. Misc. Inf., Kew **1898**: 148 (1898). —Wild in Kirkia **11**: 117 (1978). —C. Jeffrey in Kew Bull. **43**: 235 (1988). Type: Malawi, near Chitipa (Fort Hill), *Whyte* s.n. (K, holotype).
　　Vernonia homblei De Wild. in Fedde Repert. **13**: 207 (1914). Syntypes from Zaire.

A slender wiry perennial herb, 45–120 cm. tall from a thickened woody rootstock; rootcrowns densely brown-lanate, roots thong-like, spreading. Stems annual, 1–several, simple below, sparingly branched above, leafy, ribbed, glabrous; branches ascending, 2–28 cm. long. Leaves subsessile, concolorous; mid-cauline leaves up to c. 14 × 3.5 cm., decreasing in size to the stem base and apex, ± narrowly oblong-elliptic or linear-oblong, apex obtuse to acute mucronate, base cuneate to obtuse, margins serrate to remotely subserrate, lamina stiffly thinly coriaceous, glabrous or nearly so on both surfaces, ± sparsely scabridulous-hispid on the veins, often glandular-punctate beneath, venation reticulate and prominent, especially beneath. Capitula subsessile in clusters of (1)3–8; capitula-clusters numerous, sessile in the axils of upper stem- and branch-leaves. Involucres 8–11 × 5–8 mm., oblong-campanulate, later spreading-obconic to c. 12 mm. wide. Phyllaries many-seriate, ± appressed imbricate, lanate-araneose outside, apices purplish; the outer phyllaries from c. 1 mm. long, narrowly ovate and acuminate-pungent, the middle phyllaries c. 5 × up to c. 2 mm., the inner phyllaries up to c. 9 × 1 mm., linear-oblong, acuminate-apiculate. Florets c. 10 per capitulum; corollas purple, 8–10 mm. long, widening gradually upwards. Achenes 3–4 mm. long, subcylindric, narrowed at the base, obscurely 6–10-ribbed, ± appressed hispid; outer pappus of linear-lanceolate, scale-like setae c. 1 mm. long, inner pappus of copious stramineous barbellate setae 6–8 mm. long.

Zambia. N: Kasama Distr., Mungisi, 14.iii.1962, *Robinson* 5012, (K). W: Ndola, 3.v.1954, *Fanshawe* 1149 (BR; K; LISC; NDO; S; SRGH). C: Chiwefwe, c. 1340 m., 14.vii.1930, *Hutchinson & Gillett* 3658 (K). **Malawi**. N: Chitipa Distr., Kaseye Mission, c. 1250 m., 26.iv.1972, *Pawek* 5256 (K; SRGH). Also in Tanzania and Zaire. Miombo woodland and wooded grassland.

80. **Vernonia bullulata** S. Moore in Journ. Bot. **65**, Suppl. 2, Gamopet.: 44 (1927). —Mendonça, Contrib. Conhec. Fl. Angol., 1 Compositae: 10 (1943). —Wild in Kirkia **11**: 116 (1978). —Kalanda & Lisowski in Bull. Jard. Bot. Nat. Belg. **51**: 209, fig. 1 (1981). Type from Angola.
　　Vernonia polysphaera sensu S. Moore in Journ. Linn. Soc., Bot. **37**: 311 (1906) quoad *Gossweiler* 1169.
　　Vernonia kwangolana Duvign. & Hotyat in Bull. Soc. Roy. Bot. Belg. **85**: 78 (1952). Type from Zaire.

An erect perennial herb, 0.45–1(2) m. tall from a thickened woody rootstock; rootcrowns densely brown-lanate, roots thong-like, spreading. Stems annual, 1–several, simple below, sparingly ± long-branched above, leafy, pale-brownish tomentose on young growth, pubescent to glabrescent on lower stem; hairs simple, long, matted; branches ascending, 3–25 cm. long. Leaves subsessile, discolorous, 4–11(12) × 1.4–3.5(4) cm., oblong-elliptic or oblanceolate, apex subobtuse or acute, base cuneate to obtuse, margins subentire to obscurely callose-tipped denticulate; lamina coriaceous, green drying brown, ± bullate and glabrescent on upper surface, greyish tomentose with a prominent reticulate venation on lower surface. Capitula subsessile, in clusters of (1)3–11; capitula-clusters few to numerous, sessile in axils of upper stem- and branch-leaves. Involucres 7–10(15) × 4–8 mm., oblong-campanulate. Phyllaries many-seriate, imbricate, tomentose outside where exposed; outer phyllaries from c. 1 mm. long, ovate to ovate-lanceolate; the inner phyllaries up to c. 8 mm. long, narrowly oblong, apices purple, acute mucronulate. Florets c. 10 per capitulum; corollas mauve, 4–7 mm. long, widening gradually upwards, lobes c. 1 mm. long. Achenes 3–3.5 mm. long, subcylindric, narrowed at base, 5–6-sided with ± obscured ribs, appressed hispid; outer pappus of linear-lanceolate scales c. 1 mm. long, inner pappus of stramineous barbellate setae 6–8 mm. long.

Zambia. W: c. 27 km. S. of Mwinilunga on Kabompo Rd, 6.vi.1963, *Loveridge* 847 (K; M; SRGH). Also in Angola and Zaire. Miombo and mixed deciduous woodland.
The Zairian species *V. verdickii* O. Hoffm. closely approaches this species but may be distinguished by its narrowly oblong leaves rounded, not acute, at the apex and hispid, not tomentose, below.

81. **Vernonia catumbensis** Hiern, Cat. Afr. Pl. Welw. **1**, 3: 524 (1898). —O. Hoffm. in Warb., Kunene-Samb. Exped. Baum: 405 (1903). —Mendonça, Contrib. Conhec. Fl. Angol., **1** Compositae: 17 (1943). Type from Angola.
    Vernonia richardiana sensu Wild in Kirkia **11**: 86 (1978).

An erect slender perennial herb 0.8–2 m. tall; tuberous-rooted from a small woody rootcrown; tubers up to 5 × 1.5 cm., fusiform. Stems 1–several, becoming woody below, divaricately branched above, leafy, subterete, puberulous with white curly hairs, especially on the branches, glabrescent below; branches purplish. Leaves very shortly petiolate, up to c. 17 × 5.5 cm., largest on lower stem, decreasing in size up the stem, oblanceolate; apex rounded-obtuse on lower cauline leaves, becoming acute in upper leaves; tapering below to a narrow, abruptly rounded or truncate base ending in a 1–2 mm. long glabrescent petiole; margin entire towards the leaf base, becoming repand-denticulate to serrate-dentate in the upper part, the teeth callose-tipped; lamina thinly subcoriaceous, glabrescent on the upper surface, somewhat paler and puberulous on the lower surface, sometimes glabrescent. Capitula numerous, sessile or subsessile, scorpioidly cymose, the capitula secund on laxly subumbellate branches. Involucres 8–10 mm. long and 8–10 mm. wide, obconic to campanulate-obconic. Phyllaries many-seriate, appressed imbricate, increasing uniformly in length to the inside, ± coriaceous with hyaline margins and purple tips, thinly woolly outside and densely ciliate on the margins; the outer phyllaries broadly ovate, the inner oblong-lanceolate, mucronulate. Florets c. 18 per capitulum; corollas purple or whitish-mauve, up to c. 12 mm. long, narrowly tubular, slightly widened above the middle, glabrous. Achenes up to c. 3.5 mm. long, subcylindric, tapering slightly below, densely strigose; outer pappus of c. 2 mm. long bristles; inner of copious barbellate setae up to c. 9 mm. long.

Zambia. W: Solwezi, 13.v.1969, *Mutimushi* 3132 (K; NDO; SRGH). Also in Angola. Miombo woodland.
Vernonia theophrastifolia Schweinf. [*Vernonia richardiana* (O. Kuntze) Pichi Sermolli] may be distinguished from *V. catumbensis* by its smaller more numerous capitula and by its phyllaries which are glabrous or glabrescent and never woolly. The leaves are also distinctive being auriculate or hastate at the base in the former and merely narrowing to an abruptly rounded or truncate base in the latter.

82. **Vernonia adenocephala** S. Moore in Journ. Bot. **56**: 207 (1918). —Wild in Kirkia, **11**: 76 (1978). Type from Zaire.

A tough harsh perennial herb up to c. 1 m. tall, from a large woody rootstock. Stems annual, 1–several, becoming woody below, usually dark-purple, strict, simple and densely leafy, scattered whitish patent-pilose often also with short purple T-shaped hairs, or glabrescent. Leaves numerous, overlapping, thinly harshly coriaceous, subsessile, up to 11 × 5(6) cm., obovate or broadly elliptic to broadly lanceolate, apex subacute, base rounded to broadly cuneate, margins coarsely serrate-dentate to subentire, lamina ± scabrous, sparsely white pilose on the midrib and nerves beneath, glabrescent; venation prominent and reticulate beneath. Capitula numerous, sessile, congested in a very dense terminal, subglobose to ellipsoid-cylindric spike. Involucre up to 15(20) mm. long, obconic or narrowly campanulate. Phyllaries numerous, imbricate, ± cartilaginous, glabrous with large dark gland-pits outside towards the apex, finely serrate-ciliate on the margins above, the apices pungent; the outermost phyllaries c. 2 mm. long and scale-like, progressively longer towards the inside becoming lanceolate to oblong-lanceolate, the innermost c. 12 mm. long. Corollas purple, c. 13 mm. long, narrowly tubular widening slightly above. Achenes 2–3 mm. long, oblong-turbinate, obscurely 6–8 narrowly-ribbed, densely strigose; outer pappus of short narrow deeply fimbriate scales c. 0.5 mm. long; inner pappus of copious white barbellate setae c. 8 mm. long.

Zambia. W: Mwinilunga, Lisombo R., c. 15 km. SW. of Kalene Hill Mission, 11.vi.1963, *Drummond* 8292 (K; SRGH).
Also in Zaire. In mixed deciduous woodland.

83. **Vernonia turbinella** S. Moore in Journ. Linn. Soc., Bot. **47**: 266 (1925). —Wild in Kirkia **11**: 101 (1978). Type from Zaire.

A robust greyish tough perennial herb up to c. 120 cm. tall, from a large woody rootstock; rootcrowns densely shortly brown-lanate. Stems annual, 1–many, tufted, erect, simple below the synflorescence, ± densely leafy above with the leaves reduced or absent in the lower c. one third, ribbed, velutinous greyish-tomentose. Leaves closely-spaced, ascending; petioles 2–12 mm. long, very narrowly winged, tomentose or glabrescent; lamina stiffly coriaceous, up to c. 12.5 × 9 cm. in lower leaves and decreasing in size upwards, broadly ovate to ovate-oblong or elliptic, subacute to rounded at the apex, base ± abruptly cuneate to rounded sometimes asymmetrically so, decurrent on the petiole as a narrow tapering wing, margins subentire to repand, upper surface thinly sparsely tomentose or glabrescent, lower surface tomentose sometimes sparsely so; venation finely reticulate, raised but ± obscured by the indumentum beneath. Capitula numerous in a terminal panicle up to c. 22 cm. across; capitula stalks 1–15 mm. long, tomentellous; bracts 0–3 up to c. 12 mm. long, tomentellous. Involucres 6–10 × 10–15 mm., obconic-hemispheric to widely spreading, becoming almost flat. Phyllaries tightly imbricate, coriaceous, acute mucronate, grey-tomentose where exposed but glabrous and dark-purple about the tip and upper margins; the outer phyllaries from c. 2 mm. long, ovate; the inner phyllaries increasing uniformly in length to c. 6 mm. long, oblong-elliptic. Receptacle ± flat, shallowly alveolate. Corollas cream to mauve, 6–9 mm. long, funnel shaped with lobes 2–3 mm. long, glabrous. Achenes up to c. 4 mm. long, ± compressed-turbinate, somewhat 3–4-angular, obscurely ribbed, densely appressed silky hairy; pappus white or sordid, well exserted beyond the involucre, outer elements of irregular scale-like or flattened setae 1–4 mm. long, inner elements more numerous of flattened setae 6–7 mm. long and barbellate on the margins.

Zambia. N: Mbala, top of Kambole Escarpment, 1524 m., 22.iv.1969, *Richards* 24505 (K). W: 3 km. E. of Solwezi, 17.iii.1961, *Drummond & Rutherford-Smith* 6952 (BR; K; LISC; SRGH). E: Nyika Plateau, Kangampande Mt., 2130 m., 8.v.1952, *White* 2799 (FHO; K). **Malawi**. N: Viphya Plateau, Luwawa, 25.iv.1967, *Salubeni* 650 (K; SRGH).
Also in Zaire and Angola. Miombo woodland, wooded grasslands and submontane grassland, often on escarpments or rocky outcrops with tall grasses at medium altitudes.
Near *V. lampropappa* O. Hoffm. from Angola but that species has sessile narrower ± amplexicaul leaves.

84. **Vernonia helodea** Wild in Kirkia **11**: 10, fig. 5; 84 (1978). Type: Zambia, Mbala Distr., Chinakila, *Richards* 19470 (K; M; SRGH, holotype).

An erect slender subglabrous perennial herb up to c. 50 cm. tall, from a thickened base with thong-like roots. Stems annual, solitary, simple, ribbed, lanate below and sparsely pubescent above; the hairs short-stalked T-shaped. Leaves sessile, greenish, up to c. 55 × 1.5–3 mm., linear, tapering to an acute purple-tipped mucronate apex, tapering to the base, margin entire ± revolute, lamina glabrous. Capitula c. 12, laxly arranged in a terminal subpaniculate raceme, stalks slender up to c. 4 cm. long. Involucres up to c. 5 mm. long, broadly obconic spreading. Phyllaries few-seriate, becoming purple, sparsely puberulous, glabrescent outside; the outer phyllaries short subulate or linear-lanceolate, the inner up to c. 5 mm. long, oblong, abruptly contracted into a cuspidate apex. Corolla mauve, up to c. 7 mm. long, narrowly funnel-shaped, limb puberulous outside. Achenes c. 3 mm. long, subcylindric to somewhat 4-ribbed, densely long-strigose; outer pappus of short somewhat scale-like setae, inner of pale stramineous barbellate setae 4–4.5 mm. long.

Zambia. N: Mbala Distr., Chinakila, Loye Flats, c. 1200 m., 11.i.1965, *Richards* 19470 (K; M; SRGH). Known only from this gathering. Swampy grassland.

85. **Vernonia adoensis** Sch. Bip. ex Walp., Repert. Bot. Syst. **2**: 946 (1843). —Oliv. & Hiern in F.T.A. **3**: 291 (1877). —Earle Sm. in U.S. Dept. Agric. Handbook **396**: 16, figs. 1, 2A, 5, 9–11 (1971). —Hilliard in Ross, Fl. Natal: 356 (1972); Compos. Natal: 28 (1977). —Wild in Kirkia **11**: 111 (1978). —Maquet in Fl. Rwanda, Spermat. **3**: 554 (1985). —C. Jeffrey in Kew Bull. **43**: 240 (1988). Type from Ethiopia.

Ascaricidia adoensis (Sch. Bip. ex Walp.) Steetz in Peters, Reise Mozamb., Bot.: 358 (1864). Type as above.

An erect coarse perennial herb or subshrub, 0.3–3 m. tall from a woody rootstock. Stems usually annual, becoming ± woody below, 1–several, branching above and often tuberculate on the young growth, leafy, softly whitish-pubescent to tomentellous, less densely so below, or glabrescent, sometimes tawny-velvety on young growth; hairs short-stalked flagelliform, sometimes intermixed with scattered longer pilose hairs. Leaves petiolate to subsessile, petioles to c. 7 cm. long; lamina to c. 26.5 × 7.5 cm., lanceolate to oblong-elliptic, apex acute, base broadly cuneate to attenuate, or leaves ovate and rounded to subcordate at the base, margins undulate and coarsely serrate with sharply callose-tipped teeth; upper surface drying dark-green and somewhat harshly puberulous or glabrescent, lower surface greenish-grey tomentose to sparsely pilose or glabrescent and glandular. Capitula solitary on branches, or few in much reduced corymbiform cymes on short side-branches or at the stem apex, capitula shortly stalked. Involucres mostly 15–30 × 20–45 mm., broadly campanulate to cyathiform. Phyllaries apically appendaged; the outer phyllaries greenish, 8–13 mm. long including a tapering appendage 6–12 times as long as the phyllary base, or appendage ovate-lanceolate, outermost phyllaries sometimes extending onto capitulum stalk; middle phyllaries increasing from 8 to 22 mm. long (including appendage), with appendages to 12 × 12 mm. and lanceolate to broadly ovate, creamy-white, sometimes reflexed, decreasing in size towards the inside; inner phyllaries 12–20 × 2–3.5 mm., narrowly oblong with greatly reduced appendages. Florets c. 120–300 per capitulum. Corollas whitish or mauve, 12–24 mm. long, the tube long and slender before abruptly widening into a short cylindric limb exserted above the pappus. Achenes dark-brown, 4–5.5 mm. long, subcylindric to narrowly turbinate, 10-ribbed, or 20-ribbed with the development of secondary ribs, hispid-strigose, glandular; pappus consisting of an outer whorl of short linear, or minute fimbriate scales, and several inner whorls of caducous setae, setae barbellate on the margins and increasing to 10–15 mm. long inside, the innermost flattened, the outer subulate.

1. Phyllary appendages narrow, ± equalling phyllary lamina in width, tapering or linear-elliptic, often becoming expanded towards the apex; appendages narrowed below, ± constricted at junction with phyllary lamina - - - - - - - - - var. *adoensis*
– Phyllary appendages broader than phyllary lamina, ovate-lanceolate to rotundate, 4–12 mm. wide; appendages expanded below and tapering-acuminate or rounded and apiculate above - 2

2. Appendages of middle phyllaries broadly ovate-rotundate and ± strongly recurved, usually as broad as long and ± equalling the phyllary lamina in length, rounded to obtuse at the apex and shortly acuminate; outer phyllaries squarrose-recurved, ovate to broadly lanceolate often conduplicate, rarely extending onto capitulum stalk; phyllary lamina usually whitish pubescent outside - - - - - - - - - - - - - - - var. *mossambiquensis*
– Appendages of middle phyllaries ovate-lanceolate, occasionally narrowly elliptic, appressed not reflexed sometimes somewhat undulate or contorted, longer than wide and 2–7 times as long as the phyllary lamina, tapering-acuminate occasionally obtuse and apiculate apically; outer phyllaries oblong-lanceolate to subulate often sparsely scattered on capitulum stalk; phyllaries subglabrous or sparsely puberulous - - - - - - - - - var. *kotschyana*

Var. **adoensis**, TAB. **22** fig. 12.

Vernonia integra S. Moore in Journ. Bot. **46**: 39 (1918). —Eyles in Trans. Roy. Soc., S. Afr. **5**: 504 (1916). Type: Zimbabwe, Mazowe (Mazoe), *Eyles* 277 (BM, holotype; SRGH).

Leaves of upper stem subsessile or shortly petiolate; lamina lanceolate, narrowly cuneate at the base, tomentose to sparsely pubescent or glabrescent below. Phyllaries white hispidulous or puberulous; appendages linear-elliptic, as wide as phyllary (occasionally wider), ± narrowed in lower portion often constricted at junction with phyllary lamina, appendages of outer and middle phyllaries occasionally much expanded in apical portion becoming leaf-like, exceeding phyllary in length, flat or somewhat undulate.

Zimbabwe. N: Mvurwi (Umvukwe) Mts., c. 8 km. N. of Banket, 23.iv.1948, *Rodin* 4389 (K). C: Harare, v.1932, *Eyles* 7137 (K; SRGH). **Malawi**. N: Nyika Plateau, vii.1896, *Whyte* s.n. (K).
Widespread, extending from Nigeria and Ethiopia southward through Zaire and E. Africa to Zimbabwe. Grassland with scattered trees, miombo and other woodlands, often beside streams.

Var. **mossambiquensis** (Steetz) G.V. Pope in Kew Bull. **43**: 284 (1988). TAB. **22** fig. 10. Type: Mozambique in Waldern bei Boror und Rios de Sena, *Peters* s.n. (B†).
    Ascaricida mossambiquensis Steetz in Peters, Reise Mossamb., Bot.: 358 (1864) non *"mossambicensis"* Busc. & Muschl. (1913). Type as above.
    Vernonia mossambiquensis (Steetz) Oliv. & Hiern in F.T.A. **3**: 292 (1877). Type as above.
    Vernonia shirensis Oliv. & Hiern in F.T.A. **3**: 291 (1877). —Brenan in Mem. N.Y. Bot. Gard. **8**, 5: 461 (1954). Type: Malawi, lower valley of Shire River, *Meller* s.n. (K, holotype).
    Vernonia whyteana Britten in Trans. Linn. Soc., Ser. 2, **4**: 17 (1894). Type: Malawi, Zomba, x.1891, *Whyte* s.n. (BM).
    Vernonia fulviseta S. Moore in Journ. Linn. Soc., Bot. **47**: 266 (1925). Type: Malawi, *Buchanan* 1297 (BM, holotype).

Leaves, including the upper ones, usually petiolate; lamina lanceolate and ± decurrent on petiole, or ovate and truncate to rounded basally, densely tomentose below or pubescent to sparsely pilose. Phyllaries whitish pubescent outside; appendages broadly ovate to lanceolate ± equalling phyllary base in length, rounded and shortly acuminate at apex or acute, recurved or reflexed, flat or undulate, sometimes conduplicate in the outer phyllaries.

Zimbabwe. F: Nyanga (Inyanga), Pungwe R., near Mozambique border, c. 1000 m., 19.iv.1957, *Chase* 6451 (K; LISC; SRGH). **Malawi**. N: Mzimba Distr., Champira east to Lwanjati, c. 1680 m., 14.vii.1975, *Pawek* 9869 (K; MAL; MO; SRGH). C: Foot of Dedza Mt., c. 2 km. NW. of Dedza, c. 1760 m., 1.iv.1970, *Brummitt* 9575 (K). S: Thyolo Mt., c. 1200 m., 21.ix.1946, *Brass* 17701 (K). **Mozambique**. N: E. coast of L. Malawi, 1900, *Johnson* s.n. (K). Z: Namuli Peaks, near Gurué, c. 1040 m., 25.vii.1962, *Leach & Schelpe* 11466 (K; LISC; SRGH). T: Zóbuè, 17.vi.1947, *Hornby* 2760 (K; SRGH). MS: Vila de Manica - Mina André, 30.iv.1963, *Gomes e Sousa* 4802 (COI).
Not recorded outside the Flora Zambesiaca area. High rainfall miombo woodland, submontane grassland and evergreen forest margins.

Var. **kotschyana** (Sch. Bip. ex Walp.) G.V. Pope in Kew Bull. **43**: 285 (1988). TAB. **22** fig. 11. Type from Sudan.
    Vernonia kotschyana Sch. Bip. ex Walp., Repert. Bot. Syst. **2**: 947 (1843). —Oliv. & Hiern in F.T.A. **3**: 289 (1877). —Adams in F.W.T.A. ed. 2, **2**: 276 (1963). Type as above.
    Vernonia leptolepis Bak. in Bull. Misc. Inf., Kew **1898**: 147 (1898) non O. Hoffm. (1895). Type: Malawi, between Kandowe and Karonga, *Whyte* s.n. (K, holotype).
    Vernonia woodii O. Hoffm. in Engl., Bot. Jahrb. **38**: 198 (1906). —S. Moore in Journ. Linn. Soc., Bot. **40**: 106 (1911). —Eyles in Trans. Roy. Soc. S. Afr. **5**: 505 (1916). Syntypes from South Africa (Natal).
    Vernonia bequaertii De Wild. in Fedde, Repert. **13**: 206 (1914). Type from Zaire.
    Vernonia tenoreana sensu Eyles in Trans. Roy. Soc. S. Afr. **5**: 505 (1916).

Leaves on upper stem subsessile or shortly petiolate, lamina usually narrowly cuneate or attenuate to the base, sparsely pilose to glabrescent sometimes pubescent to tomentose below. Phyllaries puberulous outside; appendages lanceolate usually longer than phyllary lamina, tapering-acuminate towards the apex and ± expanded at the base, flat or somewhat undulate not recurved or conduplicate.

Zambia. N: Mporokoso, Kalungwishi R., Lumangwe Falls, 26.ii.1970, *Drummond & Williamson* 10030A (K; SRGH). W: Mwinilunga, S. of Matonchi Farm, 11.x.1937, *Milne-Redhead* 2713 (BR; K; LISC). C: Mt. Makulu Res. Sta., c. 16 km. S. of Lusaka, 29.iv.1960, *Simwanda* 130 (K; SRGH). E: Chipata Distr., Ndolo For. Res., c. 19 km. S. of Katete, 22.iv.1952, *White* 2446 (BR; FHO; K). S: c. 19 km. W. of Livingstone (17°51'S, 25°42'E), 1.iii.1967, *Earle Smith* 4674 (K). **Zimbabwe**. W: Matobo, Farm Besna Kobila, c. 1400 m., v.1957, *Miller* 4347 (K; SRGH). E: Chirinda, Rain-forest, c. 1120 m., 16.viii.1961, *Methuen* 84 (K). S: Save-Runde (Sabi-Lundi) Junction, c. 240 m., 7.vi.1950, *Wild* 3444 (BR; K). **Malawi**. N: Mzimba Distr., Chikangawa-Lusangadzi road, 26.iv.1967, *Salubeni* 669 (K; SRGH). C: Ntchisi For. Res., 1400–1590 m., 26.iii.1970, *Brummitt* 9415 (K). **Mozambique**. M: Bokissa, c. 25 km. from Maputo, 22.iv.1981, *Jansen & Macuacua* 7696 (K; LM).
Widespread from tropical W. Africa to the Sudan and Ethiopia and southwards through Zaire and eastern Africa to South Africa (Transvaal and Natal). High rainfall miombo and mixed deciduous woodland often in riverine vegetation usually in tall grass, or in evergreen rainforest.

In the NE. Flora Zambesiaca area this subspecies grades into a form with large outer phyllaries, described as *V. bracteosa* by O. Hoffmann. The specimens *Pawek* 1595 (K), 2585 (K) and 5376 (K) from the Mzimba Distr. of Malawi are intermediate between this subspecies and typical *V. bracteosa* O. Hoffm.

86. **Vernonia calvoana** (Hook.f.) Hook.f. in Bot. Mag. **94**: t. 5698 (1868). —Oliv. & Hiern in F.T.A. **3**: 293 (1877). —Adams in J.W. Afr. Sci. Assn. **3**: 117 (1957). —C. Jeffrey in Kew Bull. **43**: 235 (1988). Type from Cameroon.
  Stengelia calvoana Hook.f. in Journ. Linn. Soc., Bot. **7**: 199 (1864). Type as above.

Subsp. **meridionalis** (Wild) C. Jeffrey in Kew Bull. **43**: 237 (1988). TAB. **22** fig. 9. Type: Zimbabwe, Nyanga (Inyanga), *Chase* 1636 (K; SRGH, holotype).
  Vernonia hymenolepis subsp. *meridionalis* Wild in Kirkia **11**: 15; 113 (1978). Type as above.
  Vernonia leucocalyx sensu Goodier & Phipps in Kirkia **1**: 65 (1961).
  Vernonia bracteosa sensu Earle Sm. in U.S. Dept. Agric. Handbook **396**: 22 (1971) pro parte quoad specims. *Orpen* 11/56, *Swynnerton* 273, *Methuen* 134 a & b, *Munch* 126, *Gomes e Sousa* 4738.

An erect perennial herb or subshrub, 1–3.3 m. tall from a woody rootstock. Stems 1–several, becoming woody below, branching above, stems and branches leafy, tomentellous-velutinous to pubescent or glabrescent; hairs patent flagelliform. Leaves subsessile to ± petiolate, 3–16 × 1–6 cm., lanceolate to ovate, apex acute to somewhat acuminate, base ± broadly cuneate or abruptly narrowed onto a short tapering-winged petiole, margins coarsely serrate with sharply pointed callose-tipped teeth; upper surface drying dark-green, sparsely scabridulous-puberulous or glabrescent; lower surface pale greenish-grey, densely to sparsely woolly-tomentose; nervation somewhat prominent-reticulate beneath. Capitula numerous in ± dense clusters, terminal on branches; capitula stalks 0.4–3 cm. long. Involucre 10–16 × 10–15 mm., spreading to c. 25 mm. wide, subspherical to broadly cup-shaped. Phyllaries apically appendaged; the outer phyllaries 5–11 mm. long, including a narrowly lanceolate appendage 4–8 times as long as the phyllary base, outermost phyllaries extending onto the capitulum-stalk; middle phyllaries 8–15 mm. long (including appendage); inner phyllaries up to c. 10 × 2 mm., linear-elliptic with greatly reduced appendages; appendages creamy-white, up to c. 12 × 8(10) mm., lanceolate-ovate to very broadly ovate or rotund, up to c. 3 times as wide as lamina base, acute-obtuse to rounded-acuminate at apex, puberulous or glabrescent. Florets 25–70 per capitulum. Corollas mauve, 11–14 mm. long, tube long and slender abruptly widening above into a short cylindric limb ± exserted above the pappus. Achenes dark purple-brown when mature, 3–4 mm. long, narrowly subcylindric, tapering to the base, c. 10-ribbed (or 20-ribbed with development of secondary ribs), sparsely hispidulous; pappus several-seriate, copious, caducous, setae stramineous or pale-brownish, somewhat flattened, barbellate on margins, increasing to c. 9 mm. long on inside.

Zimbabwe. E: Chipinge, SE. Chirinda Forest, Zona River, c. 730 m., 15.v.1962, *Chase* 7716 (K; LISC; MO; SRGH). **Mozambique**. MS: Gorongosa Mt., 10.vii.1969, *Leach & Cannell* 14296 (K; LISC; SRGH).
Known only from the eastern border mountains of Zimbabwe and adjacent mountains in Mozambique. Evergreen rainforest margins and submontane grassland, often beside rivers.
Subsp. *meridionalis* is the southern-most of the 7 vicariant montane subspecies that are recognised for *V. calvoana*. The typical subspecies, subsp. *calvoana*, is confined to the mountains of west tropical Africa; subsp. *adolfi-friderici* (Muschl.) C. Jeffrey occurs in Uganda, Zaire and Rwanda; subsp. *ruwenzoriensis* C. Jeffrey occurs in Uganda and Zaire; while subsp. *oehleri* (Muschl.) C. Jeffrey, subsp. *usambarensis* C. Jeffrey, subsp. *ulugurensis* (O. Hoffm.) C. Jeffrey and subsp. *leucocalyx* (O. Hoffm.) C. Jeffrey all occur on mountains in Tanzania.
Subsp. *meridionalis* and subsp. *leucocalyx* both have smaller more numerous capitula than is usual for the other subspecies. Subsp. *leucocalyx* differs from subsp. *meridionalis* in having larger more elliptic leaves.
Vernonia hymenolepis A. Rich., which has often been confused with *V. calvoana*, replaces it in the uplands of eastern Uganda, Kenya and southern Ethiopia. It may be distinguished from *V. calvoana* by having mostly smaller narrower leaves and phyllary appendages (the leaves usually densely pubescent), and by its usually glabrous achenes.

87. **Vernonia ringoetii** De Wild., Contrib. Fl. Katanga: 228 (1921). —C. Jeffrey in Kew Bull. **43**: 237 (1988). TAB. **22** fig. 8. Type from Zaire.
  Vernonia alboviolacea De Wild. in Fedde, Repert. **13**: 205 (1914) non Muschler (1911). Type as above.
  Vernonia tuberculata Hutch. & Burtt in Rev. Zool. Bot. Africaines **23**: 39 (1932). Syntypes from Zaire.

Vernonia lasiopus sensu Earle Sm. in U.S. Dept. Agric. Handbook **396**: 39 (1971) pro parte.
Vernonia hymenolepis subsp. *leucocalyx* sensu Wild in Kirkia **11**: 16, fig. 8; 113 (1978), quoad spec. cit. *Mutimushi* 307, non (O. Hoffm.) Wild sens. str.

An erect perennial herb or subshrub, to c. 2 m. tall from a woody rootstock. Stems erect, becoming woody below, leafy, tomentellous-velutinous to pubescent or glabrescent on older growth; hairs short-stalked flagelliform. Leaves petiolate; petiole 0.1–6 mm. long; lamina 3–17 × 1–7 cm., lanceolate to narrowly oblong-lanceolate sometimes ovate, apex tapering acute to obtuse, base cuneate to somewhat abruptly narrowed and attenuate down to the petiole, margins coarsely serrate with callose-tipped teeth; upper surface drying dark-green, sparsely scabridulous-puberulous or glabrescent; lower surface pale greenish-grey, woolly-tomentose, nerve reticulation ± prominent below. Capitula numerous in large, spreading, corymbiform cymose arrangements to c. 20 cm. across, the capitula somewhat aggregated but not congested; synflorescence branches to c. 30 cm. long; capitula stalks 0.2–4 cm. long. Involucres 10–15 × 10–15 mm., spreading to c. 20 mm. wide, obconic-campanulate. Phyllaries appendaged; the outer 5–8 mm. long, including a narrowly lanceolate tapering appendage 4–7 times as long as the phyllary base, the appendage hardly exceeding the phyllary base in width, outermost phyllaries extending onto the capitulum stalk; middle phyllaries 8–12 mm. long, appendages decreasing in size; inner phyllaries to 10 × 2 mm., with greatly reduced appendages; appendages purplish to creamy-white, up to c. 12 × 4 mm., narrowly oblong to lanceolate, tapering apically, puberulous or glabrescent. Florets 15–35 per capitulum. Corollas mauve becoming whitish, 11–14 mm. long, slender tubular below abruptly widening into a short cylindric exserted limb. Achenes dark purple-brown, c. 3 mm. long, narrowly turbinate-cylindric, c. 10-ribbed (or 20-ribbed with development of secondary ribs), sparsely hispidulous; pappus setose many-seriate, copious, caducous, setae stramineous or brownish, somewhat flattened, barbellate on margins, increasing to c. 9 mm. long inside.

Zambia. W: between Mwinilunga and Matonchi Farm, 16.v.1986, *Philcox, Pope and Chisumpa* 10325 (BR; K; LISC; MO; SRGH).
Also in Burundi, Tanzania and Zaire. Mushitu (swamp forest) margins and stream sides, usually in swampy ground.

88. **Vernonia tolypophora** Mattf. in Engl., Bot. Jahrb. 59, Beibl. **133**: 6 (1924). —C. Jeffrey in Kew Bull. **43**: 238 (1988). Syntypes from Tanzania.
Vernonia hymenolepis sensu Earle Sm. in U.S. Dept. Agric. Handbook **396**: 24 (1971), quoad specim. *Brass* 17174.
Vernonia lasiopus sensu Earle Sm. in U.S. Dept. Agric. Handbook **396**: 39 (1971), quoad specim. *Fanshawe* 6732, *Richards* 9619.
Vernonia hymenolepis subsp. *tolypophora* (Mattf.) Wild in Kirkia **11**: 17; 114 (1978). Type as above.

An erect sparsely-branched subshrub, 1.2–3 m. tall. Stems to c. 3 cm. in diam. and ± woody near the base; upper stem and branches greenish to purple, leafy, sparsely pubescent to tomentose or glabrescent, the hairs each consisting of a long oblique terminal cell on a few–many-celled stalk, the indumentum a variable mixture of short and long-stalked hairs. Leaves petiolate; petiole 0.3–2 cm. long; lamina soft, up to c. 12 × 7 cm., usually smaller, ovate to lanceolate, apex acute, base ± abruptly narrowed and shortly decurrent on the petiole, sometimes broadly cuneate or occasionally subcordate, margins serrulate to serrate with callose-tipped teeth, upper surface dark-green and sparsely puberulous with scattered patent ± rigid-based hairs similar to those of the stem, lower surface pale greenish-grey sparsely pubescent to woolly-tomentose, the hairs appressed shortly-stalked and matted, interspersed with patent long-stalked hairs. Capitula in congested globose clusters; capitula-clusters numerous, 5–18-capitulate, to c. 5 cm. in diam., terminal on branches; capitula stalks 1–5 mm. long. Involucres 8–15 × 10–20 mm., broadly campanulate to subspherical. Phyllaries many-seriate, apically appendaged; outer phyllaries linear-lanceolate to lanceolate, 5–7 mm. long, including an appendage 1–2 times as long as the phyllary base, puberulous or glabrescent; middle phyllaries 7–14 mm. long (including appendage), appendages decreasing in size; inner phyllaries to c. 10 × 1.3 mm., linear-elliptic with greatly reduced appendages; appendages pale-purplish to creamy-white, 2–4.5 mm. wide, gradually tapering to the apex or ovate-lanceolate, equalling or slightly exceeding phyllary in width, wider appendages often becoming ± folded. Florets 20–40 per capitulum. Corollas mauve or white, 9–15 mm. long, slender and tubular below, abruptly widening into a short cylindric limb well exserted above pappus.

Achenes dark-brown, 2.5–3 mm. long, narrowly turbinate, 10-ribbed, ribs pale-brown, between the ribs dark purplish-brown and glandular, otherwise glabrous; pappus several-seriate, copious, caducous, setae stramineous or pale-brownish, flattened, curving outwards, barbellate on margins, c. 2 mm. long on outside increasing to c. 8 mm. inside.

Zambia. N: Mbala Distr., Vomo Gap (Fwambo side), c. 1500 m., 27.ix.1960, *Richards* 13292 (K; M; SRGH). C: Serenje, 27.ix.1961, *Fanshawe* 6732 (K). **Malawi**. N: Mzimba Distr., Mzuzu, Katoto, 1.viii.1970, *Pawek* 3645 (K). **Mozambique**. N: Lichinga (Vila Cabral), 20.v.1934, *Torre* 82 (COI; LISC).
Also in Tanzania. River and stream sides, in dambos, meadows and rainforest margins, at medium to high altitudes often massed.

89. **Vernonia filipendula** Hiern, Cat. Afr. Pl. Welw. **1**, 3: 536 (1898). —Mendonça, Contrib. Conhec. Fl. Angol., **1** Compositae: 29 (1943). —Earle Sm. in U.S. Dept. Agric. Handbook **396**: 52 (1971) pro parte. —Wild in Kirkia **11**: 114 (1978) pro parte excl. specim. *V. longipedunculata*. TAB. 21 fig. A. Type from Angola.
      Vernonia lancibracteata S. Moore in Journ. Bot. **46**: 293 (1908). Type: Zimbabwe, *Eyles* 291 (BM, holotype; SRGH).
      Vernonia graciliflora De Wild. in Fedde, Repert. **13**: 205 (1914). Syntypes from Zaire.
      Vernonia rigidifolia sensu Wild in Kirkia **11**: 122 (1978) quoad specim. *Mutimushi* 2999, excl. typ.

A stiffly erect perennial herb, 25–100 cm. tall from a small woody rootstock; roots c. 5–10, thong-like, each swelling abruptly 1.5–5 cm. from the rootstock to form a root-tuber; tubers up to c. 4 × 1.4 cm., ± fusiform. Stems annual, 1–several, simple or sparsely branched above, upper internodes ± zig-zag, stems leafy with leaves much reduced or absent near the base and increasing in size towards the apex, ± coarsely pubescent and glandular above, glabrescent below; indumentum consisting of a variable mixture of short-stalked appressed crisped hairs and long purple-stalked patent hairs with oblique elongate terminal cells. Leaves sessile, to c. 15 × 4.3 cm., mostly smaller, oblanceolate to somewhat elliptic-oblanceolate, apex subacute to rounded, base attenuate-cuneate, margins subserrate to coarsely serrate, lamina coriaceous; upper surface scabrid or scabridulous, hairs scattered patent flagelliform becoming rigid at the base; lower surface puberulous sometimes scabridulous, the indumentum similar to that of the stem; venation somewhat raised beneath. Capitula solitary and terminal, or 2–4 on short branches, occasionally to c. 30 laxly subpaniculate. Involucres 17–23 × 17–35 mm., broadly cup-shaped. Phyllaries many-seriate, apically appendaged; outer phyllaries continued onto the capitulum stalk, narrowly strap-shaped, 10–16 mm. long, including a narrow apical appendage, puberulous or glabrous; middle phyllaries from 8 to 16 mm. long (including appendage); inner phyllaries to c. 18 × 3 mm., narrowly strap-shaped with appendages greatly reduced; appendages from c. 10 × 6 mm. on outer phyllaries, decreasing in length on inner phyllaries and changing from narrowly oblong to ovate. Florets c. 30–60 per capitulum. Corollas purple to mauve fading to white, 16–23 mm. long, slender tubular below, abruptly widening into a short cylindric limb well exserted above the pappus. Achenes pale- to dark-brown when mature, 3–4 mm. long, subcylindric, hardly tapered to the base, 10-ribbed, with a thin flat annular carpopodium, strigose-hispid; pappus several-seriate, copious, ± caducous, setae barbellate on margins, c. 2 mm. long and subulate on outside increasing to c. 17 mm. long on inside, flattened and stiffly ribbon-like.

Zambia. W: c. 6 km. W. of Solwezi, 18.iii.1961, *Drummond & Rutherford-Smith* 6984 (K; LISC; SRGH). **Zimbabwe**. N: Mwami (Miami) Distr., Karoi Exp. Farm, 4.iii.1947, *Wild* 1824 (K, SRGH). **Malawi**. C: Chionjeza (Chinjiza), Dzalanyama For. Res., c. 1550 m., 9.ii.1959, *Robson* 1519 (BM; K; LISC; SRGH). S: Ntcheu Distr., Lower Kirk Range, Chipusiri, 1460 m., 17.iii.1955, *Exell, Mendonça & Wild* 954 (BM; LISC). **Mozambique**. N: Lichinga (Vila Cabral), iv.1934, *Torre* 127 (COI; LISC). T: Angónia, Msese Hill, c. 8 km. N. of Mlangeni, c. 1475 m., *Brummitt* 8610 (K; SRGH).
Also in Zaire and Angola. Miombo and mixed *Brachystegia* and *Uapaca* woodland.
V. filipendula is usually a small, 1–3-capitulate plant, with simple or very sparingly branched stems, leafless below and ± leafy on the zig-zag upper part. In the Copperbelt area of Zambia this grades into a larger, more robust form, strongly branched above with up to c. 30 capitula laxly, subpaniculately arranged (*Mutimushi* 2999). Similar specimens can be found in Zaire (*de Witte* 89) and in Angola (*Hundt* 792).

90. **Vernonia longipedunculata** De Wild. in Fedde, Repert. **13**: 207 (1914). —C. Jeffrey in Kew Bull. **43**: 241 (1988). Type from Zaire.
      Vernonia filipendula sensu Earle Sm. in U.S. Dept. Agric. Handbook **396**: 52 (1971) pro parte. —sensu Wild in Kirkia **11**: 114 (1978) pro parte quoad specim. *Richards* 2011 (BR; K; SRGH).

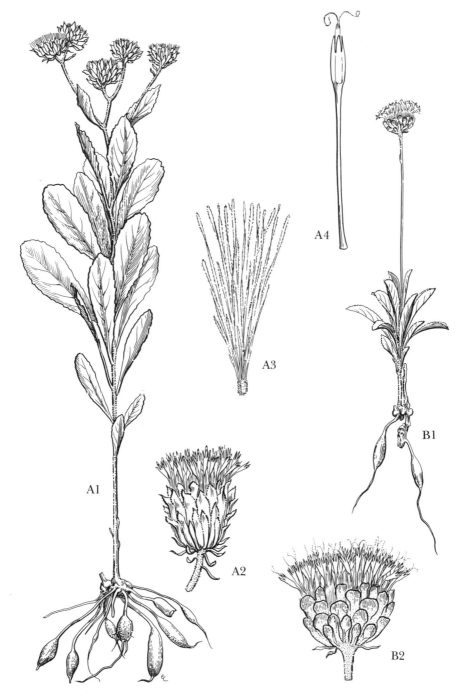

Tab. 21. A. —VERNONIA FILIPENDULA. A1, habit, note upper stem zig-zag between nodes (×⅓),
from *Philcox, Pope & Chisumpa* 10332; A2, capitulum, note outer phyllaries extending onto stalk
(× 1), from *Fanshawe* 875; A3, pappus, note ribbon-like setae (× 4); A4, corolla tube (× 2), A3 & A4
from *Mutimushi* 2477. B. —VERNONIA LONGIPEDUNCULATA var. MANIKENSIS. B1, habit
(×⅓); B2, capitulum, (× 1), B1 & B2 from *Fanshawe* 3519. Drawn by Eleanor Catherine.

An erect leafy or subscapose perennial herb, 10–90 cm. tall from a small woody rootstock; roots c. 3–10, ± thong-like, each swelling abruptly 1–8 cm. from the rootstock to form a root-tuber, or tubers sessile; tubers up to c. 6.5 × 2 cm., subfusiform to ovoid. Stems annual, 1–several, simple or branching above, ± uniformly leafy, or with leaves crowded below or subrosulate at 2–10 cm. above the stem base, pubescent to whitish-lanate or tomentose becoming less densely so above, or glabrescent; indumentum a variable mixture of white short-stalked ± appressed-crisped hairs and fewer long purple-stalked patent hairs with oblique terminal cells. Leaves sessile, 4–20 × 1.5–7 cm., obovate to oblanceolate, apex rounded to acute, base long-attenuate, margins serrulate to coarsely irregularly serrate-dentate, lamina finely whitish pubescent sometimes more densely so beneath, usually also with few to numerous long patent purplish hairs mostly on veins and the margins, densely glandular beneath, occasionally glabrescent. Capitula mostly solitary and terminal, less often up to c. 9 corymbiformly cymose. Involucres 15–20 × 25–35 mm., very broadly campanulate. Phyllaries apically appendaged, many-seriate, ± appressed-imbricate; outer phyllaries strap-shaped, 6–11 mm. long including a greenish oblong appendage slightly wider than the phyllary, puberulous or glabrous; middle phyllaries 11–14 mm. long (including appendage), stramineous in lower part, oblong-ovate, glabrous or puberulous where exposed; inner phyllaries to c. 17 mm. long, narrowly strap-shaped with appendages greatly reduced; appendages from c. 9 × 8 mm. on the outer phyllaries reducing in size on the inner, broadly ovate or oblong-ovate, greenish-white or purple-tinged, puberulous or glabrescent. Florets c. 30–70 per capitulum. Corollas purple or pale-mauve fading to white, 17–22 mm. long, slender tubular below, abruptly widening into a short cylindric limb c. 5 mm. long and well exserted above the pappus. Achenes pale- to dark-brown when mature, 4–5 mm. long, narrowly turbinate-cylindric, 10-ribbed, with a flat annular carpopodium, strigose-hispid; pappus of several-seriate, copious, ± caducous setae barbellate on the margins, the outer c. 2 mm. long and paleate, increasing to c. 14 mm. long inside, becoming flattened and ribbon-like.

1. Leaves mostly more than c. 10 × 4 cm., drying brownish - - var. *longipedunculata*
 – Leaves to c. 8 × 3 cm. mostly smaller, usually drying greenish - - - - - 2
2. Plants subscapose, appearing early in the season (August to October); leaves subrosulate or somewhat crowded on lower stem; lowermost 2–4 cm. of stem leafless and ± densely white pubescent to lanate (below leaf rosette), scapiform above the leaf rosette; capitula usually solitary - - - - - - - - - - - - - - - - var. *manikensis*
 – Stems leafy; plants appearing later in the season (October to December); leaves not subrosulate, stems ± uniformly leafy except near base where leaves are reduced in size and number; capitula (1)2–5 per stem - - - - - - - - - - - - - var. *retusa*

Var. longipedunculata
Vernonia wittei Hutch. & B.L. Burtt in Rev. Zool. Bot. Afr. **23**: 39 (1932). Type from Zaire.

Zambia. N: Mbala, Chicalungoma Road, 21.viii.1967, *Sanane* 7 (K). W: Mwinilunga Distr., c. 2–6 km. SW. of Mujibeshi River, 6.xi.1962, *Lewis* 6155 (K). **Mozambique**. N: c. 25 km. from Marrupa to Lichinga, c. 650 m., 10.viii.1981, *Jansen, de Koning & de Wilde* 179 (K; LMU).
Also from Zaire and Tanzania. A pyrophyte of miombo and mixed deciduous woodland and dambos.

Var. manikensis
(De Wild.) G.V. Pope in Kew Bull. **43**: 287 (1988). TAB. **21** fig. B. Type from Zaire.
Vernonia manikensis De Wild. in Bull. Jard. Bot. Brux. **5**: 99 (1915). Type as above.
Vernonia descampsii De Wild. in Bull. Jard. Bot. Brux. **5**: 97 (1915). Syntypes from Zaire.

Zambia. B: Masese, 11.ix.1961, *Fanshawe* 6697 (K). N: Mpulungu-Mbala Road, c. 1525 m., 11.x.1954, *Richards* 2011 (BR; K; SRGH). W: Mwinilunga Distr., c. 26 km. W. of Kabompo River, 11.ix.1930, *Milne-Redhead* 1102 (BR; K).
Also from Angola and Zaire. A pyrophyte of dambos, grassland and dry woodland.
V. pumila Kotschy & Peyr., from west and east tropical Africa, may be distinguished from this variety by its precocious flowers, and leaves grey- or greenish-brown tomentellous below.

Var. retusa
(R.E. Fr.) G.V. Pope in Kew Bull. **43**: 287 (1988). Type: Zambia, Kalungwishi River, *Fries* 1162 (UPS, holotype).
Vernonia retusa R.E. Fr., Wiss. Ergebn. Schwed. Rhod.-Kongo-Exped. 1911–1912, **1**: 323 (1916). Type as above.
Vernonia pleiotaxoides Hutch. & B.L. Burtt in Rev. Zool. Bot. Afr. **23**: 38 (1932). Type from Zaire.

Zambia. N: Kasama Distr., Mungwi, 5.xi.1960, *Robinson* 4036 (K). W: Ndola, Chichele For. Res., 9.ix.1952, *White* 3191 (BM; BR; FHO; K). C: Serenje, Mkushi River, c. 1350 m., 27.xi.1962, *Richards* 17486 (K).

Also from Zaire. A pyrophyte of dambos and dry woodland.

91. **Vernonia najas** Wild in Kirkia **11**: 1; 55, fig. 1 (1978). Type: Zambia, Mwinilunga, c. 18 km. E. of Kalene Hill, *Robinson* 6106 (K; M; SRGH, holotype).

An erect densely leafy perennial herb, to c. 45 cm. tall from a small woody rootstock; roots 2–many, thong-like, each swelling at a distance from the rootstock to form a root-tuber; tubers up to c. 5 × 2 cm., ± fusiform. Stems often purplish, annual, 1–several, long ascending-branched, ribbed to somewhat angular above, puberulous-tomentellous; branches up to c. 22 cm. long, sparsely leafy, becoming bracteate above, otherwise similar to stem. Leaves densely crowded on lower stem, ascending-overlapping, up to c. 12 × 0.7 cm., linear- to narrowly-oblanceolate, apex acute, base narrowly attenuate, margins remotely serrate with callose-tipped teeth, lamina coriaceous, glabrous on upper surface and puberulous to glabrescent beneath with reticulation prominent on both surfaces. Capitula mostly solitary and terminal on stem or branches. Involucre up to c. 15 × 20 mm., broadly campanulate. Phyllaries appendaged; the outer phyllaries from c. 3 mm. long, linear-oblong, with a greenish oblong-elliptic to ovate appendage; the middle phyllaries up to c. 12 × 4 mm., oblong with a purple broadly ovate to suborbicular appendage slightly wider than the phyllary; the inner phyllaries up to c. 14 × 2.5 mm., lorate with a smaller appendage; appendage venation prominent. Corolla purple, to c. 14 mm. long, with a long slender tube and a short cylindrical exserted limb. Achenes up to c. 4 mm. long, narrowly turbinate-cylindric, c. 10-ribbed, truncate below with a thin flat carpopodium, hispidulous and glandular between the ribs; pappus many-seriate, copious, setae pale-brownish, flattened, barbellate on the margins, c. 2 mm. long and subulate on the outside increasing to c. 10 mm. long inside.

Zambia. W: Mwinilunga-Mutshatsha road, c. 1.6 km. S. of Zambia-Zaire border, c. 1360 m., 12.xi.1962, *Lewis* 6235 (K).

Not known elsewhere. Sandy watershed grassland.

92. **Vernonia orchidorrhiza** Welw. ex Hiern, Cat. Afr. Pl. Welw. **1**, 3: 530 (1898). —Mendonça, Contrib. Conhec. Fl. Angol., 1 Compositae: 25 (1943). —Wild in Kirkia **11**: 54 (1978). Type from Angola.

An erect strict slender perennial herb, to c. 25 cm. tall from a small woody rootstock; roots 3–many, thong-like, each swelling close to the rootstock into a subfusiform root-tuber, or tubers subsessile, tubers up to c. 3 × 1 cm. Stems annual, 1–several, ?simple, leafy, strongly ribbed or angular, glabrous but tomentellous towards the apex. Leaves ascending, regularly spaced on the stem, 2–4.5 cm. long, filiform to narrowly linear, apex acute, margins entire and revolute, lamina glabrous on both surfaces or sparsely puberulous beneath, gland-dotted beneath. Capitula solitary and terminal. Involucre 8–12 × 7–15 mm., broadly campanulate. Phyllaries appendaged; the outer phyllaries from c. 5 mm. long (including appendage), linear; the middle phyllaries to c. 9 × 2.5 mm., including a short oblong appendage narrower than the phyllary base; the inner phyllaries to c. 10 × 1.5 mm., linear with a reduced appendage; appendages puberulous, glandular. Corolla purple, to c. 13 mm. long, with a long slender tube and a short cylindrical exserted limb. Achenes up to c. 3 mm. long, narrowly turbinate-cylindric, c. 10-ribbed, truncate below with a thin flat carpopodium, densely hispidulous and glandular; pappus many-seriate, copious, setae pale-brownish, flattened, barbellate on the margins, increasing to c. 8 mm. long inside.

Zambia. W: Mwinilunga Distr., Kalenda Dambo, 9.xi.1937, *Milne-Redhead* 3174 (BR; K).

Also in Angola. In seasonally wet dambos.

93. **Vernonia retifolia** S. Moore in Journ. Linn. Soc., Bot. **47**: 264 (1925). —Wild in Kirkia **11**: 115 (1978). TAB. **22** fig. 7. Type: Zambia, Luwingu, *Jelf* 44 (BM, holotype).

Vernonia venosa S. Moore in Journ. Linn. Soc., Bot. **47**: 265 (1925). Type from Zaire.

A stout erect tough perennial herb, to c. 1 m. tall from a large woody rootstock; tuberous-rooted? Stems annual, usually solitary, simple or sparingly branched near the apex, densely leafy, strongly striate, greyish-tomentellous. Leaves sessile, to c. 16 × 4 cm., ± oblanceolate to narrowly elliptic, acute at apex, cuneate at base, margins crenate-serrate;

lamina thinly coriaceous, finely pubescent on both surfaces usually more densely so beneath, soon glabrescent, glandular beneath; venation finely reticulate and markedly prominent beneath. Capitula c. 5–12 in lax clusters, or solitary on short ± leafy branches; bracts leaf-like, to c. 4 × 2 cm., subtending and equal to, or overtopping the capitula. Involucres c. 20 × 20 mm., spreading to c. 30 mm. wide, campanulate. Phyllaries many-seriate, at least the outer apically appendaged, brownish, cartilaginous, pubescent outside where exposed; the outermost ± wholly leaf-like, often overtopping the capitulum, with appendages large rotund reticulate crustaceous, enclosing shorter appressed-imbricate inner phyllaries; the inner phyllaries lorate, increasing to c. 18 mm. long while their appendages reduce to a narrow membranous brownish-purple apical tip. Florets c. 30 per capitulum. Corollas mauve fading to creamy-white, c. 22 mm. long, long slender tubular below abruptly widening into a short cylindric limb exserted above the pappus. Achenes 3–4 mm. long, broadly subcylindric and tapering slightly to the base, 10-ribbed, with a thin flat annular carpopodium, ± densely strigose-hispid, pappus several-seriate, copious somewhat caducous, setae pale-brownish, barbellate on the margins, c. 2 mm. long and subulate-paleate on outside increasing to c. 16 mm. long on inside, flattened and slightly ribbon-like.

Zambia. N: Kawambwa, 22.viii.1957, *Fanshawe* 3485 (BR; K).
Also in Zaire. Miombo woodland.
V. kapirensis De Wild. closely approaches this species but may be distinguished by its much larger capitula with more numerous phyllaries and florets.

94. **Vernonia bruceana** Wild in Kirkia **11**: 17, fig. 9; 64 (1978). Type: Zambia, Mporokoso, Kundabwika Falls, *Tyrer* 193 (BM; SRGH, holotype).
    Vernonia luaboensis Kalanda in Bull. Jard. Bot. Nat. Belg. **52**: 123 (1982). Type from Zaire.

An erect slender perennial herb, or subshrub to c. 2 m. tall from a large woody rootstock; tuberous-rooted? Stems usually solitary, becoming woody below, simple, branching only near the apex, leafy, ± appressed grey-tomentose, becoming puberulous below, glandular; hairs ± short-stalked flagelliform. Leaves with petioles to c. 3 mm. long; lamina coriaceous, up to c. 13 × 3.2 cm., oblonceolate to ± narrowly elliptic, apex acute or obtuse, base cuneate, margins irregularly serrate to subserrate, upper surface scabridulous or glabrescent, lower surface finely puberulous and prominently closely reticulate. Capitula in numerous 3–9-capitulate clusters; clusters ± scorpioidly cymose along the branches; synflorescence branches divaricate, to c. 15 cm. long, leafy with leaves bract-like above, tomentellous; capitula stalks 0–10 mm. long. Involucres 10–12 × 7–14 mm., ± narrowly cylindric-campanulate. Phyllaries lanceolate to lorate, coriaceous and thinly pubescent in upper part, glabrous and stramineous in lower part, apically appendaged; the outermost phyllaries from c. 3 mm. long (including appendage); the inner phyllaries increasing to c. 12 × 3 mm., their appendages diminishing in size; appendages up to c. 4 × 5 mm., oblong-ovate to very broadly ovate, often recurved or spreading, membranous, acute to acuminate, with nerves somewhat raised, densely tomentellous outside where exposed. Florets c. 5 per capitulum. Corollas pale mauve to white, 14–16 mm. long, slender-tubular below, abruptly dilated into a cylindric limb exserted beyond the pappus. Achenes dark-brown when mature, c. 4 mm. long, subcylindric-subfusiform, c. 10-ribbed, hispid; pappus several-seriate, copious, somewhat caducous, setae flattened barbellate increasing from c. 1 mm. long on the outside to c. 8 mm. long inside.

Zambia. N: Kaputa Distr., Kundabwika Falls on Kalungwishi R., 29°19'E, 09°12'S, 17.iv.1989, *Pope, Radcl.-Sm. & Goyder* 2170 (BR; K; LISC).
Also in Zaire. Miombo or mixed deciduous woodland on white sandy soil.

95. **Vernonia lafukensis** S. Moore in Journ. Bot. **52**: 94 (1914). TAB. **22** fig. 6. Type from Zaire.
    Vernonia sclerophylla pro parte quoad syn. *V. lafukensis* sensu Earle Sm. in U.S. Dept. Agric. Handbook **396**: 61 (1971).

An erect tough perennial herb or subshrub, to c. 75 cm. tall from a woody rootstock; tuberous-rooted? Stems annual, usually solitary, becoming woody below, simple or branching above, densely leafy, striate, silvery-white tomentose; hairs shortly-stalked with an elongate oblique terminal cell. Leaves subsessile, to c. 12.5 × 4 cm., narrowly elliptic-oblanceolate, apex acute, base tapering-cuneate, margins serrate; lamina thinly coriaceous, green and sparsely finely pilose-pubescent on the upper surface, paler and

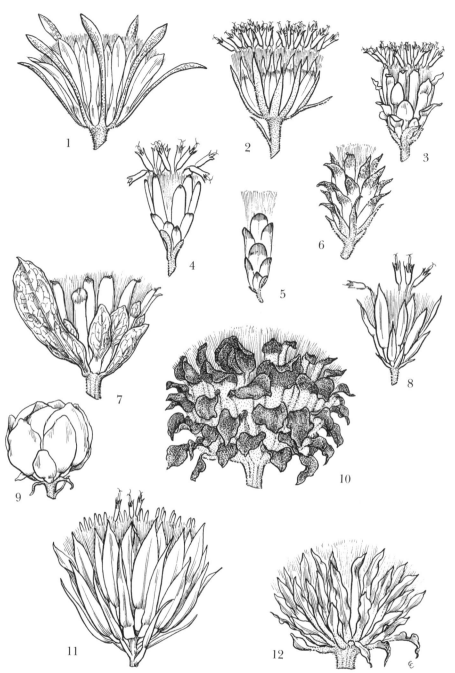

Tab. 22. VERNONIA SPP. —Capitula, with phyllaries variously appendaged. 1. —V. ANTHELMINTICA, from *Eyles* 1295. 2. —V. TUBERIFERA, from *Richards* 15440. 3. —V. INCOMPTA, from *Richards* 4469. 4. —V. SOLWEZIENSIS, appendages reduced, from *Holmes* 1449. 5. —V. SUNZUENSIS, from *Richards* 1610. 6. —V. LAFUKENSIS, from *Brummitt et al.* 17086. 7. —V. RETIFOLIA, from *Fanshawe* 3485. 8. —V. RINGOETII, from *Mutimushi* 307. 9. —V. CALVOANA subsp. MERIDIONALIS, from *Chase* 7716. 10. —V. ADOENSIS var. MOSSAMBIQUENSIS, from *Pawek* 3748. 11. —V. ADOENSIS var. KOTSCHYANA, from *Pawek* 3610. 12. —V. ADOENSIS var. ADOENSIS, from *Kelly* 44. (All capitula × 1½). Drawn by Eleanor Catherine.

more densely pilose-pubescent beneath especially on the midrib, venation closely reticulate and strongly raised on the lower surface. Capitula numerous in congested clusters up to c. 8 cm. across, terminal on stem and branches; synflorescence branches leafy, felted-tomentellous; capitula stalks 0 or up to c. 10 mm. long. Involucres 10–15 × 5–10 mm., narrowly obconic-campanulate. Phyllaries progressively longer to the inside, coriaceous and stramineous below, tapering-submembranous above and subapically aristate, finely sericeous outside where exposed, ciliolate; apicies differentiated but not petaloid-appendaged; outer phyllaries linear-lanceolate with dark long subulate apices; inner phyllaries with membranous darker tapering-acute ± recurved apices. Florets c. 7–15 per capitulum. Corollas pale-mauve to white, 14–17 mm. long, slender-tubular below, abruptly widening into a cylindric limb well exserted above the pappus. Achenes 2–3 mm. long, subcylindric, tapering somewhat below, 10-ribbed (or 20-ribbed with the development of secondary ribs), strigose-hispid; pappus several-seriate, copious, caducous, setae flattened and barbellate increasing from c. 2 mm. long on the outside to c. 10 mm. long inside.

Zambia. N: c. 8 km. from Mporokoso to Kawambwa, 1.iv.1984, *Brummitt, Chisumpa & Nshingo* 17086 (BR; K; NDO; SRGH).
Also in Zaire. Miombo woodland.
C. Earle Smith in U.S. Dept. Agric. Handbook **396**: 61 (1971) places this species in synonymy under *V. sclerophylla* O. Hoffm. However, on examining the type material it is clear that these are two distinct species readily distinguished on leaf, involucre and indumentum characters: *V. sclerophylla* lacks the prominent closely reticulate venation on the leaf lower surfaces, its involucres are twice as wide as those in *V. lafukensis* and its phyllary appendages are well developed and lack a subapical bristle. The indumentum of the stem and branches is coarsely pubescent in *V. sclerophylla*.

96. **Vernonia sunzuensis** Wild in Kirkia **11**: 13; 110 (1978). —C. Jeffrey in Kew Bull. **43**: 241 (1988). TAB. **22** fig. 5. Type: Zambia, Mbala Distr., Sunzu Mt., *Richards* 15111 (SRGH, holotype; K; PRE).
Vernonia acuta De Wild. in Bull. Jard. Bot. Brux. **5**: 90 (1915), non *V. acuta* N.E. Br. (1901). Type from Zaire.

An erect slender perennial herb or subshrub to c. 1.5 m. tall, from a large woody rootstock. Stems 1–several, becoming woody below, simple, branching only near the apex, densely leafy, striate, greyish or pale-brown shortly-woolly or felted-tomentose; hairs short-stalked with an elongate oblique terminal cell, matted. Leaves subsessile, 4–18 × 1.5–4.5 cm., ± oblong-elliptic to oblanceolate or obovate, apex acute to subobtuse or rounded, base cuneate to rounded, margins serrate to coarsely serrate, lamina coriaceous, sometimes ± discolorous; upper surface green, glabrous, and finely reticulate; lower surface appressed grey-brown tomentellous or felted; venation raised and finely reticulate beneath, somewhat obscured. Synflorescence large, capitula numerous in 15–60-capitulate clusters; capitulum-clusters terminal on stem and leafy branches; branches ± ascending, to c. 15(20) cm. long, grey-brown felted, capitula stalks 0 or to c. 5 mm. long. Involucres 9–12 × 4–5 mm., ± narrowly subcylindric, not spreading. Phyllaries tightly imbricate, longer to the inside, pubescent or felted outside where exposed, apicies ± differentiated but not appendaged, lower part coriaceous, stramineous, glabrous; tips green membranous and felted outside, apices rounded sometimes purple-tipped; the outer phyllaries from c. 2 mm. long, lanceolate-ovate; the middle phyllaries to c. 3 mm. wide, to oblong-oblanceolate; the inner phyllaries to c. 11 × 2 mm., narrowly oblong-oblanceolate and obtuse at apex. Florets 3–4 per capitulum. Corollas pale-mauve to white, 11–14 mm. long, slender tubular below abruptly widening into a cylindric limb well exserted above the pappus. Achenes pale- to dark-brown when mature, 3–4 mm. long, subcylindric to narrowly turbinate, 10-ribbed (or 20-ribbed with the development of secondary ribs), strigose-hispid; pappus several-seriate, copious, caducous, the setae narrowly-flattened and barbellate, increasing from c. 2 mm. long on outside to c. 10 mm. long inside.

Zambia. N: Mbala Distr., Chilongowelo, c. 1520 m., 9.v.1955, *Richards* 5595 (K).
Also in Tanzania and Zaire. Wooded grassland with mixed deciduous trees and tall grasses.

97. **Vernonia solweziensis** Wild in Kirkia **11**: 13; 110 (1978). TAB. **22** fig. 4. Type: Zambia, Solwezi, *Drummond & Rutherford-Smith* 7000 (SRGH, holotype; K).

An erect slender perennial herb or subshrub, to c. 1.2 m. tall from a large woody rootstock; tuberous-rooted? Stems annual, purplish, usually solitary, stout becoming woody below, branched near the apex, densely leafy, striate, shortly pale-brownish tomentose or felted becoming whitish puberulous below, glandular; hairs short-stalked with an elongate oblique terminal cell. Leaves subsessile, up to c. 12 × 3.5 cm., elliptic to lanceolate, apex acute, base cuneate, margins subentire or repand to coarsely serrate, lamina coriaceous, discolorous; upper surface glabrous and finely reticulate, lower surface appressed grey-brown tomentellous with a raised finely reticulate venation. Synflorescence of many lax capitulum-clusters; the clusters 3–9-capitulate, terminal on stem and leafy branches; branches few to many, ascending, 1–30 cm. long, felted-tomentellous; capitula stalks 0 or to c. 10 mm. long. Involucres 9–11 × 5–9 mm., spreading to c. 15 mm. wide when mature, obconic-campanulate. Phyllaries imbricate, longer to the inside, coriaceous, puberulous or glabrous, stramineous; apicies ± differentiated but not appendaged, rounded and sometimes mucronulate, convex to ± hooded and darkly-membranous tipped; the outer phyllaries c. 1.5 mm. long, ovate; the middle phyllaries up to c. 3.5 mm. wide, from ovate to oblong-obovate; the inner phyllaries up to c. 12 × 3 mm., narrowly oblong-oblanceolate. Florets c. 11 per capitulum. Corollas pale-mauve to white, 13–16 mm. long, slender-tubular below, abruptly widening into a cylindric limb exserted beyond the pappus. Achenes pale- to dark-brown when mature, 2–3 mm. long, subcylindric, tapering slightly below, 10-ribbed (or 20-ribbed with the development of secondary ribs), strigose-hispid; pappus several seriate, copious, caducous, the setae flattened barbellate increasing from c. 2 mm. long on the outside to c. 9 mm. long inside.

Zambia. W: Mwinilunga Distr., Matonchi Farm, 24.i.1938, *Milne-Redhead* 4314 (BR; K; PRE).
Not recorded from outside the Flora Zambesiaca area but would be expected to occur in Zaire. Miombo woodland and dambo margins.
V. cardiolepis O. Hoffm., from neighbouring Angola, may be distinguished by its oblanceolate leaves and by its phyllaries which have definite appendages. In the Zambian *V. solweziensis* the leaves are elliptic-lanceolate and the phyllaries lack appendages. *V. solweziensis* may be distinguished from *V. guineensis* Benth. by the presence in the latter of a distinct outer pappus of small linear-lanceolate scales and the downy-araneose, soon glabrescent, leaf upper surfaces.

98. **Vernonia incompta** S. Moore in Journ. Bot. **56**: 210 (1918). —C. Jeffrey in Kew Bull. **43**: 241 (1988). TAB. **22** fig. 3. Type: Zambia, Katuba (? Kaluba) Stream, *Kassner* 2261 (BM, holotype; E; K).
  Vernonia guineensis sensu Earle Sm. in U.S. Dept. Agric. Handbook **396**: 45 (1971) pro parte; sensu Wild in Kirkia **11**: 100 (1078) encl. typ.

An erect slender perennial herb, 20–100 cm. tall from a woody rootstock; roots 3–6, thong-like, each swelling into a fusiform root-tuber 1–5 cm. from the rootstock. Stems annual, purplish-brown, 1–several, stiff, ascending-branched in upper part, leafy, striate, tomentellous or ± felted, becoming sparsely whitish-puberulous below, glandular; indumentum a variable mixture of few-celled short-stalked and many-celled long-stalked patent hairs each with a long terminal somewhat contorted cell. Leaves sessile, 3–11 × 0.7–2 cm., ± narrowly oblanceolate, or up to c. 9 × 0.3 cm. and linear-oblanceolate, apex acute, base narrowly attenuate, margins serrate, lamina coriaceous, markedly discolorous; upper surface drying dark green-brown, glabrous or scabridous, lower surface greyish or yellowish-brown tomentellous, or felted; venation beneath ± raised reticulate, ± exposed glabrous. Capitula usually arranged in lax clusters, or sometimes solitary on branches; capitulum-clusters few to many, 2–7-capitulate, sometimes scorpiodly cymose along the synflorescence branches; branches 2–30 cm. long, leafy, tomentellous; capitula stalks 1–25 mm. long. Involucres 9–14 × 8–15 mm., spreading to c. 20 mm. wide, campanulate, lanate or tomentose outside, glabrescent. Phyllaries appendaged, appendages best developed in the middle phyllaries, sometimes absent from the others; outer phyllaries c. 2 mm. long, oblong-lanceolate or narrowly ovate, pubescent; middle phyllaries 3–9 mm. long (including appendage), greenish-stramineous and glabrous in lower part, or puberulous where exposed, expanding into a broadly ovate or oblong-ovate whitish thinly membranous appendage to 5 × 6 mm.; inner phyllaries to c. 12 × 1.5 mm., oblanceolate to linear-oblanceolate with appendage reducing in size. Florets 15–28 per capitulum. Corollas pale-mauve to white, 12–15 mm. long, slender tubular below, abruptly widening into a cylindric limb exserted beyond the pappus. Achenes pale- to dark-brown, 3–4.5 mm. long, narrowly turbinate-cylindric, 10-ribbed, strigose-hispid; pappus several-seriate, copious, ± caducous, setae barbellate, c. 2 mm. long and subulate on outside increasing to 7 mm. long inside, becoming flattened.

Zambia. N: Mbala Distr., Nkali (Kali) Dambo, c. 1520 m., 15.i.1955, *Richards* 4119 (BR; K; SRGH). W: Ndola, South Downs Road, 8.i.1962, *Linley* 258 (K; LISC; SRGH). **Malawi**. N: Chitipa Distr., Misuku Hills, c. 1220 m., 10.i.1959, *Robinson* 3141 (K; PRE; SRGH).

Also in Tanzania, Zaire and Angola. Miombo woodlands and wooded grassland, usually on sandy soil or on rocky hillsides.

V. guineensis Benth. may be distinguished by the presence of a distinct outer pappus whorl of small narrowly oblong-oblanceolate scales and also by the broader lanceolate leaves at first somewhat downy-araneose on the upper surface.

99. **Vernonia tuberifera** R.E. Fr., Wiss. Ergebn. Schwed. Rhod.-Kongo-Exped. 1911–1912, **1**: 326 (1912). —Wild in Kirkia **11**: 122 (1978). TAB. **22** fig. 2. Type: Zambia, Mukanshi R., *Fries* 1123 (UPS, holotype).

An erect slender perennial herb, to c. 37 cm. tall from a small woody rootstock. Rootstock producing 2–5 subsessile root-tubers; tubers up to 3 × 1.5 cm., subfusiform, tapering into long thong-like roots. Stems annual, 1–several, simple or sparsely branched above, leafy, ribbed, ± coarsely pubescent and glandular, more densely so towards the apex, pilose below; indumentum a mixture of ± appressed short-stalked hairs and long patent hairs, the hairs flagelliform with an often purplish stalk and an elongate terminal cell. Leaves subsessile, to 10 × 2 cm., narrowly oblanceolate, apex obtuse, base tapering-cuneate, margins subserrate with callose-tipped teeth but entire near the base, lamina scabridulous-pilose and glandular on both surfaces; the hairs scattered, patent, flagelliform, becoming rigid below; venation not raised reticulate. Capitula solitary and terminal on stems, or 2–3 on short branches. Involucres 9–11 × 11–18 mm., broadly obconic or cup-shaped. Phyllaries several-seriate, apicies ± differentiated but not appendaged, purple-tipped and glandular at the apices; the outer phyllaries 6–8 × 1–2 mm., linear-lanceolate, puberulent; the middle phyllaries to c. 9 × 3 mm., oblong-lanceolate, ± acuminate at apex, glabrous or puberulent where exposed; the inner phyllaries to c. 10 mm. long, linear-acute. Florets c. 20 per capitulum. Corollas purple or mauve, to c. 15 mm. long, slender tubular in lower two thirds abruptly widening into a cylindric limb exserted above the pappus. Achenes (immature) c. 2 mm. long, ribbed, strigose-hispid; pappus several-seriate, copious, somewhat caducous, setae increasing in length to c. 8 mm. long on inside, somewhat flattened and barbellate on the margins.

Zambia. N: Kawambwa, 3.xii.1961, *Astle* 1033 (K; SRGH). W: Mwinilunga, between Luakela and Ikelenge, 25.x.1969, *Drummond & Williamson* 9444 (K; SRGH).
Also in Angola. Watershed dambos.

100. **Vernonia anthelmintica** (L.) Willd., Sp. Pl. **3**: 1634 (1803). —C. Jeffrey in Kew Bull. **43**: 241 (1988). TAB. **22** fig. 1. Type from Sri Lanka.
   Conyza anthelmintica L., Sp. Pl. ed. 2: 1207 (1763). Type as above.
   Vernonia stenolepis Oliv. in Trans. Linn. Soc., Ser. 2, **2**: 337 (1887). —Merxm., Prodr. Fl. SW. Afr. 139: 184 (1967). —Earle Sm. in U.S. Dept. Agric. Handb. **396**: 34 (1971). —Wild in Kirkia **11**: 115 (1978). Type from Tanzania.
   Dolosanthus sylvaticus Klatt in Bull. Herb. Boiss. **4**: 473, t.5 (1896). Type from Tanzania.

A tender annual herb, 0.1–2 m. tall. Stem erect, sparingly branched above, leafy in upper part, ribbed, sparsely pilose to puberulous or glabrescent; indumentum a mixture of long patent flagelliform hairs, and short-stalked ± appressed hairs with a ± oblique terminal cell. Leaves subsessile, membranous, to c. 22 × 10 cm., elliptic or narrowly oblong-elliptic, apex acuminate, base very narrowly cuneate-attenuate, margins serrate with teeth sharply pointed, lamina sparsely pilose on upper surface and on the nerves beneath, gland-dotted especially on the lower surface; hairs flagelliform with a long many-celled stalk bearing an elongate ± oblique terminal cell. Capitula 2–24 on leafy branches; capitula stalks 0.5–11 cm. long, puberulous. Involucres 12–20 × 15–25 mm., to c. 35 mm. wide when mature, broadly campanulate to hemispheric. Phyllaries many-seriate, all except the innermost with apical appendages, the outer series usually longer than the inner; the outer phyllaries 1–2 mm. long with appendages 5–18 mm. long and linear-lorate or spathulate, green and pilose; the middle phyllaries 8–17 mm. long, with a large purplish or greenish, membranous oblong-oblanceolate appendage ± rounded at the apex, puberulous and glandular outside; the innermost phyllaries to c. 14 mm. long, strap-shaped, greenish, glabrous, purple-tipped and ± glandular at apex but without an obvious appendage. Corollas purple or mauve, 10–12 mm. long, with a long slender tube widening abruptly into a c. 3 mm. long, cylindric limb; tube ± glandular above, limb exserted beyond the pappus. Achenes dark-brown to black, 4–5.5 mm. long, subcylindric tapering somewhat to base, 10-ribbed (or 20-ribbed with the development of secondary ribs), strigose-hispid; pappus multiseriate copious, outer elements of ± persistent linear scales 0.5–1 mm. long, the other pappus elements somewhat caducous purple-tinged or stramineous, consisting of flattened barbellate setae tapering to apex, increasing to c. 9 mm. long to the inside.

Botswana. N: Mogogelo River floodplain, 19°23'S., 23°33'E. 15.iv.1974, *P.A. Smith* 888 (K; SRGH). **Zambia**. B: Shangombo (Siwelewele), near Mashi R., 7.viii.1952, *Codd* 7425 (BM; K; PRE; SRGH). C: Chisamba, 28.ii.1966, *Fanshawe* 9593 (K; NDO). E: Nyamadzi River, 25.iii.1955, *Exell, Mendonça & Wild* 1172 (BM; LISC; SRGH). S: Choma, Mapanza, 8.iii.1958, *Robinson* 2783 (K; PRE; SRGH). **Zimbabwe**. N: Makonde Distr., Mhangura (Mangula), Whindale Farm, 24.iii.1969, *Biegel* 2894 (K; SRGH). W: Victoria Falls, iv.1918, *Eyles* 1295 (BM; K; SRGH). C: Chegutu (Hartley), Poole (Farm), 3.iv.1946, *Wild* 1004 (K; SRGH). S: Chibi, Nyoni Hills, 20.iv.1967, *Müller* 597 (K; SRGH). **Malawi**. N: S. Rukuru gorge, c. 5 km. E. of Rumphi, c. 1050 m., 21.v.1970, *Brummitt* 10969 (K; LISC; MAL; PRE; SRGH; WAG). **Mozambique**. N: Entre Rios (Malema), Mutuáli, base do monte Cucuteia, c. 650 m., 16.iii.1964, *Torre & Paiva* 11201 (C; COI; EA; K; LISC; LMU; PRE). T: 17 km. de Mágoè para Mágoè Velho, para Zumbo, 43 km., c. 300 m., 3.iii.1970, *Torre & Correia* 18173 (LISC).

Also in Uganda, Kenya, Tanzania, Zaire and Namibia and in Pakistan, India, Nepal and Sri Lanka. In shady localities in riverine vegetation, miombo and mixed deciduous woodland and thickets, in Kalahari Sand and often on rocky hill slopes.

101. **Vernonia fastigiata** Oliv. & Hiern in F.T.A. **3**: 282 (1877). —Eyles in Trans. Roy. Soc. S. Afr. **5**: 503 (1916). —Merxm., Prodr. Fl. SW. Afr. 139: 182 (1967). —Hilliard in Ross. Fl. Natal: 356 (1972); Compos. Natal: 37 (1977). —Wild in Kirkia **11**: 92 (1978). Syntypes: Botswana, Kobe Pan (Koobie) to N. Shaw Valley, *Baines* s.n. (K) and from South Africa (Transvaal) *Baines* s.n. (K). *Vernonia schinzii* O. Hoffm. in Bull. Herb. Boiss. **1**: 72 (1893). Syntypes from Namibia.

A bushy annual herb, sometimes perennial with the development of a woody rootstock. Stems up to c. 1 m. tall, branched, leafy, ribbed, puberulous to scabridulous or glabrescent. Leaves subsessile, up to c. 9 × 0.8 cm., linear, rarely up to 9 × 2 cm. and oblanceolate; apex acute mucronate or acuminate; margins ± revolute and entire or remotely serrate-dentate; lamina sparsely puberulent on both surfaces becoming scabridulous on the upper surface and on the raised main nerves beneath, or glabrescent, minutely gland-pitted especially beneath. Capitula few to many, solitary and terminal on long leafy or bracteate branches. Involucres 8–14 × 12–25 mm., campanulate becoming spreading obconic. Phyllaries numerous, longer towards the inside, dark-green becoming purplish apically, white or brownish woolly-araneose, the matted hairs binding the phyllaries together, apical bristles 2–7 mm. long, usually squarrose; the outer series linear-lanceolate, attenuate-pungent or subulate, the outermost ± curved or reflexed; the middle phyllaries narrowly lanceolate and long bristle-tipped; the inner phyllaries up to c. 14 mm. long, linear or gently tapered to a pungent awn. Receptacle alveolate. Corollas purple to pale-mauve, up to c. 12 mm. long, narrowly tubular, widening slightly towards the apex. Achenes black when mature, up to c. 4 mm. long, subcylindric, tapering to the base, ± angular or faintly c. 10-ribbed, sericeous or strigose; outer pappus of short barbellate setae c. 0.5 mm. long; inner pappus of copious brownish, grey or sometimes sordid, somewhat flattened, subplumose setae 5–7 mm. long.

Botswana. N: Ngamiland, Kuke-Sehitwa road, 20.iii.1969, *de Hoogh* 178 (COI; K; M; SRGH). SW: Ghanzi, Mamuno gate, 18.i.1970, *Brown* 7924 (K; SRGH). SE: c. 1.6 km. N. of Macpe Siding, 20.i.1960, *Leach & Noel* 273 (K; SRGH). **Zimbabwe**. W: Hwange Distr., Kazuma Range, 1000 m., 12.v.1972, *Gibbs Russell* 1979 (BR; K; LISC; SRGH). C: Gweru, 11.vii.1967, *Biegel* 2200 (SRGH). E: Chipinge Distr., Save (Sabi) Valley Exp. Sta., xii.1959, *Soane* 30 (K; M; MO; SRGH). S: Mberengwa Distr., Otto Mine, S. tip of Great Dyke, 17.iii.1964, *Wild* 6398 (K; M; SRGH). **Mozambique**. GI: Chibuto, between Maniquenique and Cicacate, beside the Limpopo, 6.viii.1958, *Barbosa & Lemos* 8305 (COI; K). M: Umbeluzi, 6.iv.1949, *Myre* 437 (SRGH).

Also in South Africa (Transvaal, Natal), Swaziland and Namibia. In hot drier areas, usually in wooded grasslands on heavy black basalt soils. Also in mopani-veld, pan margins and on serpentine. Becoming a weed of cultivation and road sides.

102. **Vernonia graniticola** G.V. Pope in Kew Bull. **43**: 287 (1988). Type: Zimbabwe, Goromonzi, Domboshawa Hill, *Drummond* 11171 (K, holotype; BM; BR; CAL; COI; LISC; LM; MO; PRE; SRGH; UPS; WAG).

A bushy annual herb up to 1 m. tall, usually shorter. Stem much branched, leafy, appressed pubescent; hairs short-stalked one-armed T-shaped, the terminal cell flattened; branches up to c. 30 cm. long, similar to stem. Leaves up to c. 8 cm. long, linear to filiform, mucronulate, strongly revolute on the margins; lamina finely pubescent, sometimes sparsely so, minutely gland-pitted. Capitula numerous, laxly subpaniculately arranged with 1–5 capitula per branch; stalks 1.5–8 cm. long, appressed pubescent. Involucres up to c. 9 × 9 mm., narrowly campanulate to obconic. Phyllaries numerous, longer towards the inside, greenish to stramineous, dark-purple along the midrib, sometimes purplish apically, white pilose to slightly woolly, ciliate on the margins, apical bristles 1–3 mm. long; the outer series subulate to linear-lanceolate with pungent apices, the outermost ± curved or reflexed; the middle phyllaries narrowly lanceolate, gently tapering to a long ± straight pungent bristle; the inner phyllaries up to c. 8 mm. long, narrowly oblong-lanceolate, acuminate or abruptly narrowed, shortly bristle-tipped. Corollas purple, c. 8 mm. long, narrowly tubular, widening towards the apex. Achenes blackish when mature, up to c. 2.5 mm. long, subcylindric and tapering somewhat to the base, 4–6-angular, densely strigose; outer pappus of short oblong scales up to 0.5 mm. long; inner pappus of greyish, somewhat flattened, barbellate setae 3–4 mm. long.

Zimbabwe. C: Goromonzi Distr., E. slopes of Ngomakurira, 23.iii.1983, *Mavi* 1600 (BM; BR; COI; K; LISC; LM; MO; PRE).
Apparently confined to the shallow soil and seepage areas on granite outcrops.

103. **Vernonia kirkii** Oliv. & Hiern in F.T.A. **3**: 274 (1877). —Wild in Kirkia **11**: 66 (1978). —C. Jeffrey in Kew Bull. **43**: 229 (1988). Syntypes: Mozambique, Manica e Sofala, Chupanga (Shupanga), *Kirk* s.n. (K); Mazaro, *Stewart* s.n. (BM).
    Vernonia swynnertonii S. Moore in Journ. Linn. Soc., Bot. **40**: 107 (1911). Type: Mozambique, Manica e Sofala, lower Buzi R., *Swynnerton* 1908 (BM, holotype; K).
    Vernonia zambesiaca S. Moore in Journ. Bot. **55**: 102 (1917). Type: Zambia, Mazabuka, *Rogers* 8744 (BM, holotype; K).

An erect robust annual herb, 60–170 cm. tall. Stems straight, branching towards the apex, leafy, striate, appressed-pubescent, glandular, glabrescent; branches ± ascending, to c. 40 cm. long. Leaves sessile, to c. 16 × 2 cm., but usually less than 1 cm. wide, linear to linear-elliptic, tapering-acute at the apex or occasionally rounded-obtuse, very narrowly cuneate at the base, entire to sparsely serrulate becoming ± strongly revolute on the margins; upper surface sparsely pilose to puberulous; lower surface pubescent to tomentose, scabridulous about the margins, or glabrescent, minutely gland-pitted on both surfaces. Capitula many, corymbiform cymose at ends of branches, or laxly subpaniculate; capitula stalks stiff, to c. 11 cm. long, bracteate towards the apex. Involucres 10–13 × 15–20(25) mm., broadly campanulate to subhemispheric. Phyllaries many-seriate, longer to the inside, acute-mucronate to ± rounded and shortly acuminate, narrowly scarious on the margins becoming ± expanded about the apices of inner phyllaries, ± puberulous outside to setose about the tips of inner phyllaries, glandular outside and usually more densely so towards the apices; the outer series from c. 2 mm. long, narrowly triangular; the inner up to c. 11 mm. long, lorate to narrowly oblong-oblanceolate. Corollas purple, to c. 11 mm. long, tubular, widening slightly to a deeply-lobed narrowly funnel-shaped limb. Achenes 3–4 mm. long, narrowly cylindric-turbinate, 8–10-ribbed, densely strigose-hispid; carpopodium enlarged, being as long as it is wide; outer pappus of linear-oblong scales c. 0.5 mm. long; inner pappus of greyish-brown, subplumose setae c. 5 mm. long.

Zambia. N: Mpika Distr., Chifungwe Plain, xi.1967, *Astle* 5137 (K; SRGH). C: Luangwa Game Res., Mfuwe, 9.iv.1969, *Astle* 5685 (K). E: Chipata, 13.x.1967, *Mutimushi* 2169 (K; NDO). S: Mazabuka, 4.v.1964, *van Rensburg* 2920 (K; SRGH). **Zimbabwe.** C: near Harare, 1916, *Craster* s.n. (K). **Malawi.** C: Kasungu Game Res., Lifupa Camp, vi.1970, *Brummitt* 11613 (K). S: Chikwawa Distr., Lengwe Game Res., 100 m., 7.iii.1970, *Brummitt* 8948 (K). **Mozambique.** Z: Mocuba, Namagoa, x.1943, *Faulkner* 277 (K; PRE). T: Tete, c. 2 km. S. of Zambezi R., 11.vii.1958, *Seagrief* in CAH 3058 (BR; K; LISC; SRGH). MS: Chemba, Chiou, C.I.C.A. Expt. Sta., 12.iv.1960, *Lemos & Macuacua* 77 (COI; K; LISC).
Also in southern Tanzania. Usually on heavy black clay alluvium in floodplain grassland and in *Acacia* and *Colophospermum mopane* woodlands.

104. **Vernonia galamensis** (Cass.) Less. in Linnaea **4**: 314 (1829). —Cuf., Enum. Pl. Aethiop. Sperm.: 1069 (1963). —Wild in Kirkia **11**: 74 (1978). —M.G. Gilbert in Kew Bull. **41**: 20 (1986). —C. Jeffrey in Kew Bull. **43**: 229 (1988). Type from Senegal.
    Conyza pauciflora Willd. in L., Sp. Pl. ed. 4, **3**: 1927 (1803). Type from Senegal.
    Centrapalus galamensis Cass. in Dict. Sci. Nat. **7**: 383 (1817). Type as for *Vernonia galamensis*.
    Vernonia pauciflora (Willd.) Less., in Linnaea **4**: 292 (1829) non (Pursh.) Poir. (1817). —Oliv. & Hiern in F.T.A. **3**: 283 (1877). —Eyles in Trans. Roy. Soc. S. Afr. **5**: 504 (1916). —Adams in F.W.T.A. ed. 2, **2**: 280 (1963). —Wild in Kirkia **11**: 74 (1978). Type as for *Conyza pauciflora*.

For a full infraspecific classification and synonymy see M.G. Gilbert in Kew Bull. **41**: 19–35 (1986) and C. Jeffrey in Kew Bull. **43**: 229–231 (1988).

Var. **australis** M.G. Gilbert in Kew Bull. **41**: 26 (1986). —C. Jeffrey in Kew Bull. **43**: 230 (1988). Type: Zimbabwe, Mutare, *Chase* 7322 (K, holotype; SRGH).

An erect strict annual herb, 0.3–1.5(3) m. tall; stems straight, leafy, branching near the apex, ribbed, pubescent; branches leafy. Leaves membranous, sessile, up to 16(24) × 4 cm., very narrowly elliptic to elliptic, acuminate at the apex and tapering at the base into a very narrowly winged petiole-like midrib; margins subentire or serrulate to sharply-pointed serrate; lamina flat or sometimes ± bullate, ± sparsely pilose on both surfaces, becoming scabridulous on the upper surface, lateral nerves prominent beneath. Capitula many, cymose or sometimes solitary on the branches; capitula-stalks often leafy, widening somewhat towards the apex, pubescent. Involucres 17–19 mm. long, broadly obconic-campanulate to widely spreading. Phyllaries numerous; the outer short linear squarrose, tapering-aristate; the inner much wider and longer, increasing to c. 17 mm. long, linear-lorate, straight not squarrose, acute-obtuse and greenish-tipped, paler below, longitudinally ribbed, glabrous. Corolla blue, 13–17 mm. long, lobes 2.5–5 mm. long, linear. Achenes black, 3–5(6.5) mm. long, narrowly obovoid to fusiform, obscurely narrowly 10-ribbed, densely silvery appressed-sericeous; outer pappus of greyish-brown, very narrow setiform scales 1.5–2 mm. long; inner pappus of numerous greyish-brown barbellate setae 8.5–9 mm. long.

Zimbabwe. E: Mutare, Commonage, 1.vi.1956, *Chase* 6138 (K; LISC; M; SRGH). **Malawi**. S: Zomba Distr., Phalombe (Palombe) road, Likangala, v.1956, *Jackson* 1854 (K; SRGH). **Mozambique**. N: Cabo Delgado, Montepuez, prox. Nantulo, c. 5 km. S. Rio M'salo, c. 550 m., 9.iv.1964, *Torre & Paiva* 11822 (LISC). Z: between Nicuadala and Regulo Simogo, 29.viii.1949, *Barbosa & Carvalho* 3900 (K; SRGH). T: 40 km. de Changara para Catandica (Vila Gouveia), c. 400 m., 28.v.1971, *Torre & Correia* 18697 (LISC). MS: Chimoio, c. 16 km. N. of Vanduzi, 27.iv.1962, *Chase* 7695 (K; M; SRGH).

Also in Tanzania. Woodland and on forest margins, also a weed of cultivation.

105. **Vernonia zambiana** G.V. Pope in Kew Bull. **43**: 288 (1988). Type: Zambia, Kasama, Chishimba Falls, *Philcox, Pope & Chisumpa* 10211 (K, holotype; BR; LISC; MO; SRGH).

An erect slender or ± bushy annual herb, 10–40 cm. tall. Stems simple to much branched, sometimes branching from near the base, leafy, ribbed, densely pilose below to ± sericeous-pubescent above; the indumentum consisting of short ± appressed hairs intermixed with larger patent many-celled simple hairs, the latter fewer or absent on upper stem and branches; branches straight, stiffly ascending, up to c. 19 cm. long. Leaves subsessile, up to c. 8 × 0.8 cm., linear-oblanceolate, subobtuse at the apex, narrowly attenuate to the base, subentire to remotely serrulate on the margins, sparsely thinly pilose on both surfaces but more densely so on the midrib and nerves beneath. Capitula few to numerous, in loose 2–4-capitulate terminal clusters, sometimes solitary on long stalks; stalks mostly 1–20 mm. long, densely pubescent. Involucres 9–12 × 9–15 mm., obconic, phyllaries diverging in mature heads, densely white sericeous-lanate. Phyllaries numerous, obscurely longitudinally ribbed, densely white sericeous, long ciliate on the margins, from c. 3 mm. long and subulate outside, longer and linear-lanceolate towards the inside, gradually tapering to a finely pointed purple apex. Receptacle alveolate. Corollas creamy-white to pale-pink, 7–9 mm. long. Achenes 1.8–2.3 mm. long, narrowly angular-turbinate, 4–5-ribbed, setulose on the ribs, gland-pitted and rugulose on the faces; outer pappus of brownish or sordid narrow acuminate scales, about half as long as the achene; inner pappus of numerous brownish or sordid subplumose setae 6–7 mm. long.

Zambia. N: Between Mbala and Ndundu, c. 1590 m., 10.v.1966, *Richards* 21496 (K).

Known only from the Flora Zambesiaca area. *Brachystegia* woodland, often in sandy soil.

106. **Vernonia ambigua** Kotschy & Peyr., Pl. Tinn.: 35, t. 17B (1867). —Oliv. & Hiern in F.T.A. **3**: 272 (1877). —Mendonça, Contrib. Conhec. Fl. Angol., 1 Compositae: 12 (1943). —Adams in F.W.T.A. ed. 2, **2**: 281 (1963). —Wild in Kirkia **11**: 59 (1978). —C. Jeffrey in Kew Bull. **43**: 247 (1988). Type from Sudan.

An erect bushy, often stunted, annual herb 3–30 cm. tall. Stems diffusely branched, densely leafy, finely ribbed, densely pilose to loosely hirsute; indumentum of short ± strigose hairs intermixed with patent many-celled simple hairs; branches up to c. 14 cm. long, spreading-ascending. Leaves membranous, subsessile, ascending, the upper cauline leaves and those subtending the capitula usually overtopping the capitula; lamina up to c. 10 × 2 cm., ± narrowly oblanceolate, subacute to obtuse at the apex, cuneate to narrowly attenuate at the base, subentire to serrulate, pilose on both surfaces and especially the midribs and lateral nerves beneath. Capitula few to numerous in moderately dense clusters of 2–10 terminal on the stem and branches, or sometimes solitary; capitula stalks 0.1–2.5 cm. long, densely pilose to villous. Involucres to c. 10 mm. long, obconic, becoming hemispheric, spreading to c. 20 mm. wide. Phyllaries numerous, straight and finely tapering to a dark-tipped apex, seldom curved, densely white sericeous-lanate, ± long-ciliate on the margins, obscurely longitudinally ribbed; the outer from c. 3 mm. long, subulate; the inner series progressively longer becoming linear-lanceolate. Receptacle alveolate. Corolla light- to dark-purple, c. 9 mm. long. Achenes c. 2.3 mm. long, narrowly angular-turbinate, 4–5-ribbed, setulose on the ribs, gland-pitted and rugulose on the faces; outer pappus of brownish short narrow scales less than one quarter of the achene body in length; inner pappus of numerous brownish or sordid barbellate setae c. 7 mm. long.

Zambia. W: Mwinilunga Distr., Zambezi Source Memorial, c. 1700 m., 15.v.1986, *Philcox, Pope, Chisumpa & Ngoma* 10294 (K; MO; NDO; SRGH). C: Walamba, 22.v.1954, *Fanshawe* 1232 (K).
Also in tropical west Africa, the Sudan, Uganda, Tanzania, Zaire and Angola. High rainfall miombo woodland amongst short grasses in sandy soil, also in disturbed soil at roadsides.

107. **Vernonia petersii** Oliv. & Hiern ex Oliv. in Trans. Linn. Soc. Lond. **29**: 90 (1873); in F.T.A. **3**: 273 (1877). —Eyles in Trans. Roy. Soc. S. Afr. **5**: 504 (1916). —Mendonça, Contrib. Conhec. Fl. Angol., **1** Compositae: 10 (1943). —Merxm., Prodr. Fl. SW. Afr. 139: 183 (1967). —Wild in Kirkia **11**: 59 (1978). —Maquet in Fl. Rwanda, Spermat. **3**: 558, fig. 168, 1 (1985). —C. Jeffrey in Kew Bull. **43**: 247 (1988). Types: Mozambique, Boror, *Peters* (B†), Rios de Sena, *Peters* (B†).
Chrystallopollen latifolium Steetz in Peters, Reise Mossamb., Bot.: 364, t. 48a (1862) non *Vernonia latifolia* Lem. (1855). Type as for *Vernonia petersii*.
Vernonia eriocephala Klatt in Bull. Herb. Boiss **4**: 826 (1896). Type: Mozambique, Boruma, *Menyhart* 1112 (Z, holotype).
Vernonia karongensis Bak. in Bull. Misc. Inf., Kew **1898**: 147 (1898). Type: Malawi, between Kondowe and Karonga, *Whyte* (K, holotype).

An erect slender to divaricately branched, sometimes robust, annual herb 10–120 cm. tall. Stems simple, or more usually laxly much-branched, sometimes branching from near the base, leafy, ribbed, patent-pilose below and somewhat appressed pubescent above; branches spreading-ascending, up to c. 45 cm. long, pilose-pubescent. Leaves subsessile, up to c. 14 × 2.3 cm., linear-oblong to narrowly elliptic or narrowly oblong-oblanceolate; apex acute to obtuse; base abruptly rounded to subtruncate, or cuneate and ± attenuate; margins subentire to subserrate; lamina ± sparsely pilose on both surfaces, more densely pilose on midrib and nerves beneath. Capitula numerous, in dense 2–7-capitulate clusters, or more laxly arranged with 1–3 capitula on long stalks at the ends of branches; stalks 0.2–7 cm. long, pubescent or tomentose. Involucres 5–9 × 8–18 mm., broadly campanulate, becoming ± shallowly cup-shaped with a broad truncate base when mature, densely white sericeous-lanate below. Phyllaries numerous, ± densely white-sericeous outside, long ciliate on the margins, the apices acute to gradually tapering and dark-tipped, ± recurved at least in the outer phyllaries, lamina obscurely longitudinally ribbed, the hairs short-stalked with an appressed elongate terminal cell; the outer phyllaries from c. 2 mm. long, subulate to linear-lanceolate, the inner series longer and narrowly lanceolate. Receptacle alveolate. Corolla purple, 6–7 mm. long. Achenes 1.5–2.5 mm. long, narrowly angular-turbinate, 4–5-ribbed, setulose, especially on ribs, gland-pitted and rugulose on the faces; outer pappus of brownish or sordid, short narrow scales; inner pappus of numerous sordid or brownish barbellate setae 4–5 mm. long.

Botswana. N: Mosetse R., c. 120 km. from Francistown, 8.iii.1961, *Richards* 14598 (K; M; SRGH). **Zambia**. B: Machili, 14.iii.1961, *Fanshawe* 6442 (K; NDO; SRGH). N: Chinsali, Govt. Rest House, 13.iv.1986, *Philcox, Pope & Chisumpa* 9906 (K; NDO; SRGH). W: Kitwe, 26.iv.1966, *Mutimushi* 1384 (K; SRGH). C: c. 1.5 km. W. of Kundalila Falls, 11.iv.1986, *Philcox, Pope & Chisumpa* 9865 (K; LISC; MO; NDO; SRGH). E: Katete, St. Francis Hospital, 30.v.1956, *Wright* 101 (K). S: Mazabuka, 8.iii.1963, *van Rensburg* 1652 (K; SRGH). **Zimbabwe**. N: Gokwe Distr., Gokwe R. Fly Gate, 15.vi.1963, *Bingham* 731 (K; SRGH). W: Victoria Falls, 8.vii.1930, *Hutchinson & Gillett* 3456 (BM; K; SRGH). C: Gweru, c. 1400 m., 12.ii.1967, *Biegel* 1918 (K; SRGH). E: Mutare, 29.v.1953, *Chase* 4984 (COI; LISC; MO; SRGH). S: Mberengwa, 26.i.1931, *Norlindh & Weimarck* 5178 (LD). **Malawi**. N: Mzimba Distr., Lunyangura R., c. 1.6 km. S. of M14, c. 1160 m., 21.i.1976, *Pawek* 11384 (K; MAL; MO). C: c. 8 km. N. of Kasungu, c. 1060 m., 7.v.1970, *Brummitt* 10427 (K). S: Zomba Distr., Lower Likangala, 23.vii.1956, *Jackson* 2014 (BR; K; LISC; SRGH). **Mozambique**. N: c. 8 km. E. of Mandimba, c. 760 m., 15.v.1961, *Leach & Rutherford-Smith* 10853 (K; M; SRGH). Z: Zoa Falls on Ruo R., c. 300 m., 16.viii.1971, *Leach & Royle* 14819 (K; SRGH). T: c. 92.2 km. from Vila Mouzinho to Zóbuè, 19.vii.1949, *Barbosa & Carvalho*

3705 (K). GI: Guijá, 8.vi.1947, Pedrógão 259 (PRE). MS: Chimoio, Serra de Garuso, 2.iv.1948, *Barbosa* 1354 (LISC).

Also in SE. Tanzania, Zaire, Rwanda, Angola and Namibia. Deciduous woodland and grassland on sandy or rocky soils, often a weed of disturbed ground and old cultivation. Very widespread except at high altitudes.

In one specimen seen, *Best* 89 (K) from Kafue Mission in Zambia, the inner pappus setae are scarce and very caducous. In all other respects, however, this specimen agrees well with the species.

108. **Vernonia lundiensis** (Hutch.) Wild & G.V. Pope in Kirkia **10**: 310 (1977); **11**: 62 (1978). Type: Zimbabwe, Runde (Lundi) R., *Hutchinson & Gillett* 3270 (K, holotype).
    Triplotaxis lundiensis Hutch., Bot. S. Afr.: 465 (1946). Type as above.

An erect annual herb, 15–60 cm. tall, becoming somewhat bushy. Stems ± divaricately branched, leafy, striately ribbed, pilose and glandular; indumentum a mixture of short-stalked hairs with appressed elongate terminal cells, and few to numerous larger soft patent many-celled simple hairs with erect elongate terminal cells. Leaves subsessile, 2–12 × 0.3–2 cm., linear-elliptic to narrowly oblong, subacute at the apex, narrowly cuneate or sometimes narrowly rounded to subcordate at the base, finely serrulate on the margins, ± sparsely pilose on both surfaces, often becoming scabridulous, particularly on the margins and upper surface, glandular; indumentum of scattered patent flagelliform hairs. Capitula many, subpaniculate, usually in 2–5-capitulate clusters, or solitary on the branches; capitula stalks mostly 1–10 mm. long, longer in solitary heads, whitish pilose. Involucres 4–6 × 4–10 mm., broadly campanulate. Phyllaries many-seriate, imbricate, increasing in size to the inside, acute to acuminate at the apex, villous-setose and ciliate to glabrescent, often purple-tipped; the outer from c. 2 mm. long, linear-lanceolate; the inner to c. 6 mm. long, narrowly oblong-lanceolate. Corollas purple, exserted 2–3 mm. beyond the involucre, 5–7 mm. long, tubular, gradually widening to a narrowly funnel-shaped limb. Achenes 2–2.5 mm. long, 4–5-angled, tapering to base, angles ribbed, the faces transversely rugose with gland-dotted ridges, ± hispidulous; outer pappus of pale lacerate scales c. 1 mm. long (sometimes reduced and fringe-like), inner pappus usually absent, rarely of 1–2 short barbellate setae.

Zimbabwe. N: Bindura Distr., Kerry Farm, c. 975 m., 17°08'S, 31°20'E, 16.vii.1969, *Mogg* 34313 (LISC). E: Mutare Distr., Pounsley (near Odzi), Carolina B Farm, 27.ii.1957, *Phipps* 584 (K; SRGH). S: Chiredzi Distr., c. 4 km. from Chivirira Falls, Save (Sabi) R., 1.vi.1971, *Grosvenor* 599 (K; SRGH). **Mozambique**. MS: 9 km. from Mungári to Nhacola (Tambara), c. 200 m., 12.v.1971, *Torre & Correia* 18384 (LISC). GI: Mongorro-Panda, *Barbosa & Lemos* 8524 (SRGH).

Not known outside the Flora Zambesiaca area. Miombo and dry mixed deciduous woodlands and in *Colophospermum* tree savanna, also as a weed of cultivation in the lowveld.

109. **Vernonia miombicola** Wild in Kirkia **11**: 10; 67 (1978). —Maquet in Fl. Rwanda, Spermat. **3**: 556, fig. 168, 2 (1985). —C. Jeffrey in Kew Bull. **43**: 246 (1988). Type: Zambia, Mbala, *Richards* 22234 (M, holotype; K; MO).

An erect slender to laxly bushy annual herb, 15–60 cm. tall. Stems simple to somewhat diffusely branched, finely ribbed, densely pilose below, pubescent above; indumentum of short appressed hairs intermixed with fewer erect flagelliform hairs, the latter ± sparse or absent on the upper stem and branches; branches sometimes arising from near the stem base, spreading-ascending, up to c. 25 cm. long. Leaves subsessile, up to c. 10.5 × 4 cm., oblong or oblong-elliptic to oblanceolate, sometimes panduriform, the largest mid-cauline; apex obtuse to acute; base abruptly rounded to truncate, or tapering to a narrow abruptly rounded base with a short petiole; margins entire to denticulate; lamina pilose on both surfaces or glabrescent. Capitula numerous, corymbiform cymose at the ends of branches; stalks 0.2–7 cm. long, slender, pubescent. Involucres to c. 9 mm. long, broadly campanulate or obconic. Phyllaries numerous, straight, attenuate-pungent and dark-purple to the apices, densely to sparsely white-strigose, long ciliate on margins, ± obscurely longitudinally ribbed; the outer series short and linear-lanceolate, the inner progressively longer and narrowly lanceolate. Receptacle alveolate. Corolla dark-purple, 8–11 mm. long. Achenes 1.5–2 mm. long, narrowly angular-turbinate, 4–5-ribbed, setulose especially on the ribs, glandular and transversely rugose on the faces; outer pappus of white broad overlapping scales about one quarter the length of the achene; inner pappus of c. 10 brownish to purple-tinged barbellate setae 5–6 mm. long.

Zambia. N: Mbala Distr., Lake Chila, c. 1875 m., 25.iv.1986, *Philcox, Pope & Chisumpa* 10130 (K; LISC; MO; NDO; SRGH).

Also in tropical east Africa, Rwanda and Zaire. Locally common in open grassland or deciduous woodland, often in sandy soil, also a weed of old cultivation.

V. aemulans Vatke is distinguished from this species by its pappus, the outer elements of which are short narrow brown scales, and by the indumentum of the young growth which is often densely rufous-hairy.

110. **Vernonia acuminatissima** S. Moore in Journ. Linn. Soc., Bot. **40**: 104 (1911). —C. Jeffrey in Kew Bull. **43**: 247 (1988). Types: Mozambique, Upper Buzi, *Swynnerton* 1906 (BM, syntype); Beira, *Swynnerton* 1907 (BM; K, syntypes).
     Vernonia rogersii S. Moore in Journ. Bot. **51**: 183 (1913). —Wild in Kirkia **11**: 78 (1978). Type: Mozambique, between Beira and Nhamatanda (Vila Machado), *Rogers* 4527 (BM, holotype; K).

An erect annual herb, up to c. 120 cm. tall. Stems sparsely branched above, leafy, finely ribbed, pilose below, ± strigose above; indumentum of short-stalked flagelliform hairs with an erect or appressed elongate terminal cell; branches up to c. 30 cm. long, young growth often densely reddish-brown strigose. Leaves subsessile; mid-cauline leaves up to c. 15 × 4 cm., oblanceolate, acute at the apex and attenuate to the base; lower leaves smaller, obovate-oblanceolate, obtuse at the apex; upper leaves oblong-oblanceolate or narrowly elliptic, decreasing in size up the stem; margins subentire or serrulate; lamina thinly hispid on both surfaces. Capitula few to many, corymbiform cymose at the ends of branches, sometimes subscorpioidly cymose; stalks 0.2–4 cm. long, densely brownish-strigose. Involucres up to c. 11 mm. long, broadly obconic, widely spreading when mature. Phyllaries numerous, ± densely hispid or pilose, ± ciliate on the margins, obscurely longitudinally ribbed; the outer phyllaries up to c. 6 mm. long, linear, gradually tapering into a long ± pungent apex; the inner to c. 11 mm. long, linear-lanceolate, long acuminate to the apex, ± pungent. Receptacle shallowly alveolate. Corolla purple, 6–10 mm. long, tubular below with a slightly expanded limb. Achenes 1.5–2.5 mm. long, 4–5-sided, tapering slightly to the base, narrowly ribbed on the angles, glandular on the faces, shortly uniformly hispid; outer pappus of numerous brownish linear-lanceolate scales less than one quarter of the achene in length; inner pappus of many sordid or brownish barbellate setae c. 6 mm. long.

Zimbabwe. C: Makoni Distr., Muhonde Farm, i.1964, *Strang* 2295 (SRGH). E: Odzi, c. 1700 m., 3.vi.1936, *Eyles* 8599 (K). S: Nyoni Mts., c. 80 km. S. of Masvingo, 10.iii.1976, *Pope & Müller* 1518 (K; SRGH). **Malawi**. C: Lilongwe Distr., Dzalanyama For. Res., Chiunjiza Road, c. 5 km. SE. of Choulongwe (Chaulongwe) Falls, c. 1230 m., 22.iii.1970, *Brummitt* 9314 (K). **Mozambique**. Z: Lugela-Mocuba Distr., Namagoa Estate, ix.1946, *Faulkner* PRE 227 (BR; COI; K; SRGH). MS: Jagersberg Mts., 7.v.1948, *Munch* 61 (K; SRGH). GI: Inhambane, viii.1887, *Scott* (K). M: Maputo, 23.vii.1947, *Barbosa* 296 (SRGH).

Also in southern Tanzania. Miombo woodland and evergreen forest margins.

111. **Vernonia stellulifera** (Benth.) C. Jeffrey in Kew Bull. **43**: 225 (1988). Syntypes from Sierra Leone.
     Herderia stellulifera Benth. in Hook., Niger Fl.: 425 (1849). —Oliv. & Hiern in F.T.A. **3**: 298 (1877). —Mendonça, Contrib. Conhec. Fl. Angol., **1** Compositae: 37 (1943). Types as above.
     Triplotaxis stellulifera (Benth.) Hutch. in Bull. Misc. Inf., Kew **1914**: 356 (1914). —Adams in F.W.T.A. ed. 2, **2**: 269 (1963). —Wild in Kirkia **10**, 2: 359 (1977). Types as above.

Very similar to *V. cinerea*. A diffuse, somewhat delicate, branching annual herb to c. 75 cm. tall; branches slender, longitudinally striate, appressed pubescent; hairs short-stalked long-armed, symmetric or asymmetric, T-shaped. Leaves membranous, 1.5–6 × 0.8–3 cm., subspathulate, broadly ovate to lanceolate in upper portion and abruptly tapered below into a long narrow-winged midrib, crenate-serrate on the margins, sparsely puberulous or glabrescent on both surfaces; upper-cauline leaves narrowly lanceolate. Capitula few to numerous, small, laxly paniculate; stalks 5–25 mm. long, slender. Involucres 2–3 × c. 3 mm., spreading-campanulate, phyllaries eventually reflexed. Phyllaries narrowly oblong, pungent aristate, sometimes green-tipped, puberulous. Corollas pinkish-mauve, 1.5–2.0 mm. long, funnel-shaped. Achenes dark-brown, c. 2 mm. long, subfusiform, terete, uniformly strigulose; outer pappus very short and fringe-like of oblong-rounded fimbriate scales, inner pappus absent or of a few bristle-like scales up to c. 1.4 mm. long.

Zambia. W: Mwinilunga, Lunga River, 29.xi.1937, *Milne-Redhead* 3441 (K; LISC; PRE).

Throughout west and central tropical Africa and Angola. Moist or marshy ground usually in forest shade.

112. **Vernonia cinerea** (L.) Less. in Linnaea **4**: 291 (1829). —DC., Prodr. **5**: 24 (1836). —Oliv. & Hiern in F.T.A. **3**: 275 (1877). —Eyles in Trans. Roy. Soc. S. Afr. **5**: 503 (1916). —Mendonça, Contrib. Conhec. Fl. Angol., **1** Compositae: 14 (1943). —Adams in F.W.T.A. ed. 2, **2**: 283 (1963). —Wild in Kirkia **11**: 80 (1978). —Maquet in Fl. Rwanda, Spermat. **3**: 558 (1985). —C. Jeffrey in Kew Bull. **43**: 224 (1988). Type from Sri Lanka.

Conyza cinerea L., Sp. Pl. **2**: 862 (1753). Type as above.

Var. **cinerea** —C. Jeffrey in Kew Bull. **43**: 224 (1988).

An erect annual herb to c. 1.35 m. tall. Stems branching above, longitudinally ribbed, thinly puberulent above and coarsely pubescent below; indumentum of fine appressed T-shaped hairs intermixed, at least on lower stem, with longer patent flagelliform hairs. Leaves membranous; lower-cauline leaves 3–12 × 1.5–5 cm., subspathulate, broadly ovate to lanceolate in the upper portion and abruptly tapered below into a long narrowly-winged midrib, shortly acuminate at the apex, crenate-serrate on the margins, sparsely pubescent-scabridulous or pilose on both surfaces, somewhat paler beneath; mid-cauline leaves up to c. 10 × 2.5 cm., becoming lanceolate, attenuate at the apex and narrowly cuneate at the base; upper-cauline leaves smaller, linear-lanceolate, grading into the synflorescence bracts. Capitula small, very numerous, in a large terminal panicle, or corymbiformly cymose in clusters at ends of sparsely to much divided branches, stalks 1–15 mm. long. Involucres 4–6 mm. long, campanulate. Phyllaries ± appressed, becoming reflexed when achenes are mature; outer phyllaries from 1.5 mm. long, linear-lanccolate or subulate; inner phyllaries up to 5.5 mm. long, lorate, acuminate-apiculate, appressed pubescent. Corollas pinkish or pale-mauve, 3.5–5 mm. long, narrowly funnel-shaped. Achenes 1.2–2 mm. long, subfusiform, terete not ribbed, narrowed below, uniformly strigulose-pubescent; outer pappus of distinct short linear fimbriate scales 0.2–0.6 mm. long, inner pappus of white barbellate setae 3–4.5 mm. long and easily dislodged.

Zimbabwe. S: Save-Runde (Sabi-Lundi) Junction, Chitsa's Kraal, c. 244 m., 6.vi.1950, *Wild* 3388 (K; SRGH). **Malawi**. N: Nkhata Bay Distr., Limpasa Rice Scheme, c. 548 m., 13.viii.1977, *Pawek* 12873 (K; MAL; MO). S: c. 5 km. SW. of Nsanje, c. 65 m., 29.v.1970, *Brummitt* 11162 (K). **Mozambique**. N: Eráti, Namapa, Estação Expt. C.I.C.A., 27.iii.1961, *Balsinhas & Marrime* 316 (COI; K; LISC). Z: Mocuba, Namagoa Plantations vii.1946, *Faulkner* PRE 292 (COI; K; PRE; SRGH). MS: Inhaminga, Corone, 50 m., 20.iv.1956, *Gomes e Sousa* 4311 (COI; K). GI: Canıçadô, 18.vi.1947, *Pedrogão* 305 (K, LISC). M: Incomáti, v.1893, *Quintas* 171 (COI).

Widespread in tropical Asia, India, Australia and tropical W. Africa. Also in Zaire, Rwanda, Angola and South Africa (Transvaal). Often as a weed of cultivation and disturbed ground, also in woodland and grassland, at lower altitudes; 0–800 m.

Var. *lentii* (O. Hoffm.) C. Jeffrey is recorded from Kenya and Tanzania, and may be distinguished by its leaves silky-sericeous beneath, its larger achenes (1.6–2.8 mm. long) and its often perennial habit. Var. *ugandensis* C. Jeffrey is recorded from Uganda and Kenya at altitudes above 900 m., and may be distinguished by its shorter inner pappus (c. 3 mm. long).

113. **Vernonia meiostephana** C. Jeffrey in Kew Bull. **43**: 225 (1988). Type from Zaire.

Erlangea vernonioides Muschl. in Engl., Bot. Jahrb. **46**: 62 (1911) non *V. vernonioides* (A. Gray) Bacigal. (1931). Type as above.

Very similar to *V. cinerea* but its outer pappus is reduced to a low fringe-like rim consisting of short, distinct or ± confluent, oblong, rounded or subtruncate scales 0.1–0.2 mm. long. It is also a somewhat more slender and delicate herb with a laxer synflorescence of smaller capitula.

Zambia. B: Masese, 3.v.1961, *Fanshawe* 6530 (K; NDO; SRGH). N: Mbala, Chilongowelo, c. 1460 m., 27.v.1952, *Richards* 1843 (K). W: Mufulira, c. 1220 m., 17.iv.1948, *Cruse* 321 (K). C: Chilanga, Mt. Makulu Res. Sta., 29.iv.1960, *Simwanda* 131 (K; LISC; M; SRGH). E: Lundazi Distr., near Kalindi, 4.vii.1971, *Sayer* 1238 (SRGH). S: Mapanza Mission, c. 1066 m., 24.iv.1953, *Robinson* 181 (K). **Zimbabwe**. N: Mazowe, Henderson Res. Sta., 7.vi.1974, *Pope* 1355 (K; SRGH). W: Victoria Falls, "Rain Forest", c. 914 m., 8.vii.1930, *Hutchinson & Gillett* 3434 (BM; K; LISC; SRGH). C: Harare, Rhodesville, 4.v.1934, *Gilliland* 20 (K). E: Mutare Distr., Eastlands Farm, 10.viii.1952, *Chase* 4626 (K; MO; SRGH). S: Mwenezi, SW. Mateke Hills, Malangure R., c. 625 m., 6.v.1958, *Drummond* 5647 (K; SRGH). **Malawi**. N: Chisenga, 21.iv.1986, *Philcox, Pope & Chisumpa* 10052 (K; MO; NDO; SRGH). C: Lilongwe, Agric. Res. Sta., 1.iv.1955, *Jackson* 1562 (K; LISC; SRGH). S: Zomba Distr., Lake Chilwa, 2.v.1969, *Williams* 92 (K; SRGH).

Also in E. tropical Africa, Zaire and South Africa (Transvaal). Moist ground beside streams or water, in grassland or woodland, also as a weed of cultivation and disturbed ground; mostly above 800 m.

114. **Vernonia sylvicola** G.V. Pope in Kew Bull. **41**: 395 (1986). Type: Zambia, Solwezi, *Holmes* 1456 (K, holotype).
    Vernonia poskeana var. *poskeana* sensu Wild in Kirkia **11**, 1: 56 (1978) pro parte quoad *Robinson* 2260 (K; SRGH), *Holmes* 1456 (K), *Huchinson & Gillett* 3751 (BM; K; SRGH).

An erect annual herb. Stems c. 13–120 cm. tall, simple below, ± laxly branched above, ± strongly longitudinally striate; branches 4–30 cm. long, spreading, pubescent, ribbed or angled. Leaves of the lower stem up to c. 7 × 0.9 cm., linear-oblanceolate to linear-elliptic, lamina flat or ± revolute on the margins, tapering-attenuate to the base; upper-cauline leaves smaller, linear-lorate, revolute, puberulent to scabridulous. Capitula numerous in large open panicles, stalks slender. Involucres 8–10 mm. long, exceeding the pappus at anthesis, narrowly cylindric-campanulate, tapering to the base, not becoming widely spreading. Phyllaries numerous, progressively longer towards the inside; the outer phyllaries lanceolate, the inner narrower, gradually tapering to a ± obtuse apex, subapically mucronate or mucro absent, ± densely brownish hispid, or glabrescent. Corollas purplish, 6–11 mm. long, gradually tapering to the base. Achenes pale-brown to blackish, 1.5–2.3 mm. long, narrowly subcylindric, truncate at the apex and abruptly tapered at the base, terete or obscurely c. 5-ribbed, ± densely white-strigose, glandular; outer pappus a short fringe of narrow fimbriate scales, inner pappus of c. 14 brownish, sometimes white, barbellate or subplumose setae 5–6 mm. long.

Zambia. N: Mbala, near Kambole, 1500 m., 5.vi.1957, *Richards* 10041 (BR; K). W: Zambezi River source, c. 1700 m., 15.v.1986, *Philcox, Pope & Chisumpa* 10298 (K; MO; NDO; SRGH). C: 106 km. NE. of Serenje Corner, 16.vii.1930, *Hutchinson & Gillett* 3751 (BM; K; SRGH). S: Mochipapa, 8 km. E. of Choma, 1300 m., 29.v.1955, *Robinson* 1285 (K). **Malawi**. N: Mzimba Distr., Mzuzu, Marymount, 1390 m., 5.x.1973, *Pawek* 7328 (K; MAL; MO; SRGH). C: Ntchisi, 1350 m., 1.viii.1946, *Brass* 17087 (K). **Mozambique**. N: Lichinga (Vila Cabral), Litunde, x.1942, *Mendonça* s.n. (LISC). T: 38.6 km. from Furancungo to Ulóngué (Vila Coutinho), 15.vii.1949, *Barbosa & Carvalho* 3626 (K; SRGH).
    Also in Tanzania, Zaire and Angola. Sandy soil in high rainfall miombo woodland, evergreen forest margins, stream sides and disturbed ground, often in shade.

115. **Vernonia jelfiae** S. Moore in Journ. Linn. Soc., Bot. **47**: 262 (1925). —C. Jeffrey in Kew Bull. **43**: 242 (1988). Type: Zambia, Luwingu, *Jelf* 42 (BM, holotype).
    Vernonia chloropappa sensu Wild in Kirkia **11**: 64 (1978) pro parte quoad *van Rensburg* 1996, *Drummond* 9180, *Leach & Rutherford-Smith* 11024.

An erect slender annual herb, up to c. 85 cm. tall. Stems strict, becoming laxly branched and longitudinally ribbed or angular above; branches up to c. 30 cm. long, pubescent, ribbed. Leaves of upper stem up to 6 × 1 cm., oblanceolate to very narrowly elliptic, remotely serrulate on the margins becoming ± revolute, cuneate to the base; upper surface scabridulous, lower surface ± puberulous; lower leaves similar but decreasing in size, the lowermost soon withering. Capitula many, c. 3 per branch, corymbiform cymose; stalks sometimes dark-purple, slender, 1–8 cm. long. Involucres 8–12 mm. long, exceeding the pappus at anthesis, campanulate-obconic, later spreading. Phyllaries numerous, progressively longer towards the inside, the midribs continued as an apical bristle c. 1 mm. long, or apex finely attenuate, the lamina brownish- to whitish-pubescent below and brownish-hispid towards the apex, becoming somewhat scabridulous; the outer phyllaries subulate to lanceolate, the inner narrowly strap-shaped, narrowly acute or attenuate to the apex. Corollas purple, reddish-mauve or creamy-white, 8–10 mm. long, narrowly funnel-shaped. Achenes pale- to reddish-brown, c. 2.5 mm. long, narrowly subcylindric, truncate at apex, slightly narrowed to base, obscurely c. 5-ribbed, uniformly white strigose, glandular; outer pappus of short white scales shortly fimbriate apically or of small brownish deeply fimbriate lanceolate scales; inner pappus of 9–17 green or greenish-brown setae 6–7 mm. long, seta barbs appressed-ascending and decreasing in size and number towards the seta bases, or the barbs bristle-like and subplumose, ascending-patent and not diminishing to seta base.

Corollas usually purple; pappus setae barbellate, the barbs appressed-ascending, decreasing in size and number to the seta base; pappus scales broad, usually whitish, fimbriate-toothed on upper margin; the longest phyllaries 8–9 mm. long - - - - - var. *jelfiae*
Corollas usually creamy-white to reddish-mauve, sometimes purple; pappus setae subplumose, the bristles ascending-patent not diminishing to the seta base; pappus scales brownish, narrowly triangular, fimbriate-ciliate; the longest phyllaries 9–12 mm. long - - var. *albida*

Var. **jelfiae**, TAB. **23** fig. B.

Zambia. N: c. 16 km. E. of Kasama, Namkungu Ridge, 29.iv.1986, *Philcox, Pope & Chisumpa* 10202 (K; MO; NDO; SRGH). W: Nkana South, 9.iv.1963, *Mutimushi* 278 (K). C: c. 16 km. N. of Lusaka, 19.v.1957, *Noak* 234 (K; SRGH). S: c. 16 km. N. of Choma, 15.iv.1963, *van Rensburg* 1996, (K). **Malawi**. N: *Whyte* s.n. (K). **Mozambique**. N: c. 8 km. S. of Massangulo, c. 1070 m., 26.v.1961, *Leach & Rutherford-Smith* 11024 (K; LISC; SRGH).

Also in Zaire and Angola. High rainfall miombo woodland.

Var. **albida** G.V. Pope in Kew Bull. **41**: 394 (1986). —C. Jeffrey in Kew Bull. **43**: 242 (1988). Type: Zambia, 25 km. N. of Mpika, *Robinson* 5250 (K, holotype).

Zambia. N: Kawambwa, 26.v.1962, *Lawton* 869 (K). W: c. 12 km. W. of Chingola, Hart's Farm, 18.v.1986, *Philcox, Pope & Chisumpa* 10359 (K; LISC; MO; NDO; SRGH). C: Chiwefwe, 1.v.1957, *Fanshawe* 3243 (BR; K). **Zimbabwe**. N: 15 km. N. of Chinhoyi (Sinoia), Kapeta Ranch, 5.vi.1969, *Drummond* 9180 (BR; K; SRGH). **Malawi**. N: Mzimba Distr., between two Mzuzu junctions, 13.vi.1971, *Pawek* 4895 (K). C: Lilongwe Distr., Dzalanyama For. Res., 26.iv.1970, *Brummitt* 10161 (K). **Mozambique**. N: Lichinga (Vila Cabral), 10.v.1934, *Torre* 62 (COI).

Also in Tanzania and Zaire. In dry miombo woodland in sandy or gravelly soil, on stony ridges and rocky hillsides.

In northern Malawi, in the Chitipa District in particular, the two varieties overlap; *Pawek* 5271 & 9395 (K), for example, being intermediate between the two.

116. **Vernonia mbalensis** G.V. Pope in Kew Bull. **41**: 395 (1986). —C. Jeffrey in Kew Bull. **43**: 242 (1988). Type: Zambia, Mbala, *Bullock* 3844 (K, holotype).

An erect somewhat diffusely branched annual herb, up to c. 60 cm. tall. Stems longitudinally striate, angular, puberulous; branches up to c. 30 cm. long, spreading-ascending, often dark-purple in the lower part, shortly pubescent. Leaves largest about mid-stem, up to c. 8 × 1.3 cm., linear-oblanceolate to narrowly elliptic, acute at the apex, narrowly cuneate to the base, remotely serrulate to subentire on the margins; upper surface sparsely hispidulous-scabridulous, the lower somewhat puberulous; the lowermost leaves soon withering. Capitula numerous, in a large panicle; stalks 1–8 cm. long, slender, sometimes purple in the lower part, pubescent. Involucres c. 10 mm. long, exceeding the puppus at anthesis, narrowly cylindric-campanulate, not or hardly spreading, pubescent. Phyllaries progressively longer towards the inside; the outermost lanceolate, mucronate or shortly awned; the inner becoming narrowly strap-shaped, gently tapering to a bristle-tipped apex, brownish hispid towards the apex, pubescent below. Corollas white, cream, or mauve, narrowly funnel-shaped. Achenes pale- to reddish-brown, 2–2.5 mm. long, narrowly cylindric, truncate at the apex, slightly narrowed at the base, terete or obscurely 5-ribbed, uniformly white strigose, glandular; outer pappus of short, broadly oblong white scales, fimbriate-toothed on the upper margin; inner pappus of c. 6 greenish barbellate setae 7–8 mm. long, barbs appressed-ascending, decreasing in size and number towards the seta base.

Zambia. N: Road to Lunzua Falls, SW. of Mpulungu, c. 1425 m., 24.iv.1986, *Philcox, Pope & Chisumpa* 10100 (K; LISC; MO; NDO; SRGH).

Also in Tanzania. In miombo woodland.

117. **Vernonia chloropappa** Bak. in Bull. Misc. Inf., Kew **1898**: 146 (1898). —Brenan in Mem. N.Y. Bot. Gard. **8**, 5: 459 (1954). —Wild in Kirkia **11**: 64 (1978) pro parte. —C. Jeffrey in Kew Bull. **43**: 242 (1988). TAB. **23** fig. A. Type: Malawi, Chitipa, *Whyte* s.n. (K, mixed gathering, element A chosen as lectotype).

Vernonia kaessneri De Wild. & Muschl. in Bull. Soc. Roy. Bot. Belg. **49**: 242 (1913) nom. illegit., non S. Moore (1902). Type from Zaire.

Vernonia smaragdopappa S. Moore in Journ. Linn. Soc., Bot. **47**: 284 (1925). Type as for *V. kaessneri* De Wild. & Muschl.

An erect annual herb 30–80 cm. tall. Stems sparsely leafy, simple or becoming laxly branched, longitudinally ribbed; branches up to c. 60 cm. long, spreading, strongly ribbed, pubescent. Leaves largest about mid-stem, up to c. 8 × 1 cm., more usually c. 4 × 0.3 cm., linear-oblanceolate, obtuse and submucronate at the apex, attenuate to the base, remotely serrulate or subentire; upper surface puberulous to glabrescent; lower leaves soon withering; branch leaves c. 2 cm. long, lorate-oblanceolate. Capitula solitary on long branches, or 2(3) on stout stalks at the ends of branches, stalks often dark-purple at least below. Involucres 9–11 mm. long, broadly campanulate to obconic-spreading, ± rounded at the base, silvery-white tomentellous. Phyllaries numerous, progressively longer

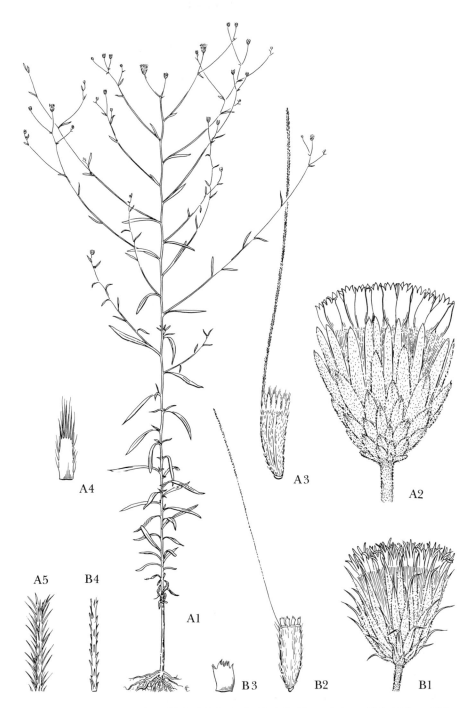

Tab. 23. A. —VERNONIA CHLOROPAPPA. A1, habit (×⅙); A2, capitulum, A1 & A2 from *Philcox, Pope & Chisumpa* 10165; A3, achene; A4, pappus scale; A5, detail of pappus seta, A3–A5 from *Pawek* 2736. B. —VERNONIA JELFIAE var. JELFIAE. B1, capitulum; B2, achene; B3, pappus scale; B4 detail of pappus seta, B1–B4 from *Richards* 16401. (Capitula × 3, achenes × 10, pappus × 30). Drawn by Eleanor Catherine.

towards the inside; the outer phyllaries lanceolate and tomentellous-pubescent, the inner narrow and tapering to a ± blunt mucronulate apex, white-sericeous becoming brownish-strigose towards the apex. Corollas white, cream or reddish-mauve, seldom purple, 10–12 mm. long, gradually tapering to the base. Achenes pale- to reddish-brown, c. 2.5 mm. long, narrowly subcylindric, truncate at the apex, ± abruptly narrowed at the base, obscurely c. 5-ribbed, densely white strigose, glandular; outer pappus of pale-brownish short paleaceous scales, ciliate-fimbriate on the margins, sometimes strigose or hispid-scabrid outside; inner pappus of emerald-green setae 7–8 mm. long, copiously uniformly bristled from apex to base, the bristles ascending-patent, not decreasing in size and number towards the base.

Zambia. N: 43 km. S. of Mbala, 18.vii.1930, *Hutchinson & Gillett* 3861 (BM; COI; K; LISC; SRGH). **Malawi**. N: Nyika, Nchenachena Spur, 1400 m., 20.viii.1946, *Brass* 17368 (K; MO; PRE; SRGH). Also in Tanzania. *Brachystegia* and *Uapaca* woodlands.

118. **Vernonia poskeana** Vatke & Hildebrandt in Oesterr. Bot. Zeit. **25**: 324 (1875). —Oliv. & Hiern in F.T.A. **3**: 274 (1877) pro parte quoad *Hildebrandt*. —Mendonça, Contrib. Conhec. Fl. Angol., **1** Compositae: 7 (1943) pro parte. —Wild in Kirkia **11**: 2; 56 (1978) pro parte. —G.V. Pope in Kew Bull. **41**, 1: 38 (1986). —C. Jeffrey in Kew Bull. **43**: 243 (1988). Type from Zanzibar.

An erect annual herb, c. 10–120 cm. tall. Stems simple or ± diffusely branched, leafy above; branches 3–40 cm. long, ribbed, puberulous. Leaves of lower stem up to 8.0 × 0.6 cm., though mostly smaller, linear, ± revolute, soon withering; upper-cauline leaves and leaves of the branches usually much narrower, c. 1–6.5 × 0.1 cm., mostly filiform-linear, ± strongly revolute, puberulous to scabridulous. Capitula numerous in open panicles, or capitula fewer and corymbiform cymose on the branches; capitula stalks slender and up to 2 cm. long, or stout and 2–6 cm. long. Involucres 6–12 mm. long, exceeding or equalling the pappus at anthesis, obconic with ± spreading phyllaries, or subfusiform to campanulate with the inner phyllaries remaining ± appressed. Phyllaries gradually narrowing to an acute or obtuse apex, not rounded or expanded above, often purple-tipped and subapically mucronate, appressed brownish- or white-hispid to pilose; the outermost short, lanceolate or narrowly ovate, the inner progressively longer becoming narrowly lanceolate to lorate. Corollas purplish, 6–11 mm. long, tapering to the base. Achenes pale- to dark-brown, 2.5–5 mm. long, narrowly obconical or obpyramidal, angular, 5–12-ribbed, hispidulous and ± glandular; outer pappus of many short overlapping oblong-lanceolate scales c. 0.5 mm. long, the inner of brownish or white barbellate setae 4–6 mm. long, not plumose.

1. Achenes 8–12-ribbed - - - - - - - - - - subsp. *botswanica*
– Achenes 5–7-ribbed - - - - - - - - - - - - 2
2. Involucre 10 mm. or more long, more than 6 mm. wide, ± campanulate or turbinate; capitula stalks not slender, the majority more than 2 cm. long; achenes 3–4 mm. long
subsp. *samfyana*
– Involucre less than 10 mm. long (sometimes to c. 10 mm. long but then less than 6 mm. wide), obconic, subfusiform to hemispheric; capitula stalks slender, the majority less than 2 cm. long; achenes 1.5–3 mm. long - - - - - - - - subsp. *poskeana*

Subsp. **poskeana**
Vernonia poskeana var. *chlorolepis* sensu Eyles in Trans. Roy. Soc. S. Afr. **5**, 4: 505 (1916) pro parte.
Vernonia poskeana var. *poskeana* sensu Wild in Kirkia **11**: 56 (1978) pro parte.

Capitula stalks slender, mostly 2 cm. or less in length. Involucre less than 10 mm. long, spreading or subfusiform, sometimes hemispheric. Phyllaries gradually tapered to the apex, or shortly tapering and ± blunt at the apex, subapically mucronate, brownish-hispid towards the apex, pilose below. Achenes 2–3 mm. long, narrowly obpyramidal, strongly 5–7-ribbed; outer pappus of brownish overlapping oblong-lanceolate scales.

Zambia. B: W. bank of Zambezi R. at Zambezi (Balovale) pontoon, 26.v.1960, *Angus* 2290 (K; SRGH). S: 16 km. N. of Choma, 15.iv.1963, *van Rensburg* 1943 (K). **Zimbabwe**. N: Sebungwe, c. 600 m., vi.1956, *Davies* 2004 (K; SRGH). W: Matobo, Besna Kobila Farm, 1470 m., 27.iii.1963, *Miller* 8418 (K; SRGH). C: between Kadoma (Gatooma) and Kwekwe (Que Que), 2.v.1958, *West* 3595 (K; SRGH). E: Odzani R. valley, 1914, *Teague* 55 (K). S: Makoholi Expt. Sta., 14.iii.1978, *Senderayi* 236 (K; SRGH). **Malawi**. C: Salima, Senga Bay, 530 m., 25.iv.1971, *Pawek* 4715 (K; MAL). **Mozambique**. N: Entre Memba e Nacala e Nampula, 17.v.1937, *Torre* 1413 (COI).

Also from Kenya, Tanzania and Angola. Deep sandy soil in grassland or deciduous woodland, on granite or serpentine rocky outcrops, stream or lake margins and roadsides.

Three variants, based on involucre shape, can be recognised: the typical form from Tanzania and Zanzibar has small obconic involucres with spreading phyllaries. The second variant, seen in southern Zambia and adjacent Zimbabwe, has subfusiform to narrowly turbinate involucres, longer than wide, with phyllaries loosely appressed-imbricate but not spreading. In central and eastern Zimbabwe these plants intergrade with a third variant in which the involucres are campanulate-hemispheric and almost as wide as long with the phyllaries remaining appressed imbricate. The involucres of the second form approach those of subsp. *samfyana* but are smaller and are borne on shorter more slender stalks.

Subsp. **samfyana** (G.V. Pope) G.V. Pope comb. et stat. nov.

    Vernonia samfyana G.V. Pope in Kew Bull. **41**: 42 (1986). Type: Zambia, Samfya, *Mutimushi* 1126 (K, holotype).

Capitula stalks mostly more than 2 cm. long. Involucre campanulate or turbinate, 10 mm. or more long. Phyllaries ovate to narrowly lanceolate, acute to obtuse mucronulate, puberulous to glabrescent. Achenes 3.2–4 mm. long, strongly c. 5-ribbed and tapering to a sharp base; outer pappus of short whitish, usually oblong, apically ± lacerate scales; inner pappus of white or brownish barbellate setae 6–7 mm. long.

    Zambia. N: N. of Samfya, Lake Bangweulu, 1200 m., 7.x.1947, *Greenway & Brenan* 8175 (K). Also in Angola. Sand dunes at lake sides and river banks.

Subsp. **botswanica** G.V. Pope in Kew Bull. **41**: 39 (1986). Type: Botswana, Khutse, *Coleman* 60 (K, holotype; LISC; SRGH).

    Vernonia poskeana var. *chlorolepis* sensu Mendonça, Contrib. Conhec. Fl. Angol., **1** Compositae: 7 (1943) pro parte.

    Vernonia poskeana sensu Merxm., Prodr. Fl. SW. Afr. 139: 184 (1967).

    Vernonia poskeana var *poskeana* sensu Wild in Kirkia **11**: 56 (1978) pro parte.

Capitula stalks mostly more than 2 cm. long. Involucre ± broadly campanulate to spreading. Phyllaries narrowly ovate to lanceolate, subapically mucronate or aristate, white-pilose or tomentellous outside. Achenes 2.5–5.0 mm. long, narrowly sub-terete, sharply tapered to the base, strongly 8–12-ribbed.

    Botswana. N: Mutsoi, 14.iv.1967, *Lambrecht* 120 (K; SRGH). SE: Gaborone, University Campus, 1000 m., 14.iii.1974, *Mott* 168c (K; UBLS). SW: 216 km. NW. of Molepolole, 15.vi.1955, *Storey* 4899 (BOL; K; PRE; SRGH). **Zimbabwe**. W: Bulilima Mangwe, 1300 m., 11.v.1972, *Norrgrann* 150 (K; SRGH). C: Gweru (Gwelo), 1400 m., 6.v.1967, *Biegel* 2142 (K; SRGH).

Also in South Africa (N. Cape Province and the Transvaal), Namibia and Angola. Sandy soil in hot dry areas, in grassland and dry deciduous woodland, also in cultivated land.

119. **Vernonia rhodesiana** S. Moore in Journ. Bot. **64**: 303 (1926). —Wild in Kirkia **11**: 58 (1978). —G.V. Pope in Kew Bull. **41**: 42 (1986). Type: Zimbabwe, Mwami (Miami), *Rand* 91 (BM, holotype).

A slender annual herb to c. 80 cm. tall. Stems erect, simple or branched above, longitudinally ribbed, puberulous. Leaves entire or some 3-lobed, subsessile, usually with a short pubescent petiole c. 3 mm. long; lamina up to c. 5.5 × 1 cm., narrowly oblong-elliptic, or 3-lobed up to c. 1.5 cm. wide and ± obovate in outline, obtuse or acute at leaf or lobe apex, cuneate at base, entire or obscurely crenate, scabridulous on upper surface and sparsely pilose or glabrescent beneath; lateral lobes 5–10 × 2–3 mm., pointing forwards; terminal lobe broader, c. 15 × 20 mm. long. Capitula numerous, laxly paniculate on many slender branches 5–9 mm. long; stalks slender up to c. 3 mm. long, pubescent. Involucre up to c. 10 × 6 mm., narrowly campanulate; phyllaries appressed, increasing in size from c. 2 mm. on the outside to c. 9 mm. long inside, lanceolate to linear-lanceolate, finely pilose-pubescent. Corollas purple, c. 7.5 mm. long, tubular in the lower half, narrowly funnel-shaped above, sparsely pilose. Achenes (immature) 2 mm. long, 5-ribbed, setulose, outer pappus of short lanceolate-fimbriate scales, inner of whitish barbellate setae c. 6 mm. long.

    Zimbabwe. N: Mwami (Miami), iv.1926, *Rand* 91 (BM). Known only from the type collection.

120. **Vernonia erinacea** Wild in Kirkia **11**: 2; 55 (1978). —G.V. Pope in Kew Bull. **41**: 40 (1986). —C. Jeffrey in Kew Bull. **43**: 243 (1988). Type: Malawi, Zomba Plateau, *Brass* 16090 (SRGH, neotype; BM; BR; K —*Wild* 1978).

Crystallopollen angustifolium Steetz in Peters, Reise Mossamb., Bot.: 366 (1864) non. *V. angustifolia* D. Don ex Hook. & Arn. (1835) nec. Michx. (1803). Type: Mozambique, Rios de Sena, *Peters* s.n. (B†).

Vernonia poskeana sensu Oliv. & Hiern in F.T.A. **3**: 274 (1877) quoad syn. et specim. Mozamb.

Vernonia poskeana var. *vulgaris* Hiern, Cat. Afr. Pl. Welw. **1**, 3: 519 (1898). —Mendonça, Contrib. Conhec. Fl. Angol., 1 Compositae: 7 (1943). Type as for *Crystallopollen angustifolium*.

An erect slender, or ± robust, slightly harsh annual herb, 0.1–2 m. tall. Stems simple or ± diffusely branched, ribbed or sulcate, sparsely appressed-pubescent. Leaves sessile, up to c. 10 × 0.5 cm., filiform to linear, less often narrowly elliptic, tapering-acute or subobtuse at the apex, often mucronate, entire or subentire, sometimes revolute, upper surface finely pubescent-scabridulous or glabrescent, gland-dotted on both surfaces. Capitula few to many, corymbiform cymose on branches or in a large paniculate arrangement; stalks slender, up to c. 5 cm. long, pubescent. Involucre up to 8 mm. long, obconic or spreading campanulate, ± as long as broad. Phyllaries c. 5-seriate, pubescent to finely pilose; outermost phyllaries spreading, 1–2 mm. long, subulate to aristate-lanceolate; middle phyllaries linear-oblong with long dark squarrose-aristate tips; inner phyllaries up to c. 8 mm. long, linear, dark-tipped and aristate or mucronate, usually not squarrose. Corollas purple, up to c. 7.5 mm. long, exserted beyond the involucre, tubular in the lower half, narrowly funnel-shaped above. Achenes 2–2.5 mm. long, narrowly ± turbinate, 5–6-angular or ribbed, finely hispid on the ribs and glandular between; outer pappus short, of oblong-lanceolate ± overlapping fimbriate scales, inner of brownish-purple or grey-brown barbellate setae up to c. 5 mm. long.

Zambia. C: Chongwe Bridge, c. 50 km. E. of Lusaka, 24.v.1962, *Best* 329 (SRGH). E: 12 km. NW. of Machinje, 30.v.1961, *Leach & Rutherford-Smith* 11079 (K; LISC; M; SRGH). S: Lochinvar Ranch, 25.iv.1962, *Angus* 3175 (LISC; MO; SRGH). **Zimbabwe**. N: Mazowe, 7.vi.1974, *Pope* 1358 (K; SRGH). E: Chipetzana R., c. 900 m., 19.iv.1907, *Swynnerton* 1828 (K). **Malawi**. C: Kasungu, 11.vii.1970, *Hall-Martin* 1739 (K; SRGH). S: Zomba Plateau, 30.v.1946, *Brass* 16090 (BM; BR; K; MO; PRE; SRGH). **Mozambique**. N: Eráti, Namapa, 16.viii.1948, *Barbosa* 1752 (LISC; SRGH). Z: Quelimane Distr., Lugela-Mocuba, Namagoa, vii.194-, *Faulkner* Kew 29 (BR; COI; K; SRGH). T: c. 33 km. SW. of Zóbuè, 1.vii.1962, *Leach & Schelpe* 11500 (K; LISC; SRGH). MS: Chimoio, 26.ii.1948, *Garcia* 399 (LISC). GI: Gaza Prov., Chibuto, Mahamba, 11.vi.1960, *Lemos & Balsinhas* 88 (PRE). M: Namaacha, *Barbosa & Lemos* 7550 (LISC).

Also in Tanzania, Zaire and South Africa (Transvaal). Tree savanna and grassland, sometimes in dambos, also in disturbed ground at roadsides and in cultivation as a weed.

121. **Vernonia steetziana** Oliv. & Hiern in F.T.A. **3**: 273 (1877). —Hilliard, Compos. Natal: 38 (1977). —G.V. Pope in Kew Bull. **41**: 41 (1986). Type: Mozambique, Kaimba Island opposite Tete, *Kirk* s.n. (K, lectotype fide Hilliard).

Crystallopollen angustifolium var. *chlorolepis* Steetz in Peters, Reise Mossamb., Bot.: 366 (1864). Type: Mozambique, Inhambane, *Peters* s.n. (B†).

Vernonia poskeana var. *chlorolepis* (Steetz) O. Hoffm. in Bol. Soc. Brot. **10**: 171 (1892): sensu Hiern, Cat. Afr. Pl. Welw. **1**, 3: 520 (1898) quoad nomina solem. Type as for *Crystallopollen angustifolium* var. *chlorolepis*.

Vernonia poskeana var. *poskeana* sensu Wild in Kirkia **11**: 3; 56 (1978) pro parte excl. typ.

An erect annual herb 0.3 to 1 m. tall. Stems unbranched below with the lower leaves soon withering, branched and ± densely leafy above; branches 5–40 cm. long, ribbed, pubescent. Leaves sessile, up to c. 10 × 0.6 cm., linear or narrowly elliptic-oblanceolate, acute to obtuse at the apex, mucronate, entire and revolute on the margins, ± gradually tapering to the base, puberulous to glabrescent, sometimes scabridulous; leaves of axillary growth much smaller and more crowded. Capitula few to numerous in lax corymbiform cymes or open panicles, capitula stalks stout, 2–10 cm. long, with a few linear bracts or ebracteate. Involucres up to c. 12 mm. long, urceolate-campanulate to hemispheric, usually rounded at the base, not or hardly spreading above, obscuring the pappus at anthesis. Phyllaries appressed imbricate, obtuse to rounded at the apices and mucronate-aristate, ± expanded membranous and often purple-tinged on the margins towards the apex, hispidulous; the outermost phyllaries c. 2 mm. long, ovate-oblong, the inner series progressively longer, changing from broadly oblong-oblanceolate to lorate. Corollas purplish, 7–11 mm. long, tapering to the base. Achenes pale-brown, 2.5–4 mm. long, narrowly obconic-turbinate, 5–7-ribbed, hispidulous on the ribs and glandular between them; outer pappus of short white lanceolate scales, the inner of brownish barbellate setae c. 6 mm. long, not plumose.

Botswana. N: c. 32 km. SW. of Samochimo at 21°48'E, 18°42'S, 2.v.1975, *Biegel, Müller & Gibbs-Russell* 5053 (K; SRGH). **Zambia**. B: Zambezi R., c. 10 km. NW. of New Sesheke Boma, 9.viii.1947, *Brenan & Keay* 7357 (K; FHO). S: Namwala, 18.iv.1963, *van Rensburg* 2053 (K). **Zimbabwe**. N: Gokwe Distr., Sengwa Res. Sta., 27.v.1976, *Guy* 2440 (BR; K; SRGH). W: Hwange Distr., 72 km. NE. of Kamativi Tin Mine, 8.v.1955, *Plowes* 1791 (K; LISC; SRGH). E: Odzi, 1060 m., 3.vi.1936, *Eyles* 8622 (K; SRGH). S: Runde (Lundi) R., 30.vi.1930, *Hutchinson & Gillett* 3279 (BM). **Malawi**. N: Nkhata Bay Distr., Old Bandawe, 52 km. S. of Nkhata Bay junction, 500 m., 25.vi.1977, *Pawek* 12806 (K; MAL; MO; SRGH; UC). S: 5 km. SW. of Nsanje, 65 m., 29.v.1970, *Brummitt* 11160 (K). **Mozambique**. N: Entre Memba e Nacala e Nampula, 17.v.1937, *Torre* 1413 (COI). Z: 91.6 km. from Mocuba on road to Namacurra, 27.viii.1949, *Barbosa & Carvalho* 3813 (K). T: Sisitso, Zambezi R., c. 290 m., 15.vii.1950, *Chase* 2775 (BM; K; LISC; SRGH). M: Maputo (Delagoa Bay), c. 50 m., viii.1886, *Bolus* 7780 (K).

Also in South Africa (Transvaal, Natal) and Swaziland. Sandy soil in open miombo or *Baikiaea* woodland, river or lake sides and in disturbed ground (roadsides) or land cleared for cultivation.

The specimens *Faulkner* 255 (K; PRE; SRGH) and *Barbosa & Carvalho* 3813 (K) differ from typical *V. steetziana* in having broadly obconic involucres, capitula in a single terminal cluster, or in several corymbiformly arranged clusters on long branches, densely leafy stems and small ± woody rootstocks (suggesting a perennial habit).

122. **Vernonia centaureoides** Klatt in Bull. Herb. Boiss. **4**: 824 (1896) "*centauroides*" non Sch. Bip. (1865–66) nom. nud. —Hilliard, Compos. Natal: 38, fig. 4 (1977). —G.V. Pope in Kew Bull. **41**: 42 (1986). Type: Mozambique, Maputo (Delagoa Bay) *Junod* 5 (BR; Z).

   Vernonia poskeana var. *centauroides* (Klatt) Wild in Kirkia **11**: 3; 58 (1978) pro parte quoad *Schlechter* 12148. Type as above.

An erect slender to somewhat bushy, stiffly-branched annual or perennial herb, up to c. 70 cm. tall; tap-root becoming woody. Stem simple in lower part or much-branched, longitudinally striate, coarsely pubescent; the hairs ± unequal-armed T-shaped; branches divaricate-ascending, mostly 3–20 cm. long, similar to the stem. Leaves subsessile, up to 45 × 4 mm., linear-oblanceolate, obtuse mucronate at the apex, entire, ± revolute, gradually tapering to the base, puberulous to ± scabridulous; much smaller and ± crowded on axillary and flowering branches. Capitula many or numerous, solitary on short leafy branches, or 2–3 corymbiformly cymose on longer branches; involucres up to c. 10 mm. long, cyathiform, width less than or ± equal to length. Phyllaries appressed imbricate, hardly spreading, becoming thinly cartilaginous, araneose-pilose to glabrescent, ciliate on upper margins, ± equalling the pappus in length; the outermost phyllaries ovate to oblong, acute mucronate; the inner increasing in length to the inside, linear-oblong, ± expanded apically and mucronate to shortly aristate at the tip. Corollas purplish, up to 10 mm. long, tapering gradually to the base. Achenes pale-brown, c. 2.5 mm. long, obconic-turbinate, 5–7-ribbed, hispidulous on the ribs, glandular between them; outer pappus of short lanceolate scales, inner pappus of brownish plumose setae 5–7 mm. long, the setae bristles fine and longer than the achene hairs.

Mozambique. M: Baía de Maputo (Delagoa Bay), 15.iii.1898, *Schlechter* 12148 (BOL; COI; E; K; PRE; S).
Also recorded from South Africa (the Transvaal and Zululand in Natal). Common in sandy soil near the coast, in grassland, open woodland and old cultivation.

123. **Vernonia perrottetii** Sch. Bip. ex Walp., Repert. **2**: 947 (1843). —Oliv. in Trans. Linn. Soc. Lond. **29**: 90, t. 56A (1873). —Oliv. & Hiern in F.T.A. **3**: 272 (1877). —Mendonça, Contrib. Conhec. Fl. Angol., **1** Compositae: 9 (1943). —Adams in F.W.T.A., ed. 2, **2**: 281 (1963). —Wild in Kirkia **11**: 52 (1978). —Maquet in Fl. Rwanda, Spermat. **3**: 556, fig. 167, 4 (1985). —C. Jeffrey in Kew Bull. **43**: 243 (1988). Type from Gambia.

   Webbia serratuloides DC., Prodr. **5**: 72 (1836) non *Vernonia serratuloides* Kunth (1820). Type as for *Vernonia perrottetii*.

An erect usually strict annual herb 20–100 cm. tall. Stems simple below, ± shortly branched above, usually branching only at the apex but sometimes with numerous short branches in the upper c. half, densely leafy, finely-ribbed, appressed-pubescent; the hairs flagelliform; branches ascending, 4–25 cm. long, less densely leafy than the stem. Leaves crowded, ± appressed-ascending, overlapping, up to c. 7 cm. long, usually shorter, filiform to linear, revolute, scabridulous. Capitula usually solitary and terminal on stems and branches, or corymbiformly cymose; capitula stalks usually gradually widening towards the apex. Involucre 10–18 × 10–20 mm., globose-cyathiform or ± turbinate. Phyllaries c. 8-seriate, closely appressed-imbricate, acute often purplish at the apex, araneose-lanate; the outer phyllaries from c. 2 mm. long, ovate-lanceolate; the inner increasing in length to c. 18 mm. long, linear-lanceolate. Corollas purple, narrowly funnel-shaped, up to c. 2

cm. long, limbs ± exserted from the involucre. Achenes to c. 5 mm. long, narrowly tapering to the base, distinctly 6–10-ribbed, densely white strigose-hispid particularly on the ribs, glandular between the ribs; pappus buff or sordid, the outer pappus of short narrow scales, the inner of long-barbellate setae c. 9 mm. long.

Zambia. B: near Angolan Border, 11.v.1925, *Pocock* 249 (BOL; PRE). N: Mbala, Ndundu, 8.v.1959, *Richards* 11380 (K; M; SRGH). W: Kitwe, 13.vii.1967, *Mutimushi* 1983 (K; NDO; SRGH). C: Lusaka, 6.iv.1956, *Noak* 197 (K; SRGH). S: Choma–Kafue R., 11.vii.1930, *Hutchinson & Gillett* 3552 (BM; K; SRGH). **Malawi**. N: Karonga, Ngala, 22.iv.1969, *Pawek* 2267 (K).

Also in tropical W. Africa, Sudan, Ethiopia, Rwanda, Uganda, Kenya, Tanzania, Angola and Zaire. In deciduous woodland or short grassland, sandy soils and laterite pans, often in disturbed ground and on roadsides.

124. **Vernonia amoena** S. Moore in Journ. Bot. **55**: 103 (1917). —Wild in Kirkia **11**: 65 (1978). —C. Jeffrey in Kew Bull. **43**: 243 (1988). Type: Zimbabwe, Hwange, *Rogers* 13300 (BM, holotype).

An erect somewhat soft annual herb, to c. 1 m. tall. Stems simple below, branched above, faintly ribbed, appressed puberulous; hairs short-stalked T-shaped. Leaves membranous, sessile or shortly petiolate, up to c. 16 × 6.5 cm., elliptic, tapering to an acuminate apex and a cuneate base, serrate crenate or subentire; upper surface with scattered small flagelliform hairs, lower surface sparsely thinly puberulous with short-stalked T-shaped hairs. Capitula few to many, ± laxly corymbiform cymose on the branches. Involucre up to c. 15 mm. long, campanulate, exceeding the pappus. Phyllaries many, stiff chartaceous, gently tapering to a pungent apex, ± narrowly membranous on the margins, hispidulous outside; outer phyllaries short, subulate-lanceolate, ± spreading, hispid; inner phyllaries to c. 15 mm. long, narrowly lanceolate, aristate. Corollas white or purplish, c. 8 mm. long, tubular but slightly wider towards the apex, pubescent and glandular except near the base of the tube; lobes linear, 1.5 mm. long. Achenes 3–4 mm. long, narrowly obovoid-cylindric, narrowly sharply 6-ribbed, strigose; outer pappus of short barbellate lanceolate-acuminate scales, inner of ± caducous white barbellate setae 6–8 mm. long.

Zambia. B: Machili, 13.iii.1961, *Fanshawe* 6410 (K; NDO; SRGH). N: Lake Tanganyika, Kumbula Island, 11.iv.1955, *Richards* 5388 (K; LISC; SRGH). C: Katondwe, 22.vi.1967, *Fanshawe* 10125 (M; NDO; SRGH). S: Namwala, Sibanzi Hill, 13.iv.1972, *van Lavieren, Sayer & Rees* 789 (SRGH). **Zimbabwe**. N: Gokwe, Copper Queen N.P.A., Muradzi, 15.iv.1964, *Bingham* 1212 (K; SRGH). W: Hwange, Gwaai-Lutope R. junction, 27.ii.1963, *Wild* 6035 (K; S; SRGH). ?C: Harare, 1906, *Flanagan* 3075 (BOL). S: Ndanga, Chitsa's Kraal, 15.vi.1950, *Chase* 2352 (SRGH). **Malawi**. N: Rumphi, Rukuru R. Gorge, 25.v.1967, *Pawek* 1125 (SRGH).

Also in Tanzania. Low altitude river valleys in the undergrowth of *Acacia, Commiphora/Colophospermum* or *Baikiaea* woodlands and in riverine thickets.

Richards 5388 (K, LISC, SRGH), from Kumbula Island in Lake Tanganyika, is unusual in that both the outer and inner pappus elements are much reduced in size - the outer to a fimbriate fringe of low scales, the inner to c. 2 mm. long setae (shorter than the achene).

12. **GUTENBERGIA** Sch. Bip.

Gutenbergia Sch. Bip. in Gedenkb. IV Jubelf. Buchdr. Mainz: 119, t.4 (1840); in Walp., Repert. **2**: 703 (1843). —Wild & G.V. Pope in Kirkia **10**: 311; 347 (1977). —C. Jeffrey in Kew Bull. **43**: 249 (1988).
Paurolepis S. Moore in Journ. Bot. **55**: 102 (1917).

Annual or perennial herbs with simple to much-branched, erect or decumbent stems. Vegetative indumentum of short-stalked, long-armed, symmetrical T-shaped hairs. Leaves opposite, alternate, or in one species in whorls of 3, sessile or petiolate, lamina upper surface sparsely pilose to scabridulous, lower surface closely whitish-silvery araneose-felted. Capitula usually small, few to numerous, laxly arranged in corymbiform cymes, or shortly-stalked in small clusters, homogamous and discoid. Involucres obconic to cyathiform; phyllaries ± diverging, increasing in size to the inside, membranous or scarious usually with subhyaline margins. Receptacle plane or alveolate. Corollas purple or mauve, regular, narrowly infundibuliform, 5-lobed, usually puberulous, the lobes sometimes with acicular bristles and/or T-shaped to Y-shaped hairs. Anthers lanceolate-appendiculate at the apex, obtuse at the base. Style-arms linear, hairy. Achenes obovoid-oblong or turbinate, rounded or truncate at the apex, ecostate, or 4–6-ribbed, or 8–10-ribbed, polished and smooth or with hooked or ± straight simple or bifid trichomes and globose-glandular trichomes. Pappus 0, or of a few short very caducous barbellate setae, or of persistent overlapping scales.

A tropical African genus of approximately 20 species.
Gutenbergia is characterised by its distinctive indumentum of T-shaped hairs, and includes 5 achene types based on the presence or absence of a pappus, the indumentum and the nature of the ribs when present.

1. Pappus of persistent, short, overlapping scales; receptacle alveolate; leaves filiform
   　　　　　　　　　　　　　　　　　　　　　　　　　　　　　　　　　　1. *filifolia*
 – Pappus absent, or of few very caducous bristles; receptacle not alveolate; leaves linear to broadly lanceolate　-　-　-　-　-　-　-　-　-　-　-　-　-　-　-　-　2
2. Leaves, at least some, in whorls of 3 or opposite, linear up to c. 1 mm. wide; stems annual from a perennial woody rootstock　-　-　-　-　-　-　-　-　-　-　-　-　2. *trifolia*
 – Leaves opposite or alternate, not in threes, linear to broadly lanceolate, 1–50 mm. wide; stems simple from an annual root or stems perennial becoming woody below　-　-　-　3
3. Achenes terete or angled but not ribbed; pappus absent　-　-　-　-　-　-　4
 – Achenes ribbed; pappus setae present or absent　-　-　-　-　-　-　-　8
4. Achenes glabrous, usually smooth and polished　-　-　-　-　-　-　-　5
 – Achenes uniformly roughly puberulous, or setose towards the base and apex　-　-　7
5. Leaves ovate-lanceolate, sessile-cordate, sometimes subauriculate, more than 5 mm. wide when mature; stems trailing/ascending; plants of sandy river banks and lake shores　8. *leiocarpa*
 – Leaves linear or narrowly strap-shaped, not or hardly widening at base, mostly less than 5 mm. wide (if more then linear-elliptic); stems erect; plants of deciduous woodland　-　-　6
6. Achenes obovoid-tubinate, ± strongly 3-angled; capitula subsessile or shortly-stalked in many 2–12-headed clusters　-　-　-　-　-　-　-　-　-　-　-　10. *kassneri*
 – Achenes subcylindric or oblong, subterete or faintly c. 4-angled; capitula more laxly arranged on stalks 3–35 mm. long　-　-　-　-　-　-　-　-　-　-　9. *adenocarpa*
7. Achenes uniformly, roughly puberulous and glandular all over, c. 2 mm. long; leaves mostly more than 1.5 cm. wide　-　-　-　-　-　-　-　-　-　7. *mweroensis*
 – Achenes glabrous about the middle, appressed-setose towards the base and apex, c. 1.5 mm. long; leaves mostly to 1 cm. wide, sometimes to 1.5 cm. wide but then phyllaries glabrescent and tapering to the apex　-　-　-　-　-　-　-　-　-　-　11. *gossweileri*
8. Achenes distinctly 8–12-ribbed, prominent ribs alternating with less prominent ribs; pappus 0; plants annual　-　-　-　-　-　-　-　-　-　-　-　-　-　-　9
 – Achenes c. 5-ribbed, ribs similar; pappus of a few very caducous barbellate setae c. 1–2 mm. long, present at the flowering stage (sometimes apparently absent in *G. eylesii* but then leaves cuneate below and shortly petiolate); plants perennial　-　-　-　-　-　-　-　10
9. Leaves 2–5 mm. wide, strap-shaped; capitula subsessile in 3–10-capitulate sub-globose clusters, subtending leaves overtopping the capitula　-　-　-　13. *spermacoceoides*
 – Leaves 5–35 mm. wide, oblong-lanceolate; capitula numerous, stalked, laxly arranged in ± diffuse compound cymes, overtopping the leaves　-　-　-　-　12. *polycephala*
10. Achene ribs swollen, rib-width equalling or exceeding the grooves in between, achenes rounded above; phyllaries with wide scarious margins not obscured by the indumentum　11
 – Achene ribs narrow, not swollen, achenes truncate above; phyllary margins usually narrow or obscured by the indumentum　-　-　-　-　-　-　-　-　-　-　-　12
11. Involucres 5–7 mm. long; corolla lobes with patent acicular bristles only, T-shaped hairs absent　-　-　-　-　-　-　-　-　-　-　-　-　-　-　-　4. *cordifolia*
 – Involucres 7–10 mm. long; corolla lobes with patent acicular bristles and T-shaped or Y-shaped hairs　-　-　-　-　-　-　-　-　-　-　3. *polytrichomata*
12. Involucres 8–10 mm. long; corolla lobes with patent acicular bristles only, T-shaped hairs absent　-　-　-　-　-　-　-　-　-　-　-　-　-　-　6. *westii*
 – Involucres 6–7 mm. long; corolla lobes with T-shaped or often Y-shaped hairs, bristles absent　-　-　-　-　-　-　-　-　-　-　-　-　-　-　5. *eylesii*

1. **Gutenbergia filifolia** (R.E. Fr.) C. Jeffrey in Kew Bull. **43**: 254 (1988). TAB. **24** fig. D. Type: Zambia, Kabwe, *R.E. Fries* 215 (UPS, holotype).
　　　Herderia filifolia R.E. Fr., Wiss. Ergebn. Schwed. Rhod.-Kongo-Exped. 1911–12, **1**: 327 (1916). Type as above.
　　　Paurolepis angusta S. Moore in Journ. Bot. **55**: 102 (1917). Type: Zambia, Kabwe, *Rogers* 7738 (BM, holotype).
　　　Paurolepis filifolia (R.E. Fr.) Wild & G.V. Pope in Kirkia **10**: 316, fig. 5, p. 332; 346, fig. 4, p.377 (1977). Type as for *Gutenbergia filifolia*.

A bushy perennial herb to c. 60 cm. tall from a woody rootstock. Stems annual, 1–many, much branched, leafy, becoming ribbed, sericeous; indumentum of short-stalked 1-

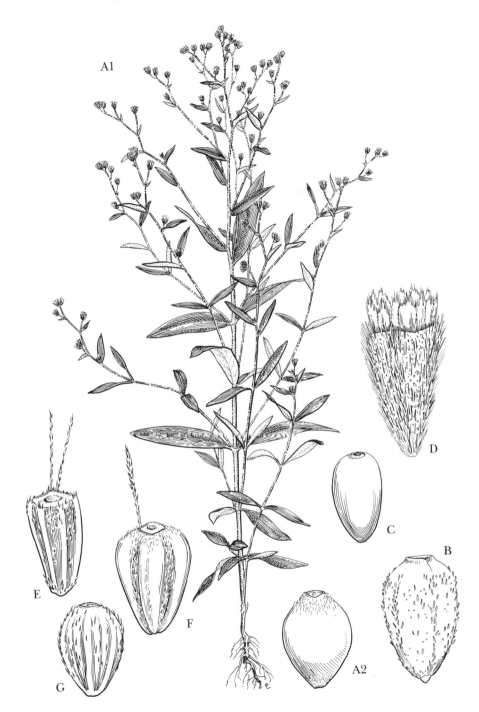

Tab. 24. A. —GUTENBERGIA GOSSWEILERI. A1, habit (× ½), from *Robinson* 2819; A2, achene, from *Pawek* 8397. B. —GUTENBERGIA MWEROENSIS, achene, from *Bullock* 1221. C. — GUTENBERGIA LEIOCARPA, achene, from *Fanshawe* 348. D. —GUTENBERGIA FILIFOLIA, achene, from *Robinson* 6676. E. —GUTENBERGIA EYLESII subsp. RETICULATA, achene, from *Richards* 5336. F. —GUTENBERGIA CORDIFOLIA, achene, from *Brummitt* 10473. G. —GUTENBERGIA POLYCEPHALA, achene, from *Wild* 1005. (All achenes × 16). Drawn by Eleanor Catherine.

shaped hairs with long dorso-ventrally flattened tapering arms. Leaves alternate, sessile, mostly 2–6 cm. long and less than 0.1 cm. wide, filiform-linear with strongly revolute margins, upper surface sparsely pilose-puberulous soon glabrescent, lower surface sericeous to floccose-lanate. Capitula many in lax, simple to compound corymbiform cymes; capitula stalks 1–6 cm. long, filiform. Involucres 4–5 × 5 mm., spreading to c. 10 mm. wide, obconic-cyathiform; phyllaries few-seriate, the outermost much reduced, the inner c. 2 series subequal and broadly oblong-oblanceolate, ± abruptly narrowed and mucronate at the apex, sericeous outside. Corollas exserted above the involucre, mauve, sometimes purplish, 6–9 mm. long, widely funnel-shaped above and shortly tubular below, limb deeply lobed, lobes to c. 4 mm. long, limb and tube appressed pilose-pubescent with T-shaped hairs. Achenes c. 2 mm. long, narrowly obpyramidal to somewhat turbinate, truncate above, tapering to a sharp base, c. 5-angular, densely hirsute with long straight minutely bifid hairs, glandular especially towards the base and apex; pappus persistent, of many often purplish or brownish, overlapping scales, c. 0.6 mm. long.

Zambia. W: Luanshya, 9.vii.1954, *Fanshawe* 1356 (K; NDO). C: Serenje Distr., Kundalila Falls, 11.iv.1986, *Philcox, Pope & Chisumpa* 9855 (K; NDO).
Also in Zaire. Rocky hillsides and outcrops in miombo woodland.

2. **Gutenbergia trifolia** Wild & G.V. Pope in Kirkia **10**: 313, fig. 4C, p. 330; 354 (1977). Type: Zambia, Mwinilunga, *Milne-Redhead* 4417 (K, holotype).

A tufted perennial herb from a woody rootstock. Stems annual, ascending, up to c. 20 cm. long, thinly white appressed-araneose; hairs short-stalked long-armed matted T-shaped. Leaves in whorls of 3, or opposite, sessile, up to c. 2.5 × 0.1 cm., linear, tapering to a pungent acuminate apex, with somewhat revolute margins, upper surface green and thinly pilose, lower surface closely white araneose-felted, midrib glabrescent. Capitula in terminal clusters of 2–6; stalks 1–30 mm. long, appressed whitish pilose. Involucres c. 6 × 5 mm., obconic-campanulate. Phyllaries many, lanceolate, tapering to a purple pungent tip, midrib prominent, densely araneose-lanate outside; the outer phyllaries from c. 2 mm. long, narrowly lanceolate, the inner to c. 6 mm. long. Corollas pale-mauve, to c. 6.5 mm. long, narrowly funnel-shaped, silky-whitish appressed-pilose in upper part, the hairs finely long-armed T-shaped. Achene (immature) c. 1 mm. long, ?narrowly obovoid, ?4-ribbed, sparsely setulose, minutely glandular; pappus 0.

Zambia. W: Mwinilunga, Kalenda Plain, 30.i.1938, *Milne-Redhead* 4417 (K).
Known only from the type specimen. In shallow peaty soil near water-hole.

3. **Gutenbergia polytrichomata** Wechuysen in Bull. Jard. Bot. Nat. Belg. **51**: 107 (1981). —C. Jeffrey in Kew Bull. **43**: 251 (1988) as '*polytrichotoma*'. Type from Tanzania.

An erect, somewhat spreading perennial herb or subshrub, 0.45–2 m. tall. Stems freely branched, leafy, appressed pubescent; indumentum of short-stalked long-armed T-shaped hairs. Leaves alternate, petiolate; petiole to c. 7 mm. long; lamina 2–12 × 1.5–5 cm., elliptic, subacute to obtuse at the apex, cuneate below, subentire to subserrate; upper surface green, drying olive-green, sparsely puberulous, the hairs short-armed T-shaped or somewhat Y-shaped, lower surface silvery-white araneose-felted, the hairs T-shaped with fine matted arms. Capitula many in somewhat lax corymbiform cymes; capitula stalks mostly 5–35 mm. long, bracteate. Involucres 7–10 × 5–10 mm., campanulate to obconic. Phyllaries many, scarious with ± broad subhyaline margins, sometimes purple-tinged, appressed-pubescent to strigose outside; the outer from c. 3 mm. long, subulate-lanceolate, the inner to c. 10 mm. long, narrowly-lanceolate, shortly pungent or bristle-tipped. Corollas purple, 7–9 mm. long, narrowly funnel-shaped, lobes shortly acicular-setose near the apex, limb and tube appressed pilose-pubescent or glabrous, the hairs short-stalked T-shaped or Y-shaped. Achenes brown, 1.2–2 mm. long, narrowly obovoid-turbinate, with c. 5 swollen-ribs, sparsely setose and glandular between the ribs; pappus of few very caducous barbellate setae 1–2 mm. long.

Zambia. N: Luwingu, 25.v.1964, *Fanshawe* 8679 (K; NDO). **Malawi**. N: South Viphya, 8.v.1970, *Brummitt* 10473 (MAL; K).
Also in Tanzania. Dry evergreen forests or thickets.

4. **Gutenbergia cordifolia** Benth. ex Oliv. in Trans. Linn. Soc. **29**: 89, t. 55 (1873). TAB. **24** fig. F. Type from Uganda.

A laxly spreading subshrub to c. 2.5 m. high, woody at the base. Stems freely branched, leafy, appressed pubescent or tomentellous; indumentum of short-stalked long-armed T-shaped hairs. Leaves alternate, petiolate; petiole to c. 17 mm. long in mature leaves; lamina up to c. 11 × 5 cm., smaller on the branches, mostly c. 3 × 1 cm., elliptic to ovate, subacute, cuneate below, margins subentire to serrulate; upper surface green drying olive-green, finely pilose-pubescent with long-stalked T-shaped hairs, the lower surface silvery-white felted with short-stalked matted T-shaped hairs. Capitula numerous, usually laxly clustered; stalks mostly 1–15 mm. long, appressed tomentellous, bracteate with a few small leaf-like bracts. Involucres 5–7 × 5–10 mm., broadly campanulate-cyathiform. Phyllaries many, shortly pungent or bristle-tipped, appressed-pubescent to glabrescent, margins broadly scarious glabrous ± hyaline or purple-tinged, or margins not very pronounced; the outer phyllaries from c. 2 mm. long, lanceolate-attenuate or ovate-apiculate; the inner to c. 9 × 2 mm., oblong-lanceolate. Corollas purple, 5–7 mm. long, narrowly funnel-shaped, lobes acicular-setose near the apex, usually glabrous below. Achenes (1)1.3–1.8 mm. long, obovoid-turbinate, with 4–5 swollen-ribs, rugulose and sparsely setose between the ribs; pappus of few very caducous barbellate setae 1–2 mm. long.

Var. **marginata** (O. Hoffm.) C. Jeffrey in Kew Bull. **43**: 252 (1988). Type: Malawi or Mozambique, *Stewart* s.n. (K, holotype).
    Vernonia marginata Oliv. & Hiern in F.T.A. **3**: 278 (1877) nom. illegit., non (Torr.) Rafin. (1832). Type as above.
    Bothriocline marginata O. Hoffm. in Engl., Pflanzenw. Ost-Afr. **C**: 403 (1895). Type as above.
    Erlangea marginata (O. Hoffm.) S. Moore in Journ. Linn. Soc., Bot. **35**: 310 (1902). Type as above.
    Gutenbergia marginata (O. Hoffm.) Wild & G.V. Pope in Kirkia **10**: 351 (1977). Type as above.

Phyllaries with pale hyaline margins, the outer acutely attenuate, the inner about 8.5 mm. long; achenes 1.5 mm. long.

Zimbabwe. E: Mutare Distr., E. Vumba Mts., Wattle Company's Coffee Plantation, c. 1372 m., 10.v.1965, *Chase* 8291 (K; SRGH). **Malawi**. N: Nkhata Bay Distr., Luwawa Dam, c. 1524 m., 20.vii.1978, *Pawek* 7250 (K; MO; SRGH). C: Dedza Mt., W. side of summit, c. 2120 m., 25.vi.1970, *Brummitt & Salubeni* 11692 (K; LISC; MAL; SRGH; UPS). S: Zomba Plateau, c. 1500 m., 5.vi.1946, *Brass* 16258 (K). **Mozambique**. N: Lichinga (Vila Cabral), May–July 1934, *Torre* 185 (COI; LISC). Z: R. Licungo, SW. of Namuli Peaks, c. 914 m., 29.vii.1962, *Leach & Schelpe* 11495 (K; LISC; M; SRGH). T: Zóbuè Mt., 3.x.1942, *Mendonça* 585A (K; LISC; LMU). MS: Vila da Manica (Macequece), 18.iii.1948, *Barbosa* 1202 (LISC).

Also in Tanzania and Zaire. Evergreen forest margins, submontane grassland and in riverine vegetation.

Var. *glanduliflora* (Wechuysen) C. Jeffrey is recorded for Uganda, Kenya, Tanzania, Rwanda, Zaire and Sudan. It is characterised by phyllaries with narrow often not very pronounced margins, and by the outer phyllaries acutely attenuate, inner phyllaries 4.7–9 mm. long, and the achenes 1.3–1.8 mm. long.

Var. *depauperata* (S. Moore) C. Jeffrey is recorded for Kenya and Tanzania, and is characterised by phyllaries with pale hyaline margins, and by the phyllaries sparsely pubescent to densely white-tomentose, the outer acutely attenuate, the inner 3.5–6.5 mm. long, and the achenes 1–1.3 mm. long.

Var. *pulchra* (B.L. Burtt) C. Jeffrey is recorded from Tanzania. It is characterised by phyllaries with pronounced hyaline margins, the phyllaries densely brown-pubescent, the outer broadly ovate and shortly apiculate, and the achenes 1.4–1.5 mm. long.

5. **Gutenbergia eylesii** (S. Moore) Wild & G.V. Pope in Kirkia **10**: 314, fig. 4E, p. 331; 350 (1977). —C. Jeffrey in Kew Bull. **43**: 253 (1988). Type: Zimbabwe, Mazowe, *Eyles* 309 (BM, holotype; BOL; K; SRGH).
    Erlangea eylesii S. Moore in Journ. Bot. **46**: 38 (1908). —Eyles in Trans. Roy. Soc. S. Afr. **5**: 502 (1916). Type as above.

A lax perennial herb to c. 60 cm. tall. Stems 1–several, becoming woody below, branched, leafy often densely so with new leaf whorls in axils of mature leaves, appressed tomentose; indumentum of short-stalked long-armed T-shaped hairs. Leaves alternate, subsessile or shortly petiolate, up to c. 8 × 5 cm., linear-lanceolate to narrowly lanceolate-oblong or elliptic, apex subacute mucronate, base rounded to cuneate, lamina undulate or flat; upper surface green drying brownish, finely pilose to pubescent, the hairs long-stalked T-shaped becoming rigid; lower surface densely whitish felted with short-stalked matted T-shaped hairs. Capitula many, often laxly clustered at the ends of branches; stalks mostly 2–15 mm. long, appressed tomentellous, with 1–2 linear bracts. Involucres 6–8 × 5–8 mm., spreading to c. 10 mm. wide, obconic to campanulate. Phyllaries many, ± narrowly lanceolate-acuminate or attenuate, aristate, appressed tomentellous outside, often purplish towards the apex and on the narrowly membranous margins; outer phyllaries c. 2 mm. long, the inner to c. 8 mm. long. Corolla purple or mauve, 6–8 mm. long, narrowly funnel-shaped, softly puberulous with T-shaped hairs especially in the upper half. Achenes c. 5 mm. long, ± narrowly turbinate, truncate above, strongly c. 5-ribbed, minutely setulose between the ribs; pappus of sparse caducous barbellate setae up to c. 3 mm. long, sometimes absent from some capitula or even entire synflorescences.

Leaf venation pinnate, lamina mostly rounded or subcordate at base and sessile; phyllaries mostly
    tapering to an attenuate-aristate apex; plants not from N. Zambia   -   -   subsp. *eylesii*
Leaf venation obviously reticulate, lamina cuneate below and subpetiolate; phyllaries acute
    mucronate; plants from the Mbala area of Zambia   -   -   -   -   subsp. *reticulata*

Subsp. **eylesii**

   Zambia. C: near Mumbwa, 1911, *Macaulay* 777 (K). **Zimbabwe**. N: Darwin, near Mutepatepa, 14.iii.1963, *Wild* 6081 (K; SRGH). C: Domboshawa, c. 1371 m., 7.iii.1946, *Wild* 909 (K; SRGH). E: Nyanga North, Mika Hill, 10.iv.1972, *Pope & Wild* 620 (K; LISC; SRGH). **Malawi**. N: Mzimba Distr., c. 100 km. from Mzuzu, Hora Mt., 30.iii.1969, *Pawek* 1902 (K). **Mozambique**. T: Tete-Harare road near Zimbabwe border, 30.vi.1947, *Hornby* 2788 (K; PRE; SRGH).
   Not known outside the Flora Zambesiaca area. Rocky soil on hillsides usually in *Brachystegia* woodland.

Subsp. **reticulata** Wild & G.V. Pope in Kirkia **10**: 315, fig. 4F, p. 331; 351 (1977). TAB. **24** fig. E. Type: Zambia, Mbala, Lake Chila, *Richards* 9914 (K, holotype; SRGH).

   Zambia. N: Mbala, c. 1676 m., 6.v.1952, *Siame* 183 (BM; K).
   Also in Tanzania. Miombo and mixed deciduous woodland and wooded grassland.

6. **Gutenbergia westii** (Wild) Wild & G.V. Pope in Kirkia **10**: 314, fig. 4D, p. 331; 349 (1977). Type: Zimbabwe, Chimanimani Mts., *West* 3621 (SRGH, holotype).
   Erlangea westii Wild in Kirkia **4**: 151 (1964). Type as above.

A perennial herb up to c. 1 m. tall. Stems leafy, branching, appressed-pilose; the hairs short-stalked long-armed T-shaped. Leaves alternate, 3–9 × 0.8–3.5 cm., ovate-lanceolate to narrowly elliptic, apices acute mucronate, bases rounded to cuneate, margins entire to remotely serrulate; upper surface green drying brownish, finely pilose to pubescent, the hairs long-stalked T-shaped becoming rigid; lower surface whitish felted with short-stalked matted T-shaped hairs. Capitula many in laxly corymbiform cymes, stalks 0.2–35 mm. long, appressed tomentellous with a few subulate bracts. Involucres 8–10 × 5–13 mm., campanulate-cyathiform. Phyllaries many, margins narrowly hyaline, minutely ciliate-serrulate, obscurely 3-nerved, appressed tomentellous to sparsely so, the outer phyllaries c. 8 mm. long and subulate, the inner longer ± subequal lorate-lanceolate and acuminate-aristate or long cuspidate. Corollas mauve-purplish, 7–8 mm. long, narrowly funnel-shaped; lobes sparsely shortly setose, T-shaped hairs absent. Achenes c. 1.5 mm. long, narrowly turbinate-cylindric, truncate above, strongly c. 6-ribbed, minutely setulose between the ribs; pappus of few very caducous barbellate setae c. 2 mm. long.

   Zimbabwe. E: Chimanimani Mts., c. 1828 m., 8.vi.1949, *Wild* 2865 (K; LISC; SRGH). **Mozambique**. MS: Serra Mocuta, c. 1200 m., 3.vi.1971, *Biegel & Pope* 3521 (K; LISC; SRGH).
   Known only from the Zimbabwe/Mozambique border area. On the quartzitic soils of the Chimanimani Range and adjacent areas, in *Uapaca sansibarica* and open *Brachystegia* woodlands.

7. **Gutenbergia mweroensis** Wild & G.V. Pope in Kirkia **10**: 312, fig. 4A, p. 330; 356 (1977). TAB. **24** fig. B. Type: Zambia, Lake Mweru, Chiengi, *Fanshawe* 4732 (K, holotype; LISC; NDO).

A procumbent perennial (? annual) herb. Stems trailing-ascending, branched, leafy, often with leaf whorls in axils of mature leaves, angular-ribbed, appressed pilose-pubescent; indumentum of short-stalked long-armed T-shaped hairs. Leaves alternate on upper stem and branches, subopposite on lower stem, sessile, 2 1.3 × 1–2.4 cm., narrowly ovate to oblong-ovate on upper stem and branches, rounded or obtuse at the apex, cordate and semi-amplexicaul below at least in upper leaves, upper surface finely pilose-pubescent with T-shaped hairs, lower surfaces silvery-white felted with short-stalked matted T-shaped hairs. Capitula small, numerous, in many 3–9-capitulate clusters corymbiform-cymose on branches 2–6 cm. long; capitula stalks 1–5 mm. long, puberulous. Involucres 3–5 × 4–6 mm., broadly cyathiform to globose. Phyllaries scarious with ± broad subhyaline margins, pilose-puberulous outside, from c. 2 mm. long and lanceolate outside to c. 4.5 mm. long and broadly oblanceolate inside, apices incurved not sharply bristle-tipped. Corollas purple, 3–6 mm. long, narrowly funnel-shaped, densely pilose-tomentellous. Achenes brown, c. 2 mm. long, oblong-obovoid, terete, minutely roughly pubescent all over and globose-glandular; pappus 0.

Zambia. N: ? Lake Mweru, 1949–1951, *Bullock* 1221 (K).
Known only from the North Province of Zambia. Sandy lake foreshore, swampy or rocky places.

8. **Gutenbergia leiocarpa** O. Hoffm. in Engl., Pflanzenw. Ost-Afr. **C**: 402 (1895). —R.E. Fr., Wiss. Ergebn. Schwed. Rhod.-Kongo-Exped. 1911–12, **1**: 319 (1916). —Wild & G.V. Pope in Kirkia **10**: 355 (1977). TAB. **24** fig. C. Type from Tanzania.
    Gutenbergia leiocarpa var. *longipedicellata* Wechuysen in Bull. Jard. Bot. Nat. Belg. **51**: 109 (1981). Type from Tanzania.
    Gutenbergia leiocarpa var. *microcarpa* Wechuysen in Bull. Jard. Bot. Nat. Belg. **51**: 110 (1981). Type from Tanzania.

A diffuse procumbent perennial herb. Stems trailing-ascending, to c. 120 cm. long, much branched, leafy often with leaf whorls in the axils of mature leaves, softly appressed pilose-puberulous; indumentum of short-stalked long-armed T-shaped hairs. Leaves opposite, becoming alternate on upper stems and branches, sessile, 1.5–8 × 0.3–2.0 cm., oblong-lanceolate or linear-elliptic, subobtuse at the apex, rounded to auriculate or cordate below, margins somewhat undulate; upper surface green drying olive-green, finely pilose-pubescent above, the hairs long-stalked T-shaped, lower surface silvery-white felted with hairs short-stalked and matted T-shaped. Capitula small, numerous, short-stalked, in many 3–8-capitulate corymbiformly cymose clusters; capitula stalks 1–5 mm. long. Involucres 3–5 × 3–5 mm., obconic-campanulate. Phyllaries many, lanceolate, scarious with narrow subhyaline margins, araneose pubescent outside to glabrescent, the outer phyllaries from c. 2 mm. long, the inner to c. 4.5 mm. long, shortly acuminate- or tapering-pungent. Corollas purple or mauve, 3–5 mm. long, narrowly funnel-shaped, densely pilose-puberulous especially in upper part. Achenes dark polished-brown or black, c. 1.3 mm. long, narrowly oblong-ovoid, terete, glabrous; pappus 0.

Zambia. N: Samfya, 3.x.1953, *Fanshawe* 348 (K; LISC; SRGH).
Also in Tanzania. Sandy shores of Lake Bangweulu and river flood plains.

9. **Gutenbergia adenocarpa** Wechuysen in Bull. Jard. Bot. Nat. Belg. **51**: 106 (1981). Type: Zambia, Chakwenga Headwaters, 100–109 km. E. of Lusaka, *Robinson* 5612 (K, holotype; M).

An erect slender to bushy annual herb, 30–60 cm. tall. Stems sparingly to much branched, densely to sparsely softly appressed-puberulous, leafy often densely so with new shoots in leaf axils; indumentum of long-armed, T-shaped hairs. Leaves opposite, becoming alternate on upper stem and branches, sessile, 2–8 × 0.1–0.4 cm., linear, mucronate at the apex, widening somewhat and subauriculate at the base, at least in the upper leaves, ± revolute; upper surface sparsely weakly pilose to scabridulous, the hairs long-stalked T-shaped becoming rigid, lower surface whitish or silvery araneose-felted with short-stalked matted T-shaped hairs. Capitula small, numerous, somewhat clustered; capitula stalks mostly 3–35 mm. long. Involucres 5–6 × 4–7 mm., obconic. Phyllaries many, membranous-scarious with broad somewhat hyaline margins, shortly acuminate to tapering pungent, often purple-tipped, finely lanate-pilose to glabrescent, the outer from c. 3 mm. long, subulate-lanceolate, the inner to c. 5.5 mm. long, narrowly oblong-lanceolate. Corollas purple, 5–6.5 mm. long, narrowly funnel-shaped, puberulous. Achenes polished brown, 1.0–1.5 mm. long, obovoid-oblong, faintly 4-angled, glabrous or sometimes glandular near the base; pappus 0.

Zambia. C: between Undaunda and Rufunsa, c. 135 km. E. of Lusaka, 1150–1200 m., 21.i.1972, *Kornás* 771 (K). **Zimbabwe**. N: Darwin Distr., Msengesi River, c. 1219 m., 8.v.1955, *Whellan* 848 (K; SRGH). **Mozambique**. T: 3 km. from Mágoè to Mágoè Velho, c. 300 m., 4.iii.1970, *Torre & Correia* 18192 (LISC; LMU). MS: Báruè, 13 km. from Changara on Catandica road, c. 400 m., 24.v.1971, *Torre & Correia* 18575 (LMU).

Not known outside the Flora Zambesiaca area. Miombo woodland, rocky hillsides and roadsides.

10. **Gutenbergia kassneri** S. Moore in Journ. Bot. **56**: 205 (1918). —Wild & G.V. Pope in Kirkia **10**: 354, fig. 5C, p. 378 (1977). Type from Zaire.
  Gutenbergia kassneri var. *angustifolia* S. Moore in Journ. Bot. **56**: 205 (1918). Type from Zaire.
  Vernonia smithiana sensu De Wild. & Muschl. in Ann. Mus. Congo Belg., sér. 4, **2**: 164 (1913).

An erect slender to bushy, often tufted, annual or perennial herb to c. 45 cm. tall. Stems sparingly- to much-branched, softly appressed-puberulous, leafy with the leaves sometimes crowded below; indumentum of short-stalked long-armed T-shaped hairs; branches to c. 30 cm. long, slender. Leaves opposite, becoming alternate on upper stem, sessile, 2–6 × 0.2–0.6 cm., linear-elliptic, acute, somewhat revolute, upper surface sparsely weakly pilose with fine T-shaped hairs, lower surface whitish or silvery araneose-felted, the midrib prominent beneath. Capitula small, few to numerous, short-stalked, in 2–12-capitulate clusters, or terminal on short branches; stalks mostly 1–3 mm. long from slender spreading branchlets 1–6 cm. long. Involucres 4–6 × 3–5 mm., obconic-campanulate. Phyllaries many, scarious, acuminate-pungent or bristle-tipped, margins somewhat hyaline, finely araneose-pilose to glabrescent, the outer phyllaries from c. 3 mm. long, subulate-lanceolate, the inner to c. 5.5 mm. long, oblong-lanceolate. Corollas purple or mauve, 4.5–6 mm. long, narrowly funnel-shaped, araneose-pilose. Achenes polished-brown, 1.0–1.5 mm. long, obovoid-turbinbate, usually 3-angled, glabrous; pappus 0.

Zambia. N: Chipili, c. 1219 m., 7.vi.1957, *Robinson* 2245 (K). W: c. 23 km. N. of Kapiri Mposhi, 10.iv.1986, *Philcox, Pope & Chisumpa* 9832 (K; NDO). C: Serenje Distr., Kundalila Falls, c. 15 km. SE. of Kanona, i.1972, *Williamson* 2136 (K; SRGH).
Also in Zaire. Shady miombo woodland, in sandy soil or rocky outcrops.

11. **Gutenbergia gossweileri** S. Moore in Journ. Linn. Soc., Bot. **47**: 257 (1925); in Journ. Bot., **65** Suppl. 2, Gamopet.: 43 (1927). —Mendonça, Contrib. Conhec. Fl. Angol., **1** Compositae: 2 (1943). —Wild & G.V. Pope in Kirkia **10**: 356, fig. 5D, p. 379 (1977). —C. Jeffrey in Kew Bull. **43**: 254 (1988). TAB. **24** fig. A. Type from Angola.
  Gutenbergia tenuis S. Moore in Journ. Linn. Soc., Bot. **47**: 258 (1925). Type: Malawi, Nyika Plateau, *Henderson* s.n. (BM, holotype).

An erect slender to bushy annual herb, 7–80 cm. tall. Stems sparsely to much branched, leafy, softly appressed-puberulous; branches to c. 40 cm. long, ± widely spreading; axils of mature leaves often with whorls of young leaves; indumentum of long-armed T-shaped hairs. Leaves opposite, becoming alternate on upper stem, sessile, mostly 2–7 × 0.5–1.5 cm., narrowly oblong, becoming oblong-lanceolate on upper stem, acute at the apex, rounded to semi-amplexicaul or sometimes broadly cuneate at the base; upper surface sparsely weakly pilose to scabridulous, the hairs T-shaped sometimes becoming rigid, the lower surface whitish or silvery araneose-felted with short-stalked matted T-shaped hairs. Capitula numerous, small, ± laxly arranged or loosely clustered; stalks mostly 1–5 mm. long, from slender branchlets to c. 2 cm. long. Involucres 3–6 × 3–6 mm., campanulate-cyathiform to somewhat spreading. Phyllaries many, green, membranous with scarious or subhyaline margins, lanceolate, shortly acuminate-pungent or tapering to a bristle-tipped apex, finely araneose-pilose outside, the outer from c. 1 mm. long, the inner to c. 6 mm. long. Corollas purple to pale-mauve, 3–5 mm. long, narrowly funnel-shaped, appressed pilose-puberulous. Achenes brown, c. 1.5 mm. long, broadly obovoid-turbinate, somewhat 2–3-sided, appressed setulose at the apex and base, glabrous about the middle; pappus 0.

Zambia. B: c. 64 km. W. of Kaoma (Mankoya), Sikelenge, Luampa R., 19.xi.1959, *Drummond & Cookson* 6623 (E; K; LISC; MO; PRE; SRGH). N: c. 29 km. from Mbala, Mukoma Escarpment, c. 1200 m., 7.iv.1962, *Richards* 16296 (K; M; SRGH). W: between Mwinilunga and Kasompe R., 16.v.1986, *Philcox, Pope, Chisumpa & Ngoma* 10344 (CAL; GA; K; LISC; MO; NDO). C: c. 10 km. E. of Lusaka, c. 1280 m., 6.ii.1956, *King* 303 (K). S: Choma, Siamambo, c. 1310 m., 20.iii.1958, *Robinson* 2819 (K; SRGH). **Zimbabwe**. N: c. 16 km. NW. of Gokwe, 21.iii.1963, *Bingham* 552 (K; SRGH). E: Odzani R. Valley, 1915, *Teague* 427 (BM; BOL; K). **Malawi**. N: Mzimba Distr., Mbawa Expt. Sta., 5.iv.1955, *Jackson* 1585 (K; PRE; SRGH). C: Lilongwe Distr., Dzalanyama For. Res., Chiunjiza Road, c. 5 km. SE. of Chaulongwe Falls, c. 1230 m., 22.iii.1970, *Brummitt* 9312 (K).
Also in Angola, Tanzania and Zaire. Miombo or mixed deciduous woodland, often in sandy soil, or in short grassland, also a weed of disturbed ground at roadsides and in cultivated fields.

12. **Gutenbergia polycephala** Oliv. & Hiern in Journ. Linn. Soc., Bot. **15**: 95 (1876); in F.T.A. **3**: 264 (1877). —Engl., Pflanzenw. Ost-Afr. **C**: 402 (1895). —Mendonça, Contrib. Conhec. Fl. Angol., **1** Compositae: 2 (1943). —Wild & G.V. Pope in Kirkia **10**: 352, fig. 5B, p.378 (1977). TAB. **24** fig. G. Type from Tanzania.

An erect slender to bushy annual herb, 15–50 cm. tall. Stems sparingly to much branched, softly appressed puberulous; indumentum of short-stalked long-armed T-shaped hairs. Leaves opposite, becoming alternate on the upper stem, subsessile, 2–6 × 0.5–3.5 cm., lanceolate to oblong-lanceolate, apex acute or subobtuse mucronate, base rounded to cordate or broadly cuneate, margins entire, upper surface sparsely weakly pilose with fine T-shaped hairs, lower surface thinly whitish araneose-felted. Capitula small, few to numerous, in a ± diffuse arrangement; stalks slender, mostly 5–15 mm. long, appressed tomentellous. Involucres 3–5 × 4–5 mm., campanulate, florets exserted. Phyllaries few-seriate, narrowly lanceolate to elliptic-oblong, blunt at the apex, narrowly scarious on the margins, sometimes serrulate near the apex, hoary appressed-pubescent outside, the outer from c. 2 mm. long, the inner to c. 4 mm. long. Corollas mauve, 4–7.5 mm. long, funnel-shaped above shortly tubular below, tube and limb appressed puberulous the hairs short-stalked T-shaped. Achenes dark-brown, 1.0–1.5 mm. long, obovoid-turbinate, c. 12-ribbed, sparsely setulose, the bristles uncinate at the apex; pappus 0.

Zambia. **C**: Luangwa Game Reserve South, S. of Lion Plain, c. 762 m., 21.iii.1966, *Astle* 4692 (K). **E**: Chipata Distr., Chizombo, c. 609 m., 20.ii.1969, *Astle* 5503 (K; SRGH). **S**: c. 20 km. from Kafue-Mazabuka road on Mapangazia road, 3.iv.1969, *Anton-Smith* in GHS 208052 (K; LISC; SRGH). **Zimbabwe**. **N**: Gokwe, Sengwa Res. Sta., 11.iv.1976, *Guy* 2418 (K; SRGH). **C**: Chegutu (Hartley), Poole Farm, 2.iv.1948, *Hornby* 2880 (K; LISC; SRGH). **Malawi**. **N**: Karonga Distr., Ngala, c. 24 km. N. of Chilumba, c. 548 m., 22.iv.1969, *Pawek* 2274 (K; SRGH). **Mozambique**. **N**: Niassa Prov., c. 26 km. E. of Malema, c. 700 m., 24.v.1961, *Leach & Rutherford-Smith* 10995 (K; LISC; SRGH). **T**: c. 7 km. on road from Changara, Monte Cameira, c. 280 m., 23.iii.1966, *Torre & Correia* 15307 (LISC). **MS**: Cheringoma Plateau, source of Mueredzi R., 18°50'S, 34°46'E, c. 210 m., 11.vii.1972, *Ward* 7818 (DBW; K).

Also in Tanzania. *Colophospermum mopane* woodland, *Adansonia* woodland and other low altitude deciduous woodland, also a weed of cultivation in these localities.

Gutenbergia petersii Steetz in Peters, Reise Mossamb., Bot. **2**: 348 (1864) may be an earlier name for this species judging by the description of the achene, but the leaves are described as being linear-lanceolate and the phyllaries acuminate-aristate. However, as the type is presumably destroyed this species must at present remain of doubtful identity.

13. **Gutenbergia spermacoceoides** Wild & G.V. Pope in Kirkia **10**: 312, fig. 4B, p. 330; 353 (1977). Type: Zambia, Kasama Distr., Mungwi, *Robinson* 4617 (K, holotype; M; MO; SRGH).

An erect slender to bushy annual herb, 10–25 cm. tall. Stems simple to much branched, softly appressed-puberulous; indumentum of short-stalked long-armed matted T-shaped hairs. Leaves opposite, becoming alternate on the upper stem, sessile, 1–5.5 × 0.2–0.4 cm., linear, apex subobtuse mucronate, margins somewhat revolute; upper surface pilose-pubescent, the hairs T-shaped with the stalks becoming rigid, or the hairs long flagelliform, lower surface white-felted with short-stalked matted T-shaped hairs. Capitula small, subsessile, in 1–numerous clusters; clusters 3–10-capitulate, ± congested, terminal on the stem and branches; capitula stalks 0–1 mm. long; subtending leaves overtopping the capitula. Involucres 3–5.5 × c. 5 mm., obconic-campanulate. Phyllaries many, subulate to linear-lanceolate, tapering to a purple pungent tip, with ± obscured narrow scarious margins, white-lanate outside, the outer from c. 2 mm. long, the inner to c. 5 mm. long. Corollas purple, 3–5 mm. long, funnel-shaped, densely white pilose-pubescent, the hairs short-stalked T-shaped. Achenes dark-brown, c. 1.3 mm. long, narrowly obovoid, c. 10-ribbed, sparsely setulose, the bristles uncinate at apex, minutely globose glandular; pappus 0.

Zambia. **N**: Kasama Distr., Chishimba Falls, c. 40 km. W. of Kasama, c. 1610 m., 30.iv.1986, *Philcox, Pope & Chisumpa* 10212 (BR; GA; K; NDO; SRGH).

Known only from the northern region of Zambia. Sandy soil often in pan-like depressions.

13. ERLANGEA Sch. Bip.

Erlangea Sch. Bip. in Flora **36**: 34 (1853). —Wild & G.V. Pope in Kirkia **10**: 316; 362 (1977). —C. Jeffrey in Kew Bull. **43**: 255 (1988).

Annual or perennial herbs. Vegetative indumentum of flagelliform hairs with an elongate terminal cell on a few to many-celled uniseriate stalk. Stems erect, simple to much-branched, becoming stout and woody at the base, strongly striately ribbed, pilose. Leaves alternate, sessile, or sometimes petiolate, subentire to serrate or ± deeply serrate-lobed. Capitula homogamous and discoid, laxly corymbiformly cymose, stalked. Involucres broadly campanulate-cyathiform; phyllaries diverging somewhat, increasing in size to the inside, membranous or scarious with a narrow subhyaline margin, obscurely 3-nerved. Corollas purple to dark-mauve, exserted, narrowly infundibuliform, regularly 5-lobed; lobe apices acicular-setose, limb and tube somewhat glandular otherwise glabrous. Anther apices appendiculate, bases subacute. Style branches subulate, hirsute. Achenes narrowly oblong-obovoid and c. 4-ribbed (prismatic and 3–6-angled), rounded at the apex, rugulose and with globose glandular trichomes sunk in pits between the ribs, short appressed setae often also present between and on the ribs, otherwise glabrous; pappus 1-seriate of caducous coarsely barbellate or subplumose setae.

A tropical African genus of approximately 5 species.

Leaves sessile or subsessile - - - - - - - - - - - - 1. *misera*
Leaves petiolate, petiole up to c. 1.7 cm. long - - - - - - - 2. *remifolia*

1. **Erlangea misera** (Oliv. & Hiern) S. Moore in Journ. Linn. Soc., Bot. **35**: 313 (1902). —Muschl. in Engl., Bot. Jahrb. **46**: 59 (1912). —R.E. Fr., Wiss. Ergebn. Schwed. Rhod.-Kongo-Exped. 1911–12, **1**: 319 (1916). —Mendonça, Contrib. Conhec. Fl. Angol., **1** Compositae: 4 (1943). —Wild & G.V. Pope in Kirkia **10**: 362, fig. 9, p. 381 (1977). TAB. **25**. Type from Zaire.
    Vernonia misera Oliv. & Hiern in F.T.A. **3**: 278 (1877). Type as above.
    Erlangea schinzii O. Hoffm. in Bull. Herb. Boiss. **1**: 71 (1893). —Muschl. in Engl., Bot. Jahrb. **46**: 59 (1912). —Eyles in Trans. Roy. Soc. S. Afr. **5**: 502 (1916). —Mendonça, Contrib. Conhec. Fl. Angol., **1** Compositae: 5 (1943). —Merxm., Prodr. Fl. SW. Afr. 139: 63 (1967). Type from Namibia.
    Bothriocline misera (Oliv. & Hiern) O. Hoffm. in Bol. Soc. Brot. **13**: 11 (1896). Type as above.
    Bothriocline schinzii (O. Hoffm.) O. Hoffm. in Warb., Kunene-Samb. Exped. Baum: 398 (1903). Type as for *Erlangea schinzii* O. Hoffm.
    Erlangea sessilifolia R.E. Fr., Wiss. Ergebn. Schwed. Rhod.-Kongo-Exped. 1911–12, **1**: 319 (1916). Type: Zimbabwe, Victoria Falls, *Fries* 168 (UPS, holotype).
    Vernonia merenskiana Dinter ex Merxm. in Mitt. Bot. Staatss. Münch. **2**: 38 (1954) nom. nud. in syn.

A slender to bushy, long-lived annual or perennial herb, 0.2–1 m. tall. Stems leafy, sparsely to much-branched in the upper part, becoming stout and woody at the base (to c. 2 cm. in diam. fide *P.A. Smith*), strongly ribbed at least above, pilose and glandular, the hairs flagelliform; branches spreading, to c. 40 cm. long. Leaves alternate, sessile or subsessile, mostly 3–14 × 0.8–4 cm., narrowly oblong to elliptic or ovate-oblong, apex obtuse, base rounded to subcordate or less often cuneate, margins subentire to serrate sometimes coarsely remotely so or dentate-lobed; shortly pilose or pubescent, ± sparsely so on the upper surface, hairs flagelliform. Capitula many, laxly corymbiform cymose; stalks slender, 0.3–9 cm. long. Involucres 5–7 × 9–15 mm., broadly campanulate. Phyllaries numerous, lanceolate, the innermost lorate, all tapering acuminate and ± bristle-tipped, margins narrowly subhyaline and finely pectinate towards the apex, pilose and glandular, ± obscurely 3-nerved, the outer from c. 2 mm. long, the inner to c. 7 mm. long. Corollas purple to dark-mauve, to c. 6 mm. long, narrowly funnel-shaped, the lobes with short acicular-setae and a few flagelliform hairs, limb and tube glandular outside. Achenes numerous, 1.2–1.8 mm. long, narrowly oblong-ovoid, c. 4-ribbed, rugulose and with globose glandular trichomes sunk in pits between the ribs, sparsely setulose or glabrous except for glands; pappus of caducous subplumose bristles 3–4 mm. long.

Caprivi Strip. Katima Mulilo, Zambezi R., 8.i.1959, *Killick & Leistner* 3318 (K; M; PRE; SRGH). **Botswana**. N: Chobe-Zambezi confluence, ll.iv.1955, *Exell, Mendonça & Wild* 1463 (BM; LISC; SRGH). SE: c. 82 km. W. of Lobatse on road to Malopo Farms, c. 1100 m., 28.ii.1977, *Mott* 1081 (K; SRGH). SW: c. 11 km. N. of Union's End, 15.iii.1969, *Rains & Yalala* 33 (K; SRGH). **Zambia**. B: Sesheke, Kazu Forest, 9.iv.1955, *Exell, Mendonça & Wild* 1449 (BM; LISC; SRGH). W: Mwinilunga,

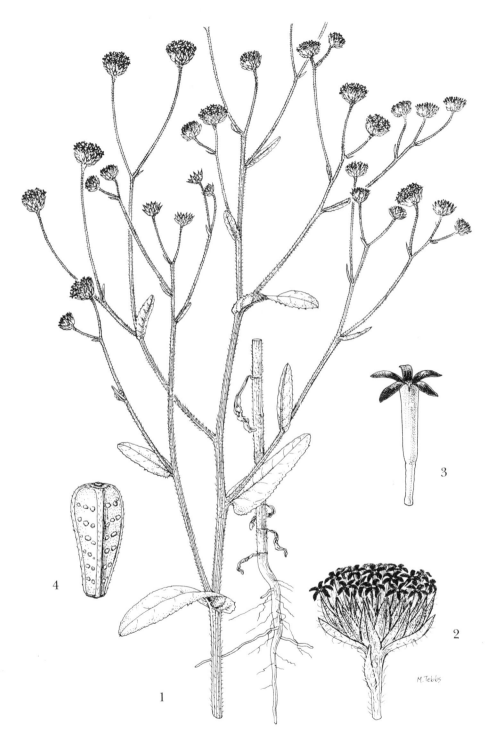

Tab. 25. **ERLANGEA MISERA**. 1, habit (× $\frac{2}{3}$); 2, capitulum (× 2); 3, corolla tube (× 6); 4, achene, showing glandular pits (× 13), 1–4 from *P.A. Smith* 2399. Drawn by M. Tebbs.

c. 96 km. S. on Kabompo Road, 1.vi.1963, *Loveridge* 711 (K; M; SRGH). S: Katambora, c. 914 m., 3.iv.1956, *Robinson* 1397 (K; SRGH). **Zimbabwe**. N: Hurungwe, Chirundu Road, 26.ii.1953, *Wild* 4043 (K; MO; SRGH). W: Hwange Nat. Park, Main Camp, 1066 m., 6.ii.1969, *Rushworth* 1499 (K; M; PRE; SRGH). **Mozambique**. T: Mágoè (Boroma), Cachomba Fort, c. 152 m., 22.vii.1950, *Chase* 2695 (COI; K; PRE; SRGH). GI: Inhambane, between Mangorre and Panda, Inhassune R., 7.iv.1959, *Barbosa & Lemos* 8524 (COI; K; LISC; PRE; SRGH). M: Maputo (Delagoa Bay), c. 30 m., 15.iii.1898, *Schlechter* 12158 (K).

Also in Angola, Namibia and northern Cape Province. Locally frequent on sand banks beside rivers, on flood plains and in pans, in deciduous woodland and grassland on Kalahari Sand, and in old cultivation.

This species is very close to *E. plumosa* Sch. Bip. from Gabon. However, the type specimen of that species is poor, and insufficient material from the type locality has been seen to make possible a decision on whether or not the two are conspecific.

2. **Erlangea remifolia** Wild & G.V. Pope in Kirkia **10**: 317, fig. 6, p. 332; 363 (1977). Type: Botswana, Tsodilo Hills, *Banks* 35 (PRE, holotype).

Very similar to *E. misera* (Oliv. & Hiern) S. Moore but readily distinguished by its petiolate leaves. Petioles up to c. 1.7 cm. long, lamina oblong-lanceolate ± abruptly narrowing to a rounded or broadly cuneate base.

Botswana. N: c. 25 km. SW. of Samochimo (Samocimo) on track to Tsodilo Hills at 21°47'E and 18°36'S, c. 1000 m., 2.v.1975, *Biegel, Müller & Gibbs-Russell* 5052 (K; SRGH).

On one of the very few isolated hill features in Botswana, west of the Okavango River. Hillsides and in *Baikiaea* woodland.

14. BOTHRIOCLINE Oliv. ex Benth.

Bothriocline Oliv. ex Benth. in Hook., Ic. Pl. **12**: 30, t. 1133 (1876). —Wild & G.V. Pope in Kirkia **10**: 317; 364 (1977). —C. Jeffrey in Kew Bull. **43**: 257–266 (1988).
        Volkensia O. Hoffm. in Engl. & Prantl, Pflanzenfam. 4(5): 387 (1894).

Annual or perennial herbs, often suffrutescent with annual stems from a woody rootstock, or shrubs. Vegetative indumentum consisting of flagelliform or ± asymmetric T-shaped hairs with uniseriate multicellular stalks each bearing an elongate erect or ± excentric terminal cell. Leaves alternate, opposite or opposite on lower stem and alternate above, or in whorls of 3, sessile or petiolate. Capitula homogamous and discoid, usually small and few to numerous, laxly arranged or clustered in corymbiform cymes, or solitary. Involucre ± campanulate to cyathiform; phyllaries increasing in size to the inside, membranous or cartilaginous or somewhat scarious, usually with a distinct subhyaline margin. Receptacle plane or alveolate, occasionally paleate on the outside. Corollas purple or mauve, regular, narrowly infundibuliform, with a deeply 5-lobed limb. Anthers lanceolate-appendiculate at the apex, subacute at the base, the connective ± slightly swollen near the base. Style arms linear-terete, hairy. Achenes subcylindric-angular not or hardly tapering above, apex ± truncate with a cupuliform depression, bluntly 3–6-angular and smooth or sparsely glandular between; or achenes narrowly obovoid-oblong, rounded above with a rimmed, shallow apical depression and with 4–5(8) low broad ribs separated by dark-brown gland-filled grooves; or achenes subcylindric-ellipsoid often somewhat curved, rounded to truncate above with a distinct apical cup or crenate cartilaginous rim, ± angular or narrowly 4–9-ribbed, glabrous or sparsely glandular between the ribs. Pappus of few very caducous barbellate setae, or pappus absent.

A tropical African genus of approximately 28 species.

1. Cauline leaves up to 0.3 cm. wide, linear; plants subaquatic, rooted in shallow water, or in marshy ground near water - - - - - - - - - - - - - - - 2
– Cauline leaves 0.5–9 cm. wide, elliptic to lanceolate or linear; plants not rooted in shallow water or marshy ground - - - - - - - - - - - - - 3
2. Capitula solitary on long branches; involucres c. 5 mm. long - - 12. *monocephala*
– Capitula many in a lax corymbiformly cymose arrangement; stalks filamentous; involucres to c. 2.5 mm. long - - - - - - - - - - - - 13. *emilioides*

3. Annual herbs, stem single from a single season taproot - - - - - - 4
- Robust perennial herbs or subshrubs; stems annual or perennial, 1–several from a woody rootstock - - - - - - - - - - - - - - - 6
4. Leaves sessile, auriculate, alternate; involucres 10–12 mm. long; phyllaries gradually tapering-pungent - - - - - - - - - - 5. *steetziana*
- Leaves petiolate or subpetiolate, bases cuneate, at least the lower ones opposite; involucres 3–8 mm. long; phyllaries shortly or abruptly tapering-pungent - - - - - - 5
5. Capitula numerous in small clusters of 3–5, capitula stalks less than 1 cm. long; phyllaries progressively longer to the inside; achenes to 1.5 mm. long - - - - 3. *laxa*
- Capitula many, laxly corymbiform-cymose on stalks 1–8.5 cm. long; phyllaries subequal; achenes 1.5–2.5 mm. long - - - - - - - - - - - 4. *mbalensis*
6. Leaves in whorls of 3, lamina less than 5 cm. long (occasionally 3-whorled in *B. inyangana* but then capitula in clusters of 20 or more, and involucres less than 7 mm. long) 2. *trifoliata*
- Leaves opposite or alternate, lamina mostly more than 5 cm. long - - - - 7
7. Phyllaries more than 5 mm. wide, cartilaginous; involucres 12–35 mm. in diam.; achenes 4–5-angular, not ribbed - - - - - - - - - - 1. *muschleriana*
- Phyllaries less than 5 mm. wide, ± scarious; involucres mostly less than 12 mm. in diam.; achenes usually (4)5–9-ribbed - - - - - - - - - - - - 8
8. Leaves all alternate - - - - - - - - - - - - - 9
- Leaves all, or at least the lower ones, opposite - - - - - - - 10
9. Involucres 7–9 mm. long; phyllaries araneose; corolla lobes acicular-setose; achenes 3–4 mm. long - - - - - - - - - - - - 11. *pectinata*
- Involucres less than 6 mm. long; phyllaries puberulous outside; corolla lobes glabrous; achenes less than 2 mm. long - - - - - - - - - 10. *ripensis*
10. Leaf upper surface hispid-strigose; leaf margins usually double-serrate, with small secondary teeth in sinuses between the main serrations - - - - - 8. *inyangana*
- Leaf upper surface puberulous to glabrescent, leaf margins not double-serrate 11
11. Leaves narrowly elliptic, narrowly tapering to a short petiole, or subsessile; upper stem and branches with few to numerous patent flagelliform hairs intermixed with smaller appressed, 1-armed T-shaped hairs; plants apparently endemic on Mt. Mulanje 9. *milanjiensis*
- Leaves lanceolate, rounded or ± abruptly narrowed to a petiole, or subsessile; upper stem and branches with ± 1-armed T-shaped hairs only, large patent flagelliform hairs absent; plants not endemic on Mt. Mulanje - - - - - - - - - - - 12
12. Leaves long-petiolate, lower surfaces sparsely to densely pubescent - - 6. *longipes*
- Leaves subsessile or shortly petiolate, lower surfaces pale-felted tomentellous (sometimes puberulous) - - - - - - - - - - - - 7. *morumbullae*

1. **Bothriocline muschleriana** Wild & G.V. Pope in Kirkia **10**: 321; 370 (1977). —C. Jeffrey in Kew Bull. **43**: 260 (1988). TAB. **26** fig. G. Type from Zaire.
  Erlangea monocephala Muschl. in Engl., Bot. Jahrb. **46**: 60 (1911) non *Bothriocline monocephala* (Hiern) Wild & G.V. Pope (1977). Type as above.

An erect suffrutescent perennial herb with annual stems to c. 90 cm. tall from a small woody rootstock; rootcrowns and stem base pale-brown lanate, roots numerous thong-like. Stems several, leafy, branching above, striate, coarsely pilose-pubescent; the hairs flagelliform. Leaves opposite, sometimes alternate on upper stem and branches, subsessile, mostly 8–18 × 1.5–6 cm., narrowly oblong-elliptic to lanceolate-elliptic, apex subobtuse, base cuneate to rounded, margins crenulate to serrate sometimes coarsely irregularly so, teeth ± callose-tipped; upper surface sparsely scabrid-pilose, the hairs flagelliform becoming rigid; lower surface paler, pilose to tomentose with a prominent reticulate venation. Capitula (1)3–12, laxly corymbiform cymose on long stalk-like branches, usually subtended by leaf-like bracts to c. 5 cm. long. Involucres 10–15 × 12–25 mm., broadly campanulate-cyathiform. Phyllaries numerous, many-seriate, purple-tipped, cartilaginous with membranous margins, glabrous or ± sparsely pubescent, the outer from c. 5 mm. long, the inner to c. 12 mm. long, ovate to broadly oblong-rotund, the innermost lorate, shortly acuminate-cuspidate at the apex. Corollas pale-mauve or purple, 9–12 mm. long, narrowly tubular widening upwards; lobes to c. 4 mm. long, linear, sparsely glandular-hairy otherwise glabrous outside. Achenes stramineous, 2.5–4 mm. long, subcylindric, bluntly 4–5-angular, ± flat or faintly grooved on the faces, truncate at the apex with a shallow rimmed depression, truncate at the base, smooth and polished; pappus of few very caducous barbellate setae 1–3 mm. long.

Zambia. N: c. 40 km. W. of Kasama, Chishimba Falls, c. 1610 m., 30.iv.1986, *Philcox, Pope & Chisumpa* 10209 (BR; K; MO; SRGH).
Also in Tanzania and Zaire. High rainfall deciduous woodland in long grass, usually beside rivers.

2. **Bothriocline trifoliata** (De Wild. & Muschl.) Wild & G.V. Pope in Kirkia **10**: 320; 369, fig. 10C, p. 382 (1977). —C. Jeffrey in Kew Bull. **43**: 260 (1988). Type from Zaire.

Erlangea trifoliata De Wild. & Muschl. in Bull. Roy. Soc. Bot. Belg. **49**: 221 (1912). Type as above.

Erlangea mooreana Alston in Bull. Misc. Inf., Kew **1925**: 363 (1925). Type from Tanzania.
Bothriocline mooreana (Alston) Wild & G.V. Pope in Kirkia **10**: 322 (1977). Type as above.

An erect perennial herb with annual stems 22–75(100) cm. tall from a small woody rootstock; rootcrowns and stem base with pale-brown lanate tufts, roots numerous thong-like. Stems 1–several, leafy, branching above, usually striately-ribbed, pilose-pubescent; the hairs flagelliform. Leaves subsessile, patent in whorls of 3, 2–5 × 0.5–2 cm., elliptic-oblong to narrowly ovate or lanceolate, apex acute mucronate, base rounded to broadly cuneate or somewhat subcordate, margins serrate-dentate with callose-tipped teeth; at first pilose on both surfaces soon glabrescent or scabridulous, glandular punctate, lower surface paler with ± prominent nervation. Capitula small, few to numerous, 1 or 3–7 corymbiformly cymose on 3-whorled stout ± ascending branches; stalks 0–30 mm. long, often subtended by small leaf-like bracts. Involucres 5–10 × 6–12 mm., campanulate, often broadly so. Phyllaries numerous, many-seriate, somewhat cartilaginous with narrow membranous or subhyaline margins, puberulous to glabrescent, margins ± ciliate; the outer from c. 2 mm. long, the inner to c. 9 mm. long, ovate to broadly oblong-rotund to lorate in the innermost, shortly acuminate-cuspidate to tapering acuminate with ± squarrose purple pungent tips. Corollas purple, 5–9 mm. long, narrowly tubular widening upwards; lobes to c. 4 mm. long, linear, sparsely glandular-hairy or glabrous. Achenes stramineous, 2–2.2 mm. long, narrowly barrel-shaped, bluntly 5-angled, faintly 1–2 dark-grooved on the faces, truncate with a shallow rimmed depression at the apex, truncate at the base, smooth and ± polished; pappus of few very caducous barbellate setae 1–2 mm. long.

Zambia. N: Isoka Distr., Nakonde, Old Fife Coffee Res. Sta., c. 1850 m., 22.iv.1986, *Philcox, Pope & Chisumpa* 10075 (K; LISC; NDO; MO). **W**: Mwinilunga, road near Musangila R., 25.i.1938, *Milne-Redhead* 4323 (K; SRGH). **C**: Serenje, Kundalila Falls, 17.ii.1970, *Drummond & Williamson* 9697 (SRGH). **E**: Nyika Nat. Park, c. 0.5 km. SW. of Govt. Rest House, 17.iv.1986, *Philcox, Pope & Chisumpa* 9966 (BR; GA; K; LISC; MAL; MO; NDO; SRGH). **Malawi. N**: Nyika Plateau, Chisanga Falls path from Nthalire road, c. 1900 m., 28.ii.1982, *Brummitt, Polhill & Banda* 16180 (BR; C; K; LISC; MAL; MO; SRGH; WAG).

Also from Tanzania, Zaire and Angola. Submontane short grassland, *Brachystegia* and *Uapaca* woodlands and high rainfall miombo.

3. **Bothriocline laxa** N.E. Br. in Bull. Misc. Inf., Kew **1894**: 388 (1894). —Wild & G.V. Pope in Kirkia **10**: 318; 370, fig. 10B, p. 382 (1977). TAB. **26** fig. F. Syntypes from the Transvaal and Malawi: Shire Highlands near Blantyre, *Last* s.n. (K, syntype).

Erlangea laxa (N.E. Br.) S. Moore in Journ. Linn. Soc., Bot. **35**: 313 (1902). —Eyles in Trans. Roy. Soc. S. Afr. **5**: 502 (1916). —Merxm. in Proc. & Trans. Rhod. Sci. Assn. **43**: 64 (1951). Syntypes as above.

An erect annual herb, 0.1–1.2 m. tall. Stems simple to widely branched above, leafy, hispid; the indumentum a mixture of appressed T-shaped hairs with larger ± patent flagelliform hairs. Leaves opposite, the uppermost alternate; petioles to c. 2 mm. long; lamina mostly 3–13 × 1–5.5 cm., elliptic, apex subacute, base cuneate, margin serrate to coarsely serrate-dentate the teeth usually callose-tipped, both surfaces finely pilose-strigose often sparsely so or glabrescent, gland-dotted, the hairs shortly-stalked flagelliform. Capitula small, in few to numerous clusters of 3–12-capitulate lax corymbiform cymes; capitula stalks 1–10 mm. long; synflorescence branches subtended by leaves or leaf-like bracts. Involucres 3–8 × 4–10 mm., campanulate. Phyllaries numerous, scarious with subhyaline margins, sparsely hispid to glabrescent, finely ciliate-pectinate on the margins towards the apex, often purple-tinged in the upper half, the outer from c. 1.5 mm. long, the inner to c. 8 mm. long, ovate to narrowly lanceolate, ± tapering acuminate to a short pungent or bristle tip. Corollas mauve or purplish, rarely whitish, mostly 4–10 mm. long, tubular, widening gradually upwards, usually glabrous except for scattered glands. Achenes pale-brown, 1.2–1.5 mm. long, narrowly obovoid-oblong, with 4–5 low broad ribs, and narrower dark-brown gland-filled grooves in between, rounded above with a shallow apical depression; pappus of a few very caducous, barbellate setae 1–3 mm. long.

Zambia. B: Kataba, 24.iv.1961, *Fanshawe* 6520 (K). **N**: c. 6.5 km. from Mbala on path to Kalala Village, near Lucheche river, c. 1500 m., 25.iv.1968, *Richards* 23230 (K; SRGH). **W**: c. 5 km. W. of

Ndola, Ndola West Nat. For., 8.iv.1986, *Philcox, Pope & Chisumpa* 9807 (K; NDO). C: Kabwe, c. 1219 m., v.1909, *Rogers* 8111 (K). E: Nyika Nat. Park, above Zambian Govt. Rest House, 19.iv.1986, *Philcox, Pope & Chisumpa* 10010 (BR; K; NDO; SRGH). S: Mapanza Mission, c. 1066 m., 3.v.1953, *Robinson* 197 (K). **Zimbabwe**. N: Mazowe Distr., Henderson Res. Sta., 7.vi.1974, *Pope* 1354 (K; SRGH). W: Bulawayo Distr., Hope Fountain Mission, c. 1400 m., 24.iv.1974, *Norrgrann* 561 (K; SRGH). C: Headlands, 1.iv.1946, *Wild* 984 (K; SRGH). **Malawi**. N: Mzimba, Mbawa Expt. Sta., 5.iv.1955, *Jackson* 1593 (K; SRGH). C: Dedza town, c. 1600 m., 23.iv.1970, *Brummitt* 10041 (K; LISC; MAL; SRGH). S: Blantyre Distr., Ndirande Mt., c. 1540 m., 17.iv.1970, *Brummitt* 9924 (K; MAL; SRGH). **Mozambique**. N: Niassa Prov., Massangulo, v.1933, *Gomes e Sousa* 1476 (COI).

Also in Tanzania, Zaire, Angola and South Africa (Transvaal). Miombo woodland, wooded grassland, submontane grassland and woodland, often as a weed of disturbed or cultivated ground and at roadsides.

4. **Bothriocline mbalensis** (Wild & G.V. Pope) C. Jeffrey in Kew Bull. **43**: 261 (1988). Type: Zambia, Mporokoso to Kawambwa road, *Richards* 9270 (K, holotype; SRGH).

    Bothriocline laxa subsp. *mbalensis* Wild & G.V. Pope in Kirkia **10**: 318, fig. 7A, p. 333 sphalm. "*mabalensis*"; 371 (1977). Type as above.

An erect annual herb. Stems 0.2–1 m. tall, simple to widely branched above, leafy, hispid; indumentum a mixture of ± patent flagelliform hairs and appressed short-stalked T-shaped hairs. Leaves opposite, the uppermost alternate; petioles to c. 1.5 cm. long; lamina mostly 3–12 × 0.7–6 cm., ± narrowly oblong-elliptic or lanceolate to ovate, apex acute, base cuneate to rounded, margins remotely serrulate to coarsely serrate-dentate, sparsely hispidulous to glabrescent and gland-dotted on both surfaces, hairs shortly-stalked flagelliform. Capitula few to many, mostly laxly corymbiform cymose on slender stalks 1–8.5 cm. long, not clustered. Involucres 5–8 × 5–14 mm., spreading campanulate. Phyllaries green, scarious with a distinct hyaline margin, sometimes expanded above, obscurely 3–6 nerved, sparsely hispidulous or glabrescent, finely ciliate-pectinate on the margins towards the apex, becoming widely diverging, the outer from c. 3 mm. long, the inner series subequal to c. 8 mm. long, narrowly oblong to oblong-lanceolate, tapering-acuminate to a pungent or bristle tip, sometimes pungent-cuspidate. Corollas purplish to mauve, fading somewhat, exserted above the involucre, 6–8 mm. long, widening gradually upwards, glabrous except for scattered glands; lobes to c. 3 mm. long, linear. Achenes pale-brown, 1.5–2.5 mm. long, narrowly obovoid-oblong with c. 5 low broad ribs and narrower dark-brown gland-filled grooves in between, rounded above with a shallow apical depression; pappus absent or of a few very caducous barbellate setae 1–3 mm. long.

Zambia. N: Road from Mbala to Lunzwa Bridge, c. 1800 m., 11.iv.1967, *Richards* 22195 (K; M; SRGH).

Also in Tanzania. Miombo and mixed deciduous woodland, often on rocky outcrops.

5. **Bothriocline steetziana** Wild & G.V. Pope in Kirkia **10**: 319 fig. 7B, p. 333; 366 (1977). Type: Mozambique, Maganja da Costa, Mulevala, *Faulkner* Pre. 221 (SRGH, neotype; K; PRE).

    Gutenbergia longipes Steetz in Peters, Reise Mossamb. Bot. **2**: 349 (1864), non *B. longipes* (Oliv. & Hiern) N.E. Br. (1894). Type: Mozambique, Niassa, Cabaceira, *Peters* s.n. (B†).
    Gutenbergia longipes var. *membranifolia* Steetz in Peters, Reise Mossamb. Bot. **2**: 349 (1864). Type as for *Gutenbergia longipes*.
    Gutenbergia longipes var. *crassifolia* Steetz in Peters, Reise Mossamb. Bot. **2**: 349 (1864). Type: Mozambique, Inhambane, *Peters* s.n. (B†).

An erect annual herb or subshrub to c. 1.2 m. tall. Stems leafy, sparsely laxly branched above, ± densely pilose-pubescent to glabrescent; indumentum of large whitish patent flagelliform hairs usually intermixed on the upper stem with smaller somewhat crispate flagelliform hairs. Leaves alternate, sessile, mostly 3–15 × 0.3–1.3 cm., linear-oblanceolate, the uppermost narrowly triangular, tapering to a narrowly subobtuse mucronate apex, ± broadly auriculate below, margins serrate; upper surface, margins and midribs below with large white multicellular hairs. Capitula few to many, laxly corymbiform cymose; capitula stalks 0.1–6 cm. long with leaf-like bracts. Involucres 10–12 × 6–10(15) mm., campanulate-cyathiform, later somewhat spreading. Phyllaries scarious with subhyaline margins, obscurely 3–5-nerved, sparsely glandular, otherwise glabrous, from c. 2 mm. long outside to c. 11 mm. long inside, lanceolate to lanceolate-acuminate, tapering to a ± pungent or short bristle-tip. Receptacle shallowly alveolate. Corollas white or mauve, 8–10 mm. long, narrowly funnel-shaped, glabrous. Achenes 1.4–2 mm. long, narrowly oblong-obovoid, rounded at the apex with a small apical depression, low and broadly 4–6-ribbed with narrower dark-brown, densely gland-filled grooves in between; pappus absent.

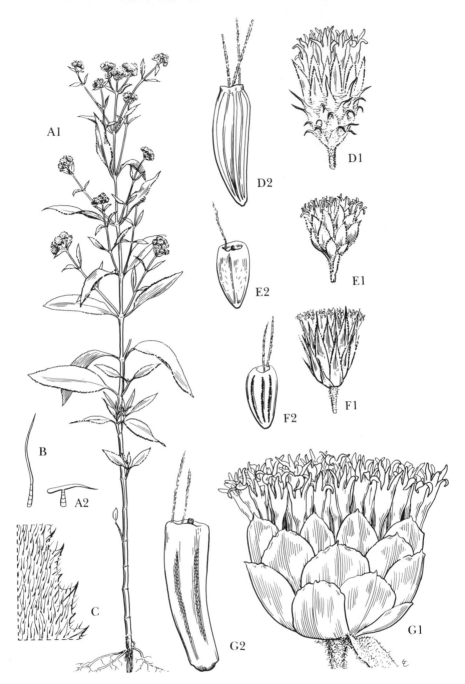

Tab. 26. A. —BOTHRIOCLINE LONGIPES. A1, habit, leaves opposite decussate (× ⅕); A2, stem trichome (× 32), A1 & A2 from *Philcox, Pope & Chisumpa* 9958. B. —BOTHRIOCLINE MILANJIENSIS, stem trichome (× 32), from *Brass* 16518. C. —BOTHRIOCLINE INYANGANA, leaf edge, upper surface strigose, margin biserrate (× 3), from *Brummitt* 11979. D. — BOTHRIOCLINE PECTINATA. D1, capitulum; D2, achene, D1 & D2 from *Brummitt* 10854. E. —BOTHRIOCLINE RIPENSIS. E1, capitulum; E2, achene, E1 & E2 from *Pawek* 5726. F. —BOTHRIOCLINE LAXA. F1, capitulum; F2, achene, note gland-filled grooves, F1 & F2 from *Robinson* 197. G. —BOTHRIOCLINE MUSCHLERIANA. G1, capitulum; G2, achene, G1 & G2 from *Richards* 1184. (Capitula × 3, achenes × 10). Drawn by Eleanor Catherine.

Mozambique. N: Niassa Prov., c. 44 km. E. of Malema, c. 640 m., 16.v.1961, *Leach & Rutherford-Smith* 10880 (K; LISC; SRGH). Z: c. 9 km. from Nicuadala to Namacurra, c. 40 m., 1.ii.1966, *Torre & Correia* 14288 (BR; EA; LD; LISC; MO; WAG).

Not known outside Mozambique. Deciduous woodland, stream sides and rocky outcrops.

6. **Bothriocline longipes** (Oliv. & Hiern) N.E. Br. in Bull. Misc. Inf., Kew **1894**: 389 (1894). —Wild & G.V. Pope in Kirkia **10**: 372, fig. 10E, p. 384 (1977). —C. Jeffrey in Kew Bull. **43**: 262 (1988). TAB. 26 fig. A. Type from Sudan.

Bothriocline schimperi var. *longipes* Oliv. & Hiern in F.T.A. **3**: 266 (1877). —O. Hoffm. in Engl., Pflanzenw. Ost-Afr. **C**: 402 (1895). Type as above.

Bothriocline schimperi var. *tomentosa* Oliv. & Hiern in F.T.A. **3**: 266 (1877). Type from Tanzania.

Erlangea longipes (Oliv. & Hiern) S. Moore in Journ. Bot. **35**: 313 (1902); in Journ. Linn. Soc., Bot. **40**: 104 (1911). —R.E. Fr., Wiss. Ergebn. Schwed. Rhod.-Kongo-Exped. 1911–1912, **1**: 319 (1916). —Robyns, Fl. Sperm. Parc Nat. Alb. **2**: 431 (1947). Type as above.

Erlangea spissa S. Moore in Journ. Bot. **35**: 313 (1902). Type from E. Africa.

Erlangea tomentosa (Oliv. & Hiern) S. Moore in Journ. Bot. **46**: 158 (1908). Type as above.

Erlangea pubescens S. Moore in Journ. Bot. **46**: 158 (1908). Types from Uganda.

Erlangea tomentosa var. *acuta* R.E. Fr. in Acta Hort. Berg. **9**: 112 (1929). Type from Kenya.

Erlangea eupatorioides Hutch. & Burtt in Rev. Bot. Zool. Afr. **23**: 36 (1932). Type a cultivated specimen from Zambia (K, holotype; BR).

Bothriocline tomentosa (Oliv. & Hiern) Wild & G.V. Pope in Kirkia **10**: 323 (1977). Type as for *Bothriocline schimperi* var. *tomentosa*.

Bothriocline eupatorioides (Hutch. & Burtt) Wild & G.V. Pope in Kirkia **10**: 321; 374 (1977).

An erect bushy suffrutex, with stems 1–2.5 m. tall from a woody rootstock. Stems herbaceous becoming woody below, leafy, branching above, puberulous, tomentellous on young growth, glabrescent below; indumentum of short-stalked asymmetric T-shaped hairs, the ± excentric terminal cell somewhat flattened and distorted. Leaves opposite; petiole to c. 4 cm. long; lamina mostly 4–20 × 1.5–9 cm., lanceolate, or the upper leaves somewhat ovate, apex tapering-acute, base rounded to broadly cuneate, margins serrate to serrate-dentate with callose-tipped teeth, upper surface sparsely puberulous to glabrescent, lower surface more densely puberulous to tomentellous, the hairs excentric T-shaped or flagelliform. Capitula small, numerous, at first in densely subglobose clusters at the ends of branches, becoming laxly corymbiform cymose; capitula stalks 0–5 mm. long. Involucres 4–7 × 3–6 mm., becoming spreading cyathiform-campanulate. Phyllaries appressed-imbricate, membranous-scarious with ± broad subhyaline margins, minutely pectinate-ciliate about the apex, puberulous soon glabrescent, sparsely glandular, from c. 1 mm. long outside to c. 6 mm. long inside, ovate-oblong to oblong-lanceolate, ± abruptly narrowed to tapering-obtuse mucronate at the apex. Corollas mauve or purplish, 4–7 mm. long, funnel-shaped, sparsely glandular otherwise glabrous. Achenes pale-brown, 1–1.8 mm. long, narrowly obovoid-oblong, with 4–5 low broad ribs, or 7–8-ribbed with the development of secondary ribs, ribs separated by narrower dark-brown, gland-filled grooves, achenes rounded above with a shallow apical depression, glabrous; pappus of few, very caducous, barbellate setae to c. 2 mm. long.

Zambia. N: Mbala, Ndundu, c. 1740 m., 15.v.1967, *Richards* 22239 (K; M; SRGH). E: Nyika Nat. Park, c. 500 m. S. of Zambian Govt. Rest House, c. 2260 m., 17.iv.1986, *Philcox, Pope & Chisumpa* 9958 (BR; K; NDO; SRGH). **Malawi.** N: Mzimba Distr., Mzuzu, Lusangadzi (Katoto Estate), c. 1371 m., 22.ix.1972, *Pawek* 5760 (K; SRGH). C: Nkhota Kota Game Res., S. of Chipata Mt., c. 1200 m., 17.vi.1970, *Brummitt* 11509 (K; MAL; SRGH).

Also in Sudan, Rwanda, Burundi, Uganda, Kenya, Tanzania, Zaire and Angola. Submontane tall grassland, usually on rocky outcrops and in evergreen forest margins at higher altitudes, in long grass in high-rainfall wooded grassland, often in riverine vegetation at lower altitudes.

7. **Bothriocline moramballae** (Oliv. & Hiern) O. Hoffm. in Engl., Pflanzenw. Ost-Afr. **C**: 403 (1895). —Wild & G.V. Pope in Kirkia **10**: 368 (1977). Type: Mozambique, Morrumbala, *Kirk* s.n. (K, holotype).

Vernonia moramballae Oliv. & Hiern in F.T.A. **3**: 278 (1877). Type as above.

Erlangea moramballae (Oliv. & Hiern) S. Moore in Journ. Linn. Soc., Bot. **35**: 313 (1902). Type as above.

Volkensia moramballae (Oliv. & Hiern) B.L. Burtt in Bull. Misc. Inf., Kew **1937**: 425 (1937). Type as above.

An erect bushy suffrutescent herb to c. 1.5 m. tall from a woody rootstock. Stems annual, herbaceous becoming somewhat woody below, leafy, branching above, puberulous; hairs short-stalked asymmetric T-shaped, or hairs with an excentric terminal cell. Leaves opposite on lower stem, alternate above, subsessile or shortly petiolate in lower leaves;

lamina 3.5–11 × 1.5–2.5 cm., lanceolate to narrowly elliptic, apex acute, base cuneate to rounded, margins coarsely serrate; upper surface puberulous, more densely so beneath, the hairs 1-armed T-shaped or flagelliform. Capitula small, numerous, ± densely clustered in corymbiform cymes; capitula stalks 0–5 mm. long. Involucres 4–6 × 4 mm., broadly ovoid to cyathiform-campanulate. Phyllaries membranous-scarious with subhyaline margins, minutely pectinate-ciliate about the apex, puberulous outside, appressed imbricate to diverging, the outer from c. 1 mm. long, the inner to c. 5 mm. long, ovate-lanceolate to oblong, the innermost lorate, apices shortly pungent and ± squarrose. Corollas purplish, c. 6 mm. long, narrowly funnel-shaped with a tubular lower portion, sparsely glandular. Achenes c. 1.2 mm. long, narrowly obovoid-oblong, with c. 5(7) low broad ribs, the ribs separated by narrower dark-brown densely gland-filled grooves, achenes rounded above with a shallow apical depression, glabrous; pappus of few very caducous barbellate setae to c. 2 mm. long.

Mozambique. N: Ribáuè, serra Mepáluè, c. 1550 m., 9.xii.1967, *Torre & Correia* 16420A (LISC). Z: Massingire, Serra da Morrumbala, l.v.1943, *Torre* 5252 (K; LISC).
Not known outside Mozambique. Hillsides on forest margins.
This species is distinguished from *B. longipes* (Oliv. & Hiern) N.E. Br. mainly by its leaves, which on the upper stem are alternate and subsessile or only shortly petiolate. It is possible that when more material is available the two taxa will prove to be conspecific.

8. **Bothriocline inyangana** N.E. Br. in Bull. Misc. Inf., Kew **1906**: 107 (1906). —C. Jeffrey in Kew Bull. **43**: 262 (1988). TAB. **26** fig. C. Type: Zimbabwe, Nyanga (Inyanga), *E. Cecil* 227A (K, holotype).
Erlangea rogersii S. Moore in Journ. Bot. **52**: 333 (1914). Type from Zaire.
Erlangea inyangana (N.E. Br.) B.L. Burtt in Bull. Misc. Inf., Kew **1937**: 423 (1937). Type as for *Bothriocline inyangana*.
Bothriocline longipes sensu Wild & G.V. Pope in Kirkia **10**: 372 (1977) non (Oliv. & Hiern) N.E. Br. (1894).

An erect bushy perennial herb or suffrutex, with annual stems 0.6–3 m. tall from a woody rootstock. Stems herbaceous becoming somewhat woody below, leafy, branching above, appressed puberulous or sometimes sparsely patent-hispid on upper stem and branches, glabrescent below; indumentum of short-stalked asymmetric, or 1-armed T-shaped hairs often intermixed with, or replaced by, long patent flagelliform hairs. Leaves opposite, sometimes 3-whorled, subsessile, sometimes shortly petiolate in older leaves, mostly 4–20 × 1.5–8 cm., elliptic to lanceolate, apex acute, base ± broadly cuneate, margins serrate to coarsely serrate-dentate, often double-serrate with small teeth alternating with larger teeth, teeth callose-tipped; lamina finely strigose-hispid on both surfaces, the lower surface sometimes also sparsely puberulous with hispid nerves. Capitula small, numcrous, ± densely clustered in corymbiform cymes; capitula stalks mostly 1–10 mm. long. Involucres 4–7 × 3–6 mm., broadly campanulate to cyathiform. Phyllaries ± appressed-imbricate, scarious with ± broad subhyaline margins, puberulous-hispidulous to glabrescent, the outer from c. 1 mm. long, the inner to c. 6 mm. long, ovate to oblong-ovate or lanceolate, ± abruptly narrowed to a mucronulate apex, or shortly acuminate. Corollas mauve or purplish, 4–7 mm. long, limb funnel-shaped and deeply linear-lobed, sparsely glandular, otherwise glabrous. Achenes pale-brown, 1–1.8 mm. long, narrowly obovoid-oblong with 4–5 low, broad, ribs and narrower dark-brown densely gland-filled grooves in between, rounded above with a shallow apical depression, glabrous; pappus of a few very caducous barbellate setae to c. 2 mm. long.

Zambia. N: Mbala, Ndundu, c. 1680 m., 2.v.1957, *Richards* 9511 (K). E: Nyika Nat. Park, c. 0.5 km. NW. of Zambian Govt. Rest House, 17.iv.1986, *Philcox, Pope & Chisumpa* 9985 (BR; K; NDO). **Zimbabwe**. E: Nyanga, 5.iv.1949, *Chase* 1289 (BM; COI; K; LISC; SRGH). **Malawi**. N: Mzimba Distr., c. 16 km. SW. of Mzuzu, Mbowe Dam, c. 1371 m., 30.vi.1974, *Pawek* 8777 (K; MO). C: c. 1 km. W. of Dedza, c. 1520 m., 22.iv.1970, *Brummitt* 10040 (K; LISC; MAL; PRE; SRGH). S: Zomba Plateau, S-facing slopes, c. 1460 m., 26.vii.1970, *Brummitt* 12231 (K; MAL; SRGH). **Mozambique**. N: Lichinga (Vila Cabral), 26.v.1934, *Torre* 110 (COI; LISC). Z: Serra do Gurué, 25.ix.1944, *Mendonça* 2286 (LISC; LMU; PRE; SRGH). MS: E. slopes of Chimanimani Mts., W. of Dombe, 26.iv.1974, *Pope & Müller* 1312 (K; SRGH).
Also in Tanzania and Zaire. Submontane tall grassland, evergreen forest margins, stream banks at higher altitudes, and in long grass in high rainfall wooded grassland.

9. **Bothriocline milanjiensis** (S. Moore) Wild & G.V. Pope in Kirkia **10**: 321; 372, fig. 10D, p. 383 (1977). TAB. **26** fig. B. Type: Malawi, Mt. Mulanje, *Whyte* s.n. (BM, holotype).
Erlangea milanjiensis S. Moore in Journ. Bot. **46**: 157 (1908). —Brenan in Mem. N.Y. Bot. Gard. **8**, 5: 458 (1954). Type as above.

An erect bushy suffrutescent perennial herb to c. 30 cm. tall from a woody rootstock. Stems annual, 1–several, herbaceous becoming somewhat woody below, leafy, branching above, puberulous, tomentose on young growth, glabrescent below; indumentum of ± patent flagelliform hairs. Leaves opposite, mostly subsessile, petiolate in larger leaves, mostly 5–21 × 1–6 cm., narrowly elliptic, tapering to an acute apex and a ± narrowly cuneate base, margins serrate with callose-tipped teeth, lamina sparsely pilose sometimes glabrescent, the hairs patent flagelliform with a bristle-like terminal cell. Capitula small, numerous in dense many-capitulate subglobose clusters; capitula mostly subsessile, or on stalks to c. 3 mm. long. Involucres 4–6 × 3–5 mm., narrowly ellipsoid- or obovoid-campanulate. Phyllaries appressed-imbricate, scarious-membranous with subhyaline margins, often minutely pectinate-ciliate about the apex, puberulous, soon glabrescent, sparsely glandular, the outer from c. 2 mm. long, the inner to c. 6 mm. long, ovate-oblong to lanceolate, ± abruptly narrowed or rounded mucronate to tapering-obtuse at the apex. Corollas purple or mauve, 6–8 mm. long, narrowly funnel-shaped, sparsely glandular, otherwise glabrous. Achenes pale-brown, 1.5–2 mm. long, narrowly obovoid-oblong, ± truncate at the apex with a distinct low rim, 4–5-ribbed, often c. 7-ribbed with the development of secondary ribs; ribs broad and low, separated by narrower dark-brown, densely gland-filled grooves, glabrous; pappus of a few very caducous barbellate setae to c. 2 mm. long.

Malawi. S: Mulanje Mt., Lichenya Plateau, c. 2150 m., 9.vii.1946, *Brass* 16747 (K; SRGH).
Apparently endemic on Mt. Mulanje. Submontane grassland amongst sheltering rocks, in scrub or on evergreen forest margins.

10. **Bothriocline ripensis** (Hutch.) Wild & G.V. Pope in Kirkia **10**: 324, fig. 7D, p. 334; 367 (1977). TAB. **26** fig. E. Type: Zambia, c. 70 km. S. of Mbala, *Hutchinson & Gillett* 3834 (K, holotype).
    Volkensia ripensis Hutch., Bot. S. Afr.: 508 (1946). Type as above.

A ± lax erect perennial herb to c. 1.5 m. tall, or stems scrambling to c. 5 m. long. Stems 1–several, leafy, branching above, shortly pilose on upper stem and branches, glabrescent below; indumentum of long-stalked 1-armed T-shaped hairs. Leaves alternate; petioles to c. 10 mm. long; lamina mostly 4–17 × 1–5 cm., narrowly ovate-lanceolate to elliptic, tapering to an acute-acuminate apex, narrowly to broadly cuneate below tapering into the petiole, margins serrate to somewhat crenulate with callose-tipped teeth; upper surface sparsely puberulous to glabrescent, lower surface paler and more densely puberulous, the hairs similar to but smaller than those on the stem. Capitula small, numerous, at first in ± dense clusters, becoming laxly corymbiform cymose; capitula stalks 0–5 mm. long. Involucres 3–5 × 3–5 mm., subglobose-cyathiform. Phyllaries relatively few, appressed imbricate, sometimes grading into the receptacular paleae, membranous-scarious with narrow subhyaline margins and ± cuspidate apices, puberulous-glandular outside, or glabrescent, sometimes pectinate-ciliate on the margins, the outer from c. 1.5 mm. long, the inner to c. 5 mm. long, ovate or oblong to narrowly lorate. Receptacular paleae sometimes present. Corollas purplish-mauve fading to white, c. 4 mm. long, with a broad funnel-shaped limb on a slender tubular stalk, sparsely glandular otherwise glabrous. Achenes 1.2–2 mm. long, subcylindric to obovoid, usually somewhat curved, c. 5-angular and usually narrowly-ribbed on the angles, the faces between ± flat with scattered brownish cell-like glands, truncate at the apex with a shallowly rimmed depression formed by the upward extension of the achene walls; pappus of a few caducous barbellate setae to c. 2 mm. long, occasionally the pappus setae somewhat paleaceous.

Zambia. N: Mbala Distr., Chitimbwa Road, Mbeya For., Chinakila path, c. 1524 m., 20.viii.1970, *Sanane* 1356 (K). W: Solwezi, Luamisambi R., 14.v.1969, *Mutimushi* 3218 (K; NDO). **Malawi.** N: Mzimba Distr., Mzuzu, Marymount towards Tung Estate, c. 1372 m., 20.ix.1972, *Pawek* 5726 (K; SRGH). C: Ntchisi Mt., c. 1400 m., 2.viii.1946, *Brass* 17101 (K; SRGH).
Not known outside the Flora Zambesiaca area. Swamp forest, gully forest and submontane evergreen forest usually beside streams.
Receptacular palea, normally absent in this genus, are sometimes to be found in plants of this species in Malawi.

11. **Bothriocline pectinata** (O. Hoffm.) Wild & G.V. Pope in Kirkia **10**: 320; 367, fig. 10A, p. 381 (1977). —Wild in Kirkia **12**: 20 (1980). —M.G. Gilbert in Kew Bull. **36**: 59 (1981). —C. Jeffrey in Kew Bull. **43**: 265 (1988). TAB. **26** fig. D. Type from Tanzania.
    Erlangea pectinata O. Hoffm. in Engl., Bot. Jahrb. **38**: 196 (1906). Type as above.

An erect soft perennial herb to c. 2 m. tall. Stems leafy, branching above, sparsely pilose soon glabrescent; hairs flagelliform. Leaves alternate, subsessile, mostly 8–26 × 1.2–6 cm., narrowly elliptic-oblanceolate, apex   tapering acuminate, base narrowly attenuate, margins finely serrate with bristle-tipped teeth, the bristles to c. 1 mm. long, lamina softly thinly herbaceous drying chartaceous, sparsely pilose soon glabrescent, hairs flagelliform. Capitula small, numerous, at first in dense subglobose clusters, becoming laxly arranged on divaricate branches or ± sessile in scorpioid cymes. Involucres 7–9 × 3–5 mm., subcylindric to somewhat campanulate. Phyllaries numerous, appressed imbricate, scarious with subhyaline margins, minutely pectinate-ciliate about the apex, araneose outside, sparsely glandular; the outer phyllaries from c. 3 mm. long, ovate with a subulate often squarrose apex; middle phyllaries lanceolate, tapering to a ± squarrose-pungent subulate apex with bristle-tips to c. 2 mm. long; the inner phyllaries to c. 8 mm. long, narrowly strap-shaped and tapering-acute at the apex. Corollas pale-lilac fading whitish, 6–7 mm. long, funnel-shaped, the lobes acicular-setose at the apex otherwise glabrous. Achenes 3–4 mm. long, narrowly ellipsoid-obovoid with a wide shallowly-rimmed depression at the apex, somewhat 4-sided at least when mature, narrowly c. 9-ribbed; pappus of caducous barbellate setae to c. 3 mm. long.

Zambia. E: Nyika, 26.vi.1966, *Fanshawe* 9749 (K; M; NDO; SRGH). **Malawi**. C: Ntchisi For. Res., south side, c. 1645 m., 25.iii.1970, *Brummitt & Evans* 9390 (EA; K; LISC; MAL; PRE; SRGH; UPS).
Also in Tanzania. Submontane evergreen forest understorey, in clearings and on forest margins.

12. **Bothriocline monocephala** (Hiern) Wild & G.V. Pope in Kirkia **10**: 319; 366 (1977). Type from Angola.
    Ethulia monocephala Hiern, Cat. Afr. Pl. Welw. **1**, 3: 514 (1898). —Mendonça, Contrib. Conhec. Fl. Angol., **1** Compositae: 1 (1943). Type as above.

A subaquatic perennial herb. Stems creeping-decumbent and rooting at the nodes, long slender and much branched, leafy, sparsely puberulous or glabrous. Leaves alternate, sessile, mostly 10–25 × 0.5–1.5 mm., linear, semi-fleshy, sparsely appressed puberulous to glabrous, gland-pitted. Capitula solitary, terminal on erect branch-ends, borne 7–15 cm. above the water. Involucres 4–5 × 5–6(8) mm., broadly obconic. Phyllaries subequal, usually purplish, sparsely puberulous outside, margins subhyaline and sparsely ciliate, the outer from c. 2 mm. long, the remainder c. 4 mm. long, lanceolate-oblong, subobtuse. Receptacular paleae sometimes present near the receptacle margins. Corollas mauve, exserted, 4–6 mm. long, narrowly funnel-shaped, minutely glandular otherwise glabrous. Achenes (immature) 1.2–1.8 mm. long, narrowly subcylindric-obovoid, ?5–8-ribbed, narrowly grooved between the ribs; pappus absent.

Zambia. N: Lake Isiba Ngandu (Lake Young), c. 1494 m., 20.vii.1938, *Greenway* 5427 (K).
Also in Zaire and Angola. Lake margins.

13. **Bothriocline emilioides** C. Jeffrey in Kew Bull. **43**: 266 (1988). Type from Tanzania.

An erect slender subaquatic annual herb to c. 45 cm. tall. Stems leafy, branching in upper part, thinly-spongy on the outside especially on the lower stem, glabrous or very sparsely puberulous, sparsely glandular; branches to c. 13 cm. long, filiform, spreading, usually purplish. Leaves alternate, sessile, soon withering on lower stem, mostly 10–45 × 2–3 mm., linear, semi-fleshy, glabrous or sparsely puberulous, gland-pitted. Capitula many, laxly paniculate, 1–8 on each branch; capitula-stalks somewhat divaricate, to c. 30 mm. long, filiform. Involucres 2–3 × 3.5–4 mm., broadly obconic. Phyllaries few-seriate, usually purplish, glandular and sparsely puberulous outside, margins subhyaline, the outer from c. 2 mm. long, the remainder c. 3 mm. long, lanceolate to ± oblong. Corollas mauve, strongly exserted, c. 3 mm. long, limb broadly campanulate, tube slender, glandular otherwise glabrous. Achenes (immature) c. 0.5–? mm. long, ?-ribbed; pappus absent.

Malawi. S: Machinga Distr., Chitundu Dambo on road to Lake Chiuta, 1.v.1982, *Patel* 902 (BR).
Also in Tanzania. Seasonally wet grassland, and pond margins.

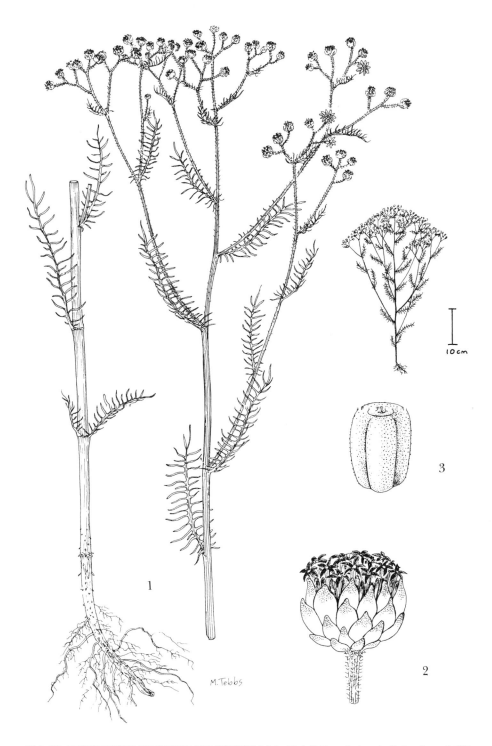

Tab. 27. RASTROPHYLLUM PINNATIPARTITUM. 1, habit (×⅔); 2, capitulum (×4); 3, achene (×20),
1–3 from *Robinson* 6602. Drawn by M. Tebbs.

15. RASTROPHYLLUM Wild & G.V. Pope.

Rastrophyllum Wild & G.V. Pope in Kirkia **10**: 324; 345 (1977). —C. Jeffrey in Kew Bull. **43**: 266 (1988).

Annual herbs. Vegetative indumentum consisting of erect flagelliform hairs, intermixed with smaller appressed ± T-shaped hairs on young growth. Stems erect, leafy, branched. Leaves alternate, sessile, or petiolate and broadly amplexicaul in *R. apiifolium* M.G. Gilbert, pinnatipartitely lobed or pinnatisect; lobes mostly simple, linear, irregular in length along a midrib, revolute (lobes oblong, deeply serrate-dentate in *R. apiifolium*). Capitula homogamous discoid, numerous in corymbiform cymes, capitula stalks short or to 2.5 cm. in *R. apiifolium*. Involucre subglobose-cupuliform (urceolate), phyllaries increasing in size to the inside, somewhat scarious or lower part ± indurated. Receptacle dentate-alveolate. Corollas purple or mauve, narrowly funnel-shaped, limb 5-lobed. Anther apices shortly ovate-appendiculate, bases obtuse-sagittate. Style arms subulate, hairy. Achenes shortly obovoid-oblong, rounded-truncate above with 4–6 low broad ribs separated by narrow grooves, surface densely beset with closely-spaced single-celled tuberculate trichomes. Pappus absent.

A ditypic tropical African genus. *Rastrophyllum apiifolium* M.G. Gilbert occurs in Tanzania.

Rastrophyllum pinnatipartitum Wild & G.V. Pope in Kirkia **10**: 325, fig. 9, p. 335; 346, fig. 3, p. 376 (1977). TAB. **27**. Type: Zambia, Mwinilunga, Ikelenge, *Robinson* 6602 (K; M; SRGH, holotype).

An erect bushy annual herb to c. 60 cm. tall. Stem leafy, branching throughout, sparsely pilose; branches to c. 28 cm. long, ascending, leafy; indumentum of scattered erect soft flagelliform hairs intermixed with smaller appressed 1–2-armed T-shaped hairs on young growth and branch-ends. Leaves alternate, sessile, 1.5–8 × 1.5–2 cm., narrowly oblong in outline, pinnatipartitely lobed; lobes 10–25 on each side, mostly simple, varying ± irregularly in length on the midrib, 1–10 mm. long, linear, revolute at least when dry, membranous, glandular-punctate, sparsely pilose. Capitula numerous, corymbiformly cymose, in clusters of 6–12, clusters terminal on branches; capitula stalks 1–10 mm. long. Involucres 3–4 × 4–5 mm., spreading to c. 10 mm. wide, subglobose-cupuliform. Phyllaries many, appressed-imbricate, somewhat scarious, purple, glandular and ± puberulent towards the apex, otherwise glabrous, margins narrowly subhyaline, the outer from c. 1 mm. long, the inner to c. 3 mm. long, ovate to oblong-lanceolate. Receptacle dentate-alveolate. Corollas purple in the upper part, c. 3 mm. long, funnel-shaped with a short tubular lower portion, lobes short, sparsely glandular outside. Achenes c. 1 mm. long, shortly obovoid-oblong, rounded-truncate above, with 4–6 broad low ribs separated by narrow grooves, densely beset (except the apical part) with minute, tuberculate trichomes; pappus absent.

Zambia. W: Mwinilunga, Ikelenge, 16.iv.1965, *Robinson* 6602 (K; M; SRGH).
Not known elsewhere. Seasonally damp grassland.

16. ETHULIA L.f.

Ethulia L.f., Decas Prima Pl. Rar. Horti Upsal.: 1 (1762). —Wild & G.V. Pope in Kirkia **10**: 324; 360 (1977). —M.G. Gilbert & C. Jeffrey in Kew Bull. **43**: 165–193 (1988). —C. Jeffrey in Kew Bull. **43**: 267–270 (1988).
Hoehnelia Schweinf. in von Höhnel, Zum Rudolf-See und Stephanie-See: 861 (1892).

Annual or short-lived perennial herbs with branching erect or decumbent stems, or stems somewhat strict; sometimes rhizomatous. Vegetative indumentum of short-stalked flagelliform hairs, or hairs short-stalked ± 1-armed T-shaped. Stems ribbed, leafy, often becoming woody below and with a pith in the upper part. Leaves alternate, sometimes opposite on the lower stem, subsessile (shortly petiolate in some Asian species), broadly-to linear-lanceolate, lamina pubescent to glabrescent, glandular-punctate. Capitula homogamous, small or very small, numerous, corymbiformly cymose sometimes subscorpioidly cymose. Involucres short, campanulate, cyathiform or subglobose; phyllaries imbricate. Receptacle flat, naked. Florets 5–100 per capitulum; corollas purple, mauve, pink or whitish in upper part, limb regularly deeply 5-lobed, ± funnel-shaped and tapering into a short tube below, or limb subcylindric and ± abruptly contracted into a short tubular lower part, glandular, lobes without apical hairs. Anthers with an apical ovate-lanceolate appendage and short obtuse basal auricles. Style arms stout, subulate, hairy. Achenes ± equalling the involucre in length, subcylindric to turbinate or tapering obpyramidal, strongly 2–6-ribbed, the ribs confluent with a pale apical rim about a ± cup-shaped depression, or apical rim-tissue expanded laterally and the achene apex truncate with a prominent lateral edge, the faces between the ribs ± flat with few to numerous unicellular glandular trichomes. Pappus absent, though apical rim sometimes produced into a pappus-like structure. n=10, 20.

1. Involucre 4–7 mm. in diam., broadly bowl-shaped; phyllaries, especially the inner, expanded membranous and ± purple-tinged about the apex; mature achenes usually cup-shaped to shortly tubular at the apex, sometimes irregularly so   -   -   4. *vernonioides* subsp. *mufindiensis*
– Involucre less than c. 4 mm. in diam., campanulate, turbinate or subglobose; phyllaries acute to rounded, not expanded at the apex; mature achenes truncate, or with a well developed, rimmed, apical depression, never with a tubular structure at the apex   -   -   -   -   -   2
2. Corollas mostly less than 2 mm. long, limb short and abruptly contracted below into a short tube   -   -   -   -   -   -   -   -   -   -   -   -   -   -   1. *conyzoides*
– Corollas 2–5 mm. long, limb funnel-shaped tapering into a short tubular lower part   3
3. Involucre more than 2.5 mm. in diam.; florets more than 20 per capitulum; corollas 3–5 mm. long; plants perennial   -   -   -   -   -   -   -   -   -   -   -   2. *rhizomata*
– Involucre less than 2.5 mm. in diam.; florets 7–15 per capitulum; corollas 2–3 mm. long; plants annual   -   -   -   -   -   -   -   -   -   -   -   -   -   3. *paucifructa*

1. **Ethulia conyzoides** L.f., Decas Prima Pl. Rar. Horti Upsal.: 1, t. 1 (1762). —Oliv. & Hiern in F.T.A. **3**: 262 (1877) pro parte. —O. Hoffm. in Warb., Kunene-Samb. Exped. Baum: 398 (1903). —Eyles in Trans. Roy. Soc. S. Afr. **5**: 502 (1916). —Mendonça, Contrib. Conhec. Fl. Angol., **1** Compositae: 1 (1943). —Merxm., Prodr. Fl. SW. Afr. 139: 63 (1967). —Wild & G.V. Pope in Kirkia **10**: 360 (1977) pro parte excl. specim. *Robinson 6773, Milne-Redhead 3406, Salubeni 1505.* —Hilliard & Burtt, Comp. Natal: 25 (1977). —Maquet in Fl. Rwanda, Spermat. **3**: 543, fig. 164 (1985). —M.G. Gilbert & C. Jeffrey in Kew Bull. **43**: 169 (1988). —C. Jeffrey in Kew Bull. **43**: 268 (1988). Type a specimen cultivated in Uppsala ex Sri Lanka (no specimen traced).

An erect or decumbent ± long-lived annual herb to c. 1.5(2) m. tall, somewhat aromatic. Stems simple or branching above, leafy, becoming somewhat woody below, ribbed or ± angular on young growth, ± appressed-puberulous, sparsely glandular; indumentum of flagelliform hairs. Leaves alternate, subsessile with the midrib narrowly-winged in the lower part and petiole-like especially in older leaves, mostly 3–12(14) × 0.5–2(3.5) cm., linear-elliptic to narrowly lanceolate, apices long-acuminate or attenuate, bases tapering and narrowly cuneate, margins subentire to coarsely serrate, thinly appressed puberulous, glandular-punctate. Capitula ± numerous in corymbiform cymes terminal on stems and branches, or in small lax or dense scorpioidly arranged clusters; capitula stalks 0–5 mm. long; synflorescence branches with small filiform bracts c. 3 mm. long. Involucres 1.5–2 × 2.5–4 mm., spreading to c. 5 mm. wide, cyathiform-campanulate becoming subglobose with mature achenes. Phyllaries few-seriate, puberulous and glandular outside, finely ciliate; the outer from c. 1 mm. long, subulate-lanceolate; the inner to c. 2 mm. long, oblong-lanceolate. Florets 19–32 per capitulum. Corollas purple, mauve or blue, up to c. 1.8(2.5) mm. long, exserted; limb cylindric or somewhat expanded above, ± abruptly contracted below into a slender tube, limb and tube ± equal in length, lobes half or less than half the limb in length, glandular otherwise glabrous. Achenes 1–1.8 mm. long, ± equalling the involucre, tapering-obpyramidal or ± turbinate, strongly 4–6-ribbed, truncate at the apex with apical tissue at first shallowly dish-shaped later flattened or patelliform and extended laterally a little beyond the achene body, in mature capitula the contiguous slightly convex achene apices produce a low dome-shaped top to the capitulum, faces between the ribs flat with few to numerous unicellular, glandular trichomes; pappus absent.

Leaves lanceolate or elliptic, sometimes narrowly so, mostly less than 6 times as long as wide, the larger usually more than 1 cm. wide, acute or shortly tapering to the apex; florets mostly 22–30 per capitulum; involucre cup-shaped at anthesis; capitula many in a somewhat laxly-branched synflorescence, or subsessile in many small clusters on lax branches   subsp. *conyzoides*
Leaves narrowly lanceolate-elliptic or very narrowly elliptic, mostly more than 6 times as long as wide, usually less than 1(1.5) cm. wide, tapering attenuate to the apex; florets c. 19 per capitulum; involucre subspherical at anthesis the phyllaries somewhat hooded; capitula numerous in a ± dense subglobose cluster   -   -   -   -   -   -   -   -   -   subsp. *kraussii*

Subsp. **conyzoides** — TAB. **28** fig. A.

Caprivi Strip. 6–8 km. from Katima Mulilo to Lisikili, c. 920 m., 24.xii.1958, *Killick & Leistner* 3076 (K; M; PRE; SRGH). **Botswana**. N: Thaoge River, 19°49'S, 22°08'E, 22.viii.1975, *P.A. Smith* 1439 (K; SRGH). **Zambia**. B: Siwelewele (Shangombo), c. 1040 m., 15.viii.1952, *Codd* 7558 (K; PRE). N: Mbereshi-Luapula R. Swamp, c. 900 m., 14.i.1960, *Richards* 12348 (K). C: c. 48 km. S. of Lusaka, 8.ii.1957, *Noak* 93 (K; SRGH). E: Petauke, Mvuvye R., c. 850 m., 5.vii.1958, *Robson* 838 (BR; K; LISC; SRGH). S: Livingstone Distr., Katambora, 13.i.1956, *Gilges* 537 (K; SRGH). **Zimbabwe**. N: Makonde Distr., Mhangura (Mangula), c. 865 m., 28.ix.1969, *W.B.G. Jacobsen* 4001 (K; PRE). W: Victoria Falls, xii.1904, *Allen* 83 (K; SRGH). E: Odzi, c. 1065 m., 30.v.1936, *Eyles* 8623 (K). S: Birchenough Bridge, Save (Sabi) R., i.1938, *Obermeyer* 2450 (K; PRE; SRGH). **Malawi**. N: Mwenembwe (Mwanemba), 2.iii.1903, *McClounie* 61 (K). **Mozambique**. Z: c. 15 km. from Mopeia to Campo, c. 70 m., 28.xii.1967, *Torre & Correia* 16773 (LISC). MS: Chibabava, Lower Buzi, c. 120 m., 28.xi.1906, *Swynnerton* 1815 (BM; K; SRGH). GI: Xai-Xai (Vila de João Belo), 21.i.1942, *Torre* 3900 (LISC). M: Namaacha, 29°59'S, 32°01'E, ii.1931, *Gomes e Sousa* 424 (K).

Also in tropical Asia and eastwards to Indochina, and in Egypt, Ethiopia and tropical Africa. On moist ground, in shallow water or swampy places, on river and stream banks, in reed-beds sometimes rooted in *Salvinia sp.* mats.

Subsp. **kraussii** (Sch. Bip. ex Walp.) M.G. Gilbert & C. Jeffrey in Kew Bull. **43**: 173 (1988). Type from South Africa (Natal).
    Ethulia kraussii Sch. Bip. ex Walp., Rep. Bot. Syst. 2, Suppl. 1: 945 (1843). Type as above.

Mozambique. M: Inhaca Island, sea level to 90 m., 29.xii.1956, *Mogg* 29056 (J; K).
Also in South Africa (Natal) and Brazil. Fresh water swamp.

2. **Ethulia rhizomata** M.G. Gilbert in Kew Bull. **43**: 175 (1988). —C. Jeffrey in Kew Bull. **43**: 268 (1988). TAB. **28** fig. B. Type: Zambia, Mporokoso Distr., Kalungwishi R., Lumangwe Falls, *Richards* 12100 (K, holotype).

This species is very similar to *E. conyzoides* L.f., though slightly larger. Plants are typically rhizomatous perennial herbs; rhizomes slender with long stout roots. Stems woody in the lower part. Leaves mostly similar in shape and size to those in *E. conyzoides* but can be up to c. 16 cm. long. Involucres usually 1.5–2 mm. long, but sometimes up to c. 2.8 mm. long; capitulum stalks up to c. 10 mm. long. Corollas 3–4.5 mm. long, narrowly funnel-shaped, with the limb ± tapering into a short tube (not abruptly contracted), lobes more than half the limb in length, almost linear, apically acute.

Zambia. B: Mongu, Lealui, 6.i.1966, *Robinson* 6773 (COI; K; M; SRGH). N: Mbala, Chinakila, Loye Flats, c. 1200m., 11.i.1965, *Richards* 19464 (K). W: Mwinilunga Distr., Lunga R. just below Mudjanyama R., 25.xi.1937, *Milne-Redhead* 3406 (BR; K; LISC). C: c. 51 km. NW. of Kabwe, Mufukushi R. near Chipepo, c. 1130 m., 20.i.1973, *Kornaś* 3052 (K).

Also in Tanzania, Zaire and Angola. In swampy places, sandy beaches beside rivers and lakes, flood-plains and seasonally wet grassland, often in shallow water, in reed beds or stands of papyrus.

3. **Ethulia paucifructa** M.G. Gilbert in Kew Bull. **43**: 181 (1988). —C. Jeffrey in Kew Bull. **43**: 269 (1988). TAB. **28** fig. C. Type from Tanzania.

An erect annual herb to c. 1(1.3) m. tall (sometimes apparently perennial). Stems branched above, becoming somewhat woody below, leafy, ribbed, puberulous to tomentellous especially on young growth, sparsely glandular; hairs flagelliform. Leaves alternate, subsessile, mostly 4–10 × 0.6–2 cm., up to c. 17 × 3 cm. in well-grown plants, narrowly lanceolate-elliptic, apex ± tapering acute, base narrowly cuneate, margin obscurely dentate or serrulate, lamina puberulous with fine appressed hairs, glandular-punctate. Capitula numerous, short-stalked or subsessile in corymbiform-cymose clusters; synflorescence branches densely pubescent, bracteate with filiform bracts. Involucres 2–2.6 × 1.5–2 mm., to c. 2.5 mm. wide in fruiting capitula, turbinate-campanulate. Phyllaries few-seriate, apices subacute to rounded, margins subhyaline, puberulous or glabrescent, ciliate, glandular above; outer phyllaries c. 1 mm. long and narrowly triangular, inner phyllaries to c. 2.3 mm. long and broadly lanceolate to elliptic-oblong. Florets 7–11(13) per capitulum. Corollas pale-purplish or lilac, 2–3 mm. long, exserted, narrowly funnel-shaped, limb deeply lobed, glandular otherwise glabrous. Achenes c. 1 mm. long, very narrowly turbinate, narrowly to sharply 3–4(5)-ribbed, the apex truncate with a low rim confluent with the ribs and forming a distinct apical cup-like depression, the faces between the ribs somewhat flat and ± densely covered in glandular trichomes; pappus absent.

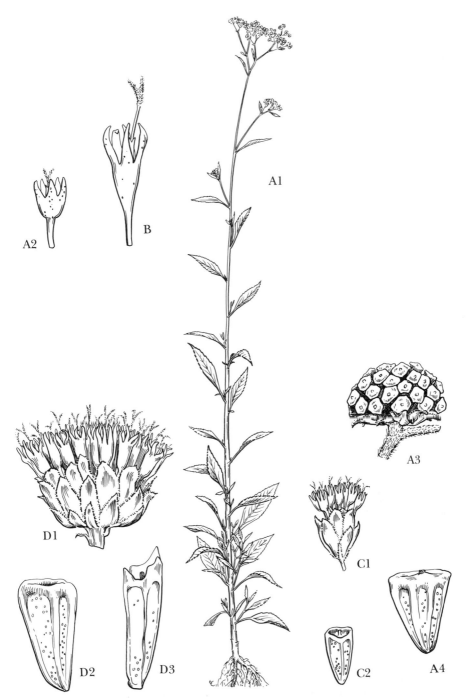

Tab. 28. A. —ETHULIA CONYZOIDES subsp. CONYZOIDES. A1, habit (×⅕), from *Robson* 838; A2, corolla, from *Richards* 14825; A3, mature capitulum; A4, achene, A3 & A4 from *P.A. Smith* 1439. B. —ETHULIA RHIZOMATA, corolla, from *Richards* 19464. C. —ETHULIA PAUCIFRUCTA. C1, capitulum; C2, achene, C1 & C2 from *Gomes e Sousa* 719. D. —ETHULIA VERNONOIDES subsp. MUFINDIENSIS. D1, capitulum; D2, achene, D1 & D2 from *Salubeni* 1505; D3, achene, note cup-shaped apical rim, from *Perdue & Kibuya* 11079. (Capitula × 6, corollas × 8, achenes × 15). Drawn by Eleanor Catherine.

Mozambique. N: Ribáuè, 14°50'S, 38°20'E, c. 600 m., ix.1931, *Gomes e Sousa* 719 (K).

Also in Tanzania. Flood-plain grassland on black cotton clay soils or alluvium, moist ground near water, and a weed of disturbed ground and cultivation.

4. **Ethulia vernonioides** (Schweinf.) M.G. Gilbert in Kew Bull. **43**: 187 (1988). —C. Jeffrey in Kew Bull. **43**: 269 (1988). Type from Kenya.

   Hoehnelia vernonioides Schweinf. in von Höhnel, Zum Rudolf-See und Stephanie-See, Appendix: 861 (1892). —O. Hoffm. in Engl. & Prantl, Nat. Pflanzenfam. **4**: 388 (1892). —S. Moore in Journ. Linn. Soc., Bot. **35**: 306, pl. 8 (1902). —R.E. Fr. in Acta Hort. Berg. **9**: 110 (1929). —Agnew, Upland Kenya Wild Fl.: 420 (1974). —Wild & G.V. Pope in Kirkia **11**: 140 (1978). Type as above.

An erect or decumbent perennial herb to c. 1.3 m. tall. Stems woody at the base, strict or sparsely- to much-branched in the upper part, ± densely leafy on the upper stem and branches, closely appressed puberulous above, glabrescent below or glabrous; indumentum hairs short-stalked with an excentric apical cell; branches strict, to c. 45 cm. long. Leaves alternate or spirally arranged, subsessile, crowded, 2–6.5 × 0.3–1 cm., narrowly oblong-elliptic, grading into synflorescence bracts, acute to subobtuse at the apex, cuneate below, margins serrulate to serrate or subentire, lamina subcoriaceous, glabrous or finely sparsely puberulous, glandular-punctate. Capitula many, loosely clustered in terminal, corymbiform cymes; capitula stalks 0–10 mm. long. Involucres 3–5 × 5–6(8) mm. in flowering capitula, to c. 7 mm. wide in fruiting capitula, broadly campanulate to bowl-shaped. Phyllaries many, imbricate, margins membranous becoming much expanded and ± appendage-like about the apex, usually purplish-tinged, puberulous and ± ciliolate; the outer phyllaries from c. 1.5 mm. long and narrowly ovate-lanceolate, the inner to c. 4 mm. long and narrowly oblong to narrowly oblanceolate-spathulate. Florets 40–100 per capitulum. Corollas purple or lilac, exserted, 5–6 mm. long, narrowly funnel-shaped; the limb deeply narrowly-lobed, scattered glandular with the lobes sometimes sparsely short-bristled. Achenes 1.5–3 mm. long, subcylindric, tapering slightly to the base, strongly 4–5-ribbed or angular, the apical rim usually produced into a thin-walled short apical tube, often irregular or asymmetrical on its upper margin, the faces between the ribs flat with few to numerous glandular trichomes; pappus absent.

Subsp. **mufindiensis** M.G. Gilbert in Kew Bull. **43**: 188 (1988). —C. Jeffrey in Kew Bull. **43**: 269 (1988). TAB. **28** fig. D. Type from Tanzania.

Stems often finely and sparsely puberulous; involucres of fruiting capitula 4.5–6 mm. wide; phyllaries 1.5–2 mm. wide; achenes mostly 45–50 per capitulum, 1.5–2.2 mm. long.

   Malawi. C: Dedza Distr., Namanungu stream, near Chongoni Forest, 23.xi.1970, *Salubeni* 1505 (K; SRGH). **Mozambique**. N: Lichinga (Vila Cabral), 23.v.1934, *Torre* 95 (COI; LISC).

Also in Tanzania. Submontane grassland, on river banks and in swampy ground.

   Development of the tubular apex to the achene is apparently variable. The apical tube, typical for this species, is greatly reduced leaving the achene low-rimmed and somewhat truncate in the Malawi material from Dedza district (*Pawek* 4108 (K) and *Salubeni* 1505 (K)).

   Subsp. *vernonioides* is recorded from Kenya and Tanzania. It is glabrous with larger fruiting capitula (6–8 mm. in diam.) containing up to c. 100 achenes; achenes up to 3 mm. long.

17. BRACHYTHRIX Wild & G.V. Pope

Brachythrix Wild & G.V. Pope in Kirkia **11**: 25; 135 (1978). —C. Jeffrey in Kew Bull. **43**: 270 (1988).

Perennial herbs with annual stems from small woody rootstocks; rootcrowns with or without lanate tufts; roots numerous, thong-like and usually swelling to form root-tubers. Vegetative indumentum a varying mixture of large long-stalked flagelliform hairs intermixed with short-stalked T-shaped hairs, the elongate terminal cell of the T-shaped hairs transverse and ± equal-armed, or ± excentric-ascending. Stems erect usually branching above, leafy, sometimes scapiform. Leaves alternate, sessile or subsessile, bases cuneate less often auriculate; upper surface ± scabrous, sparsely pubescent or glabrescent, the hairs scattered flagelliform often becoming somewhat rigid; lower surface pubescent or felted, sometimes finely pilose especially on the nerves, or glabrescent, glandular-punctate. Capitula homogamous, solitary or few–many in lax corymbiform cymes, or aggregated in ± dense 3–many-capitulate terminal clusters, or clusters lateral and ± scorpioidly cymose. Involucres campanulate to broadly cup-shaped, occasionally subglobose or narrowly ovoid. Phyllaries imbricate, several-seriate. Receptacle ± flat. Corollas purple, narrowly funnel-shaped, regularly deeply 5-lobed; lobes apically setulose or glabrous. Anthers with a narrowly triangular hyaline appendage at the apex and short subobtuse tails at the base. Style-arms subulate, hairy outside. Achenes narrowly oblong-obovoid, 4–5-angular with narrow ribs on the angles, truncate to slightly rounded at the apex, faces between the ribs with large scattered brownish glandular cells, very sparsely hispid or glabrous. Pappus 2-seriate, the outer of short lacerate free or ± united scales or reduced to a low rim, the inner of short very caducous barbellate setae. Pollen described as tricolporate, subechinolophate to echinate and micropunctate.

A small south-central and east tropical African genus of 5 species.

1. Capitula solitary, terminal; pappus setae twice as long as the achene   5. *lugarensis*
   - Capitula many, ± subsessile in dense clusters, or long-stalked in lax corymbiform cymes; pappus setae ± equalling the achene in length   - - - - - - - - - 2
2. Leaf bases auriculate, lamina widest in upper half, oblanceolate or obovate-oblong (at least on lower stem); phyllaries ± oblong and abruptly narrowed or rounded mucronulate at the apex
             3. *sonchoides*
   - Leaf bases cuneate, lamina widest about the middle, oblong-elliptic; phyllaries finely tapering or narrowly acute-pungent at the apex (if obtuse mucronate then leaves scabrid)   - - 3
3. Mature leaves less than 1.5 cm. wide, upper surface sparsely appressed-puberulous or glabrescent; roots tuberous; rootcrowns lacking brown lanate tufts   - - 4. *pawekiae*
   - Mature leaves 2–7 cm. wide, upper surface ± scabridulous; roots numerous thong-like; rootcrowns brown tufted-lanate   - - - - - - - - - - - 4
4. Leaf lower surface ± sparsely stiffly pubescent or glabrescent; phyllaries progressively longer to the inside, shortly bristle-tipped   - - - - - - - - 1. *glomerata*
   - Leaf lower surface densely grey-felted; phyllaries subequal, or the outer at least more than half the involucre in length and tapering to a long bristle tip   2. *malawiensis*

1. **Brachythrix glomerata** (Mattf.) C. Jeffrey in Kew Bull. **43**: 270 (1988). TAB. **29** fig. A. Type from Tanzania.
   Ageratinastrum glomeratum Mattf. in Notizbl. Bot. Gart. Berl. **11**: 412 (1932). Type as above.
   Brachythrix brevipapposa subsp. *brevipapposa* Wild & G.V. Pope in Kirkia **11**: 27; 138, tab. 1A (1978). —Kalanda in Bull. Jard. Bot. Nat. Belg. **56**: 386 (1986). Type: Zambia, Mbala Distr., Mambwe (Mwambe), *Richards* 12749 (SRGH, holotype; K; M).

An erect perennial suffrutescent herb with annual stems, up to c. 1.7 m. tall from a small woody rootstock; rootcrowns pale-brownish tufted-lanate; roots numerous, thong-like. Stems usually solitary, branching above, leafy, pubescent; indumentum a mixture of large long-stalked flagelliform hairs and hairs with an elongate excentric apical cell on a short stalk. Leaves subsessile, the largest shortly petiolate, mostly 5–14 × 2–7 cm., broadly elliptic to narrowly oblong-elliptic or lanceolate, apex acute, base cuneate to somewhat rounded, margins serrate; upper surface ± scabrous, rarely glabrescent, the hairs scattered becoming rigid at the base; the lower surface paler and stiffly puberulous particularly on the nerves; venation laxly reticulate and ± prominent beneath. Capitula numerous, aggregated in many 3–20-capitulate clusters; the capitula-clusters corymbiform or somewhat scorpioidly cymose, on long branches; capitula stalks 0–10(20) mm. long. Involucres 6–10 × 4–10 mm., spreading to c. 12 mm. wide, narrowly to broadly obconic-campanulate with ± diverging phyllaries, tapering or rounded at the base, occasionally subglobose or subfusiform with phyllaries appressed imbricate. Phyllaries numerous in many series, stiffly scarious with subhyaline margins, pilose-pubescent sometimes araneose, glabrescent; the outer from c. 2 mm. long, the inner to c. 8 mm. long, lanceolate to lorate, acute mucronate sometimes tapering pungent to the apex, occasionally ± recurved above. Florets 40–120 per capitulum. Corollas purple, 6–9 mm. long, very narrowly funnel-shaped, lobes ± densely setulose at the apex, sparsely glandular on the limb and tube, otherwise glabrous. Achenes 2–2.5 mm. long, narrowly obovoid-subcylindric, truncate to slightly rounded at the apex, 4–5-angular and narrowly ribbed on the angles, the faces between with scattered enlarged brownish cells; pappus 2-seriate, the outer of reduced free to ± united lacerate scales, the inner of a few tawny caducous barbellate setae 0.75–2 mm. long.

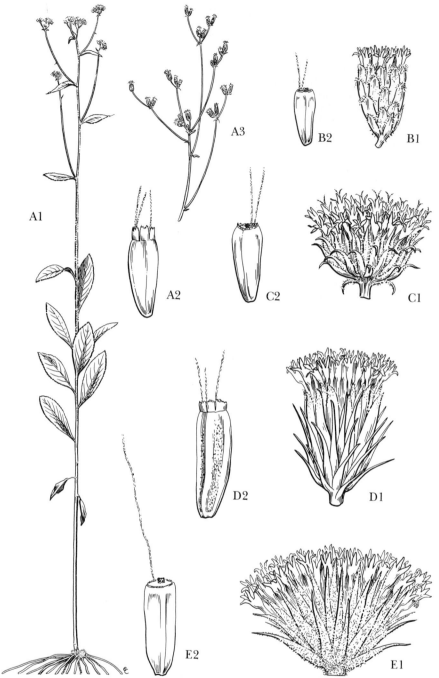

Tab. 29. A. —BRACHYTHRIX GLOMERATA. A1, habit (×$\frac{1}{5}$); A2, achene, A1 & A2 from *Pawek* 4807; A3, subscorpioid form of synflorescence (×$\frac{1}{3}$), from *Richards* 11318. B. —BRACHYTHRIX SONCHOIDES. B1, capitulum; B2, achene, outer pappus a low rim, B1 & B2 from *White* 2756. C. —BRACHYTHRIX PAWEKIAE. C1, capitulum; C2, achene, outer pappus a low rim, C1 & C2 from *Pawek* 3385. D. —BRACHYTHRIX MALAWIENSIS. D1, capitulum; D2, achene, D1 & D2 from *Brummitt et al.* 16201. E. —BRACHYTHRIX LUGARENSIS. E1, capitulum; E2, achene, outer pappus a low rim, E1 & E2 from *Richards* 3984. (Capitula × 3, achenes × 10). Drawn by Eleanor Catherine.

Zambia. N: Mbala Distr., Mombashi River, c. 1500 m., 7.iv.1967, *Richards* 22187 (K; M). W: Kitwe, 10.ii.1968, *Fanshawe* 10250 (K; NDO; SRGH). C: Serenje Distr., Kundalila Falls, 29.iii.1984, *Brummitt, Chisumpa & Nshingo* 16969 (K). **Malawi**. N: Mzimba Distr., Champira Forest, c. 1676 m., 20.iv.1974, *Pawek* 8425 (K). C: Dowa Distr., Ngara Hill, c. 1250 m., 7.ii.1959, *Robson* 1509 (BM; K; LISC; PRE; SRGH). S: Ntcheu, Chirobwe, 18.iii.1955, *Exell, Mendonça & Wild* 1017 (BM; LISC; SRGH). **Mozambique**. N: Maniamba, 40 km. from Lichinga (Vila Cabral) to Unango, c. 1100 m., 2.iii.1964, *Torre & Paiva* 10937 (C; COI; EA; K; LISC; LMU; PRE).

Also in Tanzania and Zaire. Submontane grassland and with tall grasses at lower altitudes, in *Brachystegia* woodland, on hillsides, often on termitaria.

2. **Brachythrix malawiensis** (Wild & G.V. Pope) G.V. Pope in Kew Bull. **45**: 698 (1990). TAB. **29** fig. D.
   Type: Malawi c. 50 km. from Rumphi on Chisenga Road, *Richards* 14373 (K, holotype; SRGH).
   Brachythrix brevipapposa subsp. *malawiensis* Wild & G.V. Pope in Kirkia **11**: 28; 139, tab. 1B (1978). Type as above.

This species is very similar to *B. glomerata* (Mattf.) C. Jeffrey, but differs from it in the following features: Leaves narrowly oblong-elliptic, densely grey-felted on the lower surface; the hairs of the leaf lower surface short-stalked T-shaped consisting of a 1–2-celled stalk bearing an elongate transverse terminal cell with long-tapering matted arms, the indumentum ± obscuring the lax prominent venation. Phyllaries linear-lanceolate, subequal or the outer phyllaries 5–7 mm. long, mostly more than half as long as the involucre, their apices finely tapering, subulate.

Malawi. N: Nyika Plateau, c. 6 km. downstream from Lake Kaulime, c. 2133 m., 11.ii.1968, *Simon, Williamson & Ball* 1754 (K; M; SRGH).
Known only from the Nyika Plateau. Submontane grassland and woodland.

3. **Brachythrix sonchoides** Wild & G.V. Pope in Kirkia **11**: 29; 137, tab. 1D (1978). TAB. **29** fig. B. Type: Malawi, Nyika Plateau, Chelinda Bridge, *Hilliard & Burtt* 4398 (E, holotype; K).

An erect perennial suffrutescent herb with annual stems to c. 1.3 m. tall; tuberous-rooted from a small woody rootstock, sometimes with woody stoloniferous rhizomes; rootcrowns pale-brownish tufted-lanate; tubers to c. 10 × 0.7 cm., slender fusiform. Stems usually solitary, branching above, leafy, puberulous; indumentum a mixture of long-stalked flagelliform hairs and short-stalked unequal-armed, or excentric, T-shaped hairs. Leaves sessile, mostly 4–9 × 0.7–3 cm., oblanceolate-oblong sometimes obovate-oblong, acute to ± rounded at the apex, auriculate-cordate below, margins subentire to serrate, upper surface scabridulous with scattered ± rigid hairs, lower surface stiffly puberulous with patent long-stalked T-shaped hairs, most numerous on the nerves; venation prominent beneath. Capitula numerous, aggregated in dense subglobose 4–many-capitulate clusters; capitula-clusters terminal, racemose or corymbiformly cymose, 1–3 cm. in diam. Involucres 6–9 ×× 2.5–4 mm., narrowly obovoid-turbinate the phyllaries remaining ± appressed imbricate. Phyllaries several-seriate, thinly scarious, purplish where exposed, araneose to white- or brownish-lanate towards the apex, glabrescent below; the outer phyllaries from c. 2 mm. long, ovate-lanceolate; the inner to c. 7 mm. long, linear, apices subacute mucronate. Florets c. 20 per capitulum. Corollas purple, 5–7 mm. long, very narrowly funnel-shaped, setulose at lobe apices, sparsely glandular on limb and tube. Achenes c. 1.5 mm. long (immature), narrowly obovoid-subcylindric, truncate at the apex, 4–5-angular and narrowly ribbed on the angles, faces between the ribs with scattered enlarged brownish cells; outer pappus reduced to a low rim, inner pappus of caducous brown barbellate setae 1–1.5 mm. long.

Zambia. E: Nyika Plateau, Kangampande Mt., c. 2133 m., 7.v.1952, *White* 2756 (FHO; K). **Malawi**. N: Nyika Plateau, between Chelinda and Dembo R., c. 2300 m., 13.v.1970, *Brummitt* 10661 (K; SRGH).
Known only from the Nyika Plateau. Submontane grassland, evergreen forest margins and stream-bank vegetation.

4. **Brachythrix pawekiae** Wild & G.V. Pope in Kirkia **11**: 29; 139, tab. 1C (1978) "*pawekii*". TAB. **29** fig. C. Type: Malawi, Nyika Plateau, Chelinda Bridge, *Pawek* 3385 (K, holotype).

An erect perennial herb 20–40 cm. tall, with annual stems from a small woody rootstock; roots short thong-like, each swelling abruptly into a root tuber; tubers 1.5–2.5 × 0.5–1 cm., ovoid-fusiform. Stems purple, usually solitary, branching sparsely above or throughout, leafy except below, puberulous; indumentum a mixture of relatively few long-stalked flagelliform hairs, and more numerous short-stalked appressed ± unequal-armed T-shaped hairs. Leaves mostly 3–9 × 0.4–1.3 cm., decreasing in size to the stem apex and base, narrowly oblong-elliptic, acute or subobtuse mucronate at the apex, cuneate below, margins subentire to remotely serrulate; lamina thinly coriaceous, upper surface puberulous with sparse appressed T-shaped hairs, sometimes with patent flagelliform hairs or glabrescent, lower surface ± strigose with T-shaped hairs, these often more numerous on the nerves, sometimes intermixed with flagelliform hairs. Capitula few to many, solitary on branches up to c. 6 cm. long and somewhat racemose, or in ± lax capitulum-clusters cymosely arranged. Involucres 5–10 × 9–11 mm., very broadly obconic-campanulate to broadly cup-shaped, wider than long. Phyllaries many-seriate, narrowly acute-pungent or finely tapering to a bristle tip, pilose with appressed short-stalked hairs intermixed with few to numerous larger patent flagelliform hairs; the outer phyllaries from c. 2 mm. long, very narrowly lanceolate, the inner to c. 6 mm. long, lanceolate, or phyllaries subequal to c. 8 mm. long and linear-lanceolate or subulate. Florets c. 70 per capitulum. Corollas purple, 5–6 mm. long, narrowly funnel-shaped. Achenes 1.5–2 mm. long, subcylindric to narrowly turbinate, truncate at the apex, c. 4-angular and narrowly ribbed on the angles, with scattered brownish cells on the faces between, otherwise glabrous; pappus 2-seriate, the outer reduced to a low rim of small ± free or united lacerate scales, inner pappus of caducous brown barbellate setae 1–2 mm. long.

Malawi. N: Nyika Plateau, c. 2 km. N. of Chosi Peak, 19.iv.1986, *Philcox, Pope & Chisumpa* 10018 (BR; GA; LISC; MAL; MO; NDO; SRGH).
Known only from the Nyika Plateau. Submontane short grassland.
Specimens from Nganda Mt. (*Grosvenor & Renz* 1156 and *Pawek* 2062 (K)) differ from the typical material in that the phyllaries are long ± subulate and subequal with an indumentum in which long-stalked patent flagelliform hairs predominate.
Brachythrix stolzii (S. Moore) Wild & G.V. Pope, from adjacent Tanzania, is closely related to this species but may be distinguished by its subscapose habit in which the largest leaves are ± clustered on the lower stem (and not midcauline) and by its outer pappus of free membranous scales.

5. **Brachythrix lugarensis** (O. Hoffm.) Wild & G.V. Pope in Kirkia **11**: 26; 136, tab. 1E (1978). —C. Jeffrey in Kew Bull. **43**: 270 (1988). TAB. **29** fig. E. Type from Tanzania.
  Vernonia lugarensis O. Hoffm. in Engl., Bot. Jahrb. **30**: 125 (1901). Type as above.

A stiffly erect perennial herb to c. 40 cm. tall, with annual stems from a small woody rootstock or slender woody rhizome; roots thong-like, each usually swelling into a root-tuber. Stems solitary, or 2–3 together, simple, ± densely leafy hispid-pubescent; indumentum a mixture of long-stalked flagelliform hairs and short-stalked appressed unequal-armed T-shaped hairs. Leaves sessile, mostly 2–7 × 0.5–3 cm., ovate to narrowly elliptic, acute to ± acuminate at the apex, cuneate to rounded or subcordate below, margins subentire to sharply serrate; lamina membranous and sparsely puberulous, sometimes scabridulous on the upper surface, the hairs flagelliform ± scattered and becoming rigid on the leaf upper surface, usually confined to main veins on lower surface, appressed short-stalked T-shaped hairs sometimes present, glandular-punctate; main nerves prominent below. Capitula solitary, terminal; involucres 9–12 × 14–20 mm., very broadly obconic-campanulate to cup-shaped, ± truncate at the base. Phyllaries many-seriate, tapering-pungent at the apex, finely sparsely puberulous, ciliolate; the outer from c. 5 mm. long, subulate, the inner to 10 mm. long becoming linear-lanceolate. Corollas purple, 7–8 mm. long, funnel-shaped, deeply-lobed. Achenes 2–2.5 mm. long, subcylindric to narrowly obovoid, truncate at the apex, 4–5-angular and narrowly ribbed on the angles, with scattered brownish cells on the faces between, otherwise glabrous; pappus 2-seriate, the outer series reduced to a low rim-like structure, the inner of many brown caducous somewhat flattened barbellate setae 0.75–5 mm. long.

Zambia. N: Mbala Distr., Nkali Dambo, c. 1590 m., 26.xii.1967, *Richards* 22806 (K; SRGH).
Also in Tanzania. Open and wooded grassland, and in dambos.

18. AGERATINASTRUM Mattf.

Ageratinastrum Mattf. in Notizbl. Bot. Gart. Berl. **11**: 412 (1932). —Wild & G.V. Pope in Kirkia **10**: 326; 357 (1977). —C. Jeffrey in Kew Bull. **43**: 270–271 (1988).
  Ageratina O. Hoffm. in Engl., Bot. Jahrb. **28**: 503 (1900) non Spach (1841).

Perennial herbs with annual stems from woody rootstocks. Vegetative indumentum of short-stalked flagelliform hairs. Stems many, densely leafy, with short internodes, branching above. Leaves alternate or spirally arranged, opposite or whorled, sessile to shortly petiolate, felted-tomentose, sparsely pilose or glabrescent, glandular-punctate. Capitula homogamous, numerous, in lax or ± dense clusters at the ends of branches, or the capitulum-clusters arranged in a large corymbiform cyme, occasionally the capitula somewhat scorpioidly arranged. Involucres campanulate to subglobose, florets exserted; phyllaries several-seriate, progressively larger to the inside. Receptacle flat, shallowly alveolate. Corollas purple or mauve, darker in the upper half, regularly 5-lobed, lobes lanate or pilose. Anthers with a broad hyaline apical appendage and short acute tails at the base. Style-arms subulate, hairy outside. Achenes obovoid-oblong or turbinate, 4–5-sided and obscurely ribbed on the angles, truncate at the apex, smooth or strongly rugose on the faces between the ribs, glabrous or sparsely setulose, minutely brown gland-dotted. Pappus persistent, of ± united broad shiny scales lacerate on the upper margin, or of scales sometimes free and alternating with paleaceous setae. Pollen described as triporate, lophate and emicropunctate.

A small east-central tropical African genus of two species.

Leaves tomentellous on the upper surface; achenes smooth not rugose between the ribs
1. *polyphyllum*
Leaves glabrous or sparsely thinly pilose on the upper surface; achenes rugose or transversely ridged between the ribs - - - - - - - - - - - - - 2. *palustre*

1. **Ageratinastrum polyphyllum** (Bak.) Mattf. in Notizbl. Bot. Gart. Berl. **11**: 412 (1932). —Brenan in Mem. N.Y. Bot. Gard. **8**, 5: 462 (1954). —Wild & G.V. Pope in Kirkia **10**: 358, fig. 6 (1977). —Kalanda & Lisowski in Bull. Jard. Bot. Nat. Belg. **51**: 457, fig. 1, 1–5 (1981). —C. Jeffrey in Kew Bull. **43**: 271 (1988). TAB. **30** fig. A. Type: Malawi, Nyika Plateau, *Whyte* 252 (K, holotype).
    Ageratum polyphyllum Bak. in Bull. Misc. Inf., Kew **1898**: 148 (1898). Type as above.
    Ageratina polyphylla (Bak.) O. Hoffm. in Engl., Bot. Jahrb., **28**: 504 (1900). Type as above.
    Ageratina goetzeana O. Hoffm. in Engl., Bot. Jahrb., **28**: 504 (1900). Type from Tanzania.
    Ageratinastrum goetzeanum (O. Hoffm.) Mattf. in Notizbl. Bot. Gart. Berl. **11**: 412 (1932). Type as for *Ageratina goetzeana*.
    Ageratinastrum virgatum Mattf. in Notizbl. Bot. Gart. Berl. **11**: 413 (1932). Type from Tanzania.

An erect tufted perennial herb to c. 1 m. tall, with annual stems from a woody rootstock; rootstock with a woody taproot. Stems many, strict, branching near the apex, densely leafy with internodes mostly less than 10 mm. long, densely pilose-tomentose with glands between the hairs, becoming pubescent or glabrescent below; indumentum of short-stalked flagelliform hairs. Leaves sessile or subsessile, overlapping, alternate or spirally arranged, sometimes subopposite or whorled, 2–4.5 × 0.4–1.2 cm., narrowly oblong-lanceolate to narrowly-elliptic or ovate, acute mucronate at the apex, cuneate to rounded below, margins ± revolute, upper surface densely tomentellous to sparsely pilose, lower surface paler, tomentose or pubescent, glandular punctate. Capitula numerous, in few to many 3–15-capitulate clusters, or the capitulum-clusters ± corymbiformly cymose on branches 1.5–9 cm. long, occasionally the capitula somewhat scorpioidly arranged along branches. Involucres 3–5 × 4–6 mm., campanulate to subglobose, florets exserted. Phyllaries several-seriate, purple above, acute, tomentose to woolly outside; the outer from c. 1 mm. long, narrowly ovate; the inner to c. 5 mm. long, narrowly oblong to lanceolate. Receptacle shallowly alveolate. Corollas purple, darker in the upper half, 4–6 mm. long, narrowly funnel-shaped, puberulous to tomentellous outside or only the lobes hispidulous, glandular. Achenes 1.5–2 mm. long, narrowly oblong-obovoid, often curved, truncate at the apex, 4–5-angular, obscurely ribbed on the angles and ± sparsely setulose, the faces smooth with scattered brown minute gland-dots, otherwise glabrous; pappus of 4–7 broad ± united shiny scales, 0.5–1 mm. long and lacerate on the upper margin, or sometimes the scales free and alternating with 4–7-narrower paleaceous setae which exceed or ± equal the scales in length.

Zambia. E: Nyika Nat. Park, c. 0.5 km. SW. of Zambian Govt. Rest House, 17.iv.1986, *Philcox, Pope & Chisumpa* 9978 (BR; K; LISC; MO; NDO; SRGH). **Malawi**. N: Nyika Plateau, Nchenachena Spur, c. 2000 m., 20.viii.1946, *Brass* 17345 (BM; K; SRGH).
Also in Tanzania. Submontane grassland.

2. **Ageratinastrum palustre** Wild & G.V. Pope in Kirkia **10**: 326; 358 (1977). TAB. **30** fig. B. Type: Zambia, Mbala, Ndundu Swamp, *Richards* 16151 (K, holotype; M).

An erect stiff perennial herb to c. 1 m. tall, with annual stems from a woody rootstock. Stems many, strict, branching above, densely leafy with internodes less than 2 cm. long,

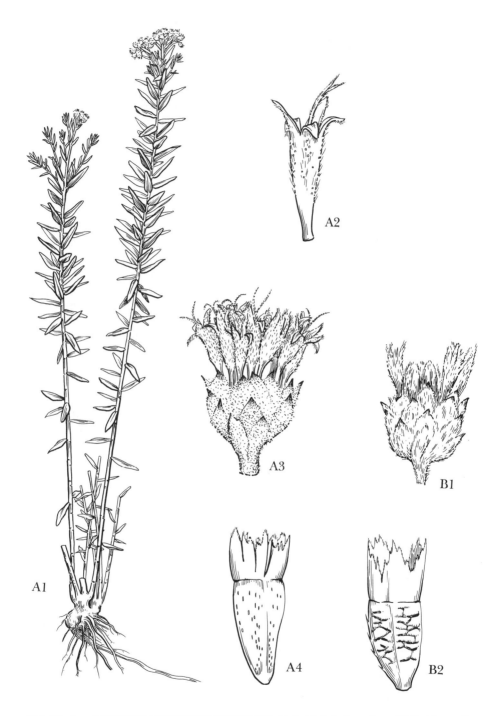

Tab. 30. A. —AGERATINASTRUM POLYPHYLLUM. A1, habit (×¼), from *Brass* 17345; A2, corolla
tube (× 6); A3, capitulum (× 5), A2 & A3 from *Pawek* 3403; A4, achene (× 15), from *Brummitt*
10823. B.—AGERATINASTRUM PALUSTRE. B1, capitulum (× 5); B2, achene (× 15), B1 & B2
from *Sanane* 458. Drawn by Eleanor Catherine.

whitish pilose-pubescent, glabrescent below; indumentum of short-stalked flagelliform hairs. Leaves alternate or spirally arranged sometimes opposite or whorled, subsessile or with petioles to c. 2 mm. long; lamina 1–3.3 × 0.4–1.2 cm., narrowly oblong-lanceolate to narrowly ovate, acute mucronulate at the apex, cuneate to rounded below, margins ± revolute, upper surface glabrous or thinly sparsely pilose-puberulous with nerves ± impressed, the lower surface paler and sparsely pilose especially on the nerves, glandular-punctate. Capitula numerous, singly or in 2–10-capitulate clusters at the ends of branches, or capitulum-clusters ± corymbiformly cymose at the stem apex. Involucres 3–4 × 3–4.5 mm., campanulate to subglobose; florets exserted. Phyllaries several-seriate, purple and acute above, tomentellous to woolly outside; the outer phyllaries from c. 1 mm. long, ovate; the inner to c. 4 mm. long, narrowly oblong-lanceolate. Receptacle shallowly alveolate. Corollas mauve, darker in the upper half, 5–6 mm. long, narrowly funnel-shaped, puberulous or thinly pilose outside at least on the lobes, glandular, the lobes purple-tipped. Achenes 1–1.5 mm. long, turbinate, truncate at the apex, 4–5-sided with obscure narrow ribs on the angles, rugose or transversely ridged on the faces between the ribs, glabrous or sparsely setulose, minutely brown gland-dotted; pappus of broad ± united shiny, white brownish or sometimes purplish scales, 0.5–1 mm. long and ± lacerate on the upper margin, occasionally scales ± free and alternating with 1–3 narrow paleaceous setae which ± exceed the scales in length.

Zambia. N: Mporokoso-Sanga Hill Road, c. 1350 m., 19.i.1960, *Richards* 12436 (K; M; SRGH). Not known outside the Flora Zambesiaca area. Swampy or dambo areas with tall grasses.

19. DEWILDEMANIA O. Hoffm.

Dewildemania O. Hoffm. in Ann. Mus. Congo, Sér. 4, **1**: x (1903). —Wild & G.V. Pope in Kirkia **10**: 327; 343 (1977). —C. Jeffrey in Kew Bull. **43**: 271 (1988).

Perennial herbs with erect, usually tufted, annual stems from small woody rootstocks; roots radiating, thong-like. Stems ± strict, branching near the apex, glabrous, sparsely pilose or pubescent, densely leafy with leaves often grading into the phyllaries. Leaves spirally arranged, subsessile, filiform to ± elliptic or lanceolate, glabrous. Capitula homogamous, 1–many, solitary or in lax terminal corymbiform cymes. Involucres broadly cup-shaped to turbinate-campanulate, florets exserted. Phyllaries several seriate, larger to the inside, appressed imbricate to somewhat spreading, the outer often grading into the bracts or leaves below, glabrous to pubescent. Receptacle flat or ± dome-shaped, paleaceous with the paleae grading from the inner phyllaries and usually exceeding them in length. Corollas purple, narrowly funnel-shaped, regularly 5-lobed, the lobes sparsely to densely setose. Anthers with a small triangular hyaline apical appendage, anther thecae brownish and shortly acute at the base. Style arms terete tapering to the apex, hairy. Achenes oblong-turbinate and 4–5-ribbed sometimes somewhat angular and 4–5-sided, setulose to hispid, or glabrous, with scattered large brown cells on the sides; pappus 2-seriate, the outer series of 5–15 broad or narrow scales, the inner of somewhat larger narrow paleaceous setae.

A small south-central and east tropical African genus of three species.

1. Leaves lanceolate or elliptic-lanceolate, more than 3 mm. wide; involucres usually more than 12 mm. wide; capitula often solitary on the stem - - - - - - 1. *platycephala*
 – Leaves linear or filiform, less than 2 mm. wide; involucres mostly less than 12 mm. wide; capitula usually 3–21 per stem - - - - - - - - - - - - - - - 2
2. Synflorescence branches simple stiff and leafy, bearing a single terminal capitulum; phyllaries more than 50 per involucre; outer pappus of c. 5 broad stramineous scales 2. *filifolia*
 – Synflorescence branches usually divided, ± filiform, sparsely leafy or bracteate, with 1–3 capitula per branch; phyllaries fewer than 40 per involucre; outer pappus of 10–15 brownish or purplish narrowly lanceolate scales - - - - - - - - - - - 3. *stenophylla*

1. **Dewildemania platycephala** B.L. Burtt in Kew Bull. **4**: 496 (1950). —Wild & G.V. Pope in Kirkia **10**: 345 (1977). —C. Jeffrey in Kew Bull. **43**: 271 (1988). TAB. **31** fig. A. Type from Tanzania.

An erect tufted perennial herb to c. 70 cm. tall, with annual stems from a small woody rootstock; roots thong-like. Stems strict, simple (rarely branched at the apex), densely

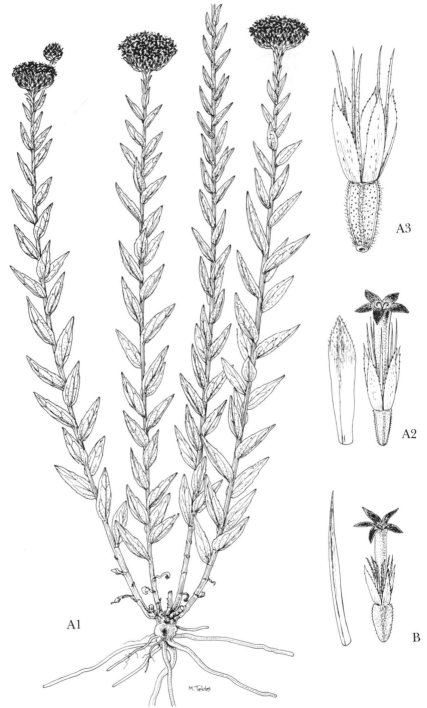

Tab. 31. A. —DEWILDEMANIA PLATYCEPHALA. A1, habit, annual stems from woody rootstock (×⅔), from *Philcox, Pope & Chisumpa* 10112; A2, floret and subtending receptacular palea (× 5); A3, achene with pappus (× 8), A2 & A3 from *Richards* 1147. B.—DEWILDEMANIA FILIFOLIA, floret and receptacular palea (× 5), from *Fanshawe* 10043. Drawn by M. Tebbs.

leafy, glabrous, or sparsely pilose just below the insertion of the leaves. Leaves spirally arranged, ± grading into the phyllaries, subsessile, mostly 1.5–3 × 0.3–1.4 cm., lanceolate to narrowly elliptic-lanceolate, acute at the apex, cuneate to rounded below, margins subentire or obscurely denticulate, somewhat revolute, lamina glabrous glandular-punctate and obscurely 3-nerved from the base. Capitula usually solitary and terminal, sometimes 2–3(6) corymbosely cymose at the stem apex. Involucres 7–9 × 12–20 mm., broadly cup-shaped. Phyllaries numerous, appressed-imbricate, narrowly lanceolate to lorate ± differentiated above, glabrous to pubescent, finely long-ciliolate; the outer from c. 2 mm. long, the inner to c. 8 mm. long. Receptacle flat, paleaceous; paleae purplish, 7–9 mm. long, lorate, somewhat pilose. Corollas purple, 5–8 mm. long, narrowly funnel-shaped, deeply lobed, densely to sparsely setose on the lobes. Achenes 1–2 mm. long, angular oblong-turbinate, c. 5-ribbed, setulose on the ribs, with scattered large brown cells on the sides; pappus 2-seriate, the outer series of 5–6 chartaceous oblong-elliptic scales c. 2.5 mm. long, the inner of 5–6 narrower paleaceous setae to 4 mm. long and barbellate-laciniate on the margins.

Zambia. N: Mbala Distr., road to Chinakila from Mpulungu, c. 1800 m., 24.iv.1986, *Philcox, Pope & Chisumpa* 10112 (BR; K; MO; NDO; SRGH).
Also in Tanzania. Short grassland usually in deciduous woodland on rocky outcrops and dambo margins.

2. **Dewildemania filifolia** O. Hoffm. in Ann. Mus. Congo, Sér. 4, **1**: x (1903). —B.L. Burtt in Kew Bull. **4**: 496 (1950). —Wild & G.V. Pope in Kirkia **10**: 343, fig. 2 (1977). TAB. **31** fig. B. Type from Zaire.

An erect tufted perennial herb to c. 50 cm. tall, with annual stems from a small woody rootstock; roots numerous, thong-like. Stems several, wiry, branching above, densely leafy, glabrous or sparsely puberulous becoming pubescent at the base. Leaves spirally arranged, ascending, subsessile, mostly 1–3.5 × 0.15 cm., linear, decreasing in size on the upper stem and branches usually grading into the phyllaries, acute at the apex, tapering slightly below, margins entire somewhat revolute, lamina ± stiffly coriaceous, glabrous on both surfaces. Capitula 3–9, solitary and terminal on leafy branches 3–13 cm. long. Involucres 7–11 × 10–12 mm., obconic-turbinate. Phyllaries numerous, glabrous, linear-lanceolate, stiffly subulate or pungent to the apex, margins hyaline in the lower part, the outer phyllaries from c. 4 mm. long, the inner to c. 8 mm. long. Receptacle conical or dome-shaped, paleaceous; paleae becoming purplish above, 7–9 mm. long, narrowly lanceolate, stiffly subulate-pungent. Corollas purple, 5–7 mm. long, narrowly funnel-shaped, glabrous. Achenes 1.5–1.75 mm. long, oblong-turbinate, angular and 4–5-sided to subterete and 4–5-ribbed, setulose or hispid with scattered large brown cells on the sides; pappus 2-seriate, the outer series of 5–7 chartaceous oblong-elliptic scales 0.5–1.5 mm. long, the inner of 5–7 narrower paleaceous setae to c. 2 mm. long.

Zambia. W: c. 5 km. S. of Solwezi, near Lufubwa Nat. Monument, 17.v.1986, *Philcox, Pope, Chisumpa & Ngoma* 10354 (BR; CAL; LISC; MO; NDO; SRGH). **Mozambique.** N: Lichinga (Vila Cabral), ii.1934, *Torre* 17 (COI; LISC).
Also in Zaire. Miombo woodland often on dry rocky hillsides.

3. **Dewildemania stenophylla** (Bak.) B.L. Burtt in Kew Bull. **4**: 496 (1950). —Wild & G.V. Pope in Kirkia **10**: 344 (1977). —C. Jeffrey in Kew Bull. **43**: 271 (1988). Type: Malawi, Nyika Plateau, *Whyte* 219 (K, holotype).
Athrixia stenophylla Bak. in Bull. Misc. Inf., Kew **1897**: 270 (1897). Type as above.

A slender erect perennial herb to c. 75 cm. tall, with annual stems from a small woody rootstock; roots numerous, thong-like. Stems 1–several, branching above, densely leafy, glabrous, or sparsely puberulous below. Leaves spirally arranged, decreasing in size on the upper stem, absent or few on the branches, sessile, mostly 1.5–4.5 cm. long, filiform, acute at the apex, margins ± revolute, glabrous. Capitula many, solitary and terminal, or several at the ends of branches; capitula stalks 1–6 cm. long, slender, few-bracteate. Involucres 4–9 × 5–8 mm., spreading to c. 12 mm. wide, broadly turbinate-campanulate. Phyllaries many, glabrous, subulate-pungent, margins ± broadly hyaline and somewhat ciliolate; the outer phyllaries from c. 2 mm. long and subulate-lanceolate, the inner to c. 6 mm. long and lanceolate to oblong-ovate. Receptacle slightly dome-shaped, paleaceous; paleae 6–7 mm. long, narrowly lanceolate, acute-pungent at the apex, glabrous. Corollas purple, 4–6 mm. long, narrowly funnel-shaped, glabrous. Achenes 2–2.5 mm. long, angular oblong-turbinate, obscurely 4–5-ribbed, setulose, with scattered large brown cells sometimes visible on the sides; pappus 2-seriate, the outer series of 10–15 brownish-purple narrowly lanceolate scales c. 1 mm. long, the inner of paleaceous setae fewer and slightly exceeding the outer scales.

Zambia. N: c. 30 km. WNW. of Kasama, Chishimba Falls, 31.iii.1984, *Brummitt & Chisumpa* 17051 (BR; C; K; LISC; MO; NDO; SRGH; UPS; WAG). **Malawi**. N: Nyika Plateau, vii.1896, *Whyte* 219 (K). Also in Tanzania. *Brachystegia* woodland usually on rocky hillsides.

20. ELEPHANTOPUS L.

Elephantopus L., Sp. Pl.: 814 (1753); Gen. Pl. ed. 5: 355 (1754). —Philipson in Journ. Bot. **76**: 299–305 (1938). —Wild & G.V. Pope in Kirkia **10**: 340–342 (1977). —C. Jeffrey in Kew Bull. **43**: 272–274 (1988).

Perennial ± hirsute herbs with annual stems from woody rootstocks; roots numerous, thong-like. Vegetative indumentum sparse to dense, consisting of patent to appressed stiff bristle-like hairs. Stems 1–several, leafy or scapiform, branching above. Leaves alternate, often basal, sessile, or narrowly attenuate and petiole-like before widening into a ± stem-clasping, ± stem-sheathing base; upper leaves grading into leaf-like bracts. Capitula homogamous, few-flowered, numerous, syncephalous in bracteate glomerules; glomerules terminal on stiff synflorescence branches, laxly corymbiform-cymose tending to be subscorpioid-cymose in arrangement. Involucres narrowly ovoid-cylindric; phyllaries few-seriate, ± cartilaginous with pungent apices, strigose to densely hispid or glabrescent. Receptacle small, plane. Corollas 5-lobed, asymmetric, more deeply cleft on one side. Anthers sagittate at the base. Style arms linear-terete, hairy. Achenes narrowly turbinate-cylindric, c. 10-ribbed, usually glandular between the ribs, setulose; pappus 1-seriate, setae 5–10 with bases broad scale-like and usually overlapping, gradually or abruptly tapering and subterete barbellate above.

A tropical genus of c. 32 species occurring in Asia, Australia, Madagascar and N. and S. America, with c. 9 species from Africa.

Stems scapiform, the lower leaves rosette-forming; pappus elements gradually tapering above a
    scale-like lower c. one half - - - - - - - - - - - 1. *scaber*
Stems leafy, the lower leaves separated by short internodes; pappus elements abruptly narrowed
    above a dilated short scale-like base - - - - - - - - 2. *mollis*

1. **Elephantopus scaber** L., Sp. Pl.: 814 (1753). Type from India.

An erect tough hirsute perennial herb to c. 130 cm. tall, with scapiform annual stems from a woody rootstock; roots numerous, thong-like. Stems 1–several, branching above, leafy below, densely hirsute, or ± strigose on upper stem and branches. Basal leaves mostly 7–62 × 1.5–10 cm., narrowly oblanceolate or ovate-elliptic, subacute or rounded at the apex, narrowly cuneate to long-attenuate below with the midrib becoming very narrowly winged or petiole-like, the base ± stem-clasping or shortly stem-sheathing, margins entire or sometimes serrulate-crenulate, lamina somewhat scabridulous to glabrescent with scattered patent 1–2 mm. long bristle-like hairs; cauline leaves 1–5, smaller than the basal leaves, sessile and ± stem-clasping, grading into ± conduplicate synflorescence bracts. Capitula sessile, crowded in glomerules 1.2–3 cm. in diam.; glomerules terminal on few–many stiff synflorescence branches 3–11 cm. long, or sometimes several glomerules ± scorpioidly arranged along a branch, glomerule-subtending bracts 3–5, ovate, leaf-like, ± equalling the glomerules in length. Florets c. 4 per capitulum. Involucres 7–10 × 2–3 mm., narrowly ovoid-cylindric. Phyllaries usually 2-seriate with c. 4 phyllaries in each series, oblong-lanceolate, cartilaginous with pungent apices and membranous margins, strigose or ± densely hispid where exposed, or glabrous; the inner phyllaries 7–9 mm. long, nearly twice as long as the outer. Corollas mauve, 4.5–7 mm. long, narrowly funnel-shaped, 5-lobed but more deeply cleft down one side, glandular at the lobe apices otherwise glabrous. Achenes 3–4 mm. long, narrowly turbinate-cylindric, c. 10-ribbed, setulose mainly on the ribs; pappus of 7–10 elements 3.5–6 mm. long, overlapping scale-like below and gradually tapering into a subterete barbellate seta.

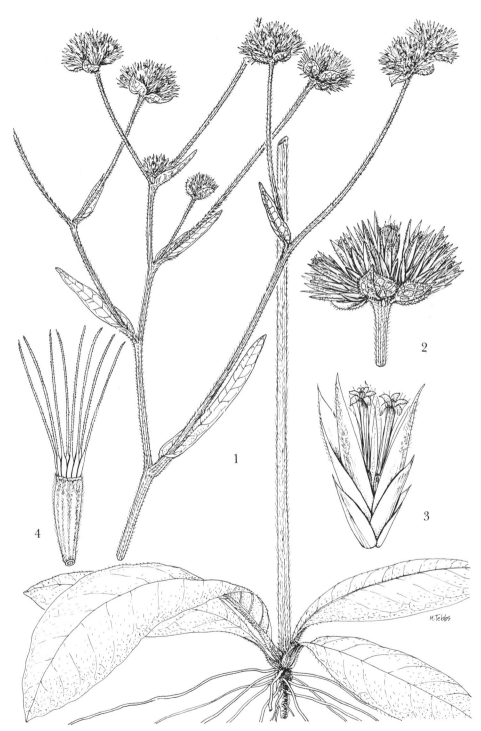

Tab. 32. ELEPHANTOPUS SCABER var. PLURISETUS. 1, habit (× ⅔), from *Biegel* 2869; 2, glomerule of capitula (× 1); 3, single capitulum from glomerule (× 4); 4, achene (× 6), 2–4 from *Richards* 5522. Drawn by M. Tebbs.

Subsp. **plurisetus** (O. Hoffm.) Philipson in Journ. Bot. **76**: 303 (1938). —Mendonça, Contrib. Conhec. Fl. Angol., **1** Compositae: 37 (1943). —Brenan in Mem. N.Y. Bot. Gard. **8**, 5: 462 (1954). —Wild & G.V. Pope in Kirkia **10**: 341 (1977). —C. Jeffrey in Kew Bull. **43**: 273 (1988). Type from Tanzania. *Elephantopus welwitschii* Hiern, Cat. Afr. Pl. Welw. **1**, 3: 540 (1898). Type from Angola.

Key to the varieties of subsp. plurisetus

1. Branches and upper stem densely patent-pilose or tomentose, the hairs 2–3 mm. long
   var. *hirsutus*
 - Branches and upper stem appressed-hispid or strigose, the hairs 1–2 mm. long   2
2. Plants robust; glomerules mostly more than 20 mm. in diam.; bracts subtending the synflorescence branches and glomerules mostly 10 mm. or more wide; phyllaries often densely hispid-sericeous, sometimes glabrescent in the lower part - - - - var. *argenteus*
 - Plants not robust; glomerules usually up to c. 15 mm. in diam., rarely more than 20 mm. in diam.; bracts subtending the synflorescence branches and glomerules mostly less than 10 mm. wide; phyllaries mostly sparsely strigose in the upper part otherwise glabrescent, sometimes wholly glabrescent - - - - - - - - - - - - - var. *plurisetus*

Var. **plurisetus** O. Hoffm. in Engl., Bot. Jahrb. **30**: 426 (1902). —Wild & G.V. Pope in Kirkia **10**: 341 (1977). —C. Jeffrey in Kew Bull. **43**: 273 (1988). TAB. **32**. Type as above.

Zambia. B: Zambezi (Balovale), 64 km. on Kabompo Road, *Drummond & Rutherford-Smith* 7362 (SRGH). N: Mbala, c. 1676 m., 13.iii.1950, *Bullock* 2638 (K). W: Solwezi Distr., Mutanda Bridge, 20.vi.1930, *Milne-Redhead* 551 (K). C: c. 16 km. E. of Lusaka on Great East Road, c. 1219 m., 19.iii.1955, *Best* 71 (K; LISC). E: c. 4.8 km. from Lundazi on road to Mzimba, 13.ii.1957, *Angus* 1520 (COI; K; SRGH). S: Kalomo, 21.i.1962, *Mitchell* 12/100 (K; NDO). **Zimbabwe**. N: Makonde Distr., Susuje River on Kildonan Road, 29.iii.1969, *Pope* 12 (COI; K; SRGH). C: Harare, Warren Hills, Tyndale Road, c. 1524 m., 5.ii.1969, *Biegel* 2869 (COI; K; LISC; M; SRGH). **Malawi**. N: c. 43 km. from Kagaza Gate on Nyika-Chisenga road, c. 1600 m., 21.iv.1986, *Philcox, Pope & Chisumpa* 10049 (BR; K; MAL; NDO). C: Lilongwe Distr., Dzalanyama For. Res., Chiunjiza Road, c. 5 km. SE. of Chaulongwe Falls, c. 1230 m., 22.iii.1970, *Brummitt* 9288 (K; SRGH). S: Liwonde, Chaoni between Chikala and Malosa Mts., c. 609 m., 20.iv.1937, *Lawrence* 376 (K). **Mozambique**. N Massangulo Mts., 13°55′S, 35°35′E, c. 1100 m., iii.1933, *Gomes e Sousa* 1353 (K). Z: Gurué, Mt. Currarre, R. Louissi, 11.ii.1964, *Torre & Paiva* 10537 (LISC; LMU).

Also in Tanzania, Zaire and Angola. Miombo woodland and wooded grassland, often on rocky hillsides.

Var. **argenteus** C. Jeffrey in Kew Bull. **43**: 273 (1988). Type from Tanzania.

Zambia. N: Mansa, 11.v.1964, *Fanshawe* 8604 (K; NDO). W: Kitwe Distr., Mwekera For. Training School, 28°21′E, 12°51′S, 21.iv.1989, *Pope, Radcl.-Sm. & Goyder* 2218 (BR; K; LISC; NDO).

Also in Tanzania. *Brachystegia*, "Chipya", woodland and grassland.

Var. **hirsutus** Philipson in Journ. Bot. **76**: 303 (1938). —Wild & G.V. Pope in Kirkia **10**: 342 (1977). —C. Jeffrey in Kew Bull. **43**: 273 (1988). Type from Tanzania.

Zambia. N: Isoka Distr., 12 km. from Mukombe School on Nakonde road, c. 1500 m., 22.iv.1986, *Philcox, Pope & Chisumpa* 10064 (BR; K; LISC). **Malawi**. N: Chitipa Distr., above Chisenga Village, towards Mafinga Hills, c. 1585 m., 12.vii.1970, *Brummitt* 12060 (K).

Also in Tanzania. Seasonally wet grassland, dambos.

Var. *brevisetus* Philipson is recorded from Uganda, Kenya, Tanzania, Rwanda, Burundi and Zaire, and is characterised by its phyllaries almost glabrous to somewhat densely pilose; its stems and capitula stalks ascending-pilose; and its pappus bristles 3.5–4.5 mm. long.

Subsp. *scaber*, with 5(6) pappus setae, is confined to tropical Asia.

2. **Elephantopus mollis** Kunth in Humb., Bonpl. & Kunth, Nov. Gen. et Sp. Pl., ed. fol. **4**: 20 (1818); ed. quatro **4**: 26 (1820). —Philipson in Journ. Bot. **76**: 303 (1938). —Mendonça, Contrib. Conhec. Fl. Angol., **1** Compositae: 38 (1943). —Adams in F.W.T.A. ed. 2, **2**: 268, fig. 258 (1963). —Wild & G.V. Pope in Kirkia **10**: 342, fig. 1 (1977). —C. Jeffrey in Kew Bull. **43**: 273 (1988). Type from Venezuela.

An erect hirsute perennial herb to c. 40 cm. tall, with annual stems from a small vertical woody rootstock. Stems mostly solitary, shortly branched above, striate, leafy throughout but with leaves usually more crowded below, sparsely to densely hispid with patent hairs 1–2 mm. long, ± strigose on upper stem and branches. Leaves usually subsessile, mostly 7–15(26) × 2–5(7) cm., obovate-oblanceolate to oblong-oblanceolate, narrowly cuneate or sometimes the midrib narrowly winged and petiole-like, ± stem-clasping and ± stem-sheathing at the base, margins ± crenulate with callose-tipped teeth, scabridulous with scattered bristle-like hairs on both surfaces, the hairs usually most numerous on the prominent midrib and venation beneath, lamina somewhat chartaceous sometimes bullate in older leaves, minutely glandular beneath; upper cauline leaves grading into foliaceous bracts, bracts not conduplicate. Capitula sessile, crowded in glomerules 1–2 cm. in diam.; glomerules terminal on few–many short synflorescence branches, or sometimes several glomerules ± scorpioidly arranged along a branch, glomerule-subtending bracts 2–4, ovate, leaf-like, ± equalling the glomerules in length. Florets c. 4 per capitulum. Involucres to c. 10 × 2–3 mm., narrowly ovoid-cylindric; phyllaries few-seriate, ovate to oblong-lanceolate, chartaceous to ± cartilaginous with pungent apices, margins membranous, sparsely hispid in the upper part or glabrescent, the inner 4 phyllaries 7–9 mm. long and nearly twice as long as the outer. Corollas white, 5–6 mm. long, consisting of a slender tube and a short campanulate deeply-lobed limb, more deeply cleft on one side, glabrous. Achenes 3–4 mm. long, narrowly turbinate-cylindric, c. 10-ribbed, setulose; pappus elements usually 5, c. 4 mm. long, very shortly scale-like at the base and abruptly tapered into a long barbellate apical seta.

Zambia. W: Mwinilunga Distr., Kasombosombo River, 14.vi.1963, *Edwards* 777 (K; SRGH).
Also in Uganda, Tanzania, Zaire and Angola; a native of the moist neotropics widely naturalised in the Old World. Marshy ground fringing swamp forest (mushitu).

Tribe 4. LACTUCEAE Cass.

Lactuceae Cass. in Journ. Phys. **88**: 151 (1819). —C. Jeffrey in Kew Bull. **18**: 427–486 (1966). —A.S. Tomb, *Lactuceae*, Syst. Rev. in Heywood et al., Biol. & Chem. Comp.: 1067–1079 (1977) "Cichorieae".

Annual, perennial or biennial herbs (occasionally subshrubs) with milky juice. Leaves alternate or radical, entire, lobed, runcinate or dentate-pinnatisect, sometimes spinulose. Capitula homogamous; florets perfect, ligulate. Receptacle mostly epaleate, paleate in *Hypochaeris*. Corollas yellow, rarely blue, purplish or white; ligules 5-toothed. Anther-bases sagittate, auricles acute or setaceous acuminate, not tailed. Style branches usually slender, apically tapered or obtuse, inner surfaces usually flattened, papillose. Achenes subcylindric or ellipsoid-fusiform and ± laterally compressed, often rostrate (beaked) above, usually variously ribbed, surfaces smooth to rugose or ± tuberculate, glabrous or pubescent; pappus 1–several-seriate, of setae or setae intermixed with down-like hairs, sometimes of plumose bristles or scales (or wanting). Pollen echinolophate or echinate, usually tricolporate. n = 9 is the most frequently recorded haploid chromosome number, but also recorded are n = 11, 10, 8, 7, 6, 5, 4, 3.

A tribe of about 71 genera, mostly in the northern hemisphere. The tribal name *Lactuceae* has priority over the more often used name *Cichorieae*.

Key to the genera after C. Jeffrey in Kew Bull. **18**: 430 (1966).

1. Pappus of scales up to 2 mm. long; capitula stalks conspicuously expanded upwards
   23. Cichorium
   – Pappus of hairs or setae at least 4 mm. long; capitula stalks not conspicuously expanded
   upwards - - - - - - - - - - - - - 2
2. Pappus setae plumose, or at least the inner ones plumose; receptacular paleae present
   22. Hypochaeris
   – Pappus setae usually barbellate-setose, not plumose; receptacle epaleate - - - 3
3. Achenes 2–3.5 mm. long, obconic, truncate above; pappus of comparatively few coarse scabrid
   setae; ligules drying greenish - - - - - - - - - **-21. Tolpis**
   – Achenes, pappus and ligules not as above - - - - - - - - 4
4. Achenes tuberculate in the upper part, with a long slender beak; scapes 1-capitulate, hollow
   26. Taraxacum
   – Achenes and scapes not as combined above, or plants not scapigerous - - - 5
5. Pappus dimorphic, of setae and fine down-like hairs intermixed - - - - 6
   – Pappus of setae only, monomorphic, rarely with an outer ring of very short setae 7
6. Achenes compressed, not beaked; involucres of mature capitula usually more than 8 mm.
   in diam. - - - - - - - - - - - - - **28. Sonchus**
   – Achenes subcylindrical, beaked; involucres less than 5 mm. in diam. *Launaea intybacea*

7. Achenes truncate, without ascending hairs, or without ascending rough projections on the ribs or margins - - - - - - - - - - - - - **27. Launaea**
- Achenes beaked or at least somewhat attenuate above, with ascending hairs or rough projections on the ribs or at least on the margins when mature - - - - - - - - 8
8. Achenes ± compressed and unequal-ribbed, with ascending evident or submicroscopic hairs on the ribs or at least the margins; involucres mostly cylindrical; plants sometimes acaulescent or scandent; flowers yellow, blue, purple or white - - - - - - **24. Lactuca**
- Achenes ± terete and equal-ribbed, rough with ascending projections on the ribs; involucres usually obconic; plants never acaulescent or scandent; flowers yellow **25. Crepis**

21. TOLPIS Adans.

Tolpis Adans., Fam. Pl. **2**: 112 (1763). —C. Jeffrey in Kew Bull. **18**: 432 (1966).

Perennial or annual herbs, usually scapigerous, ± pubescent. Leaves radical, or cauline and alternate, entire to dentate (lobed). Capitula few–numerous in lax panicles. Receptacles epaleate, alveolate; alveolae ± fimbriate. Phyllaries 2–3-seriate. Corollas pubescent outside, ligules yellow or the innermost purplish-brown, usually becoming greenish when dry. Anthers sagittate, auricles acute or shortly setaceous-acuminate. Style branches short, obtuse; collecting hairs short. Achenes obconic, truncate at the apex not narrowed or beaked, finely-ribbed, glabrous; pappus setose, mostly 2-seriate, persistent.

A genus of about 20 species, mainly in the Mediterranean region.

Leaves radical, 4–5(16) cm. long, ± pubescent; phyllaries 6–8 × 0.5 mm., linear-lanceolate
1. *capensis*
Leaves cauline, 7–15(19.5) cm. long, scabridulous; phyllaries 8–10 × 1–1.25 mm., narrowly lanceolate
2. *mbalensis*

1. **Tolpis capensis** (L.) Sch. Bip. in Bonplandia **9**: 172 (1861). —C. Jeffrey in Kew Bull. **18**: 432 (1966). —Guillarmod, Fl. Lesotho: 303 (1971). —Hilliard in Ross, Fl. Natal: 377 (1972). —Agnew, Upland Kenya Wild Fls.: 494 (1974). —Hilliard, Comp. Natal: 598 (1977). —Lawalrée in Fl. Rwanda, Spermat. **3**: 684, fig. 215(1) (1985). —Lawalrée, Dethier & Gilissen in Fl. Afr. Centr., Compos., Cichorioideae: 6, pl. 1 (1986). TAB. **33**. Type from South Africa (Cape Province).
    Hieracium capense L., Pl. Afr. Rar.: 17 (1760). —Harv. in Harv. & Sond., F.C. **3**: 529 (1865). —Humbert., Comp. Madag.: 158 (1923); Fl. Madag., Comp., 189: 893, t. CLXVI (1963). Type as above.
    Hieracium madagascariense DC., Prodr. **7**, 1: 218 (1838). Type from Madagascar.
    Tolpis madagascariensis (DC.) Sch. Bip. in Bonplandia **9**: 172 (1861). Type as for *Hieracium madagascariense*.
    Crepis ephemera Hiern, Cat. Afr. Pl. Welw. **1**, 3: 616 (1898). —Mendonça, Contrib. Conhec. Fl. Angol., **1** Compositae: 148 (1943). Types from Angola.
    Crepis ephemerioides S. Moore in Journ. Bot. **54**, Suppl. 2, Gamopet.: 286 (1916). Type: Zimbabwe, Mt. Chirinda, *Swynnerton* s.n. (BM, holotype).
    Tolpis ephemera R.E. Fr. in Acta Hort. Berg. **8**: 271, t. 1 (1925). Type from Kenya.

An erect scapose perennial herb from a stout semi-woody taproot; rootcrowns lanate. Scapes up to c. 80 cm. tall, solitary or several, branching above, slender, finely ribbed, ± sparsely pilose below, glabrous above. Leaves 2–8, sub-rosulate, rarely also 1–2 cauline, mostly 4–5 × 1–1.5 cm., oblanceolate to narrowly elliptic, sometimes up to 15 cm. long and narrowly oblanceolate, or up to 3.5 cm. wide and narrowly obovate, apex acute to rounded, base narrowly cuneate to attenuate, margins remotely denticulate, lamina glabrous to sparsely pubescent. Capitula many, laxly corymbiform cymose or paniculate, stalks up to 26 cm. long. Involucres up to c. 11 × 7 mm. in flower, obconic-campanulate, cylindric in bud with hooded phyllaries, subtended by 2–21 subulate bracteoles which extend down the capitulum stalk. Phyllaries 2-seriate, sub-equal, 6–8(10) mm. long, linear to narrowly lanceolate, attenuate at the apex, pubescent to tomentose and papillose-glandular outside, the outer phyllaries often with a thickened ± verruculose midrib, phyllaries changing abruptly into bracteoles. Florets numerous, corollas 9–10.5 mm. long, corolla tube c. 5 mm. long, cylindrical, crisped-pilose outside; ligule erect yellow, often drying greenish, equalling or exceeding the tube in length, strap-shaped, pubescent on the back, apical lobes papillose, ligules often longer and more deeply lobed in the outer florets. Achenes 2–3.5 mm. long, narrowly obconic-turbinate, 4–5-angled, finely 12–15-ribbed; pappus of 20–30 persistent rigid barbellate setae 5–7 mm. long, often with an outer whorl of minute setae.

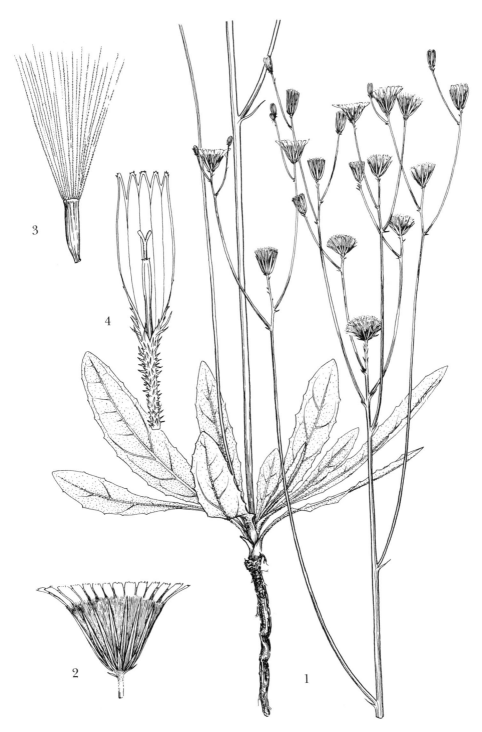

Tab. 33. TOLPIS CAPENSIS. 1, habit (× ⅔); 2, capitulum (× 2); 3, achene (× 8); 4, corolla (× 10), 1–4 from *Brummitt* 9742. Drawn by Eleanor Catherine.

Zambia. N: Mbala Distr., Fwambo area, c. 1680 m., 11.ix.1956, *Richards* 6151 (K). W: Solwezi to Kasempa, 30.ix.1947, *Greenway & Brenan* 8135 (K). E: Nyika Plateau, 2250 m., 10.xi.1967, *Richards* 22462 (K). **Zimbabwe**. N: Guruve (Sipolilo), Nyamunyeche Estate, 19.x.1978, *Nyariri* 434 (SRGH). W: Matobo Distr., Besna Kobila Farm, c. 1500 m., i.1961, *Miller* 7678 (K; SRGH). C: Harare, 1.i.1937, *Eyles* 8897 (K). E: Nyanga (Inyanga), c. 2000 m., 18.x.1946, *Wild* 1372 (K; SRGH). S: Mberengwe Distr., Mt. Buhwa, ridge c. 2 km. E. of summit, c. 1500 m., 30.x.1973, *Biegel, Pope & Gosden* 4332 (K; SRGH). **Malawi**. N: Rumphi Distr., Nyika Plateau, c. 1870 m., 17.x.1975, *Pawek* 10228 (K; MAL; MO; SRGH; UC). S: Mulanje Mt., 1 km. S. of Tuchila Hut, c. 1930 m., 8.iv.1970, *Brummitt* 9742 (K; SRGH). **Mozambique**. Z: Gurué, near Nuirre R., *Mendonça* 2248 (LISC). MS: Beira Distr., Mt. Gorongosa summit area, iii.1972, *Tinley* 2404 (K; LISC; SRGH).

Also in the Sudan, Uganda, Rwanda, Burundi, Kenya, Tanzania, Zaire, Angola, South Africa and Madagascar. An infrequent pyrophyte of grassland, seasonally moist areas, and open deciduous woodland.

2. **Tolpis mbalensis** G.V. Pope in Kew Bull. **39**: 167 (1984). Type: Zambia, Mbala Distr., Kambole Escarpment, *Richards* 24540 (K, holotype).

An erect perennial herb up to c. 100 cm. tall from a semi-woody taproot; root-crown lanate. Stems annual, solitary, often flexuous, simple below and branching above, striate, ± densely pilose below and sparsely so or glabrescent above. Leaves cauline (4)7–15.5(19.5) × (0.5)1–3.5(5) cm., narrowly oblanceolate to narrowly obovate, upper leaves becoming linear, margins acute to ± attenuate, bases attenuate, margins remotely dentate to denticulate, both surfaces pilose to somewhat strigose and scabridulous, rarely glabrescent. Capitula few to many, laxly corymbiform cymose or paniculate, capitula stalks up to c. 40 cm. long, ± flexuous, with 2–23 subulate bracts. Involucres up to c. 14 × 10 mm., obconic; phyllaries 2-seriate subequal 8–10 mm. long, narrowly lanceolate, attenuate to the apex, araneose-lanate and papillose-glandular outside; outer phyllaries often with a thickened verruculose midrib, changing ± abruptly into the bracteoles. Florets numerous; corolla tube 5–7 mm. long, cylindrical, crisped-pubescent on the outside; ligule 7–10 mm. long, erect, strap-shaped, strigose to pilose on the outside, drying greenish, apical lobes up to 2 mm. long papillose. Achenes 2–3.5 mm. long, obconic-turbinate, truncate above, 4–5-angled, finely 12–15-ribbed, scabridulous or glabrous; pappus of numerous persistent barbellate setae 5–7 mm. long, equalling or slightly exceeding the involucre in length, sometimes with an outer series of minute setae.

Zambia. N: c. 3 km. from Mbala, 1700 m., 24.iv.1959, *McCallum-Webster* 626 (K; SRGH). **Malawi**. N: Nyika Plateau, Chisanga Falls, 10°32'S, 33°42'E, 21.v.1989, *Pope, Radcl.-Sm. & Goyder* 2311 (K; LISC).

So far recorded only from N. Zambia and N. Malawi. A pyrophyte of short sub-montane grassland, high rainfall plateau grasslands and woodlands, in sandy and rocky soils and escarpments.

22. HYPOCHAERIS L.

Hypochaeris L., Sp. Pl. **2**: 810 (1753); Gen. Pl. ed. 5: 352 (1754). —C. Jeffrey in Kew Bull. **18**: 432 (1966).

Perennial or annual, usually scapose herbs. Stems solitary or few, often branched. Leaves mostly basal or radical, entire or sinuate-dentate to pinnatifid. Capitula solitary or several in sparingly-branched corymbiform cymes. Receptacles flat, paleate. Involucres sub-cylindrical or campanulate, phyllaries several-seriate. Corollas yellow or white; anthers sagittate; style branches long, obtuse, sweeping-hairs short (barbellate). Achenes cylindric-ellipsoid, at least the inner ones beaked, ribbed often scabridulous; pappus 1–2-seriate of barbellate or plumose setae.

A genus of 50–100 species, native to Europe, the Near East, N. Africa and South America. One species introduced as a weed in the Flora Zambesiaca area.

Hypochaeris radicata L., Sp. Pl. **2**: 811 (1753). —Ross-Craig, Draw. Brit. Pl.: 18, pl. 21 (1963). —Humbert, Fl. Madag., Comp., III: 872 (1963). —Henderson & Anderson, Common Weeds in S. Afr.: 378, t. 188 (1966). —Guillarmod, Fl. Lesotho: 303 (1971). —Hilliard in Ross, Fl. Natal: 378 (1972). —Hilliard, Comp. Natal: 599 (1977). TAB. **34**. Type from Europe.

A scapose perennial herb up to c. 60 cm. tall, from a stout semi-woody taproot. Leaves numerous in a basal rosette, (4)5–12(20) × 1–4 cm., oblanceolate to narrowly elliptic in

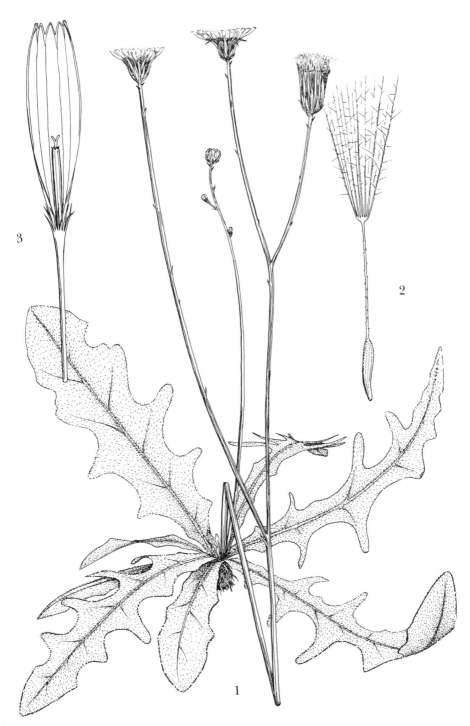

Tab. 34. **HYPOCHAERIS RADICATA.** 1, habit (×⅔); 2, achene (× 4), 1 & 2 from *Chase* 8487; 3, corolla (× 6), from *Goldsmith* 158/68. Drawn by Eleanor Catherine.

outline, sinuate-dentate to pinnatilobed, obtuse to rounded at the apex, attenuate below, ± densely hispid. Scapes 1 to several, usually branched, striate, glabrous or hispid; branches long ascending, remotely bracteate. Capitula solitary erect, up to c. 2.5 cm. long, obconic or deeply cup-shaped. Phyllaries 3–4-seriate, the outer c. 4 × 1.5 mm., the inner to c. 18 × 2.5 mm., lanceolate, acute, ciliate above, setose-papillose along the exposed midribs otherwise glabrous, eventually reflexed. Receptacular paleae up to c. 22 mm. long, conduplicate and enclosing the achenes, subulate above, apex drying to a brownish arista exserted above the mature pappus, glabrous. Florets numerous, corollas c. 15 mm. long; corolla tube 5–6 mm. long, somewhat 5-ribbed and with an incomplete ciliate fringe where the tube splits to form the ligule, otherwise glabrous; ligule ± erect 8–9 × 2.5 mm., strap-shaped, exserted above the involucre, yellow sometimes with a greenish or reddish stripe outside, glabrous. Achenes all with long slender beaks; achene body 4–5 mm. long, cylindric-fusiform, finely c. 14-ribbed, scabridulous, tapering ± abruptly into a 4.5–7 mm. long beak; beaks of the outer achenes shorter than the inner; pappus 2-seriate, the outer pappus of short barbellate setae, the inner of plumose setae to c. 9 mm. long.

Zimbabwe. E: Chimanimani Distr., Charter Forest Estate, 1000 m., x.1968, *Goldsmith* 158/68 (COI; K; SRGH). **Mozambique**. MS: Himalaya Mts., Carvalho's Estate, 28.xi.1966, *Dale* SKF 438 (K; SRGH).

A cosmopolitan weed, recorded from sub-montane localities in the Flora Zambesiaca area, usually in disturbed areas such as roadsides.

23. CICHORIUM L.

Cichorium L., Sp. Pl. **2**: 813 (1753); Gen. Pl., ed. 5: 354 (1754).

Erect annual or perennial herbs with stout taproots. Stems usually solitary, divaricately branched. Leaves alternate, pinnatifid to coarsely dentate. Capitula numerous, in small clusters spicately arranged along branches, or terminal on short branches. Involucres cylindrical, phyllaries 2-seriate. Receptacles ± flat, epaleate. Corollas blue, rarely pink or white; anthers sagittate at the base; style branches long slender obtuse, with long sweeping-hairs (hispid). Achenes obconic, obscurely 5-angled, smooth, glabrous; pappus of minute scales.

A genus of approximately 10 species, native to Europe, W. Asia and N. Africa, one species is introduced in the Flora Zambesiaca area.

Cichorium intybus L., Sp. Pl. **2**: 813 (1753). —Harv. in Harv. & Sond., F.C. **3**: 609 (1865). —Ross-Craig, Draw. Brit. Pl. 17, pl. 26 (1962). —Humbert, Fl. Madag., Comp., III: 871 (1963). —C. Jeffrey in Kew Bull. **18**: 433 (1966). —Guillarmod, Fl. Lesotho: 303 (1971). —Biegel & Mavi in Wild, Bot. Dict. Afr. & English Plant Names: 121 (1972). —Hilliard in Ross, Fl. Natal: 377 (1972). —Sell in Tutin et al., Fl. Europ. **4**: 304 (1976). —Hilliard, Comp. Natal: 597 (1977). —Lawalrée, Dethier & Gilissen in Fl. Afr. Centr., Compos., Cichorioideae: 4 (1986). TAB. **35**. Type from Europe.

A stiffly erect ± robust perennial herb up to c. 120 cm. tall, from a long stout taproot. Stems annual, usually divaricately branched above, striate, densely hispid-setose to glabrescent. Basal leaves 7–30 × 1–12 cm., oblanceolate in outline, runcinate-pinnatifid, obtuse-acute at the apex, semi-amplexicaul below, dentate to remotely denticulate on the margins, hispid or pilose to glabrescent; cauline leaves smaller, mostly oblong-lanceolate and undivided, sessile and amplexicaul, dentate to entire, hispid or glabrescent. Capitula 2.5–3 cm. in diam. measured across the expanded ligules, 2–4 sessile in axillary clusters on the usually very long ± flexuous patent-ascending branches, the clusters subtended by large foliaceous bracts, or capitula solitary and terminal on apically thickened stalks up to c. 9(14) cm. long. Involucres to c. 11(14) mm. long, cylindric. Phyllaries 2-seriate, coriaceous, glandular-ciliate on the margins and midrib, or glabrous; outer phyllaries up to 8 mm. long, narrowly ovate, patelliform-thickened below, apices acute becoming recurved; inner phyllaries up to 11(14) mm. long, lanceolate, erect, apically pubescent. Corollas c. 18 mm. long; the lower portion c. 2 mm. long, cylindric, pubescent or pilose-hairy above; ligule c. 16 mm. long, widening to c. 4 mm. at about the middle, lorate, bright blue rarely pink or white, pubescent or glabrous outside, densely pubescent on the apical lobes, ligules spreading. Achenes c. 2.5 mm. long, obconic, truncate above, somewhat 5-angled, rugose-scabridulous; pappus several-seriate of numerous persistent short scales up to c. one quarter of the length of the achene.

Tab. 35. CICHORIUM INTYBUS. 1, habit (×$\frac{1}{15}$), from *Hubbard* 11476; 2, flowering axis (×$\frac{2}{3}$); 3, achene (× 15); 4, phyllary, outer surface (× 4), 2–4 from *Whellan* 2099. Drawn by Eleanor Catherine.

Zimbabwe. C: Harare, Arcadia, ii.1964, *Whellan* 2099 (K; SRGH). E: Mutare, near railway station, 1100 m., 14.xi.1952, *Chase* 4715 (BM; COI; K; LISC; SRGH).

Recorded in the Flora Zambesiaca area as a cultivated herb and also as a garden escape, occurring on roadsides and in disturbed areas.

"Chicory", native of northern Europe. Cultivated as a garden ornamental, or as a medicinal herb or salad plant, an extract of the root is used as a substitute for, or an additive to, coffee.

24. LACTUCA L.

Lactuca L., Sp. Pl. **2**: 795 (1753); Gen. Pl., ed. 5: 348 (1754). —C. Jeffrey in Kew Bull. **18**: 436–457, figs. 2–8 (1966).

Annual, perennial or biennial herbs, erect often scapigerous, sometimes scandent. Stems mostly solitary and branched above, glabrous, crispate-pubescent or setulose, rarely spinescent. Leaves radical, or cauline and alternate, often auriculate-amplexicaul, undivided to deeply runcinate-pinnatisect, or rarely pinnately compound, margins entire or remotely denticulate to coarsely dentate, glabrous or crispate-pubescent to setulose, rarely spinose-setose on midribs beneath, often glaucous. Capitula few to numerous, corymbose or thyrsoid-paniculate, sometimes solitary. Phyllaries 3–5-seriate, elongating in the fruiting stage. Receptacle epaleate, alveolate; alveolae ± fimbriate. Corollas yellow whitish purplish or blue; anthers sagittate at the base, auricles acute or shortly setaceous-acuminate; style branches long with conspicuous sweeping hairs. Achenes homomorphic, ± ellipsoidal, ± compressed, apex constricted and produced as a distinct filamentous beak, or beak absent; achene body ribbed, minutely rugose-scabridulous, ascending-hispidulous; pappus several-seriate of uniform white or straw-coloured, barbellate setae, rarely with an outer ring of minute setae.

A genus of about 100 species mainly from temperate Eurasia and Africa. A few species are widespread weeds, while *L. sativa* L., known only in its cultivated form, is the salad plant "lettuce".

Pterocypsela Shih in Acta Phytotax. Sin. **26**: 385 (1988). Type: *Lactuca indica* L.

The genus *Pterocypsela* Shih in Asia is clearly distinguished from *Lactuca* (in the strict sense) by its dark achenes broadly winged on the margins, 1–3-ribbed on the faces and with a short thick rarely filiform beak. In the African species, however, achene variation is such that *Pterocypsela* cannot be satisfactorily separated, and for the time being at least the new combination *Pterocypsela indica* (L.) Shih, made for *Lactuca indica* L. cannot be used. Until the position of the African species is elucidated the traditional treatment of *Lactuca* is maintained here.

1. Plants not scandent nor scrambling - - - - - - - - - - 2
– Plants scandent or scrambling - - - - - - - - - - 18
2. Capitula arranged in a tight cushion immediately above the densely woolly apex of the taproot - - - - - - - - - - - - - - 1. *lasiorhiza*
– Capitula borne on evident stems, or on scapes usually more than 10 cm. tall; taproot crown not densely woolly - - - - - - - - - - - - - - 3
3. Capitula large; involucres at time of flowering at least 14 × 5 mm., pappus setae c. 11 mm. long 10. *imbricata*
– Capitula small; involucres at time of flowering not exceeding 13 × 5 mm., pappus setae 4–8 mm. long - - - - - - - - - - - - - - - 4
4. Achene beak absent or hardly developed, at the most a stalk less than 1 mm. long (TAB. **36** figs. B, C, D); leaves usually simple sometimes lobed - - - - - - - 5
– Achenes with a distinct ± filiform beak 1 mm. or more long (TAB. **36** fig. J); leaves, at least the lowermost, variously lobed to pinnatipartite - - - - - - - 10
5. Achenes 2–5-ribbed on each face, somewhat laterally compressed or subfusiform-ellipsoid 6
– Achenes usually 1-ribbed on each face, strongly flattened-ellipsoid with thickened margins 8
6. Stem leaves present, well developed; achenes narrowly ellipsoid 4. *setosa*
– Stem leaves usually absent, if present then reduced and bract-like, rarely well developed; achenes broadly ellipsoid - - - - - - - - - - - - 7
7. Stems glandular-setose; setae 1.5–4 mm. long, well developed, purplish; leaves appearing with the flowers - - - - - - - - - - - - - 2. *calophylla*
– Stems glabrous or ± setulose below; leaves appearing after the flowers 3. *mwinilungensis*
8. Plants ± glabrous; lower leaves seldom exceeding 11 cm. in length; achenes reddish-brown, conspicuously hairy; mostly a herb of miombo woodland - - - - - 8. *homblei*
– Plants glandular-setose, at least on leaves and lower stems; lower leaves up to 25 cm. long; achenes dark-brown to black, not conspicuously hairy; mostly herbs of grassland or wooded grassland - - - - - - - - - - - - - - - 9

9. Capitula numerous, mostly stalked, arranged in large panicles; involucres at time of flowering up to 8 mm. long increasing to c. 10 mm. in fruiting capitula; phyllaries often with a sub-apical horn-like papilla - - - - - - - - - - - - - *6. schweinfurthii*
- Capitula in few–many subsessile clusters of 2–4 along a simple or sometimes branched axis; involucres at time of flowering 9–10 mm. long increasing to c. 13 mm. long in fruiting capitula; phyllaries without apical papillae - - - - - - - - - - *7. longispicata*

10. Perennial herbs with variously swollen semi-woody taproots; florets c. 5 per capitulum (up to c. 15 in *L. inermis*) - - - - - - - - - - - - - - - - - 11
- Annual or biennial herbs without swollen taproots; florets 10 or more per capitulum 14

11. Subscapose herbs; flowers appearing before the leaves - - - - - *9. praecox*
- Herbs with leafy stems; flowers and leaves appearing together - - - - 12

12. Stem leaves and bracts pointed-auriculate at the base; outer phyllaries ± lanceolate; corollas blue purplish or white less often yellow, pappus yellowish-white or tawny *12. inermis*
- Stem leaves and bracts exauriculate; outer phyllaries ± ovate; corollas yellow, pappus white 13

13. Stem leaves linear-lanceolate, usually 2–4-lobed; synflorescence paniculate *11. zambeziaca*
- Stem leaves oblanceolate-elliptic, not lobed; capitula in subsessile clusters along a simple, or sometimes branched axis - - - - - - - - - - - - *7. longispicata*

14. Achene beak shorter than body of achene; achenes more than 1 mm. wide, dark-brown to black - - - - - - - - - - - - - - - - - - - 15
- Achene beak exceeding the body in length; achenes up to 1 mm. wide, pale reddish-brown 16

15. Corollas blue; lower leaves sinuate-runcinate with broadly triangular apical lobes; florets c. 10 per capitulum - - - - - - - - - - - - - *5. schulzeana*
- Corollas yellow; lower leaves pinnatifid with lanceolate apical lobes; florets c. 20 or more per capitulum - - - - - - - - - - - - - - - *13. indica*

16. Stem leaves patent; leaves and stems espinulose; synflorescences corymbose; phyllaries erect in fruit; cultivated - - - - - - - - - - - - - - - *L. sativa*
- Stem leaves held vertically; spinulose on leaf midribs beneath and on lower part of stems; synflorescence a pyramidal panicle; phyllaries patent or deflexed in fruit - - - 17

17. Stem leaves narrowly lanceolate, the lowermost sometimes pinnately lobed; lobes narrowly triangular becoming recurved, up to 4 × 0.7 cm. - - - - - - *L. dregeana*
- Stem leaves ± broadly elliptic in outline and runcinate; lobes mostly recurved-oblong 1–3 × 0.5–2 cm., or leaves undivided and oblong-oblanceolate - - - - - *14. serriola*

18. Leaves pinnately compound with at least the lateral lobes on short petiolules; outer phyllaries broadly ovate; involucres conical in fruiting capitula - - - - - *16. paradoxa*
- Leaves simple, pinnatipartite with 2–4(6) ± recurved lateral lobes; lobes usually widest at the bases, upper leaves subentire; phyllaries lanceolate; involucres subcylindric to narrowly conical in fruiting capitula - - - - - - - - - - - - - - - 19

19. Terminal lobe of leaf broadly ovate-sagittate; achenes broadly elliptic, compressed, middle ribs wing-like - - - - - - - - - - - - - *17. glandulifera*
- Terminal lobe of leaf lanceolate-sagittate; achenes ellipsoid-subfusiform, ribs not wing-like *15. attenuata*

1. **Lactuca lasiorhiza** (O. Hoffm.) C. Jeffrey in Kew Bull. **18**: 438, fig. 2(2), (1966). —Lawalrée, Dethier & Gilissen in Fl. Afr. Centr., Compos., Cichorioideae: 48 (1986). TAB. **36** fig. A. Type from Tanzania.

    Sonchus lasiorhizus O. Hoffm. in Engl., Bot. Jahrb. **30**: 444 (1901). —R.E. Fr. in Acta Hort. Berg. **8**: 117 (1924). Type as above.

    Sonchus ledermannii R.E. Fr. in Acta Hort. Berg. **8**: 118 (1924). —Adams in F.W.T.A., ed. 2, **2**: 296 (1963). Type from Cameroon.

    Sonchus quercifolius Phillipson in Journ. Bot. **77**: 88 (1939). Type from Central African Republic.

    Launaea ledermannii (R.E. Fr.) L. Boulos in Bot. Not. **115**: 59 (1962). Type as for *Sonchus ledermannii*.

An acaulescent perennial herb rarely more than 7 cm. tall; taproot large, semi-woody, up to c. 3 cm. in diam. at the apex; rootcrown very densely brown sericeous-lanate with hairs c. 2 cm. long. Leaves radical, appearing after flowering, c. 9(25) cm. long, mostly oblanceolate in outline, acute at the apex, attenuate to cuneate below, remotely denticulate to runcinate-pinnatifid, ± densely pilose to glabrescent, midribs lanate below. Capitula erect, few to numerous, densely aggregated in a pulviniform synflorescence up to 10 cm. in diam.; flowering stalks congested, usually shorter than the capitula but occasionally up to c. 3.5 cm. long, partly or completely enveloped in the rootcrown hairs, simple or shortly branched, bracteate, glabrous to pilose, lanate in the axils; bracts 4–12 mm. long, lanceolate. Involucre 11–18 × 4–8 mm., cylindric to narrowly campanulate. Phyllaries 3–4-seriate, from c. 4 mm. long outside to c. 15 mm. long inside, lanceolate acute to acuminate, margins hyaline entire or remotely serrulate. Corollas c. 16 mm. long; tubes narrowly-cylindric, rugulose, puberulous above; ligules yellow, sometimes becoming reddish-purple outside, c. 10 mm. long, lorate, erect, glabrous, apical lobes densely papillose outside. Achenes dark-brown c. 9 mm. long including the beak, fusiform, apically attenuate into a beak c. 3 mm. long, narrowly-ribbed, scabridulous, glabrous; pappus copious several-seriate, of numerous barbellate setae up to 10 mm. long.

Zambia. N: Mbala Distr., Lumi R., c. 1500 m., 17.viii.1956, *Richards* 5846 (K). W: Chingola Distr., Luano Forest, 21.viii.1969, *Mutimushi* 3530 (K). E: Nyika Plateau, c. 3 km. SW. of Rest House, 2150 m., 21.x.1958, *Robson* 230 (K; LISC; SRGH). **Zimbabwe**. E: Chimanimani Distr., Musapa Mt., 1.x.1966, *Grosvenor* 258 (K; SRGH). **Malawi**. N: Nyika Plateau, road to Kasaramba, c. 2250 m., 17.xi.1967, *Richards* 22581 (K). C: Dedza Distr., Chongoni For. Res., 18.ix.1968, *Salubeni* 1143 (COI; K; SRGH). S: Mt. Mulanje, c. 1820 m., 1896, *McClounie* s.n. (K).

Also in Ghana, Nigeria, Cameroon, Central African Republic, Zaire and Tanzania. A pyrophyte of montane grassland, high rainfall woodland or seasonally waterlogged grassland (dambos), flowering after grass has burnt.

2. **Lactuca calophylla** C. Jeffrey in Kew Bull. **18**: 439, figs. 2(4), 4, (1966). TAB. **36** fig. B. Type from Tanzania.

A subscapose perennial herb up to c. 75 cm. tall, from a large semi-woody taproot. Flowering stalks erect, somewhat robust, usually solitary, mostly branching above, sparsely to densely glandular-setose; setae 1.5–4 mm. long, patent soft purple. Leaves 1–6 ± prostrate in a basal rosette, arising from a sheath of leaf bases persisting from the previous season, 5.5–17.5(22) × 2–12 cm., elliptic to rotund rarely oblate; apices obtuse, often mucronate in young leaves; bases rounded to cuneate with the midrib winged petiole-like; margins denticulate, often ciliate between the teeth; upper surface green glabrous or sparsely purple-setose particularly along and near the margins and midrib, lower surface purplish or green with a prominent purple reticulation, purple-setose on midrib and main nerves, occasionally also puberulent; cauline leaves absent, or if present reduced and bract-like. Synflorescence up to c. 25 cm. long, narrowly cylindric-pyramidal, bracteate; bracts up to c. 2 mm. long, ovate. Capitula numerous, stalked; involucres cylindric, 8–10 × 1.75–2 mm. at time of flowering, increasing to c. 12 × 3 mm. in fruiting capitula. Phyllaries 2–3-seriate, green, minutely tuberculate-rugose outside and puberulous above; the outermost ovate, usually less than half the length of the inner; the innermost up to c. 12 mm. long, lanceolate-lorate, obtuse. Florets 4–6 per capitulum; corollas yellow, 11.5–16 mm. long, tube cylindric, ligule 9–12 × 2.5–4 mm. lorate with lobes 1–2 mm. long and apically papillose. Achenes reddish-brown, 4.5–5.5 mm. long, ellipsoid, not or hardly beaked, broadly 2–3-ribbed with 2–3 smaller ribs between, minutely strigose; pappus several-seriate of barbellate setae 4–7 mm. long.

Zambia. N: Mbala Distr., Mbala-Kasama road, c.1500 m., 30.xii.1956, *Richards* 7420 (K; SRGH). E: Nyika Plateau, 27.xii.1962, *Fanshawe* 7261 (K; SRGH). **Malawi**. N: Nyika Plateau, Lake Kaulime, c. 2250 m., 4.i.1959, *Richards* 10456 (K). C: Ntchisi Mt., c. 1450 m., 21.ii.1959, *Robson & Steele* 1709 (K; SRGH).

Also in Tanzania. A pyrophyte locally frequent in mountain grassland and high rainfall woodland.

3. **Lactuca mwinilungensis** G.V. Pope in Kew Bull. **39**: 168 (1984). —Lawalrée, Dethier & Gilissen in Fl. Afr. Centr., Compos., Cichorioideae: 54 (1986). TAB. **36** fig. C. Type: Zambia, Mwinilunga, *Milne-Redhead* 3522 (K, holotype; BR).

Lactuca praecox sensu C. Jeffrey in Kew Bull. **18**: 449 (1966) pro parte quoad *Quarre* 4761; sensu Dethier in Bull. Jard. Bot. Nat. Belg. **52**: 375 (1982) pro parte quoad *Quarre* 4761.

A subscapose perennial herb up to c. 70 cm. tall, from a semi-woody fusiform-turbinate rootstock. Flowering stalks erect striate much-branched glabrous, or sparsely setulose below; branches mostly simple up to c. 35 cm. long, divaricate-ascending, glabrous, bracteate. Leaves often reddish-purple on the margins and veins, 1–6 in a ± prostrate basal rosette, appearing after the flowers; 6–15 × 3–7 cm., broadly elliptic to narrowly obovate, rounded to obtuse at the apices, ± broadly cuneate below with a petiole-like winged midrib up to c. 2.5 cm. long, denticulate on the margins, glabrous; cauline leaves absent, or if present, sessile and bract-like, up to c. 4 cm. long, narrowly oblanceolate to narrowly triangular. Synflorescences paniculate, diffuse. Capitula precocious, numerous, subsessile or stalked, solitary or in clusters of 2–4 spaced along the branches; capitula stalks up to c. 3 cm. long, glabrous. Involucres cylindric 8–10 × 2 mm. at time of flowering, increasing to c. 12 × 3 mm. in fruiting capitula. Phyllaries 3–4-seriate glabrous green, often brown or purple-tinged towards the apices, ciliate above, spreading with age; the outermost ovate, less than half the length of the inner; the innermost up to c. 11 mm. long,

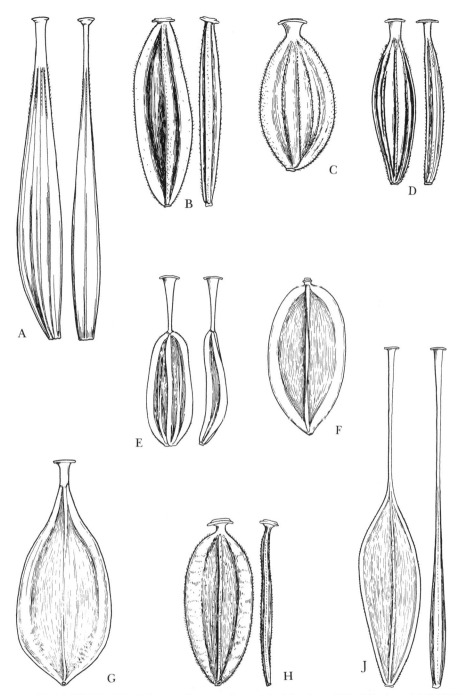

Tab. 36. LACTUCA SPP. Achenes, face or face and side views (all ×10). A. —LACTUCA
LASIORHIZA, from *Robson & Angus* 230. B. —LACTUCA CALOPHYLLA, from *Richards*
10394. C. —LACTUCA MWINILUNGENSIS, from *Milne-Redhead* 3522. D. —LACTUCA
SETOSA, from *Milne-Redhead* 2822. E. —LACTUCA SCHULZEANA, from *Milne-Redhead*
4144. F. —LACTUCA SCHWEINFURTHII, from *Richards* 451. G. —LACTUCA
LONGISPICATA, from *Richards* 4694. H. —LACTUCA HOMBLEI, from *Robinson* 5942.
J. —LACTUCA PRAECOX, from *Robson & Angus* 201. Drawn by Eleanor Catherine.

lanceolate-lorate, ± cucullate in bud. Florets c. 5 per capitulum; corollas yellow, 11–15 mm. long, the tube narrowly cylindric and puberulus above; ligule 8–11 × 2.5–3.5 mm., lorate, glabrous. Achenes reddish-brown, 4–5 × 1.5–2 mm., ellipsoid, somewhat compressed, not or hardly beaked, broadly-ribbed on the margins or angles, with 1–3 narrower ribs between, minutely hispidulous; pappus several-seriate, copious of fine, white barbellate setae c. 5 mm. long.

Zambia. W: Mwinilunga Distr., c. 0.8 km. S. of Matonchi Farm, 6.xii.1937, *Milne-Redhead* 3522 (K). Also in Zaire. In *Brachystegia* woodland.

4. **Lactuca setosa** Stebbins ex C. Jeffrey in Kew Bull. **18**: 444, figs. 2(5), 5 (1966). —Dethier in Bull. Jard. Bot. Nat. Belg. **52**: 375 (1982). —Lawalrée, Dethier & Gilissen in Fl. Afr. Centr., Compos., Cichorioideae: 55, pl. 13, (not pl. 11, *Richards* 11108), (1986). TAB. **36** fig. D. Type: Zambia, Ndola, *Fanshawe* 1747 (BR; K, holotype; SRGH).

An erect perennial herb up to 95(120) cm. tall, from a stout fusiform taproot. Stems solitary, simple below branching above, leafy with leaves ± crowded below, often purplish towards the base, sparsely to densely glandular-setose, or glabrescent; setae 0.5–2 mm. long, patent, gland-tipped. Lower leaves crowded or sub-rosulate, 3.5–16 × 2–6.5 cm., ± broadly elliptic or oblanceolate, acute to obtuse mucronate at the apex, cuneate below with a tapering-winged petiole-like midrib, irregularly denticulate, ciliate between the teeth, glabrous, or densely setose on the midrib, occasionally sparsely puberulent beneath, lateral nerves ± prominent beneath; upper leaves more spaced, decreasing in size towards the stem apex, becoming oblong-lanceolate sessile and semi-amplexicaul. Synflorescences corymbose-paniculate with spreading-ascending branches up to c. 25 cm. long, or contracted and ± globose; bracts subtending the lower branches foliaceous up to 3 cm. long and narrowly lanceolate. Capitula shortly stalked or subsessile, often ± aggregated. Involucres 5–8 × 1.5–2 mm. at time of flowering, cylindric, increasing to c. 12 × 2.5 mm. in fruiting capitula. Phyllaries 3–4-seriate somewhat rugulose outside and ciliolate above, glabrous or rarely glandular-setulose on the midrib; the outermost c. 3 mm. long, ovate; the innermost up to c. 12 mm. long and lanceolate-lorate, obtuse, margins entire and narrowly hyaline. Florets c. 5 per capitulum. Corollas yellow, 8–11.5 mm. long; tube narrowly cylindric; ligule 4–7 mm. long, lorate, lobes c. 1 mm. long apically papillose. Achenes not or hardly beaked, reddish-brown, 4.5–5.8 mm. long, narrowly ellipsoid, or obovoid and ± truncate above, somewhat laterally compressed, broadly ribbed on the margins with 2–5 narrower ribs between, hispidulous on ribs; pappus several-seriate, of numerous white barbellate setae c. 5.5 mm. long.

Zambia. N: Mbala Distr., Lungu For., 12 km. from Kalambo Falls, 1560 m., 11.i.1975, *Brummitt & Polhill* 13716 (K; SRGH). W: Mwinilunga Distr., Kalenda Ridge, W. of Matonchi Farm, 18.x.1937, *Milne-Redhead* 2822 (BM; K; LISC). **Malawi**. N: Chitipa Distr., Misuku Hills, Mughesse, c. 1780 m., 5.i.1974, *Pawek* 7773 (K; MAL; SRGH). C: Dedza Distr., Chongoni For. School, 4.iii.1967, *Salubeni* 565 (SRGH).
Also from Tanzania and Zaire. Herbs of swampy plateau grassland and high rainfall woodland, or montane grassland and forest margins.
The northern Malawi specimens, *Pawek* 4354, 7773, & 8020, differ from typical *L. setosa* in that the achenes are apically truncate with broad marginal ribs ending abruptly rather than tapering above, and the lower leaves taper into petiole-like bases and are not sub-sessile.

5. **Lactuca schulzeana** Büttner in Verhandl. Bot. Ver. Brandenb. **31**: 72 (1890). —S. Moore in Journ. Linn. Soc., Bot. **37**: 328 (1906). —Mendonça, Contrib. Conhec. Fl. Angol., **1** Compositae: 150 (1943). —Adams in F.W.T.A. ed. 2, **2**: 293 (1963). —C. Jeffrey in Kew Bull. **18**: 446, fig. 2(10), (1966). —Dethier in Bull. Jard. Bot. Nat. Belg. **52**: 374 (1982). —Lawalrée, Dethier & Gilissen in Fl. Afr. Centr., Compos., Cichorioideae: 49 (1986). TAB. **36** fig. E. Type from Zaire.
Lactuca macroseris Hiern, Cat. Afr. Pl. Welw. **1**, 3: 620 (1898). —Mendonça, op. cit.: 149 (1943). Types from Angola.

An erect long-lived giant annual herb, up to 3(3.5) m. tall. Taproot with several thick ± fleshy lateral roots. Stem solitary simple up to c. 1.5 cm. in diam., striately ribbed hollow glaucous, sparsely setose below, glabrous above. Leaves cauline somewhat crowded below; the lowermost up to c. 32 × 12 cm., elliptic-lanceolate in outline, grossly lyrate-runcinate, apices acute to obtuse, bases cuneate-attenuate with midrib ± narrowly winged petiole-like and semi-amplexicaul, denticulate on the margins, glabrous or sparsely setose particularly on the midrib and main nerves beneath; the upper leaves decreasing in size towards the stem apex, becoming lanceolate, undivided and sessile, ± sparsely setose. Synflorescence elongate-paniculate; branches short, patent and much thinner than the stem axis, sparsely glandular setulose, bracteate; bracts foliaceous. Capitula numerous stalked; involucres 6–8 mm. long at time of flowering, increasing to c. 12 mm. in fruiting capitula, cylindric to very narrowly campanulate. Phyllaries 4–5-seriate, purplish, glabrous, ciliolate above; the outermost c. 1.5 mm. long, ovate; the innermost up to c. 12 mm. long, lorate-lanceolate, obtuse to acute. Florets c. 10 per capitulum. Corollas 10–12.5 mm. long, blue to purplish; tube c. 5 mm. long, narrowly cylindric, sparsely setulose about the outside of the throat; ligule c. 6.5 mm. long, oblanceolate-lorate, lobes c. 1 mm. long, apically papillose. Achenes dark-brown, c. 4.5 mm. long including a distinct 1 mm. long beak, c. 1.25 mm. wide, fusiform, 3–4-angled, 1–several-ribbed between the angles, minutely hispidulous; beak slender, discolorous; pappus several-seriate of white barbellate setae c. 5 mm. long.

Zambia. W: Mwinilunga Distr., West Lunga River at Mwinilunga, 11°44'S, 24°26'E, c. 1280 m., 23.i.1975, *Brummitt, Chisumpa & Polhill* 14036 (K).
Also in Cameroon, Zaire and Angola. A herb of moist, tall grassland, stream sides and riverine forest.

6. **Lactuca schweinfurthii** Oliv. & Hiern in F.T.A. **3**: 452 (1877). —Robyns, Fl. Sperm. Parc Nat. Alb. **2**: 602 (1947). —F.W. Andrews, Fl. Pl. Sudan, **3**: 38 (1956). —Adams in F.W.T.A. ed. 2, **2**: 293 (1963). —C. Jeffrey in Kew Bull. **18**: 448, fig. 2(13), (1966). —Dethier in Bull. Jard. Bot. Nat. Belg. **52**: 381 (1982). —Lawalrée, Dethier & Gilissen in Fl. Afr. Centr., Compos., Cichorioideae: 58 (1986). TAB. **36** fig F. Type from the Sudan.

An erect perennial herb to c. 2.4 m. tall, from a semi-woody taproot. Stems solitary simple striately ribbed, glaucous often purple-tinged below, sparsely glandular-setose; setae up to c. 3 mm. long, patent, usually dark-purple. Leaves cauline, somewhat crowded below; the lowermost up to c. 20 × 4.5 cm., lanceolate to narrowly elliptic undivided, acute, ± shallowly sinuate-dentate and irregularly denticulate on the margins, attenuate to rounded and shortly petiole-like at the base, glabrous or setose on the midrib and nerves beneath; upper leaves decreasing in size towards the stem apex becoming lanceolate sessile amplexicaul and sparsely setulose or glabrescent. Synflorescence an elongate terminal panicle; branches up to c. 15 cm. long, slender, patent-ascending, sparsely setulose or glabrescent, bracteate. Capitula stalked, + aggregated in loose clusters; involucres 6–8 × 1.5–2 mm. at time of flowering, increasing to c. 10 × 4 mm. in fruiting capitula, cylindric to narrowly campanulate, glaucous, often purple-tinged. Phyllaries 2–3-seriate, increasing in length towards the inside, ciliolate-pubescent above and often subapically corniculate; the outermost ovate, the innermost up to 10 mm. long, narrowly lanceolate, obtuse. Florets c. 5 per capitulum. Corollas 8–10 mm. long, yellow; tubes slender, glabrous; ligule erect, c. 5.5 mm. long, narrowly obovate-lorate, lobes up to 1 mm. long, minutely papillose above. Achenes dark-brown to black, hardly or very shortly beaked, c. 4.5 × 2 mm., broadly ellipsoidal, strongly flattened, margins thickened, narrowly 1-ribbed or occasionally 2-ribbed on each face, minutely hispidulous; pappus several-seriate of white barbellate setae c. 5.5 mm. long.

Zambia. N: Mbala Distr., Old Isanya Road, c. 1540 m., 12.i.1952, *Richards* 451 (K).
Also in the Cameroon, Zaire, Angola, Tanzania and Sudan. A herb of open grassland.

7. **Lactuca longispicata** De Wild. in Ann. Mus. Congo, Bot., sér. 5, **1**: 87, pl. 16 (1904), as "longespicata"; op. cit. **2**: 219 (1907). —Th. & H. Durand, Syll. Fl. Congol.: 316 (1909). —C. Jeffrey in Kew Bull. **18**: 448, fig. 2(14), (1966). —Dethier in Bull. Jard. Bot. Nat. Belg. **52**: 382 (1982). —Lawalrée, Dethier & Gilissen in Fl. Afr. Centr., Compos., Cichorioideae: 60, pl. 12 (1986). TAB. **36** fig. G. Type from Zaire.

An erect perennial herb to c. 1.3 m. tall, from a woody taproot. Stems solitary simple striate, ± densely glandular-pilose below, sparsely glandular-setose above, glabrescent; hairs up to c. 3 mm. long. Leaves cauline; the lowermost up to c. 25 × 6 cm., oblanceolate to elliptic, undivided, obtuse mucronate, shallowly sinuate-dentate and irregularly denticulate, attenuate with midrib narrowly winged petiole-like at the base, sparsely pilose-setose particularly on the midrib beneath; upper leaves decreasing in size towards the stem apex, becoming narrowly-elliptic to lanceolate, sessile and semi-amplexicaul. Synflorescence racemose to sparingly-branched divaricate-paniculate, up to c. 50 cm. long, branches much reduced (sometimes to c. 20 cm. long), sparsely glandular-setose, glabrescent, bracteate. Capitula subsessile to shortly stalked, 2–6 aggregated in remote dense clusters; involucres 8–10 × 2 mm. at time of flowering, to c. 14 × 4 mm. in fruiting capitula, cylindric to campanulate, often purple-tinged. Phyllaries 2–3-seriate, ciliolate above otherwise glabrous; the outermost phyllaries usually less than half the length of the inner, narrowly ovate; inner phyllaries 10–13 × 1.5-3 mm., narrowly lanceolate. Florets c. 5 per capitulum. Corollas 10–11 mm. long, yellow; tube c. 4.5 mm. long, very narrowly cylindric, glabrous; ligule erect c. 6 mm. long, oblanceolate, lobes apically papillose. Achenes dark-brown to black, 5–6 mm. long (including 0.5–1 mm. long beak), broadly flattened-ellipsoid, margins swollen, narrowly 1-ribbed on each face, ± sparsely hispidulous; pappus many-seriate of white barbellate setae c. 5–6 mm. long, outermost setae much smaller.

Zambia. N: Mbala Distr., Inono River Source, top of Escarpment, c. 1540 m., 26.ii.1955, *Richards* 4694 (K).
Also in Zaire and Tanzania. High rainfall wooded grassland.

8. **Lactuca homblei** De Wild. in Fedde, Repert. **13**: 210 (1914). —C. Jeffrey in Kew Bull. **18**: 448, fig. 3(15), (1966). —Dethier in Bull. Jard. Bot. Nat. Belg. **52**: 382 (1982). —Lawalrée, Dethier & Gilissen in Fl. Afr. Centr., Compos., Cichorioideae: 59 (1986). TAB. **36**fig. H. Type from Zaire.

An erect perennial herb to c. 120 cm. tall, from a semi-woody narrowly turbinate taproot. Stems solitary, branched above, glabrous or sparsely glandular-setulose, often purple-tinged. Leaves contemporaneous with flowers or sometimes delayed, cauline, up to c. 10 × 2.5 cm., oblanceolate-elliptic, undivided, obtuse mucronate, remotely denticulate and ciliate on the margins, bases attenuate to semi-amplexicaul, glabrous or sparsely setulose on the midrib and main nerves; upper leaves decreasing in size towards the stem apex becoming linear-lanceolate. Synflorescences paniculate up to 40 cm. long, occasionally much reduced and spicate; branches up to c. 15 cm. long, patent-ascending, bracteate, or much reduced. Capitula stalked or subsessile, solitary or in clusters of 2–4. Involucres 7–9 × 1.5–2 mm. at time of flowering, increasing to 12 × 2.5 mm. in fruiting capitula, cylindric, glaucous to purple-tinged. Phyllaries ciliolate above otherwise glabrous, the outermost ± narrowly ovate, the innermost up to 11 × 2.5 mm., narrowly lanceolate. Florets c. 5 per capitulum; corollas yellow, up to c. 11 mm. long, the tube c. 5 mm. long, cylindric, glabrous; ligule c. 6 mm. long, lorate, erect. Achenes reddish-brown, 4.5–5 × 1.75–2 mm. (including the c. 0.5 mm. long beak) flattened-ellipsoid, margins swollen, narrowly 1-ribbed, rarely 2–3-ribbed on each face, hispidulous; pappus several-seriate of white, barbellate setae c. 5–6 mm. long.

Zambia. W: Mwinilunga Distr., 6 km. N. of Kalene Hill, 12.xii.1963, *Robinson* 5942 (K; SRGH).
Also in Zaire. *Brachystegia* woodland.
Milne-Redhead 3139 differs from typical *L. homblei* in that the synflorescence is reduced and spiciform, and the young flowers are developed before the leaves. However, it agrees well with this species in other respects; in particular its habit and achene features, and is therefore included here. A similar variation in the synflorescence arrangement is to be found in the closely related *L. longispicata*.

9. **Lactuca praecox** R.E. Fr. in Wiss. Ergebn. Schwed. Rhod.-Kongo-Exped. 1911–12, **1**: 352 (1916). —Brenan in Mem. N.Y. Bot. Gard. **8**, 5: 489 (1954). —C. Jeffrey in Kew Bull. **18**: 449, fig. 3(16), (1966) excl. *Quarre* 4761. —Dethier in Bull. Jard. Bot. Nat. Belg. **52**: 375 (1982) excl. *Quarre* 4761. —Lawalrée, Dethier & Gilissen in Fl. Afr. Centr., Compos., Cichorioideae: 49 (1986). TAB. **36** fig. J. Types: Zambia, Malolo, N. of Luwingu, Kalungwishi R., *R.E. Fries* 1111, 1111a (UPS, syntypes).

An erect subscapose perennial herb 7–55 cm. tall, from a narrowly turbinate rootstock. Flowering stalks 1–several ± diffusely branched, or much contracted and subpulviniform, glabrous or ± sparsely glandular-setose, bracteate; bracts foliaceous, sessile, up to c. 4 × 0.6 cm., very narrowly triangular, glabrous; branches mostly simple, up to c. 30 cm. long, divaricate-ascending, or much reduced, glabrous. Leaves poorly known, appearing after flowering, radical, dark-brown lanate in the axils, up to c. 10 cm. long, linear-oblong and sessile, or elliptic to narrowly ovate and broadly cuneate to attenuate at the base, margins ciliate-denticulate with teeth up to c. 1.5 mm. long; petiole-like midribs up to 4 cm. long, persisting as polished golden-brown scales. Capitula precocious, numerous, sessile or stalked, solitary or in clusters of 2–4; stalks up to c. 5 cm. long, glabrous. Involucre 7–10 × 2.5 mm. at time of flowering, increasing to c. 17 × 3.5 mm. in fruiting capitula, cylindric to narrowly conical, later narrowly campanulate. Phyllaries glabrous, guttulate, green sometimes purple-tinged, ciliolate above, the outer from c. 3 mm. long and narrowly ovate, the inner to c. 16 mm. long and lanceolate. Florets c. 5 per capitulum; corollas yellow, 12–16 mm. long, tube narrowly cylindric, ± densely pilose near the throat; ligules

10–14 × 3–4.5 mm., oblanceolate, ± pilose below, lobes papillose above. Achenes dark-brown to black, 8–9 mm. including a 2.5–3.5 mm. long paler beak, 1.5–1.75 mm. wide, narrowly ellipsoid, swollen-ribbed on the margins with 1 or sometimes 2–3 smaller ribs between, minutely antrorsely hispidulous; pappus several-seriate, copious, of fine white barbellate setae, c. 5 mm. long.

Zambia. N: Mbala Distr., Uningi (Ningi) Pans, c. 1500 m., 26.viii.1960, *Richards* 13149 (K; SRGH). W: Kitwe, 13.ix.1968, *Mutimushi* 2677 (K; SRGH). E: Nyika Plateau, c. 3.2 km. SW. of Rest House, 2150 m., 21.x.1958, *Robson & Angus* 201 (BM; K; LISC; SRGH). **Malawi**. N: Chitipa Distr., N. of Nganda, 1860 m., 5.ix.1972, *Synge* WC 384 (K). C: Nkhota Kota Distr., Chintembwe, c. 1400 m., 9.ix.1946, *Brass* 17581 (K; PRE; SRGH).

Also from Angola, Zaire and Tanzania. A pyrophyte of medium to high altitude grassland and *Brachystegia* woodland.

Fanshawe 11510, from the Mukutus Mts. of NE. Zambia, and similar material from Mbisi Mt. in the Ufipa district of Tanzania (*Bullock* 1850, 3416 and *Napper* 993) differ somewhat from typical *L. praecox* in habit and achene features. The synflorescences are much contracted and seldom more than 15 cm. tall – the branches congested and usually setose, whilst the achene faces are usually more than 1-ribbed. These differences are, however, not sufficiently discontinuous to warrant separate taxonomic recognition of this material.

10. **Lactuca imbricata** Hiern, Cat. Afr. Pl. Welw. **1**, 3: 620 (1898). —Mendonça, Contrib. Conhec. Fl. Angol., **1** Compositae: 149 (1943). —C. Jeffrey in Kew Bull. **18**: 449, fig. 3(17), (1966). — Lawalrée, Dethier & Gilissen in Fl. Afr. Centr., Compos., Cichorioideae: 68, pl. 14 (1986) excl. *Lactuca hockii* De Wild. TAB. **37** fig. A. Type from Angola.

An erect ± robust perennial herb to 1 m. tall, exceptionally to c. 2 m. tall; rootstock semi-woody, narrowly turbinate. Stems solitary, sometimes several, branching above, ± crowded leafy below, green occasionally purple-tinged, ± glaucous, glabrous or sometimes sparsely setulose. Leaves ascending or ± appressed to the stem, sessile membranous mostly pinnatipartite with a linear rhachis, up to c. 30 cm. long, elliptic in outline, remotely denticulate on the margins, glabrous or the midrib setose beneath; terminal lobe up to c. 12 × 0.7 cm., linear-lanceolate; lateral lobes in c. 5 pairs, remote, up to c. 8 × 0.6 cm., linear-lanceolate; basal leaves sometimes smaller and undivided, linear-lanceolate, remotely dentate, their lower portions often persisting as chartaceous scales, dark-brown lanate in the axils. Synflorescences terminal, laxly paniculate or corymbiform; branches up to c. 50 cm. long, divaricate, glabrous. Capitula stalked; involucres 15–23 × 6–8 mm. at time of flowering, increasing to c. 30 × 10 mm. in fruiting capitula, conical-cylindric becoming campanulate. Phyllaries imbricate increasing uniformly in length towards the inside, the outermost broadly ovate and rounded-obtuse, the innermost narrowly lanceolate and attenuate, ciliolate above otherwise glabrous, guttulate on exposed areas, usually glaucous, purple-tinged above and on the margins. Florets c. 20 per capitulum. Corollas yellowish-green, 14–23 mm. long, overtopping the involucres; tube narrowly cylindric, pilose at the throat outside, ligule c. 11 × 1.5 mm., lorate. Achenes reddish-brown, c. 12 × 2.5 mm. including the 4–5 mm. long beak, narrowly flattened-ellipsoid, broadly-ribbed on the margins with 1–2 smaller ribs between, hispidulous; pappus several-seriate, of fine white, barbellate setae, c. 11 mm. long.

Herbs 50–175 cm. tall; lower leaves up to c. 30 × 16 cm., pinnatifid; capitula usually more
than 10 - - - - - - - - - - - - - - var. *imbricata*
Herbs not exceeding 35 cm. in height; lower leaves c. 10 × 0.8 cm., usually undivided; capitula
up to 10 - - - - - - - - - - - - - var. *hockii*

Var. **imbricata** —Dethier in Bull. Jard. Bot. Nat. Belg. **52**: 381 (1982). Type from Angola.
    Lactuca gilletii De Wild. in Ann. Mus. Congo, Bot. sér. 5, **1**: 86, pl. 14 (1904). Type from Zaire.

Herbs 50–175(200) cm. tall; leaves somewhat crowded on lower stem, lower leaves up to c. 30 × 16 cm., mostly pinnatifid, lobes remote up to c. 8 × 0.6 cm., linear-lanceolate; capitula usually more than 10, outer phyllaries broadly ovate.

Zambia. N: Old road to Stembewe near Mbala, c. 1540 m., 14.i.1955, *Richards* 4081 (K; SRGH). W: Kitwe, 9.i.1968, *Mutimushi* 2414 (K; NDO). C: 10 km. E. of Lusaka, c. 1200 m., ll.xi.1955, *King* 203 (K). E: Lundazi, 17.x.1967, *Mutimushi* 2265 (K; SRGH). S: Between Kafue Hook Pontoon and Mumbwa, 21.xi.1959, *Drummond & Cookson* 6753 (K; LISC; SRGH). **Zimbabwe**. N: Mazowe, University College Farm, xii.1964, *Smith* 744 (SRGH). C: Harare, 12.ix.1932, *Eyles* 7173 (K). **Malawi**. N: Mzimba Distr., Mzuzu, Marymount, c. 1400 m., 26.ii.1975, *Pawek* 9109 (K; MAL; MO; SRGH). C: Kasungu to Chamama, c. 1000 m., 16.i.1959, *Robson & Jackson* 1206 (K). **Mozambique**. N: Lichinga (Vila Cabral), l.viii.1934, *Torre* 294 (COI; LISC). Near Lake Malawi, 1902, *Johnson* 490 (K).

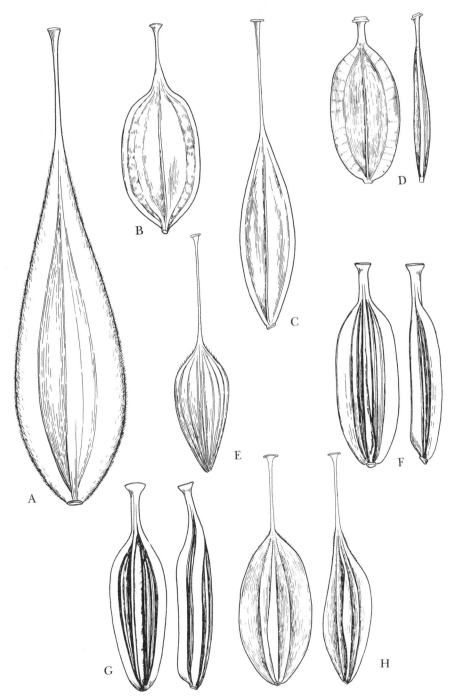

Tab. 37. LACTUCA SPP. Achenes, face and side views (all × 10). A. —LACTUCA IMBRICATA, from *Mutimushi* 2414. B. —LACTUCA ZAMBEZIACA, from *Brummitt, Chisumpa & Polhill* 14115. C. —LACTUCA INERMIS, from *Loveridge* 1155. D. —LACTUCA INDICA, from *Gomes e Sousa* 3738. E. —LACTUCA SERRIOLA, from *Best* 1375. F. —LACTUCA ATTENUATA, from *Brummitt* 12291. G. —LACTUCA PARADOXA, from *Pawek* 10051. H. —LACTUCA GLANDULIFERA, from *Mendonça* 2132. Drawn by Eleanor Catherine.

Also in Angola, Zaire and Tanzania. A pyrophyte of miombo woodland and grassland.

Var. **hockii** (De Wild.) Dethier in Bull. Jard. Bot. Nat. Belg. **52**: 381 (1982). Type from Zaire.
Lactuca andongensis Hiern, Cat. Afr. Pl. Welw. **1**, 3: 622 (1898). —Mendonça, Contrib. Conhec.
Fl. Angol., **1** Compositae: 151 (1943). Type from Angola.
Lactuca hockii De Wild. in Fedde, Repert. **13**: 211 (1914). —Lawalrée, Dethier & Gilissen in Fl.
Afr. Centr., Compos., Cichorioideae: 67 (1986). Type as above.

Herbs up to c. 35 cm. tall; leaves mostly in a crowded basal rosette, mostly undivided,
up to c. 12 × 0.8 cm., linear-lanceolate; capitula rarely more than 10, outer phyllaries
lanceolate.

Zambia. N: Mbala, 8.ix.1969, *Fanshawe* 10644 (K; SRGH).
Also from Angola and Zaire. A pyrophyte of grassland and miombo woodland.

11. **Lactuca zambeziaca** C. Jeffrey in Kew Bull. **18**: 450, figs. 3(20), 8 (1966). —Dethier in Bull. Jard.
Bot. Nat. Belg. **52**: 380 (1982). —Lawalrée, Dethier & Gilissen in Fl. Afr. Centr., Compos.,
Cichorioideae: 60, pl. 11 (not pl. 13, *Fanshawe* 1747), (1986). TAB. **37** fig. B. Type: Zambia, Lake
Chila, *Richards* 11108 (K, holotype).

An erect perennial herb, often viscid, up to c. 150(200) cm. tall; taproot semi-woody
turbinate. Stems solitary, branching above, glabrous, green and often purple-tinged.
Leaves contemporaneous with flowers, ± membranous, glabrous, 18–32 × 0.25–1.5 cm.,
narrowly elliptic to linear-lanceolate, usually with 2–4 well developed lobes in the lower
half, ± attenuate and acute at the apex, narrowly attenuate below and usually semi-
amplexicaul at the base, subentire; terminal lobe 12–23 cm. long, elliptic-linear; lateral
lobes 0.5–6 cm. long, very narrowly triangular. Synflorescences paniculate, up to c. 40 × 15
cm., cylindrical; branches divaricate slender glabrous. Capitula numerous, stalked; stalks
1.5–12 mm. long; involucres 8–9 × 1.5–2 mm. at time of flowering, increasing to c. 12 × 2.5
mm. in fruiting capitula, cylindric. Phyllaries ciliolate above otherwise glabrous, guttulate,
green and often purple-tinged where exposed; the outermost less than half the length of
the inner, broadly ovate; the innermost 9–11 × 1–2 mm., linear-lanceolate, margins
becoming involute except near the apices. Florets c. 5 per capitulum; corollas yellow, up to
c. 12 mm. long, tube narrowly cylindric, sparsely pilose near the throat; ligule c. 6 mm.
long, erect. Achenes dark-brown to black, 5–6.5 × 1.5–2 mm. including a 1–1.5 mm. long
paler-coloured beak, compressed-ellipsoid, broadly thickened at the margins, narrowly
1-ribbed on each face, hispidulous or glabrous; pappus several-seriate, of white
barbellate setae c. 6 mm. long.

Zambia. N: 48 km. E. of Mporokoso, c. 1400 m., 13.iv.1961, *Phipps & Vesey-FitzGerald* 3109 (COI; K;
LISC; SRGH). W: 60 km. S. of Mwinilunga on Kabompo road, 25.i.1975, *Brummitt, Chisumpa &
Polhill* 14115 (K; SRGH).
Also in Angola and Zaire. Miombo or mixed deciduous woodland with grass understorey.

12. **Lactuca inermis** Forssk., Fl. Aegypt.-Arab.: 144 (1775). —J.R.I. Wood in Kew Bull., **39**: 132 (1984).
—Lawalrée in Fl. Rwanda, Spermat. **3**: 688, fig. 216(1), (1985). —Lawalrée, Dethier & Gilissen in
Fl. Afr. Centr., Compos., Cichorioideae: 62 (1986). TAB. **37** fig. C. Type from Arabia, Yemen.
Lactuca capensis Thunb., Prodr. Pl. Cap.: 139 (1800). —Harv. in Harv. & Sond., F.C. **3**: 526
(1865). —Oliv. & Hiern in F.T.A. **3**: 452 (1877). —Hiern, Cat. Afr. Pl. Welw. **1**, 3: 621 (1898). —De
Wild., Bull. Jard. Bot. Brux. **3**: 315 (1912). —R.E. Fr., Wiss. Ergebn. Schwed. Rhod.-Kongo-Exped.
1911–12, **1**: 352 (1916). —Eyles in Trans. Roy. Soc. S. Afr. **5**, 4: 523 (1916). —S. Moore in Journ.
Bot. **65**, Suppl. 2: 66 (1927). —Mendonça, Contrib. Conhec. Fl. Angol., **1** Compositae: 150
(1943). —Robyns, Fl. Sperm. Parc Nat. Alb. **2**: 601 (1947). —C. Jeffrey in Kew Bull. **18**: 450, t. 3,
fig. 21, (1966). —Hilliard, Comp. Natal: 632 (1977). —Dethier in Bull. Jard. Bot. Nat. Belg. **52**:
377 (1982). Type from South Africa (Cape Province).
Lactuca abyssinica Fresen. in Mus. Senckenb. **3**: 72 (1839). —Oliv. & Hiern in F.T.A. **3**: 453
(1877). —Hiern, Cat. Afr. Pl. Welw. **1**, 3: 621 (1898). —Mendonça, Contrib. Conhec. Fl. Angol., **1**
Compositae: 151 (1943). Type from Ethiopia.
Lactuca capensis var. *duruensis* De Wild. in Ann. Mus. Congo, Bot. sér. 5, **2**: 217 (1907). Type
from Zaire.
Lactuca seretii De Wild. in Ann. Mus. Congo, Bot. sér. 5, **2**: 218 (1907). Type from Zaire.
Lactuca vanderystii De Wild. in Bull. Jard. Bot. Brux. **3**: 315 (1912). Type from Zaire.
Lactuca pallidocoerulea Dinter in Fedde, Repert. **30**: 189 (1932). Type from Namibia.
Lactuca leptocephala Stebbins in Bull. Jard. Bot. Brux. **14**: 223 (1936). Type from Zaire.
Lactuca kenyaensis Stebbins in Bull. Jard. Bot. Brux. **14**: 224, t. 19, fig. c–d (1936). Type from
Kenya.

Lactuca lebrunii Robyns in Bull. Jard. Bot. Brux. **17**: 105 (1943); Fl. Sperm. Parc Nat. Alb. **2**: 603 (1947). Type from Zaire.

Lactuca schweinfurthii sensu Robyns, Fl. Sperm. Parc Nat. Alb. **2**: 602 (1947) saltem pro parte, non Oliv. & Hiern (1877).

See C. Jeffrey in Kew Bull. **18**: 450 (1966) for full synonymy.

An erect perennial herb up to 1(1.5) m. tall, with a long thick semi-woody taproot. Stems solitary, or proliferating at the rootcrown, branched above, glabrous or setulose, glaucous, scattered leafy or leaves ± crowded on the lower stem, sometimes rosulate. Leaves contemporaneous with the flowers, sessile membranous glabrous, or lower surface sparsely hispid with midrib and main nerves glandular-setose, entire or remotely denticulate, 8–20 × 0.3–1.6 cm., ± narrowly lanceolate, undivided to ± runcinate with lateral lobes up to c. 1.5 cm. long, bases semi-amplexicaul in basal leaves to sagittate-auriculate in cauline leaves, auricle lobes up to c. 2 cm. long and very narrowly triangular. Synflorescence laxly or congested paniculate, rarely somewhat racemose; branches usually c. 8–10 cm. long and diffusely subdivided, or up to c. 25 cm. long ascending and sparsely or much subdivided, glabrous, bracteate; bracts leaf-like, narrowly triangular, sagittate-auriculate. Capitula numerous, stalked, solitary or loosely aggregated; involucres 8–10 × 1.5–2 mm. at time of flowering, increasing to c. 18 × 4 mm. in fruiting capitula, cylindric becoming narrowly conical in fruit. Phyllaries glabrous, sometimes guttulate, green and often purple-edged or spotted; outer phyllaries 2–6 mm. long, lanceolate; the innermost up to c. 17 mm. long, lanceolate; margins becoming involute except near the apex. Florets 4–14 per capitulum; corollas blue purplish or white, less often yellow, 0–17 mm. long, tube tufted-setulose near the throat; ligules c. 6–11 × 2 mm., oblanceolate-lorate, glabrous. Achenes dark-brown to black 4.5–8.5 × 1.25–2 mm. including a 1.5–2.5 mm. long paler-coloured beak, flattened-ellipsoid, broadly margined, narrowly 1-ribbed on each face, hispidulous or glabrous; pappus several-seriate, of persistent white to yellowish-white barbellate setae 6–7 mm. long.

Zambia. B: Kaoma (Mankoya), near resthouse, 20.xi.1959, *Drummond & Cookson* 6655 (K; SRGH). N: Mbala Distr., Fwambo area, Ninina hill, near Kawimbe, c. 1500 m., 6.ix.1956, *Richards* 6115 (K; SRGH). W: Ndola, 23.vii.1953, *Fanshawe* 166 (K; LISC; SRGH). C: Mt. Makulu Res. Stat., 20 km. S. of Lusaka, 14.i.1957, *Angus* 1494 (K; LISC; SRGH). E: Chimutengo/Petauke, 3.x.1966, *Mutimushi* 1460 (K; NDO). S: c. 48 km. NE. of Mazabuka, c. 1000 m., 12.vii.1930, *Hutchinson & Gillett* 3568 (BM; K; SRGH). **Zimbabwe**. N: Makonde, Silverside Mine, 3.ii.1965, *Wild* 6802 (K; SRGH). W: Matobo, Farm Besna Kobila, c. 1500 m., ii.1955, *Miller* 2679 (K; LISC; SRGH). C: Harare, Borrowdale, 8.xi.1953, *Wild* 4149 (K; LISC; SRGH). E: Chimanimani Distr., Kasipiti, c. 1300 m., 20.ix.1964, *Loveridge* 1155 (K; SRGH). S: Makoholi Experiment Station, 14.iii.1978, *Senderayi* 233 (K; SRGH). **Malawi**. N: Mzimba Distr., Mzuzu, c. 1400 m., 12.x.1973, *Pawek* 7379 (K; MO; SRGH; UC). C: Near Kasungu Hill, c. 1100 m., 14.i.1959, *Robson & Jackson* 1150 (BM; K; LISC; SRGH). S: Blantyre, Chancellor College campus, Chichiri, c. 1170 m., 12.iii.1970, *Brummitt* 9025 (K; SRGH). **Mozambique**. N: Lichinga (Vila Cabral), 1.vii.1934, *Torre* 180 (COI; LISC). Z: Gurué, between Nuirre and Lolae, 24.ix.1944, *Mendonça* 2256 (LISC). MS: Serra Zuira, planalto Tsetserra, c. 1840 m., 12.xi.1965, *Torre & Pereira* 12911 (LISC). M: Malola near Maracuene, i.1946, *Pimenta* 43101 (LISC).

Widely distributed in sub-Saharan Africa, also in the Yemen and Madagascar. In grassland or woodland in higher rainfall areas, often in moist localities and seasonally wet depressions. Becoming a weed of disturbed areas.

A species of great variability but characterised by the combination of the perennial habit, the sagittate-auriculate leaves, the blue or white rarely yellow flowers, and the narrow outer phyllaries.

13. **Lactuca indica** L., Mant. **2**: 278 (1771). —Bak., Fl. Maurit. Seych.: 180 (1877). —Bews, Fl. Natal and Zululand: 227 (1921). —Merrill in Bot. Mag. Tokyo **51**: 194, t. 3 (1937); in Journ. Arn. Arb. **19**: 373 (1938). —Humbert, Fl. Madag., Comp., III: 874, t. 162, fig. 1–3 (1963). —C. Jeffrey in Kew Bull. **18**: 454, fig. 3(22), (1966). —Hilliard, Comp. Natal: 632 (1977). TAB. **37** fig. D. Type from Indonesia.

Pterocypsela indica (L.) Shih in Acta Phytotax. Sin. **26**: 387 (1988). Type as above. (See note under the genus *Lactuca*).

A stout erect leafy annual or biennial herb up to c. 1.5(4) m. tall. Stems solitary, branching above, glabrous. Leaves cauline sessile, not or scarcely auriculate, glabrous; lower leaves up to c. 22 × 14 cm., pinnatipartite, narrowly ovate to broadly elliptic in outline, remotely 2–5-lobed, the terminal lobe up to c. 9.5 × 2.5 cm. lanceolate, lateral lobes up to c. 5.5 × 1 cm. lorate; upper leaves smaller, undivided and lanceolate to linear-lanceolate, sometimes lobed. Synflorescence paniculate; branches ascending, glabrous with ovate to narrowly lanceolate bracts. Capitula numerous, stalked; stalks up to

c. 2.5 cm. long, imbricately bracteolate particularly when young; involucres 8–10 × 3.5–4 mm., cylindric at time of flowering, increasing to c. 14 × 7 mm. and becoming pyriform in fruiting capitula. Phyllaries 4–5-seriate, glabrous; the outermost 2–5 mm. long, ovate; the innermost to c. 14 mm. long, lorate-lanceolate. Florets 20–28 per capitulum; corollas yellow 13–15 mm. long, tube slender, cylindric, pubescent about the mouth; ligule c. 7–9 × 1.5–2 mm., lorate, glabrous. Achenes dark-brown or black 4.5–5 × 1.5–2 mm. including a c. 1 mm. long paler-coloured beak, flattened-ellipsoid, ± winged on the margins and narrowly 1(2)-ribbed on each face, glabrous; pappus several-seriate, of white barbellate setae c. 7 mm. long.

Mozambique. M: Maputo R., 90 km. S. of Maputo, Salamanga, 7.vi.1948, *Gomes e Sousa* 3738 (COI; K; LISC).
Also in Asia, from India to China and Indonesia, and recorded from South Africa (Natal), Madagascar, Mauritius, Reunion and the Seychelles. Mainly coastal below 1200 m., on river banks and roadsides, or as a weed of cultivation.

14. **Lactuca serriola** L., Cent. Pl. **2**: 29 (1756). —C. Jeffrey in Kew Bull. **18**: 455 (1966). —Hilliard, Comp. Natal: 633 (1977). TAB. **37** fig. E. Type from Europe.
   Lactuca virosa sensu Oliv. & Hiern in F.T.A. **3**: 453 (1877); sensu F.W. Andrews, Fl. Pl. Sudan **3**: 39 (1956) non L.

An erect annual or biennial herb up to 2.5 m. tall. Stems solitary, robust, simple below and branching above, glabrous to ± densely spinulose in the lower part, sometimes strongly so, leafy. Leaves ± appressed to the stem, sessile, usually up to 10 × 6 cm., exceptionally to 20 × 10 cm., broadly elliptic in outline and runcinate-pinnate, or some leaves undivided and oblong-oblanceolate; lateral lobes 2–6, narrowly to broadly oblong, recurved; apices rounded to obtuse mucronate; margins irregularly denticulate and ± sinuate-dentate; bases sagittate-auriculate, the auricle lobes up to c. 2 cm. long and narrowly pointed; lamina glabrous but midribs beneath often weakly to strongly spinulose. Synflorescences paniculate, divaricately branched; branches up to 23 cm. long, glabrous; capitula on slender, bracteate stalks up to c. 4 cm. long. Involucre 6–8 mm. long and cylindric at time of flowering, increasing to 13 × 3.5 mm. and becoming conical in fruiting capitula. Phyllaries glabrous, c. 2 mm. long outside increasing to c. 13 mm. long inside becoming narrowly lanceolate, margins becoming involute. Florets c. 14 per capitulum; corollas yellow, c. 9 mm. long, tubes slender, ± densely pilose about the mouth; ligules erect, c. 5 mm. long, lorate. Achenes pale- or greyish-brown; beaks paler, equalling or exceeding the achene in length; achene 3 × 1 mm., oblanceolate, somewhat laterally compressed, broadly-ribbed on the margins and narrowly c. 5-ribbed on each face, hispidulous towards the apices; pappus several-seriate of numerous white, barbellate setae c. 4 mm. long, free on the apically expanded beak.

Botswana. N: Ngamiland, *Curson* 56 (PRE). SE: S. of Otse (Ootze), 25°05'S, 25°40'E, 17.viii.1978, *Hansen* 3457 (C; GAB; SRGH). **Zimbabwe**. W: Bulawayo, Hillside, garden, c. 1400 m., 3.i.1980, *Best* 1375 (K; SRGH).
A native of the E. Mediterranean and N. Africa from the Canary Islands to Ethiopia, now a widely spread weed, introduced into southern Africa. A weed of cultivation and moist places, also a ruderal on roadsides and waste places.
Lactuca serriola can be distinguished from the closely related *L. dregeana* DC. by the shape of the undivided leaves. In the latter, these are narrowly lanceolate and ± attenuate at the apices, not oblong-oblanceolate. *L. dregeana* is apparently known only from the drier parts of the northern Cape Province and Orange Free State in South Africa.

15. **Lactuca attenuata** Stebbins in Bull. Jard. Bot. Brux. **14**: 346, pl. 8(f–j) (1937). —Robyns, Fl. Sperm. Parc Nat. Alb. **2**: 604 (1947). —C. Jeffrey in Kew Bull. **18**: 455, fig. 3(29), (1966). —Dethier in Bull. Jard. Bot. Nat. Belg. **52**: 372 (1982). —Lawalrée in Fl. Rwanda, Spermat. **3**: 690, fig. 216(3), (1985). —Lawalrée, Dethier & Gilissen in Fl. Afr. Centr., Compos., Cichorioideae: 46, pl. 8 (1986). TAB. **37** fig. F. Type from Zaire.
   Lactuca semibarbata Stebbins in Bull. Jard. Bot. Brux. **14**: 348, pl. 8(k–o), (1937). —Robyns, Fl. Sperm. Parc Nat. Alb. **2**: 604 (1947). —C. Jeffrey in Kew Bull. **18**: 455, fig. 3(31), (1966). Type from Zaire.
   Lactuca attenuatissima Robyns in Bull. Jard. Bot. Brux. **17**: 106 (1943); Fl. Sperm. Parc Nat. Alb. **2**: 604, pl. 54 (1947). —C. Jeffrey in Kew Bull. **18**: 455 (1966). Type from Zaire.

A vigorous scandent short-lived perennial herb up to c. 2 m. high. Stems terete, ± densely leafy, setose, or glabrescent above. Leaves up to c. 15 × 3 cm., ± narrowly lanceolate in outline, ± runcinate, 3–7-lobed, narrowly-winged petiole-like and dentate-auriculate at the base, subentire to denticulate often ciliate on the margins, setose particularly on the midrib and reticulation beneath; terminal lobe up to 6 cm. long, narrowly triangular-sagittate, attenuate above; lateral lobes up to 2 × 1 cm., recurved acute-mucronate. Synflorescences paniculate, axillary or terminal, usually borne clear of the foliage; branches glabrous and with leaf-like bracts. Capitula numerous; stalks up to c. 3 cm. long with 1–several c. 2 mm. long lanceolate bracteoles. Involucres c. 10 × 2 mm. and cylindric at time of flowering, increasing to c. 15 × 2.5 mm. and becoming narrowly conical in fruiting capitula; phyllaries narrowly lanceolate, the outer less than half the length of the innermost, obtuse and ciliolate above, glabrous or sometimes with 1–5 conspicuous setae along the midrib, green often purple-tinged where exposed. Florets c. 5 per capitulum; corollas 13–14 mm. long, tube slender pubescent about the mouth, ligule yellow often with a purple central band outside, 7–8 × 3–3.5 mm., lorate, spreading. Achenes pale-brown, up to c. 6 × 1.5 mm. including a 1 mm. long beak, narrowly ellipsoid, somewhat compressed with swollen ribs on the margins and 4–5 corky wing-like ribs on each face, hispidulous or glabrous; pappus several-seriate, of equal white barbellate setae up to c. 7.5 mm. long.

Malawi. S: Mulanje Mt., Nayawani Shelf, near top of pathway to Madzeka Basin, c. 2070 m., 28.vii.1970, *Brummitt* 12291 (K; SRGH).

Also in Burundi, Rwanda, Uganda and Zaire. In *Widdringtonia* forest, montane woodland or scrubland above 2000 m.

16. **Lactuca paradoxa** Sch. Bip. ex A. Rich., Tent. Fl. Abyss. **1**: 461 (1848). —Oliv. & Hiern in F.T.A. **3**: 454 (1877). —Engl., Hochgebirgsfl. Trop. Afr.: 455 (1892). —O. Hoffm. in Engl., Pflanzenw. Ost-Afr. **C**: 421 (1895). —R.E. Fr. in Acta Hort. Berg. **9**: 161 (1929). —Stebbins in Bull. Jard. Bot. Brux. **14**: 344 (1937). —Robyns, Fl. Sperm. Parc Nat. Alb. **2**: 606 (1947). —C. Jeffrey in Kew Bull. **18**: 456, fig. 3(32), (1966). —Agnew, Upland Kenya Wild Fls.: 500 (1974). —Dethier in Bull. Jard. Bot. Nat. Belg. **52**: 370, fig. 1 pro parte excl. A & B (1982). —Lawalrée, Dethier & Gilissen in Fl. Afr. Centr., Compos., Cichorioideae: 43, pl. 7 pro parte excl. A & B (1986). TAB. **37** fig. G. Type from Ethiopia.

　　Pertya paradoxa (A. Rich.) Schweinf., Beitr. Fl. Aethiopia: 153 (1867). Type as above.
　　Lactuca paradoxa var. *pedicellato-foliolata* De Wild., Pl. Bequaert. **5**: 458 (1932). Types from Zaire.

A scandent diffuse short-lived perennial herb up to 2 m. high. Stems up to c. 5 m. long, slender terete, ± leafless and stramineous below, leafy and much entangled above, glabrous or rarely sparsely setulose. Leaves imparipinnately compound, up to c. 14.5 × 8 cm., narrowly ovate in outline, with 3–5 pinnae on a narrowly-winged midrib, the midrib narrowly-winged petiole-like and auriculate below, denticulate or entire, glabrous or sparsely setose beneath; the terminal pinna ± sessile on the winged midrib, 2–5.5 × 1.3–3 cm., narrowly ovate, acute-acuminate above, cordate below; lateral pinnae subopposite, petiolules 1–10 mm. long and not or hardly winged, 2–4.5 × 0.8–2.6 cm., oblong-ovate, ± acute mucronate above and cuneate to cordate or asymmetric at the base. Synflorescences paniculate, axillary and terminal, mostly amongst the foliage; branches diffusely dichotomously subdivided, bracteate; bracts leaf-like ± oblanceolate denticulate auriculate. Capitula numerous; stalks up to c. 2.5(4.5) cm. long with 1–several c. 2 mm. long ovate-lanceolate bracteoles. Involucres c. 9 × 2 mm. and cylindric at time of flowering, increasing to c. 12 × 2.5 mm. and becoming conical in fruiting capitula; phyllaries c. 2–5 mm. long and broadly ovate on the outside, c. 12 mm. long and narrowly lanceolate on the inside, obtuse and ciliate above, glabrous. Florets c. 5 per capitulum; corollas up to c. 14 mm. long, tube slender cylindric and pilose above, ligule yellow 6–8 × 1.5–2 mm., lorate. Achenes pale-brown, up to c. 6 × 1.5 mm. including a c. 1.5 mm. long beak, narrowly ellipsoid, somewhat compressed with swollen ribs on the margins and 4–5 narrower wing-like ribs on each face, hispidulous; pappus several-seriate, of white barbellate setae c. 7 mm. long.

Malawi. N: Rumphi Distr., Nyika Plateau, Kafwimba Rain Forest, c. 1900 m., 15.viii.1975, *Pawek* 10051 (K; MAL; MO; SRGH; UC).

Also in Zaire, Ethiopia, Uganda, Kenya and Tanzania. In high altitude evergreen forests, often as a pioneer component of cleared forest.

17. **Lactuca glandulifera** Hook.f. in Journ. Linn. Soc., Bot. **7**: 203 (1864). —Oliv. & Hiern in F.T.A. **3**: 454 (1877). —Engl., Hochgebirgsfl. Trop. Afr.: 455 (1892). —O. Hoffm. in Engl., Pflanzenw. Ost-Afr. **C**: 421 (1895). —R.E. Fr. in Acta Hort. Berg. **9**: 161 (1929). —Stebbins in Bull. Jard. Bot. Brux. **14**: 350, pl. 7(e–f), (1937). —Robyns, Fl. Sperm. Parc Nat. Alb. **2**: 604 (1947). —Brenan in

Mem. N.Y. Bot. Gard. **8**, 5: 489 (1954). —Adams in F.W.T.A., ed. 2, **2**: 293 (1963). —C. Jeffrey in Kew Bull. **18**: 456, fig. 3(33), (1966). —Agnew, Upland Kenya Wild Fl.: 500 (1974). —Dethier in Bull. Jard. Bot. Nat. Belg. **52**: 370 (1982). —Lawalrée in Fl. Rwanda, Spermat. **3**: 690. fig. 216(2), (1985). —Lawalrée, Dethier & Gilissen in Fl. Afr. Centr., Compos., Cichorioideae: 43 (1986). TAB. **37**fig. H. Type from the Cameroons.

Lactuca glandulifera forma *calva* R.E. Fr. in Acta Hort. Berg. **9**: 162 (1929). —Stebbins in Bull. Jard. Bot. Brux. **14**: 350 (1937). Types from Kenya.

Lactuca integrifolia De Wild., Pl. Bequaert. **5**: 456 (1932), non Bigel. (1824) nom. illegit. Type from Zaire.

Lactuca wildemaniana Stebbins in Bull. Jard. Bot. Brux. **14**: 349 (1937). —Robyns, Fl. Sperm. Parc Nat. Alb. **2**: 603 (1947). Type as for *L. integrifolia* De Wild.

Lactuca glandulifera var. *calva* (R.E. Fr.) Robyns, Fl. Sperm. Parc Nat. Alb. **2**: 606 (1947). —Brenan in Mem. N.Y. Bot. Gard. **8**, 5: 489 (1954). —Adams in F.W.T.A., ed. 2, **2**: 293 (1963). Type as for *L. glandulifera* forma *calva*.

A creeping or scandent, somewhat viscid, short-lived perennial herb up to 1.5(3) m. high. Stems up to 2(6) m. long, becoming leafless stramineous and often peeling-papyraceous below, ± densely leafy above with young growth tinged reddish-purple, ± densely glandular-setose or glabrescent. Leaves up to c. 10.5(12) × 5 cm., narrowly ovate-elliptic in outline, runcinate-pinnatipartite, usually 3-lobed, sometimes 5-lobed or simple, acicular-denticulate to subentire, ± densely glandular-setose especially on the midribs and reticulation beneath, midrib narrowly-winged and petiole-like towards the base becoming auriculate; terminal lobe up to c. 5(7) cm. long, ovate-sagittate, acuminate-mucronate rarely obtuse; lateral lobes up to c. 2.5 × 1.5 cm., recurved oblong-lanceolate, rounded to acute-mucronate, sessile or somewhat asymmetrically and abruptly narrowed at the base. Synflorescences paniculate, mostly terminal; branches dichotomous, densely glandular-setose, or glabrescent, bracteate. Capitula numerous, somewhat aggregated when mature; stalks up to c. 4 cm. long, filiform, densely glandular setose to glabrescent, with 1–several bracteoles. Involucres up to c. 10 × 2 mm. and narrowly cylindrical at time of flowering, increasing to c. 14 × 3 mm. and becoming narrowly conical in fruiting capitula; phyllaries increasing somewhat abruptly in length towards the inside, narrowly lanceolate, obtuse-acute, glabrous or sparsely glandular setose, usually purplish outside, ± spirally twisted or with involute margins when dry. Florets c. 5 per capitulum; corollas c. 12 mm. long, tube slender and minutely pilose above, ligule yellow often purplish outside, c. 6 × 1.5 mm., lorate. Achenes pale-brown, up to c. 6 mm. long including a c. 2 mm. beak, narrowly ellipsoid and somewhat compressed, with conspicuously swollen ribs on the margins and 4–6 corky wing-like ribs on each face, glabrous or hispidulous; pappus several-seriate of equal white barbellate setae, c. 6.5 mm. long.

Malawi. S: Thyolo (Cholo) Mt., c. 1200 m., 19.ix.1946, *Brass* 17656 (K; PRE; SRGH). **Mozambique**. Z: Gurué, between Marraquelo and Muchiouiua, 20.ix.1944, *Mendonça* 2132 (LISC).

Also from tropical W. Africa, Ethiopia, Zaire, Rwanda, Burundi and E. Africa. In montane evergreen forests, often as a pioneer in disturbed areas, and in montane grassland.

25. CREPIS L.

Crepis L., Sp. Pl. **2**: 805 (1753); Gen. Pl., ed. 5: 350 (1754). —Babcock, The Genus Crepis [Univ. Calif. Publ. Bot., **21–22**: 1–1030 (1947)]. —C. Jeffrey in Kew Bull. **18**: 457–462 (1966).

Hieraciodes Kuntze, Rev. Gen. **1**: 344 (1891) nom. illegit.

Perennial (annual or biennial) herbs, subscapose from semi-woody rootstocks in the Flora Zambesiaca area, or caulescent, hispid or glandular-setose sometimes also tomentellous, or glabrous. Stems or flowering stalks erect, 1–several. Leaves radical and rosulate, or cauline and alternate, entire to coarsely toothed (or pinnatifid). Capitula solitary, or few to many corymbosely (paniculately) arranged. Involucres cylindric-campanulate. Phyllaries 2–several-seriate, the inner often thickening or carinate and longer than the outer in fruiting capitula. Receptacle epaleate, pitted, fimbrilliferous. Florets numerous, yellow, rarely white or pink; ligules usually longer than the tube; anthers sagittate at the base, auricles acute or shortly setaceous-acuminate; style branches long slender, sweeping-hairs medium to long. Achenes homomorphic, subterete, sometimes attenuate into an apical beak, uniformly many-ribbed. Pappus 1–many-seriate of shortly barbellate, persistent (deciduous) setae.

A genus of about 200 species, chiefly in temperate and subtropical Europe and Asia, a few in N. America and in tropical and southern Africa.

Achenes not beaked, subterete, tapering to the apex, usually 5–6 mm. long; inner phyllaries mostly glabrous, sometimes pubescent or setulose; immature capitula obovoid; leaves ± broadly oblanceolate-spathulate, often more than 6 cm. long   -   -   -   -   -   -   1. *newii*
Achenes ± distinctly beaked, subfusiform and constricted-attenuate above, usually 7–10 mm. long; inner phyllaries usually pubescent and often also glandular-setose; immature capitula obconic; leaves oblanceolate usually less than 6 cm. long   -   -   -   -   -   2. *hypochaeridea*

1. **Crepis newii** Oliv. & Hiern in F.T.A. **3**: 449 (1877). —O. Hoffm. in Engl., Pflanzenw. Ost-Afr. **C**: 422 (1895). —Muschl. in Mildbraed, Deutsch Zentr.-Afr.-Exped. 1907–1908, **2**: 409 (1910). —R.E. Fr. in Svensk. Bot. Tidskr. **22**: 355 (1928). —Babcock in Univ. Calif. Publ. Bot. **22**: 369, t. 74, 75 (1947). —Brenan in Mem. N.Y. Bot. Gard. **8**, 5: 489 (1954). —Adams in F.W.T.A. ed. 2, **2**: 294 (1963). —C. Jeffrey in Kew Bull. **18**: 459 (1966). —Agnew, Upland Kenya Wild Fl.: 501 (1974). —Lawalrée in Fl. Rwanda, Spermat. **3**: 695 (1985). —Lawalrée, Dethier & Gilissen in Fl. Afr. Centr., Compos., Cichorioideae: 22 (1986). TAB. **38** fig. A. Type from Tanzania.
  Crepis swynnertonii S. Moore in Journ. Bot. **54**: 285 (1916). —Babcock in Univ. Calif. Publ. Bot. **22**: 388, fig. 85 (1947). Type: Zimbabwe, *Swynnerton* s.n. (BM, holotype).
  See C. Jeffrey in Kew Bull. **18**: 459 (1966) for full synonymy.

An erect subscapose perennial herb, up to c. 90 cm. tall from a large semi-woody rootstock. Flowering stalks erect often solitary, branching above, markedly striate, usually glabrous, sometimes sparsely glandular-setulose becoming hispid. Leaves mostly radical in a rosette, usually c. 10 × 2 cm. sometimes up to c. 36 × 6 cm., oblanceolate-spathulate, acute or rounded mucronate at the apex, narrowly cuneate below, ± coarsely recurved-dentate on the margins, less often denticulate or lobed, glabrous, or sparsely glandular-setose beneath, midribs usually ± densely setose; cauline leaves when present smaller and linear-lanceolate. Capitula erect, few to many in corymbose clusters on the branches; stalks 0.3–4 cm. long, ± densely pubescent. Involucres obovoid in bud, ± cylindric later. Phyllaries few-seriate, becoming swollen at the base, glabrous outside or puberulent, sometimes dark glandular-setulose on the midribs; the outer 2–6 × 0.5–1 mm. and subulate, extending briefly onto the capitulum stalk; the inner up to c. 12 × 1.5 mm. becoming linear-lanceolate. Corollas yellow, up to c. 15 mm. long; ligule c. 10 × 2.5 mm., strap-shaped. Achenes brown ripening to a bright orange-red, 4.5–7 × 0.5–0.75 mm., subterete, tapering above but not drawn out into a beak, finely uniformly ribbed, ribs antrorsely muricate; pappus of numerous, barbellate setae, up to 6 mm. long.

Zimbabwe. E: Chimanimani Distr., Tarka For. Res., c. 1200 m., x.1968, *Goldsmith* 147/68 (K; COI; SRGH). **Malawi**. N: Viphya Plateau, c. 30 km. SW. of Mzuzu, c. 1680 m., 12.xi.1972, *Pawek* 5937 (K; SRGH). C: Ntchisi, c. 1350 m., 9.ix.1946, *Brass* 17573 (K; SRGH). S: Mt. Mulanje, Likhubula (Likabula) Valley, c. 1200 m., 17.x.1941, *Greenway* 6336 (K). **Mozambique**. N: Lichinga (Vila Cabral), xii.1934, *Torre* 430 (LISC).
  Also from Rwanda, Burundi, Zaire, Uganda, Kenya and Tanzania. Locally frequent in montane grassland, on forest margins and in grassy dambos, sometimes in cultivated lands.
  Usually glabrous or somewhat pubescent in the Flora Zambesiaca area, this species becomes hispid-setose on the involucre and synflorescence branches further north.
  C. oliveriana (Kuntze) C. Jeffrey, from Kenya, Tanzania and the Cameroons, may be distinguished from *C. newii* and *C. hypochaeridea* by its larger capitula, its setose involucres and capitulum stalks and by its darkly pigmented phyllaries (olive-green or blue-black, at least when dry).

2. **Crepis hypochaeridea** (DC.) Thell. in Viert. Nat. Ges. Zürich, **66**: 255 (1921). —Babcock in Univ. Calif. Publ. Bot. **22**: 357, t. 68 (1947). —Sussenguth & Merxm. in Proc. & Trans. Rhod. Sci. Assn. **43**: 64 (1951). —C. Jeffrey in Kew Bull. **18**: 460 (1966). —Hilliard, Comp. Natal: 634 (1977). —Lawalrée, Dethier & Gilissen in Fl. Afr. Centr., Compos., Cichorioideae: 23, pl. 4 (1986). TAB. **38** fig. B. Type from South Africa (Cape Province).
  Anisoramphus hypochoerideus DC., Prodr. **7**: 251 (1838). —Sond. in Harv. & Sond., F.C. **3**: 530 (1865). Type as above.
  Crepis bumbensis Hiern, Cat. Afr. Pl. Welw. **1**, 3: 616 (1898). —S. Moore in Journ. Bot. **65**, Suppl. 2: 66 (1927). —Mendonça, Contrib. Conhec. Fl. Angol., **1** Compositae: 148 (1943). Type from Angola.
  Crepis simulans S. Moore in Journ. Bot. **54**: 286 (1916). —Babcock in Univ. Calif. Publ. Bot. **22**: 393, t. 88 (1947). Type: Zimbabwe, *Swynnerton* s.n. (BM, holotype).
  Crepis chirindica S. Moore in Journ. Bot. **54**: 286 (1916). —Babcock in Univ. Calif. Publ. Bot. **22**: 363, t. 71 (1947). Type: Zimbabwe, *Swynnerton* s.n. (BM, holotype).
  Crepis congoensis Babcock in Bull. Jard. Bot. Brux. **14**: 301 (1937); in Univ. Calif. Publ. Bot. **22**: 365, t. 72 (1947). Type from Zaire.

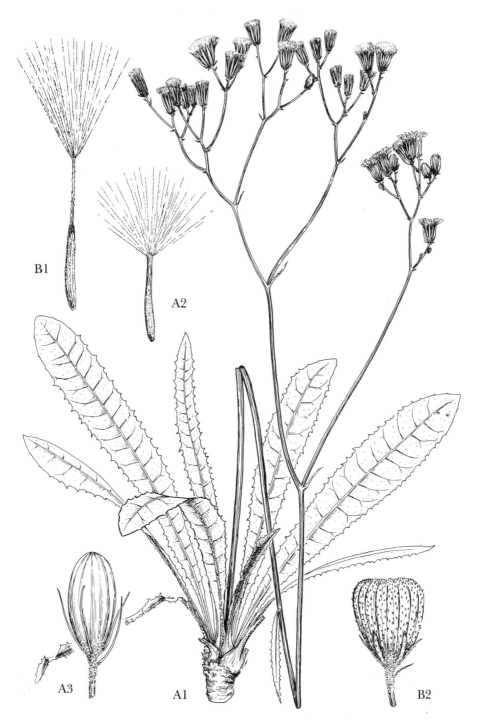

Tab. 38. A. —CREPIS NEWII. A1, habit (× $\frac{2}{3}$); A2, achene (× 4), A1 & A2 from *Pawek* 2975; A3, immature capitulum (×3), from *Richards* 10390. B. —CREPIS HYPOCHAERIDEA. B1, achene (× 4), from *Richards* 6110; B2, immature capitulum (× 3), from *Richards* 5982. Drawn by Eleanor Catherine.

Crepis hypochaeridea subsp. *rhodesica* Babcock in Univ. Calif. Publ. Bot. **19**: 400 (1941); op. cit., **22**: 360, t. 69 (1947). Type: Zimbabwe, *Walters* 2322 (K, holotype).

An erect subscapose perennial herb, up to c. 50 cm. tall from a stout semi-woody taproot. Flowering stalks 1 to several, simple or sparingly branched, markedly striate, glabrous to stiffly glandular-setose, often also white-pubescent above. Leaves numerous, mostly radical in a rosette, up to c. 10 × 1.5 cm., exceptionally to c. 15 × 3 cm., linear-oblanceolate to oblanceolate, rounded to acute mucronate at the apices and narrowly tapering below, sinuate denticulate to ± recurved-dentate, sometimes coarsely toothed, glabrous to ± densely hispid or glandular-setose; upper leaves, when present, usually bract-like and sessile. Capitula erect, solitary at ends of branches, or few in lax corymbs. Involucres obconical and somewhat flat-topped in bud becoming broadly cylindric-campanulate. Phyllaries few-seriate, glabrous or pubescent and ± densely glandular-setose outside, the setae sometimes black and up to c. 1.5 mm. long, the midribs swelling at the base; the outer phyllaries 2–6 × 0.5–1 mm., linear, loosely arranged and extending briefly onto the capitulum stalk; the inner phyllaries up to c. 13 mm. long becoming linear-lanceolate. Corollas yellow, up to 12 mm. long, tubes short, ligules strap-shaped. Achenes brown ripening to a bright orange-red, distinctly beaked, 7–10 mm. long including the 1.5–4 mm. long attenuate beak, fusiform or subterete, finely uniformly ribbed, the ribs antrorsely muricate; pappus of numerous barbellate setae c. 5 mm. long.

Zambia. N: Mbala Distr., Mbala-Mbeya road, c. 1500 m., 4.ix.1956, *Richards* 6110 (K). W: Chingola, 30.ix.1955, *Fanshawe* 2471 (K; SRGH). C: Munshiwemba, x.1940, *Stohr* 325 (BOL; PRE). **Zimbabwe**. N: High veld, 1200–1500 m., *Walters* 2322 (K; SRGH). C: Makoni Distr., 5 km. NW. of Headlands, c. 1550 m., 10.x.1965, *Chase* 8312 (K; SRGH). E: Chirinda, 1150 m., 20.x.1947, *Wild* 2071 (COI; K; SRGH). **Malawi**. N: Nyika Plateau, Kasaramba road, c. 1 km. from view, c. 2400 m., 4.iii.1977, *Pawek* 12452B (K; MO; MAL).

Also in Nigeria, Cameroon, Zaire, Angola, Tanzania and South Africa. A pyrophyte of open often seasonally water-logged grassland, or miombo woodland.

C. hypochaeridea is variable in facies, leaf size and achene shape over its range of distribution; in material from Natal (South Africa) the leaves can be larger and deeply lobed, and the involucres densely hispid-setose. However, intermediates occur and the material is treated as a single species.

26. TARAXACUM Weber ex Wiggers

Taraxacum Weber ex Wiggers, Prim. Fl. Hols.: 56 (1780) nom. conserv.

Perennial scapigerous herbs with simple or branched taproots. Leaves radical, rosulate, entire, sinuate-dentate or runcinate. Scapes 1–many, simple, hollow; capitula large solitary terminal. Involucres oblong-campanulate; phyllaries 2-seriate, the inner series erect, the outer series shorter and spreading or recurved. Receptacle ± flat, epaleate, pitted. Corollas yellow (white), ligulate; anthers sagittate at the base with shortly setaceous-acuminate auricles; style branches long slender, sweeping hairs long. Achenes fusiform-oblanceolate, many-ribbed, narrowed and shortly conical above before tapering ± abruptly into a slender stalk-like beak (rostrum), beaks rarely absent, the ribs antrorsely muricate or echinate above. Pappus many-seriate consisting of numerous slender, unequal setae.

A genus of the north temperate regions and temperate South America, with some species now widely distributed as weeds: "dandelions".

Apomixis, a common feature of the genus, has given rise to numerous genetically invariable taxa, some of which are adventive in Africa. *Taraxacum* material from the Flora Zambesiaca area probably represents a number of apomictic forms and for the time being is treated as belonging to a single aggregate species, as yet unnamed. *Taraxacum officinale*, the name that has been widely used for these plants, is now seen to be correctly applied to plants from the Scandanavian arctic — A.J. Richards [in Taxon **34**: 633–644 (1985)] having lectotypified *Leontodon taraxacum* L. by the specimen collected by Linneaus in Lapland. [See also A.J. Richards in Watsonia **9**, suppl., (1972)].

Taraxacum sp. agg.

Taraxacum officinale sensu Humbert, Fl. Madag., Comp. III: 873 (1963). —C. Jeffrey in Kew Bull. **18**: 463 (1966). —Henderson & Anderson, Common Weeds in S. Africa: 402, t. 200 (1966). —Hilliard in Ross, Fl. Natal: 378 (1973). —Agnew, Upland Kenya Wild Fls.: 496 (1974). —Blundell, Wild Fls. of Kenya: 87, t. 17 (1982). —Lawalrée in Fl. Rwanda, Spermat. **3**: 684, fig. 215(2), (1985). —Lawalrée, Dethier & Gilissen in Fl. Afr. Centr., Compos., Cichorioideae: 10 (1986). TAB. **39**.

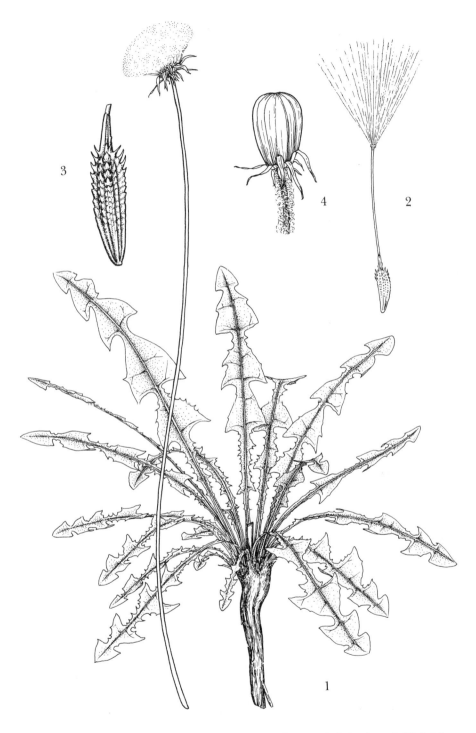

Tab. 39. TARAXACUM SP. AGG. 1, habit (×⅔); 2, achene and pappus (× 4); 3, achene (× 10), 1–3 from *Fanshawe* 11419; 4, immature capitulum (× 2), from *Biegel* 3174. Drawn by Eleanor Catherine.

A scapigerous perennial herb with a simple or branched taproot. Leaves many radical rosulate, 5–11 × 1.5–3.5 cm., exceptionally to c. 34 × 6 cm., oblanceolate in outline, recurved-dentate to runcinate-pinnatifid, glabrous or somewhat pubescent; apical lobe ovate-hastate; lateral lobes of 2–6 pairs, spreading recurved-falcate to broadly triangular, acute, margins entire to laciniate-dentate, sinuses sometimes strongly toothed. Scapes 1–several stout simple, 1–40 cm. tall, glabrous often ± densely lanate at first. Capitula solitary, terminal; involucres up to c. 15 × 9 mm. in fruiting heads, oblong-campanulate. Phyllaries glabrous, sometimes subapically corniculate; the inner series olive-green outside, up to c. 14 mm. long, linear-lanceolate with scarious margins; the outer series brownish-purple when dry, shorter, spreading or reflexed. Corollas yellow, up to c. 10 mm. long; ligule with a darker strip outside; style arms yellowish when dry. Achenes pale- to dark-brown, 2.5–3 mm. long with a beak 4–7 mm. long, oblanceolate, 10–14-ribbed, ± abruptly narrowed above into a cone 0.75–1 mm. long which tapers into the slender beak; ribs uniform narrow antrorsely muricate or echinate above; pappus white, of numerous fine minutely barbellate setae 4–6 mm. long.

Zambia. W: Kitwe, 21.iv.1972, *Fanshawe* 11419 (K; NDO). **Zimbabwe**. W: Matobo Distr., Matopos Research Station, 20.xii.1968., *Mangona* s.n. (SRGH). C: Harare, traffic island, Second Street Extension — East Road, 1460 m., 19.viii.1969, *Biegel* 3174 (SRGH). **Mozambique**. MS: Quinta da Fronteira (Border Farm), 25.i.1966, *Chase* 8365 (SRGH).

Introduced and usually naturalised in the Cameroons, Rwanda, Burundi, Zaire, Kenya, Tanzania and South Africa. A weed of lawns and plant nurseries.

27. LAUNAEA Cass.

Launaea Cass. in Dict. Sci. Nat. **25**: 321 (1822). —C. Jeffrey in Kew Bull. **18**: 463–477 (1966).

Annual perennial or biennial herbs (spiny subshrubs), glabrous, with or without stems, sometimes caespitose, occasionally stoloniferous. Stems or flowering stalks 1–several, branching above or simple. Leaves cauline and alternate, or radical and rosulate, entire or sinuate-dentate to runcinate-pinnatipartite, often acicular-denticulate on the margins. Capitula few to numerous, subsessile or shortly stalked, often in lax irregular corymbs, sometimes solitary, or in clusters along the branches. Receptacle epaleate. Involucres narrowly cylindric-campanulate. Phyllaries 2–several-seriate, becoming ± swollen and corky towards the base; the innermost subequal, usually greatly exceeding the outer. Florets few to many, yellow, or the ligule sometimes purplish or red outside; anthers sagittate at the base with the auricles shortly setaceous-acuminate; style branches long slender, sweeping hairs long. Achenes pale-brown to black, subterete or somewhat angular, or compressed, sometimes apically attenuate or beaked, ribbed; ribs 5–many, uniformly narrow or somewhat inflated, smooth or muricate, glabrous. Pappus many-seriate, of setae only or of setae intermixed with numerous down-like hairs.

A genus of 30–40 species tropical and subtropical in distribution, extending from the Canary Islands through the Mediterranean region to E. Asia and down through Africa to South Africa and to Madagascar.

1. Stem prostrate; plant stoloniferous with a leaf rosette at each node; achene ribs ± spongy and inflated; leaves coriaceous   -   -   -   -   -   -   -   -   -   - 5. *sarmentosa*
- Stem or scape erect, plants sometimes tufted, leaf rosettes 1 per plant; achene ribs usually narrow, becoming narrowly spongy; leaves membranous to thinly fleshy   -   -   - 2
2. Plants scapigerous, sometimes ± caespitose, perennial; flowers usually precocious   3
- Plants with leafy stems, annual or biennial; flowers and leaves appearing together   6
3. Involucres not exceeding 10 × 2 mm.; florets c. 3–5 per capitulum   -   -   - 4. *rogersii*
- Involucres more than 10 × 2 mm.; florets mostly more than 5 per capitulum   -   - 4
4. Florets 20–25 per capitulum   -   -   -   -   -   -   -   -   -   - 2. *nana*
- Florets 6–12 per capitulum   -   -   -   -   -   -   -   -   -   -   - 5
5. Phyllaries mostly 4–6-seriate, imbricate, with middle phyllaries up to half as long as the involucre, the inner phyllaries 7–9 subequal, in 2–3 series   -   -   -   -   -   - 3. *violacea*
- Phyllaries mostly 2–3-seriate, the inner phyllaries markedly longer than the outer 1–2 series with intermediate sized phyllaries absent from between them, the inner phyllaries c. 5 subequal in a single series   -   -   -   -   -   -   -   -   -   -   -   - 1. *rarifolia*
6. Pappus homogeneous, of setae only; mature achenes pale creamy-brown, rounded to ± abruptly tapered at the apex   -   -   -   -   -   -   -   -   -   -   - 6. *cornuta*
- Pappus heterogeneous, of setae intermixed with numerous down-like hairs; mature achenes greyish-black, constricted and ± distinctly beaked at the apex   -   -   - 7. *intybacea*

1. **Launaea rarifolia** (Oliv. & Hiern) L. Boulos in Bot. Not. **115**: 59 (1962). —C. Jeffrey in Kew Bull. **18**: 464 (1966). —Merxm., Prodr. Fl. SW. Afr. 140: 3 (1967). —Hilliard, Comp. Natal: 620 (1977). —Vollesen in Opera Bot.: 86 (1980). —Lawalrée in Fl. Rwanda, Spermat. **3**: 687, fig. 215(3), (1985). —Lawalrée, Dethier & Gilissen in Fl. Afr. Centr., Compos., Cichorioideae: 16 (1986). TAB. **40** fig. A. Types: Malawi, *Meller* s.n. (K, syntypes) and from Nigeria.

    Sonchus rarifolius Oliv. & Hiern in F.T.A. **3**: 460 (1877). —O. Hoffm. in Engl., Pflanzenw. Ost-Afr. **C**: 421 (1895). —R.E. Fr., Wiss. Ergebn. Schwed. Rhod.-Kongo-Exped. 1911–1912, **1**, 2: 350 (1916); in Acta Hort. Berg. **8**: 113 (1924). —Sussenguth & Merxm. in Proc. & Trans. Rhod. Sci. Assn. **43**: 71 (1951). —Adams in F.W.T.A. ed. 2, **2**: 296 (1963). Types as above.

    Lactuca welwitschii Scott-Elliot in Journ. Linn. Soc., Bot. **29**: 30 (1891). —Humbert, Comp. Madag.: 156 (1923). Types from Angola and Madagascar.

    Sonchus macer S. Moore in Journ. Bot. **37**: 404 (1899). —Eyles in Trans. Roy. Soc. S. Afr. **5**, 4: 523 (1916). —R.E. Fr. in Acta Hort. Berg. **8**: 117 (1924). Type: Zimbabwe, *Rand 627* (BM, holotype).

    Lactuca verdickii De Wild. in Ann. Mus. Congo, sér 4, **1**: 170 (1903). Type from Zaire.

    Sonchus violaceus sensu Eyles in Trans. Roy. Soc. S. Afr. **5**, 4: 523 (1916) quoad *Rogers* 4062.

    Sonchus verdickii (De Wild.) R.E. Fr. in Acta Hort. Berg. **8**: 113 (1924). Type as for *Lactuca verdickii*.

    Lactuca varianii I.M. Johnst. in Contrib. Gray Herb., n.s., **75**: 26 (1925). Type from Angola.

    Sonchus welwitschii (Scott-Elliot) Chiov. ex S. Moore in Journ. Bot. **65**, Suppl. 2: 67 (1927). —Mendonça, Contrib. Conhec. Fl. Angol., **1** Compositae: 152 (1943). Type as for *Lactuca welwitschii*.

    Launaea macer (S. Moore) L. Boulos in Bot. Not. **115**: 58 (1962). Type as for *Sonchus macer*.

    Launaea verdickii (De Wild.) L. Boulos in Bot. Not. **115**: 58 (1962). Type as for *Lactuca verdickii*.

A scapose strict wiry perennial herb up to c. 95 cm. tall, or sometimes a contracted small caespitose pulviniform herb 3–15 cm. high, often glaucous; taproot slender terete semi-woody; rootcrowns 1–several, usually shortly brown-lanate. Leaves in basal rosettes, usually appearing after the flowers, up to c. 25 × 4 cm., exceptionally to 35 × 5.5 cm., obovate or narrowly oblanceolate to subspathulate, linear when young, obtuse or acute at the apex, attenuate below, entire to callose-tipped denticulate or sometimes coarsely dentate-lobulate on the margins, glabrous. Scapes often purple-tinged below, 1–several, wiry, ± sparsely branched above, glabrous; or scapes much reduced, diffusely-branched and aggregated in dense clusters, c. 3–15 cm. tall. Capitula few to many, erect; capitula stalks bracteate, bracts ± ovate, lanate in the axils. Involucres in flowering capitula c. 12–15 × 2 mm. and very narrowly ovoid-cylindric, up to c. 19 × 3.5 mm. in mature capitula. Phyllaries green, often purple-tinged, mostly 2–3(4)-seriate, glabrous, occasionally puberulous outside, the inner often ciliolate about the apex, each usually with a short subapical wing-like glandular outgrowth on the midrib; outer phyllaries mostly 1–2-seriate imbricate, several extending onto the capitulum stalk, 2–3 mm. long, ovate to lanceolate, margins broadly hyaline; inner phyllaries c. 5, markedly longer than the outer, subequal, linear-lanceolate, accrescent, eventually reflexed. Florets c. 6–12 per capitulum; corollas yellow, 13–15 mm. long, tube puberulous outside near the throat, ligule drying purplish c. 6–8 mm. long, lorate, glabrous. Achenes 5–8 mm. long, narrowly subfusiform, somewhat 4–5-angular or ± compressed, finely 12–14-ribbed, glabrous or hispidulous, greyish-green; ribs often minutely barbellate, becoming narrowly spongy; pappus many-seriate of barbellate setae 7–10 mm. long.

Phyllaries greenish, sometimes purple-tinged; outer phyllaries glabrous; scapes usually up to 95 cm. tall, sometimes contracted and ± aggregated, 3–15 cm. tall, much branched   var. *rarifolia*
Phyllaries brown to reddish-purple; outer phyllaries usually ciliolate; scapes greatly abbreviated, 0–6 cm. tall, ± branched   -   -   -   -   -   -   -   -   -   -   var. *nanella*

Var. **rarifolia**

Plants usually scapose, sometimes caespitose-pulviniform; scapes 1–several, wiry up to 95 cm. tall, or sometimes scapes few–many aggregated, contracted and ± much-branched, 3–15 cm. high. Phyllaries greenish, sometimes purple-tinged; outer phyllaries glabrous not ciliolate.

Botswana. N: Chobe Nat. Park, between Serondella and Ngwezumba R., 17.x.1972, *Pope, Biegel & Russell* 803 (SRGH). **Zambia**. B: Kalabo, 16.x.1963, *Fanshawe* 8095 (K; NDO). N: Mbala, golf course, 22.ix.1949, *Bullock* 1048 (K). W: Kitwe, 13.ix.1968, *Mutimushi* 2686 (K; NDO). C: 10 km. E. of Lusaka, 1300 m., 30.ix.1955, *King* 159 (K). E: Lundazi, 18.x.1967, *Mutimushi* 2244 (K; NDO). S: Choma, 28.ix.1961, *Fanshawe* 6738 (K). **Zimbabwe**. N: Mwami (Miami), c. 1380 m., 4.x.1946, *Wild* 1265 (K; SRGH). W: Matobo Distr., Farm Besna Kobila, c. 1460 m., x.1957, *Miller* 4570 (K; SRGH). C: Harare, Hatfield, 9.ix.1969, *Pope* 173 (K; SRGH). E: Chirinda, c. 1160 m., 20.x.1947, *Wild* 2062 (K; SRGH).

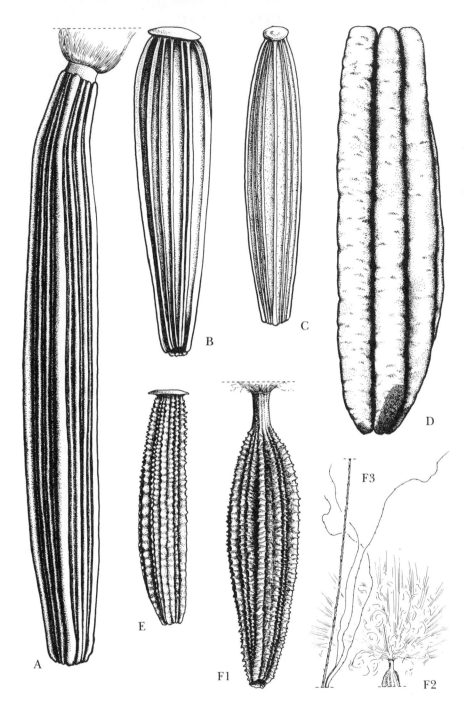

Tab. 40. LAUNAEA SPP. Achenes (all × 20). A. —LAUNAEA RARIFOLIA, from *King* 159. B. —LAUNAEA NANA, from *Bingham* 969. C. —LAUNAEA ROGERSII, from *Milne-Redhead* 2642. D. —LAUNAEA SARMENTOSA, from *Mendonça* 3285. E. —LAUNAEA CORNUTA, from *Plowes* 2068. F. —LAUNAEA INTYBACEA. F1, achene (× 20); F2, pappus (× 4); F3, dimorphic pappus elements (× 28), F1–F3 from *Wild & Drummond* 6819. Drawn by Eleanor Catherine.

Malawi. N: Mzimba Distr., Mzuzu, Marymount, c. 1370 m., 7.xi.1976, *Pawek* 11924 (K; MAL; MO; SRGH). C: Dedza Distr., Chongoni For. Res., 21.ix.1967, *Salubeni* 843 (COI; SRGH). S: Thondwe (Ntondwe), 21.x.1905, *Cameron* 121 (K). **Mozambique**. N: Lichinga (Vila Cabral), 28.x.1934, *Torre* 429 (COI; LISC). Z: Entre o Alto Molócuè e o cruzamento para Ile, a 25 km. do Alto Molócuè, 16.x.1949, *Barbosa & Carvalho* 4450 (K). MS: Manica, Rotanda near Messambuzi, c. 1000m., 17.xi.1965, *Torre & Correia* 12989 (LISC).

Widespread in subsaharan Africa and in Madagascar. A pyrophyte of grassland and miombo woodland.

For the most part these plants are scapose with erect wiry, often solitary, flowering stalks. However, ecological variants occur and are usually short, caespitose and pulviniform in habit. Such specimens are recorded throughout the distribution range of the species. On the Nyika and Viphya plateaux in Malawi, at higher altitudes, these plants often have dark-green phyllaries, *Richards* 22605 (K), *Pawek* 2795, 11792, 11878 (K).

A variant found in Barotseland (Zambia) with the flowers and young leaves ± contemporaneous, is characterised by the scapes being somewhat decumbent, and the subapical wing-like glandular outgrowth on the phyllary midrib being well developed. Examples of this are seen in *Fanshawe* 5807, 8095 and 8976 (K).

Var. **nanella** (R.E. Fr.) G.V. Pope comb. et stat. nov. Type: Zambia, between Samfya (Fort Rosebury) and Lwela (Luera) River, *Fries* 510a (UPS, holotype).
    Sonchus nanellus R.E. Fr., Wiss. Ergebn. Schwed. Rhod.-Kongo-Exped. 1911–1912, **1**: 351 (1916); in Acta Hort. Berg. **8**: 117 (1924). Type as above.
    Launaea nanella (R.E. Fr.) L. Boulos, in Bot. Not. **115**: 58 (1962). Type as above.

Plants pulviniform; scapes greatly abbreviated, 0–6 cm. long, bearing few–many ± unbranched capitulum stalks. Phyllaries brown to reddish-purple; outer phyllaries usually ciliolate.

Zambia. W: Mwinilunga Distr., by R. Mwanamitowa, c. 16 km. W. of R. Lungu and c. 64 km. S. of the Boma, 15.viii.1930, *Milne-Readhead* 912 (K); Solwezi, 16.viii.1953, *Fanshawe* 246 (K). **Malawi**. N: 40 km. N. of Mzimba, 1.xi.1966, *Gillett* 17525 (K).

Also in Tanzania. An infrequently collected pyrophyte of *Brachystegia* woodlands and dambos.

2. **Launaea nana** (Bak.) Chiov., Result. Sc. Miss. Stef.-Paoli **1**: 108 (1916). —C. Jeffrey in Kew Bull. **18**: 466 (1966) pro parte. Agnew, Upland Kenya Wild Fl.: 496 (1974) —Hilliard, Comp. Natal: 620 (1977). —Vollesen in Opera Bot. **59**: 86 (1980). —Lawalrée in Fl. Rwanda, Spermat. **3**: 687, fig. 212 (1985). —Lawalrée, Dethier & Gilissen in Fl. Afr. Centr., Compos., Cichorioideae: 18 (1986). TAB. **40** fig. B. Types from Nigeria and Mozambique (K, syntypes not traced at K).
    Lactuca nana Bak. in Bull. Misc. Inf., Kew **1895**: 17 (1895). Type as above.
    Sonchus nanus O. Hoffm. in Engl., Pflanzenw. Ost-Afr. C: 421 (1895). —R.E. Fr. in Acta Hort. Berg. **8**: 108 (1924), non Sond. (1865) nom. illegit. Types: Mozambique, Beira, *Braga* 64, and also from Sierra Leone, Nigeria, Togo, Tanzania, Zaire and Angola (B†, syntypes).
    Sonchus elliotianus Hiern, Cat. Afr. Pl. Welw. **1**, 3: 623 (1898). —Eyles in Trans. Roy. Soc. S. Afr. **5**, 4: 523 (1916). —R.E. Fr., Wiss. Ergebn. Schwed. Rhod.-Kongo-Exped. 1911–1912, **1**: 351 (1916); in Acta Hort. Berg. **8**: 115 (1924). —S. Moore in Journ. Bot. **65**, Suppl. 2: 67 (1927). —Mendonça, Contrib. Conhec. Fl. Angol., 1 Compositae: 153 (1943). —F.W. Andr., Fl. Pl. Sudan **3**: 53 (1956). —Adams in F.W.T.A., ed. 2, **2**: 296 (1963). Type as for *Launaea nana* (Bak.) Chiov.
    Launaea elliotiana (Hiern) L. Boulos in Bot. Not. **115**: 58 (1962). Type as for *Sonchus elliotianus*.

An acaulescent perennial herb with a terete semi-woody taproot; rootcrown sometimes lanate. Leaves 2–6 radical, appearing after the flowers, glabrous, up to c. 20 × 6 cm., oblanceolate, linear when young, sinuate-dentate to pinnately-lobed, apices acute-obtuse, margins subciliate-denticulate, bases narrowly cuneate, midribs prominent beneath. Scapes rarely more than c. 8 cm. tall, much branched, glabrous; branches slender, bracteate. Capitula erect, numerous, densely aggregated in a pulviniform cluster up to c. 10 cm. in diam. Involucres 10–13 × 3–4 mm. in flowering capitula increasing to c. 18 × 5 mm. in fruiting capitula, cylindric, somewhat spreading later. Phyllaries green to dark-purple, 3–4-seriate, each usually with a short subapical wing-like outgrowth on the midrib, eventually swollen and corky towards the base, glabrous except for the ± densely ciliate apex, margins relatively narrowly hyaline; outer phyllaries few, imbricate, 2–6 mm. long, lanceolate; inner phyllaries 7–8, subequal, accrescent, eventually up to c. 18 mm. long, narrowly lanceolate. Florets c. 23 per capitulum; corollas yellow, 10–15 mm. long, ligule becoming purplish-brown c. 4–5 mm. long, lorate glabrous. Achenes reddish-brown, 4–5 × 0.75–1 mm., subfusiform-oblanceolate, somewhat 4-angular to ± compressed, narrowly 10–14-ribbed, glabrous; pappus copious, several-seriate of barbellate setae 8–10 mm. long.

Zambia. W: Luanshya, 6.ix.1963, *Fanshawe* 7981 (K; NDO). C: 8 km. E. of Lusaka, c. 1280 m., 26.x.1955, *King* 181 (K). S: Mazabuka, Central Res. Sta., c. 1000 m., 14.viii.1931, *Trapnell* in CRS. 420 (PRE). **Zimbabwe**. N: Gokwe River Fly Gate, 20.xi.1963, *Bingham* 969 (K; SRGH). W: Bulawayo, Bellevue, ix.1908, *Chubb* 348 (BM). C: Harare, 31.viii.1960, *Rutherford-Smith* 36 (K). E: Gungunyana For. Res., c. 1060 m., x.1961, *Goldsmith* 90/61 (SRGH). **Malawi**. N: Nkhata Bay Distr., Viphya, 42 km. SW. of Mzuzu, c. 1860 m., 1.xi.1967, *Pawek* 1499 (SRGH). C: Kasungu Game Res., c. 1080 m., 23.ix.1969, *Hall-Martin* 419 (PRE). S: Zomba, Matokoteza, 8.ix.1978, *Msiska* 201 (SRGH). **Mozambique**. N: Belém (Amaramba), Mandimba, 10 km. from Lichinga (Vila Cabral), 8.x.1942, *Mendonça* 640 (BM; COI; LISC). Z: Mocuba, Namagoa, 19.ix.1948, *Faulkner* Kew 303 (K).

Also from W. Africa, Sudan, Zaire, Rwanda, Uganda, Kenya, Tanzania, Angola and South Africa. A pyrophyte of grassland and open miombo woodland.

3. **Launaea violacea** (O. Hoffm.) L. Boulos in Bot. Not. **115**: 58 (1962). Type from Tanzania.
    Sonchus violaceus O. Hoffm. in Engl., Bot. Jahrb. **30**: 443, figs. A–G (1901). Type as above.

Similar to *L. nana* (Bak.) Chiov. Plants pulviniform, up to c. 6 cm. high, flowering before the leaves appear. Immature leaves oblanceolate, recurved callose-tipped denticulate, edges and midribs red; mature leaves not known. Scapes much abbreviated, branched, glabrous, glaucescent, bracteate; bracts ± attenuate at the apex and lanate in the axils. Involucres 12–16 × 2–4 mm. in flowering capitula, deep-purple, drying olive-green. Phyllaries 4–6-seriate, imbricate, glabrous except for the ± lanate-ciliate apices of the inner phyllaries, margins relatively narrowly hyaline; the outer phyllaries up to c. 3 mm. long, ovate to lanceolate, not ciliate on the margins; the middle phyllaries up to half the involucre in length; the innermost 7–9 phyllaries subequal and in 2–3 series. Florets 8–10 per capitulum. Corollas white pinkish or yellow, tubes pubescent outside near the throat, ligules erect. Achenes up to c. 5 mm. long, subcylindric, many-ribbed with ribs minutely barbellate, narrow at first later swelling, narrowly-spongy; pappus copious, several seriate of barbellate setae 8–10 mm. long.

Malawi. N: Rumphi Distr., Nyika, c. 2011 m., 1.x.1969, *Pawek* 2879 (K).
Also in Tanzania. A rare pyrophyte of montane grassland, mostly above 2000 m.
L. violacea differs from *L. rarifolia* mainly in its involucre, with phyllaries more numerous and clearly 4–6-seriate, the inner series containing 7–9 subequal phyllaries. In *L. rarifolia* the phyllaries are 2–3-seriate with a marked difference in length between the outer and inner series, and with the inner series containing c. 5 ± linear phyllaries.

4. **Launaea rogersii** (Humbert) Humbert & Boulos in Humbert, Fl. Madag., Comp. III: 882 (1963).
    —C. Jeffrey in Kew Bull. **18**: 467 (1966). —Lawalrée, Dethier & Gilissen in Fl. Afr. Centr., Compos., Cichorioideae: 14, pl. 3 (1986). TAB. **40** fig. C. Type from Zaire.
    Lactuca rogersii Humbert, Comp. Madag.: 156, 307 (1923). Syntypes from Zaire and Madagascar.
    Sonchus pycnocephalus R.E. Fr. in Acta Hort. Berg. **8**: 114, pl. 3(3–4), (1924). Type: Zambia, Bwana Mkubwa, *Fries* 489 (UPS, holotype).
    Launaea pycnocephala (R.E. Fr.) L. Boulos in Bot. Not. **115**: 58 (1962). Type as for *Sonchus pycnocephalus*.

A scapose, glabrous somewhat glaucous perennial herb up to c. 60 cm. tall, from a semi-woody rootstock. Leaves appearing after the flowers, 2–5 radical ± prostrate, up to c. 13 × 8 cm., elliptic-obovate to rotund, apices obtuse to rounded, margins entire or acicular-denticulate, bases ± broadly cuneate, lamina thinly succulent, glabrous and ± glaucous beneath. Flower stalks green or purple-tinged, usually solitary, wiry, branched above with divaricate-ascending slender branches, or flower stalks shorter and more densely branched. Capitula numerous, subsessile or shortly stalked, singly or in groups of 2–4 along the branches. Involucres c. 7 × 1.5 mm., narrowly cylindric, accrescent, becoming c. 10 mm. long and spreading; phyllaries green to purplish, glabrous, each often subapically corniculate, eventually swollen and corky towards the base; inner phyllaries 5, subequal, 7–10 mm. long, lorate-lanceolate, 4–5 times longer than the outer. Florets 3–4(6) per capitulum; corollas yellow becoming reddish-purple, up to 10.5 mm. long, ligules c. 5 mm. long, glabrous. Achenes light- or reddish-brown, c. 4.5 mm. long, narrowly subfusiform, finely ribbed, glabrous or minutely hispidulous; pappus many-seriate, composed of white barbellate setae 4–6 mm. long.

Zambia. W: Solwezi Distr., Jiwundu R., 14.ix.1930, *Milne-Redhead* 1122 (K). C: Mumbwa, *Macaulay* 887 (K).
Also in Angola and Zaire. A pyrophyte of seasonally waterlogged grassland (dambos).
The name *Lactuca rogersii* Humbert was based on two specimens – *Rogers* 10035 from Zaire and *Perrier de la Bathie* 2976 from Madagascar, with subsequent implied lectotypification on the former by

Humbert & Boulos in 1963 for *Launaea rogersii*. However, the latter specimen with its narrow leaves, lanate rootcrown and larger capitula and achenes is more correctly referred to *L. rarifolia* (Oliv. & Hiern) Boulos. Furthermore all material agreeing with *Rogers* 10035 has a western central African distribution, being found only in Angola, Zaire and Zambia.

5. **Launaea sarmentosa** (Willd.) Sch. Bip. ex Kuntze, Rev. Gen. **1**: 350 (1891). —O. Hoffm. in Engl., Pflanzenw. Ost-Afr. **C**: 421 (1895). —C. Jeffrey in Kew Bull. **18**: 468 (1966). —Hilliard, Comp. Natal: 621 (1977). TAB. **40** fig. D. Type from India.
　　Prenanthes sarmentosa Willd., Phyt.: 10, t. 6, fig. 2 (1794). Type as above.
　　Launaea bellidifolia Cass. in Dict. Sci. Nat. **25**: 321 (1822). —Oliv. & Hiern in F.T.A. **3**: 460 (1877). —Humbert, Comp. Madag.: 157 (1923); Fl. Madag., Comp. III: 883 t. 164, figs. 4–7 (1963). Type from Madagascar.
　　Microrhynchus dregeanus DC., Prodr. **7**: 181 (1838). —Harv. in Harv. & Sond., F.C. **3**: 527 (1865). Type from South Africa.

A glabrous perennial herb with a prostrate creeping stem; rootstock semi-woody. Stems up to c. 1.4 m. long, flagelliform, usually solitary, simple or sometimes branched, often rooting at the nodes, internodes c. 5–15 cm. long. Leaves few to numerous, ± densely crowded at the stem base and also in fascicles at the nodes, the leaf fascicles often subtended by 2 broadly ovate-reniform foliaceous bracts; basal leaves subcoriaceous, up to 11(14)–2.5(3) cm., spathulate to oblanceolate, apices rounded-obtuse, margins subentire to denticulate, rarely also sinuate-dentate to lobed, bases attenuate often subauriculate below; fascicle leaves similar but usually smaller. Capitula solitary or few from the rootcrown and at the nodes; stalks 1–3 cm. long, simple, bracteate; involucres 10–13 × 3–4 mm. in flowering capitula, up to c. 15 × 5 mm. long in fruiting capitula, narrowly oblong-cylindrical; phyllaries imbricate, increasing uniformly from 3–4 mm. long and ovate on the outside to 13–15 mm. long and lanceolate on the inside, all eventually swollen and corky towards the base. Florets 12–20 per capitulum; corollas yellow, c. 15 mm. long, ligule c. 7 mm. long, lorate. Achenes greyish-brown, 4–5 × 1.5 mm., subfusiform-cylindric, with c. 5 ± inflated spongy ribs; pappus copious of fine, white, minutely barbellate setae 9 mm. long.

Mozambique. N: Ilha de Moçambique, near Fostaleza, 30.x.1942, *Mendonça* 1161 (LISC). Z: Maganja da Costa, Raraga beach, c. 5 m., 15.ii.1966, *Torre & Correia* 14695 (LISC). MS: Beira, near Macuti Lighthouse, 18.ii.1961, *Goodier* 1042 (K; SRGH). GI: Inhambane, Zavala, Quissico, 6.xii.1944, *Mendonça* 3285 (LISC). M: Maputo, Costa do Sol, 2.iii.1960, *Balsinhas* 125 (BM; COI; K; LISC; PRE).
Along the east coast of Africa from Egypt to South Africa, and on the coasts of Madagascar, the Seychelles and Mauritius. Also in India, Sri Lanka and eastwards to Indo-China and Java. On sandy beaches and dunes.

6. **Launaea cornuta** (Hochst. ex Oliv. & Hiern) C. Jeffrey in Kew Bull. **18**: 468 (1966). —Agnew, Upland Kenya Wild Fl.: 496 (1974). —Vollesen in Opera Bot.: 86 (1980). TAB. **40** fig. E. Type from the Sudan.
　　Sonchus cornutus Hochst. ex Oliv. & Hiern in F.T.A. **3**: 459 (1877). —O. Hoffm. in Engl., Pflanzenw. Ost-Afr. **C**: 421 (1895). —R.E. Fr. in Acta Hort. Berg. **8**: 111 (1924). —F.W. Andr., Fl. Pl. Sudan **3**: 51 (1956). Type as above.
　　Sonchus bipontini var. *exauriculatus* Oliv. & Hiern in F.T.A. **3**: 459 (1877). Type from the Sudan.
　　Sonchus exauriculatus (Oliv. & Hiern) O. Hoffm. in Engl., Pflanzenw. Ost-Afr. **C**: 421 (1895). —R.E. Fr. in Acta Hort. Berg. **8**: 110 (1924). —Robyns, Fl. Sperm. Parc Nat. Alb. **2**: 598, pl. 53 (1947). —Brenan in Mem. N.Y. Bot. Gard. **8**, 5: 489 (1954). —F.W. Andr., Fl. Pl. Sudan **3**: 51 (1956). —Adams in F.W.T.A., ed. 2, **2**: 296 (1963). Type as for *Sonchus bipontini* var. *exauriculatus*.
　　Sonchus kabarensis De Wild., Pl. Bequaert. **5**: 463 (1932). Type from Zaire.
　　Launaea exauriculata (Oliv. & Hiern) Amin ex L. Boulos in Bot. Not. **115**: 59 (1962). —Lawalrée in Fl. Rwanda, Spermat. **3**: 687 (1985). —Lawalrée, Dethier & Gilissen in Fl. Afr. Centr., Compos., Cichorioideae: 11, pl. 2 (1986). Type as for *Sonchus exauriculatus*.

An erect glabrous long-lived annual herb, or perennial herb from a semi-woody rootstock. Stem usually solitary up to 1.7 m. tall, hollow leafy and branched above, or sometimes divaricately branched from near the base; branches up to c. 50 cm. long. Leaves sessile, mostly 5–10 × 1 cm., up to 25 × 3 cm. in well-grown plants, linear to narrowly elliptic, entire or with 1–3 pairs of lobes near the base, leaf and lobe apices acute, margins entire to acicular-denticulate, midrib prominent beneath; lobes 0.3–5.5 × 0.1–1 cm. Capitula numerous, in a lax ± diffusely branched panicle; capitula stalks up to c. 2.5 cm. long, with 0 to many reddish-tinged bracteoles. Involucres 7–10 × 3–4 mm., not markedly accrescent, ovoid-cylindric, later spreading, glabrous, rarely shortly white puberulous outside; phyllaries 2–3-seriate, 2–4 mm. long and narrowly ovate on the outside, up to 7–10 mm. long and narrowly lanceolate on the inside, green and often reddish outside, sometimes subapically corniculate, midribs becoming swollen and corky below. Florets 10–25 per capitulum; corollas yellow, up to c. 11 mm. long, ligule often becoming reddish outside, 7–8 mm. long, lorate. Achenes buff or pale-brown, 2.5–4 mm. long, subfusiform to narrowly obovoid, uniformly narrowly 6–14-ribbed, ribs muricate; pappus of white barbellate setae 4.5–5.5 mm. long, not exceeding the involucre.

Zambia. N: Mkupa above Mbuga, 8.x.1949, *Bullock* 1177 (K). C: Chilanga, Quien Sabe Ranch, c. 1100 m., 24.viii.1929, *Sandwith* 130 (K). E: Jumbe near Chikowa turn-off, c. 1000 m., 12.x.1958, *Robson & Angus* 62 (BM; LISC; SRGH). S: Kafue Basin, 26.iii.1963, *van Rensburg* 1804 (K; SRGH). **Zimbabwe**. N: Darwin Distr., SE. corner of Chiswiti Reserve, c. 800 m., 18.i.1960, *Phipps* 2336 (K; SRGH). E: Chipinge Distr., Chisumbanje, Lower Save Valley, c. 420 m., 4.ii.1959, *Plowes* 2068 (K; SRGH). S: Save-Runde (Sabi-Lundi) Junction, Chitsa's Kraal, 240 m., 6.vi.1950, *Wild* 3385 (K; LISC; SRGH). **Malawi**. N: Rumphi Distr., 7 km. west of Rumphi on road to Katumbi, Runyina bridge, 1070 m., 20.v.1970, *Brummitt* 10932 (K; SRGH). C: Nkhota Kota Distr., 6 km. N. of Benga turn-off by Lipyodzi bridge, 485 m., 15.vi.1970, *Brummitt* 11418 (K; SRGH). S: Nsanje Distr., between Muona and Shire R., 80 m., 20.iii.1960, *Phipps* 2567 (K; SRGH). **Mozambique**. N: Eráti, Namapa, C.I.C.A. Experiment Station, 27.iii.1961, *Balsinhas & Marrime* 311 (K; PRE; SRGH). Z: 7.3 km. from Megaza on road to Aguas Quentes, 14.vi.1949, *Barbosa & Carvalho* 3076 (SRGH). T: Cahora Bassa, 380 m., 17.ii.1973, *Correia, Marques & Belo-Correia* 3857 (LMU; SRGH). MS: Gorongosa Nat. Park, 5.xi.1963, *Torre & Paiva* 9050 (LISC).

Also from Nigeria, Cameroon, Central African Republic, Sudan, Ethiopia, Somali Republic, Rwanda, Zaire, Uganda, Kenya, Tanzania and Zanzibar. On alluvial soils and in disturbed ground at roadsides and in old cultivation near rivers.

7. **Launaea intybacea** (Jacq.) Beauverd in Bull. Soc. Bot. Genève sér 2, **2**: 114 (1910). —C. Jeffrey in Kew Bull. **18**: 472 (1966). —Merxm., Prodr. Fl. SW. Afr. 140: 3 (1967). —Agnew, Upland Kenya Wild Fl.: 496 (1974). TAB. **40** fig. F. Type cultivated in Vienna from seed from Central America.
    Lactuca intybacea Jacq., Icon. Pl. Rar. **1**: 16, t. 162 (1784). —Adams in F.W.T.A., ed. 2, **2**: 294 (1963). Type as above.
    Scorzonera pinnatifida Lour., Fl. Cochin. **2**: 479 (1790). Type: Mozambique, *Loureiro* (type not traced).
    Sonchus goreënsis Lam., Encycl. **3**: 397 (1791). Type cultivated at Paris from seed from Senegal.
    Lactuca goreënsis (Lam.) Sch. Bip. in Flora **25**: 422 (1842). —Oliv. & Hiern in F.T.A. **3**: 452 (1877). —Engl., Hochgebirgsfl. Trop. Afr.: 454 (1892). —Hiern, Cat. Afr. Pl. Welw. **1**, 3: 619 (1898). —Humbert, Comp. Madag.: 155 (1923). —S. Moore in Journ. Bot. **65**, Suppl. 2, Gamopet.: 66 (1917). —Mendonça, Contrib. Conhec. Fl. Angol., 1 Compositae: 149 (1943). Type as for *Sonchus goreënsis*.
    Launaea goreënsis (Lam.) O. Hoffm. in Engl., Pflanzenfam. IV, **5**: 370 (1893). —Humbert, Fl. Madag., Comp. III: 878 (1963). Type as for *Sonchus goreënsis*.
    Lactuca pinnatifida (Lour.) Merr., Comm. Lour. Fl. Cochin.: 397 (1935). —F.W. Andr., Fl. Pl. Sudan **3**: 38 (1956). Type as for *Scorzonera pinnatifida*.
    See C. Jeffrey in Kew Bull. **18**: 472 (1966) for full synonymy.

An erect glabrous annual herb. Stems up to c. 1(2) m. tall, hollow, leafy throughout or leaves ± crowded below, divaricately branched above; branches up to c. 50(65) cm. long, ascending. Leaves up to c. 24 × 9.5 cm., decreasing in size towards the stem apex, broadly oblanceolate-elliptic in outline, runcinate or remotely pinnately-lobed, apices obtuse-acute to rounded, margins strongly acicular-denticulate, bases sessile to subauriculate, midrib prominent beneath. Capitula numerous, subsessile to shortly-stalked, singly or in clusters of 2–8 spaced along the branches. Involucres c. 10 × 2.5 mm. in flowering capitula, up to c. 12 × 4 mm. in fruiting capitula, conical-cylindric at first later spreading. Phyllaries several-seriate, grading into the bracteoles of the capitulum stalk; the outer series imbricate increasing ± uniformly in length from c. 1.5–5 mm., then abruptly to 9–12 mm. long in the innermost, broadly ovate-acuminate on the outside and narrowly lanceolate-acute inside, all with a distinct hyaline margin, the midribs becoming swollen and corky towards the base. Florets 20–26 per capitulum; corollas yellow, 11–13 mm. long, ligule c. 5 mm. long and lorate becoming purplish. Achenes grey-black when ripe, 3–4 mm. long, subcylindric-ellipsoid somewhat flattened, constricted above or drawn out into a short distinct apical beak, ± 4-angular with ribs on the angles and 2 secondary ribs between the angles, ribs sharply rugose-muricate; pappus heterogeneous, white, copious, c. 7 mm. long, composed of stiff barbellate setae intermixed with numerous filamentous hairs.

Botswana. N: Makadikadi (Makarikari) Pan, between Nata and Odiakwe on Francistown road, 8.iii.1965, *Wild & Drummond* 6819 (K; LISC; SRGH). SE: Mosu Village near Sowa (Soa) Pan,

16.i.1974, *Ngoni* 342 (K; SRGH). **Mozambique**. T: Cahora Bassa, S. bank of Zambezi R., 225–330 m., 28.iv.1972, *Pereira & Correia* 2273 (LMU). GI: Inhambane, Vilanculos, 126 km. from Massinga on Save R., c. 100 m., 21.x.1967, *Torre & Correia* 15803 (LISC). M: Road between Chinguanine (Chinhanguanine) and Magude, c. 50 m., 5.vii.1948, *Myre* 20 (COI; PRE; SRGH).

Of scattered occurrence in semi-arid areas throughout the Old and New World tropics, including the islands. An infrequent herb, often in sandy soil on pan edges and on banks of dry water courses, also on coastal plains.

Jeffrey in Kew Bull. **18**: 474 (1966) notes that *Lactuca flava* Forssk., Fl. Aegypt.-Arab.: 143 (1775), described from the Yemen, may possibly provide the oldest epithet for *L. massauensis* but the description does not exclude the possibility of its representing *L. intybacea*, but in the absence of a type specimen it has not been possible to adopt the epithet for either of these species.

28. SONCHUS L.

Sonchus L., Sp. Pl. **2**: 793 (1753); Gen. Pl., ed. 5: 347 (1754). —C. Jeffrey in Kew Bull. **18**: 478–482 (1966). —Boulos in Bot. Not. **125**: 287–319 (1972); **126**: 155–196 (1973); **127**: 7–37, 402–451 (1974).

Annual, perennial or biennial herbs. Stems erect, sometimes scrambling, mostly solitary, branched above, leafy throughout or leaves ± crowded below, hollow (at least in the Flora Zambesiaca area). Leaves alternate, cauline, undivided or runcinate to pinnately-lobed, ± auriculate to sagittate, margins subentire or acicular-denticulate to irregularly and sharply dentate; basal leaves smaller, less dissected. Capitula few in terminal clusters, or many in lax compound corymbs, sometimes solitary, stalked or some subsessile. Receptacle pitted, epaleate. Involucres 1–2 times as long as broad, cylindric-campanulate to spreading, ± persistently tomentose below, sometimes glabrous. Phyllaries several-seriate, imbricate, densely tomentose to glabrescent, sometimes glandular-setose outside, the outer eventually ± swollen at the base. Florets numerous, yellow, the ligule sometimes reddish outside with age; anthers sagittate at the base with the auricles shortly setaceous-acuminate; style branches medium to long, sweeping hairs small. Achenes reddish-brown, narrowly obovoid to ellipsoid, ± compressed, glabrous, without a beak, ribbed; ribs 1–4, often with secondary ribs between, uniformly narrow, eventually ± swollen, or marginal ribs wing-like, smooth, rugulose or retrorsely and minutely tuberculate. Pappus several-seriate, dimorphic of stiff straight setae intermixed with down-like hairs.

A genus of about 50 species, mostly native to Africa, extending from the Canary and Cape Verde Islands and Madeira, through the Mediterranean countries and Eurasia to Australia and New Zealand. A few species are cosmopolitan weeds.

1. Plants perennial; achene ribs subterete sometimes rugulose but never wing-like, or muricate
   - 2
 - Plants annual; achene marginal ribs wing-like, or ribs all subterete and muricate - - 8
2. Mature achenes c. 4 mm. or more long; plants up to 1 m. tall, erect; axillary shoots when present glabrous - - - - - - - - - - - - - - - - 3
 - Mature achenes 2.5–4 mm. long; plants 1 m. or more tall erect, or weak-stemmed and ± scrambling; axillary shoots usually present, densely brownish-lanate or glabrous - - 6
3. Involucres glabrous; phyllaries not glandular-setose, the outermost broadest just above the base; leaves entire or remotely pinnately-lobed, lobes (when present) spreading and strap-shaped not recurved - - - - - - - - - - - - - - - - - 3. *integrifolius*
 - Involucres ± persistently woolly below; outer phyllaries often fleshy glandular-setose on the midribs, broadest at the base; leaves all runcinate, or the lowermost at least shallowly runcinately-lobed, or leaves pinnately-lobed, the lobes acute-recurved, or leaves sometimes entire - - - - - - - - - - - - - - - - - 4
4. Leaves linear-elliptic or narrowly lanceolate in outline, lamina length:width = c. 10:1, ± attenuate at the apices with few to many remote, spreading lateral lobes, margins entire to remotely denticulate; stems usually more leafy towards the base - - - - 4. *dregeanus*
 - Leaves elliptic to oblanceolate (oblong-oblanceolate) in outline, lamina length:width = c. 5:1, acute at the apices, ± shallowly runcinate to pinnately-lobed or leaves subentire, margins usually sharply and strongly denticulate; stems leafy throughout - - - - - 5
5. Involucre c. 15 mm. long; outer phyllaries somewhat narrowly oblong-lanceolate, obtuse-acute at the apex, ± equalling the innermost in width; all phyllaries membranous, not becoming distinctly thickened at the base, sometimes setose on the midribs - - - 2. *friesii*
 - Involucre c. 20 mm. long; outer phyllaries ovate-lanceolate, acute-acuminate, wider than the innermost, becoming coriaceous and thickened at the base and along the midribs; midribs not setose - - - - - - - - - - - - - - 1. *wilmsii*

6. Leaves runcinate-pinnatipartite, at least on the lower stem; lobes recurved and unequally tapered, the lobe bases equal to or exceeding the midrib lamina in width; leaf bases ± broadly auriculate, wider than the midrib lamina; phyllaries in flowering capitula ± blunt apically, usually not setose outside   -   -   -   -   -   -   -   -   -   -   -   7. *luxurians*
- Leaves entire, or if lobed then lobes spreading-falcate, mostly narrower than the midrib lamina; leaf bases not markedly auriculate; phyllaries ± tapered or blunt apically, often glandular-setose outside   -   -   -   -   -   -   -   -   -   -   -   -   -   7
7. Involucres 10–13(14) mm. long; phyllaries in flowering capitula ± blunt apically; wool at the base of the involucres whitish or absent; capitula stalks often glandular-setose; leaf bases shortly sagittate; plants mostly weak-stemmed scramblers   -   -   -   -   -   6. *bipontini*
- Involucres 13–18 mm. long; phyllaries gently tapered to the apex, at least in the outer 1–2 series; wool at the base of the involucres brownish; capitula stalks eglandular, often tomentose; leaves with large sagittate bases; tall, erect herbs   -   -   -   -   -   -   5. *schweinfurthii*
8. Achenes ellipsoid, strongly compressed; marginal ribs broadly wing-like, smooth or rugulose; leaf auricles usually rounded   -   -   -   -   -   -   -   -   -   -   8. *asper*
- Achenes narrowly obovoid, ± compressed, ribs all subterete and muricate with retrorse projections; leaf auricles usually ± triangular-acute   -   -   -   -   -   9. *oleraceus*

1. **Sonchus wilmsii** R.E. Fr. in Acta Hort. Berg. **8**: 107, t. 2, fig. 1 (1924). —Boulos in Bot. Not. **127**: 422, fig. 3G (1974). —Hilliard, Comp. Natal: 626, fig. 24 G (1977). Type from South Africa (Transvaal).

An erect perennial herb, up to c. 1 m. tall. Stems simple, somewhat woody below, ± dilated hollow and stout above, glabrous sometimes glaucous, leafy throughout. Leaves sessile, 5–15 × 2.5–6(8) cm., oblanceolate to panduriform in outline, lanceolate-ovate and smaller towards the stem apex, the basal few leaves smaller and evanescent; cauline leaves variable, runcinate-dentate throughout, or remotely and narrowly-lobed pinnatisect throughout, acute mucronate at the apex, semi-amplexicaul or auriculate-sagittate below, denticulate to sharply and irregularly dentate on the margins, glabrous sometimes glaucous. Capitula 2–7 in a much reduced corymbiform-cymose arrangement, or somewhat aggregated with stalked capitula subtended by subsessile immature capitula, or sometimes the capitula solitary on long branches. Involucres c. 20 × 12 mm. in fruiting capitula, cylindric-campanulate to spreading, ± persistently white-lanate below. Phyllaries imbricate, increasing from c. 4 mm. long on the outside to c. 20 × 2 mm. on the inside; the outermost narrowly ovate, becoming subcoriaceous with a thickened midrib, the basal portion eventually swollen; the inner phyllaries narrower, lanceolate-lorate, membranous, puberulous or glabrescent. Florets numerous. Corollas yellow, c. 18 mm. long, pubescent about the junction of the tube and ligule; ligule often becoming pinkish outside, 4–5 mm. long, oblong. Achenes pale-brown, c. 4.5 mm. long, narrowly ellipsoid, slightly flattened, glabrous, ribbed; marginal and middle ribs well developed with c. 2 secondary ribs between, the main ribs becoming swollen and rugulose; pappus white copious, 12–14 mm. long, composed of minute barbellate setae and fine down-like hairs.

Mozambique. M: Maputo, Namaacha, 4.xi.1966, *Balsinhas* 1110 (COI).
Also from South Africa (Transvaal and Natal) and Lesotho. Open grassland from sea-level to c. 2000 m., also in disturbed or cultivated ground. A "native vegetable" fide *Borle* 293.

2. **Sonchus friesii** L. Boulos in Bot. Not. **120**: 456, fig. 1 (1967); **127**: 421, figs. 3F, 16 (1974). —Hilliard, Comp. Natal: 626 (1977). Type: Zimbabwe, Inyanga Downs, *Norlindh & Weimarck* 4656 (LD, holotype & isotype).

An erect perennial herb up to 90(120) cm. tall, from a semi-woody taproot. Stems mostly solitary, stout, leafy, hollow, glabrous. Leaves runcinate-dentate or entire (coarsely pinnatilobed-runcinate in Natal), glabrous; basal leaves 6–7 in a rosette, 2–8.5 × 0.5–2.5 cm., oblanceolate, evanescent; cauline leaves few to many, somewhat crowded on the lower stem, 4–14(18) × 1–4(10) cm., narrowly oblong-oblanceolate to somewhat panduriform in outline, the uppermost becoming linear-lanceolate, apices obtuse or acuminate, margins coarsely sharply and irregularly-toothed, bases semi-amplexicaul sometimes auriculate-sagittate with narrow lobes up to 1.5 cm. long. Capitula up to 7(40) in lax reduced corymbiform cymes. Involucres c. 15 × 5–10 mm. in fruiting capitula, ± broadly cylindric-campanulate, persistent white-lanate below. Phyllaries imbricate, 3–4 × 1–2 mm. on the outside, increasing to 12 mm. long on the inside, to c. 15 mm. long in fruiting capitula, lanceolate-lorate, obtuse-acute, membranous, midribs of the outermost often setose becoming thickened towards the base, white- or brownish-lanate, glabrescent towards the apex. Florets numerous. Corollas yellow, c. 12–17 mm. long, pubescent about the junction of the tube and ligule; ligule often purplish outside, c. 5 mm. long, oblong. Achenes pale-brown, c. 4–5 mm. long, narrowly ellipsoid, slightly flattened glabrous ribbed, the marginal ribs and main rib on each face well developed with c. 2 secondary ribs between, the main rib eventually swollen and rugulose; pappus white, c. 10 mm. long, composed of minutely barbellate setae intermixed with down-like hairs.

Stems stout, usually 5–8 mm. in diam.; leaves mostly runcinate; involucres broadly cylindric, usually more than 8 mm. in diam.; phyllaries usually setose along the midrib   -   - var. *friesii*
Stems relatively slender, usually c. 3 mm. in diam.; leaves mostly unlobed; involucres ± narrowly cylindric, c. 5 mm. in diam. in fruiting capitula; phyllaries not setose on the midrib   -   -   -   -   -   -   -   -   -   -   -   -   -   var. *integer*

Var. **friesii**

Zimbabwe. E: Nyanga (Inyanga), c. 3 km. W. of Mt. Inyangani, c. 2100 m., 6.xii.1930, *Fries, Norlindh & Weimarck* 3487 (K; LD).
Also in South Africa (Natal, recorded from coastal and midland areas, along roadsides in secondary grassland and in forest). In montane grassland in the Flora Zambesiaca area.
The Zambian distribution of this species, as indicated by Boulos, and also by Hilliard, loc. cit. was based on the misidentification of *Rogers* 8645 (K); a specimen more correctly assigned to *S. asper* (L.) Hill. *S. friesii* is not known to occur in Zambia.

Var. **integer** G.V. Pope var. nov.* Type: Zimbabwe, Chimanimani, Gungunyana For. Res., *Goldsmith* 81/61 (K, holotype; SRGH).

Zimbabwe. E: Mutare Distr., Tsetsera Mts., Butler North, 1500 m., 20.xi.1957, *Chase* 6766 (K; SRGH). **Malawi**. S: Zomba Plateau, Chitinji, 22.vi.1984, *Tawakali & Nachamba* 263 (MAL).
So far recorded only from the eastern border mountains of Zimbabwe. Montane grassland or scrubland and ground cleared for cultivation.
The specimens *Norlindh & Weimarck* 3753 (LD) and 4526 (LD) from the lower slopes of Mt. Inyangani are intermediate between the two varieties.

3. **Sonchus integrifolius** Harv. in Harv. & Sond., F.C. **3**: 528 (1865). —R.F. Fr. in Acta Hort. Berg. **8**: 108 (1924). —L. Boulos in Bot. Not. **127**: 419, figs. 3E, 15 (1974). —Hilliard, Comp. Natal: 627 (1977). Syntypes from South Africa (Natal).
Sonchus deluguensis Thell. in Viert. Nat. Ges. Zürich **66**: 252 (1921). Syntypes. Mozambique, Maputo (Delagoa Bay), *Junod* 174 (BR; G, not seen); *Schlechter* 11971 (BM).

An erect glabrous perennial herb up to c. 55 cm. tall, from a semi-woody taproot. Stems usually simple, terete and finely striately-ribbed, leafy throughout but with leaves more crowded in the lower part. Leaves numerous sessile patent-ascending, usually undivided and entire to sinuate-dentate, sometimes with remote spreading (not recurved) lorate lobes up to c. 3.5 × 0.5 cm.; basal leaves c. 10 × 1 cm., narrowly oblanceolate; cauline leaves up to c. 17 × 2 cm., linear-elliptic to narrowly-elliptic becoming lanceolate towards the stem apex, attenuate-acute, margins subentire to denticulate, semi-amplexicaul and sagittate at the base. Capitula few in lax corymbiform cymes, or solitary and terminal on long synflorescence branches, glabrous at least when mature. Involucres up to c. 23 × 12 mm. in fruiting capitula, cylindric-campanulate later spreading. Phyllaries imbricate, increasing in length towards the inside, becoming thickened below; outermost c. 4 mm. long, ovate to lanceolate, sub-coriaceous; innermost to c. 17 × 3 mm. long, increasing to c. 23 cm. long in fruiting capitula, narrowly lanceolate, membranous, eventually involute and reflexed. Florets numerous. Corollas yellow, 15–18 mm. long, ± pubescent about the junction of tube and ligule; ligule 4–5 mm. long, oblong, reddish outside. Achenes pale-brown, 4–5.5 mm. long, narrowly ellipsoid, slightly laterally compressed, strongly 4-ribbed with c. 2 secondary ribs between, the larger ribs eventually thickened, glabrous; pappus white, c. 12 mm. long, composed of minutely barbellate setae and down-like hairs.

Zimbabwe. C: Marondera, Grasslands Res. Station, 29.i.1954, *Corby* 788 (K; SRGH). E: Mutare, Odzani R. Valley, *Teague* 590 (BOL). **Mozambique**. GI: Inhamissa, on the road from Machongo to Chongoanine, 17.xii.1957, *de Aguiar Macedo* 50 (SRGH).
Also from South Africa (Cape Province, Natal, and the Transvaal) and Lesotho. Open, usually seasonally waterlogged grassland, and in disturbed ground.

* Var. **integer** G.V. Pope var. nov., a varietate typica caulibus pro ratione exilibus plerumque c. 3 mm. diam., foliis pro parte maxima elobatis, involucris ± anguste cylindricis in capitulis fructificantibus c. 5 mm. diam., phyllariis non in costa setosis differt. Typus: Zimbabwe, Chimanimani, Gungunyana For. Res., *Goldsmith* 81/61 (K, holotype; SRGH).

Specimens of this species with distinctly lobed leaves were described as *S. delagoeensis* by Thellung (the leaf lobes being strap-shaped, spreading and apically obtuse). Intermediate leaf forms occur and Thellung's plant cannot be maintained as a separate species. However, the specimens with lobed leaves appear to be recorded only from coastal localities and may therefore represent an infra-specific taxon.

4. **Sonchus dregeanus** DC., Prodr. **7**: 184 (1838). —Harv. in Harv. & Sond., F.C. **3**: 528 (1865). —R.E. Fr. in Acta Hort. Berg. **8**: 109 (1924). —Humbert in Fl. Madag., Comp., III: 889 (1963). —L. Boulos in Bot. Not. **120**: 456 (1967); **127**: 418, figs. 3D, 14 (1974). —Hilliard, Comp. Natal: 627 (1977). Syntypes from South Africa (Cape Province).

Sonchus ecklonianus DC., Prodr. **7**: 184 (1838). —Harv. in Harv. & Sond., F.C. **3**: 528 (1865). —R.E. Fr. in Acta Hort. Berg. **8**: 109 (1924). Type from South Africa (Cape Province).

An erect glaucous perennial herb 12–90 cm. tall, from a semi-woody taproot. Stems usually scapiform, 1–several-branched from near the base, sometimes simple, striately-ribbed. Leaves ± numerous, mostly basal, often rosulate, patent-ascending, up to c. 12(19) × 1(2.3) cm. Lower leaves very narrowly-oblanceolate to linear-elliptic, remotely lobed to sinuate-dentate or sharply denticulate, sometimes entire, attenuate-acute apically, glaucous; lobes up to c. 2.5 × 0.5 cm., recurved, narrowly lanceolate. Cauline leaves few, grading into foliaceous bracts, or absent; lanceolate entire, attenuate above, semi-amplexicaul and sagittate-auriculate below. Capitula few in lax corymbiform cymes, sometimes solitary; mature capitula often subtended by 1–several subsessile buds; capitula bases and subtending buds densely white woolly. Involucres up to c. 19 × 9 mm. in fruiting capitula, cylindric-campanulate, white-lanate below, the receptacle becoming somewhat swollen. Phyllaries olive-green, glaucous, imbricate; the outermost c. 3 mm. long, ovate, lanate and often also patent-setose outside, semi-coriaceous; the innermost c. 15 × 2 mm. in flowering capitula, increasing to c. 19 mm. in fruiting capitula, glabrous, or setose on the midribs and apices, membranous, becoming involute. Florets numerous. Corollas yellow, c. 15 mm. long, pubescent about the junction between the tube and ligule; ligule c. 4 mm. long, strap-shaped. Achenes pale-brown, c. 5 mm. long, narrowly ellipsoid, slightly compressed, strongly c. 4-ribbed with 1–2 secondary ribs between, ribs eventually swollen and ± rugulose; pappus white, c. 10 mm. long, composed of few minutely barbellate setae intermixed with down-like hairs.

Zimbabwe. W: Matobo, Pomongwe, 4.i.1948, *West* 2555 (SRGH). S: Great Zimbabwe Nat. Park, 28.iii.1973, *Chiparawasha* 638 (K; SRGH).

Also from South Africa (Cape Province, Natal, Transvaal) Lesotho and Madagascar. Open grassland in moist localities, often on sandy soil, also in disturbed ground.

5. **Sonchus schweinfurthii** Oliv. & Hiern in F.T.A. **3**: 458 (1877). —O. Hoffm. in Engl., Pflanzenw. Ost-Afr. **C**: 421 (1895). —Hiern, Cat. Afr. Pl. Welw. **1**, 3: 623 (1898). —R.E. Fr., Wiss. Ergebn. Schwed. Rhod.-Kongo-Exped. 1911–1912, **1**: 350 (1916); in Acta Hort. Berg. **8**: 98 (1924). —Mendonça, Contrib. Conhec. Fl. Angol., **1** Compositae: 152 (1943). —Robyns, Fl. Sperm. Parc Nat. Alb. **2**: 595 (1947). —F.W. Andr., Fl. Pl. Sudan **3**: 51 (1956). —Adams in F.W.T.A. ed. 2, **2**: 296 (1963). —C. Jeffrey in Kew Bull. **18**: 479 (1966). —Boulos in Bot. Not. **127**: 402, figs. 1, 2A (1974). —Agnew, Upland Kenya Wild Fls.: 498 (1974). —Lawalrée in Fl. Rwanda, Spermat. **3**: 694 (1985). —Lawalrée, Dethier & Gilissen in Fl. Afr. Centr., Compos., Cichorioideae: 38 (1986). TAB. **41** fig. A. Type from Zaire.

Sonchus bipontini var. *pinnatifidus* sensu Eyles in Trans. Roy. Soc. S. Afr. **5**, 4: 523 (1916). —sensu Mendonça, Contrib. Conhec. Fl. Angol., **1** Compositae: 151 (1943) non Oliv. & Hiern (1877).

An erect robust perennial herb, usually 1–1.5 m. tall, exceptionally to c. 2 m. tall, from a terete semi-woody taproot. Stems solitary, branching above, green or purplish often glaucous, striately ribbed, hollow, glabrous, leafy. Leaves ascending, glabrous often glaucous, thinly coriaceous, the lowermost small and narrowly elliptic. Cauline leaves sessile up to c. 36 × 3 cm., linear-elliptic to narrowly-lanceolate, entire or dentately- to pinnately-lobed, attenuate to the apex, ± revolute and subentire to acicular-denticulate on the margins, sagittate-auriculate with auricles c. 2–5 cm. long, acute; lobes when present usually in the basal portion of the leaf, few to many ± remote, patent or ± reflexed, c. 7 × 1.2 cm.; upper leaves auriculate and stem clasping, becoming narrowly lanceolate and grading into the synflorescence bracts. Capitula solitary, or more often in few to many 2–5-capitulate clusters; mature capitula usually subtended by one or more buds; the bases of the capitula as well as the subtending buds and the upper synflorescence branches enveloped in a brownish densely woolly indumentum. Involucres up to c. 18 × 11 mm. in

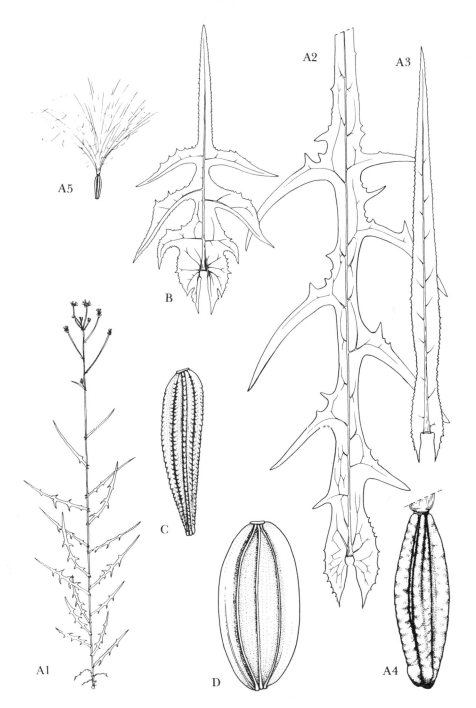

Tab. 41. SONCHUS SCHWEINFURTHII. A1, habit (×$\frac{1}{12}$), from *Brummitt* 9164; A2, mid-cauline leaf, from *Richards* 452; A3, mid-cauline leaf, from *Fries, Norlindh & Weimarck* 2432; A4, achene; A5, achene and pappus (× 2), A4 & A5 from *Richards* 1245. B. —SONCHUS LUXURIANS, upper-cauline leaf, from *Perdue & Kibuya* 8311. C. —SONCHUS OLERACEUS, achene, from *Brummitt* 8736. D. —SONCHUS ASPER, achene, from *Mott* 741. (Leaves × $\frac{1}{2}$, achenes ×15). Drawn by Eleanor Catherine.

fruiting capitula, broadly cylindric to campanulate, brownish-lanate below to glabrescent, becoming swollen and corky below. Phyllaries olive-green to purplish, imbricate; the outermost from c. 4 mm. long, ovate-lanceolate, woolly at first often also with brownish-purple setae on the back; the innermost 13–15 mm. long, up to c. 18 mm. long in fruiting capitula, puberulent to glabrescent or setose towards the apex, ± ciliate above, involute when dry. Florets very numerous. Corollas yellow, 12–16 mm. long, pubescent about the junction between tube and ligule; ligule 3–4 mm. long, shortly strap-shaped, reddish-tinged outside, glabrous. Achenes pale- to reddish-brown, 3–3.5 × 0.75–1 mm., narrowly ellipsoid, somewhat compressed, c. 4-angular and ribbed on the angles with c. 2 smaller ribs on the faces between, ribs becoming swollen corky and ± rugulose; pappus white, dimorphic, c. 11 mm. long, composed of minutely barbellate slightly flattened setae intermixed with down-like hairs.

Zambia. N: Old Isanya Road, c. 1520 m., 18.i.1955, *Richards* 4145 (K; SRGH). W: Kitwe, 19.ix.1967, *Mutimushi* 2108 (K; SRGH; NDO). C: Chakwenga Headwaters, 100–129 km. E. of Lusaka, 28.x.1963, *Robinson* 5794 (K; SRGH). E: Nyika Plateau, c. 2250 m., 10.xi.1967, *Richards* 22452 (K). **Zimbabwe**. C: Harare, Borrowdale, 8.xi.1953, *Wild* 4150 (K; SRGH). E: Tarka For. Res., c. 1160 m., x.1968, *Goldsmith* 148/68 (COI; K; SRGH). **Malawi**. N: Mzimba Distr., Mzuzu, St. John's, c. 1380 m., 14.ii.1975, *Pawek* 9079 (SRGH). C: Ntchisi Distr., 11 km. S. of Ntchisi, c. 1580 m., 18.vi.1970, *Brummitt* 11550 (K). S: Zomba Plateau, Chingwe's Hole, c. 1890 m., 16.iii.1970, *Brummitt* 9164 (K; SRGH). **Mozambique**. N: Lichinga (Vila Cabral), 17.vi.1934, *Torre* 132 (LISC; COI). MS: Border Farm, 25.i.1966, *Chase* 8364 (K; SRGH).

Also in Guinea, Nigeria, Cameroon, Central African Republic, Sudan, Ethiopia, Zaire, Rwanda, Burundi, Uganda, Kenya, Tanzania and Angola. In seasonally or permanently wet soil, beside streams, dams or irrigation furrows; often as a ruderal or in cultivated or disturbed ground on roadsides in moist localities.

6. **Sonchus bipontini** Aschers. in Schweinf., Beitr. Fl. Aethiop.: 160 (1867). —Oliv. & Hiern in F.T.A. **3**: 458 (1877). —R.E. Fr. in Acta Hort. Berg. **8**: 100 (1924). —S. Moore in Journ. Bot. **65**, Suppl. 2, Gamopet.: 67 (1927). —Mendonça, Contrib. Conhec. Fl. Angol., **1** Compositae: 151 (1943). —C. Jeffrey in Kew Bull. **18**: 479 (1966). —Boulos in Bot. Not. **125**: 300 (1972); **127**: 411, figs. 2F, 8 (1974). —Agnew, Upland Kenya Wild Fls.: 498 (1974). —Lawalrée in Fl. Rwanda, Spermat. **3**: 692 (1985). —Lawalrée, Dethier & Gilissen in Fl. Afr. Centr., Compos., Cichorioideae: 34 (1986). Types from Ethiopia.
Sonchus bipontini forma *glanduligerus* R.E. Fr. in Acta Hort. Berg. **8**: 101 (1924). Type from Zaire.
Sonchus bequaertii De Wild., Pl. Bequaert. **5**: 461 (1932). Syntypes from Zaire.
Sonchus bipontini var. *louisii* Robyns in Bull. Jard. Bot. Brux. **17**: 105 (1943); Fl. Sperm. Parc Nat. Alb. **2**: 596 (1947). —Boulos in Bot. Not. **125**: 300 (1972); **127**: 414 (1974). —Lawalrée, Dethier & Gilissen in Fl. Afr. Centr., Compos., Cichorioideae: 35 (1986). Type from Zaire.
Sonchus bipontini var. *glanduligerus* (R.E. Fr.) Robyns, Fl. Sperm. Parc Nat. Alb. **2**: 596 (1947). —Boulos in Bot. Not. **125**: 300 (1972); **127**: 412 (1974). —Lawalrée, Dethier & Gilissen in Fl. Afr. Centr., Compos., Cichorioideae: 35 (1986). Type as for *S. bipontini* forma *glanduligerus*.

A weak-stemmed perennial herb with a semi-woody rootstock. Stems trailing or scrambling to c. 3 m. long, or erect and up to c. 1.2 m. tall, mostly solitary, branched above, very leafy throughout becoming leafless below, terete hollow, sometimes ± glandular-setose above otherwise glabrous. Leaves glabrous, largest and crowded on the upper stem, 5–20 × 0.25–1.6 cm., linear-elliptic becoming linear-lanceolate and sessile towards the stem apex, attenuate above, entire or less often ± recurved acicular-dentate to lobed on the margins, attenuate and shortly sagittate at the base in lower leaves, becoming amplexicaul and subauriculate-sagittate in upper leaves; lobes up to c. 2 cm. long narrowly runcinate-triangular; ± well developed leafy axillary shoots usually present, at least in upper leaves, the young axillary growth densely brown-lanate later glabrescent. Capitula subsessile or shortly stalked, 2–many ± aggregated in terminal clusters or in lax corymbiform cymes; mature capitula subtended by 1–several subsessile buds; the bases of the capitula as well as the subtending buds and the capitula stalks enveloped in a white persistent lanate indumentum, synflorescence branches and capitula stalks densely glandular-setose or glabrous. Involucres c. 10–13 mm. long in fruiting capitula, cylindric later spreading, phyllaries eventually reflexed. Phyllaries few-seriate, imbricate, white-lanate outside where exposed, glabrescent towards the apices, usually also glandular-setose outside; the outermost c. 4 mm. long, narrowly ovate; the innermost 10–13 × 1.5–2.5 mm., lanceolate, ± obtuse at the apices. Florets numerous. Corollas yellow, c. 10–13 mm. long, pubescent about the junction of the tube and ligule; ligule c. 3–4 mm. long, oblong. Achenes reddish-brown, 2.5–3 mm. long, narrowly ellipsoid, somewhat flattened, c.

4-angular with ribs on the angles, smaller ribs developing in between, the main ribs becoming thickened and transversely rugulose, glabrous; pappus white, dimorphic, c. 11 mm. long composed of minutely barbellate setae intermixed with down-like hairs.

Malawi. N: Mzimba Distr., Viphya Plateau, Kasangadzi, 5.xi.1968, *Salubeni* 1196 (K; SRGH).

Also from Ethiopia, Sudan, Rwanda, Zaire, Uganda, Kenya and Tanzania. In montane forest, or forest margins, in dense vegetation or thickets, also in disturbed ground.

7. **Sonchus luxurians** (R.E. Fr.) C. Jeffrey in Kew Bull. **18**: 480 (1966). —L. Boulos in Bot. Not. **127**: 414, figs. 3A, 11 (1974). —Agnew, Upland Kenya Wild Fls.: 498 (1974). —Lawalrée in Fl. Rwanda, Spermat. **3**: 694, fig. 217(4), (1985). —Lawalrée, Dethier & Gilissen in Fl. Afr. Centr., Compos., Cichorioideae: 36, pl. 6 (1986). TAB. **41** fig. B. Type from Tanzania.

Sonchus bipontini var. *pinnatifidus* forma *luxurians* R.E. Fr. in Acta Hort. Berg. **8**: 102 (1924). Type as above.

Sonchus bipontini var. *pinnatifidus* sensu R.E. Fr. in Acta Hort. Berg **8**: 102 (1924); sensu Robyns, Fl. Sperm. Parc Nat. Alb. **2**: 597 (1947), non Oliv. & Hiern (1877).

Sonchus schweinfurthii sensu Brenan in Mem. N.Y. Bot. Gard. **8**, 5: 489 (1953) non Oliv. & Hiern (1877).

Sonchus oliveri-hiernii var. *luxurians* (R.E. Fr.) L. Boulos in Bull. Jard. Bot. Brux. **32**: 106 (1962). Type as for *Sonchus luxurians*.

A somewhat robust perennial herb from a semi-woody rootstock. Stems trailing or scrambling to c. 2(6) m. long, or erect and up to c. 1 m. tall, 1–several, branched above, leafy throughout or the lowermost leaves withered and the stem leafless below, green or purple-tinged, terete hollow glabrous. Leaves glabrous, 5–25 cm. long, the upper cauline being the largest, all leaves runcinate, or pinnately-lobed with a narrowly-winged midrib, or upper leaves and those of the axillary shoots undivided and linear-lanceolate; lateral lobes 2–8, mostly 1–3.5(6.5) × 0.5–1.5(2.3) cm., recurved oblong-lanceolate, asymmetrically tapered, less often 1.5–4 × 0.3–0.5 cm. and narrowly-triangular; midrib lamina 3–14 mm. wide, linear-elliptic; lobe and leaf apices acute to ± attenuate; margins recurved acicular-denticulate; leaf bases amplexicaul and narrowly auriculate-sagittate in lower leaves to rounded-auriculate with sagittate auricle lobes in the upper leaves; ± well developed axillary shoots often present at least in the upper leaves, the young axillary growth densely brown lanate later glabrescent. Capitula many, subsessile or shortly-stalked, 2–10 ± densely aggregated in solitary or corymbiformly cymose globose clusters, or clusters much reduced with fewer capitula laxly arranged, mature capitula subtended by 1–several subsessile buds; the bases of the capitula as well as the subtending buds and the capitula stalks covered in a white persistent lanate indumentum, the synflorescence branches and capitula stalks not glandular-setose. Involucres up to c. 16 × 9 mm. in fruiting capitula, ovoid-cylindric to spreading. Phyllaries olive-green, few-seriate, imbricate; densely white-lanate outside, later glabrescent towards the apices, rarely also setose outside; the outermost c. 4 mm. long, ovate-lanceolate, eventually swollen below; the innermost 10–15 × 1.5–2.5 mm., oblong-lanceolate, obtuse to rounded at the apices, glabrous, eventually involute. Florets numerous. Corollas yellow c. 12–14 mm. long, pubescent about the junction of tube and ligule; ligule 4.5–5 mm. long, oblong becoming orange above. Achenes pale-brown, 3–4 mm. long, narrowly ellipsoid, ± flattened, c. 4-angular, with ribs on the angles and with smaller ribs in between, the larger ribs becoming thickened and transversely rugulose, glabrous; pappus white, dimorphic, 8–10 mm. long, composed of minutely barbellate setae intermixed with down-like hairs.

Malawi. N: Rumphi Distr., Uzumara Rain Forest, 48 km. NE. of Njakwa, 1820 m., 29.xi.1970, *Pawek* 4036 (K; MAL). S: Mulanje Mt., Sombani Basin, 2 km. above Hut, 2070 m., 5.vi.1970, *Brummitt* 11273 (K; SRGH).

Also in Zaire, Rwanda, Burundi, Ethiopia, Uganda, Kenya and Tanzania. Above 1000 m., in thickets or dense vegetation on forest margins or on forest stream banks, also in disturbed ground.

The leaves are not all typically runcinate in the Flora Zambesiaca area. The uppermost leaves are often entire and linear-lanceolate with attenuate apices, and the lower leaves somewhat narrowly runcinately-lobed. These plants intergrade with the more typical, broadly-lobed runcinate forms found further north.

Specimens in which the upper leaves are undivided and narrow can be distinguished from *S. bipontini* and *S. schweinfurthii* by their ± broadly auriculate leaf bases, and the lateral lobes of the lower leaves which although narrow are recurved and asymmetrically tapered (in *S. luxurians*). *S. bipontini* has smaller capitula, the stalks and phyllaries of which are often glandular-setose, and leaves broadest about the middle and not markedly auriculate at the base.

8. **Sonchus asper** (L.) Hill, Herb. Brit. **1**: 47 (1769). —O. Hoffm. in Engl., Pflanzenw. Ost-Afr. **C**: 421 (1895). —Humbert, Comp. Madag.: 156 (1923). —R.E. Fr. in Acta. Hort. Berg. **8**: 97 (1924). —S. Moore in Journ. Bot. **65**, Suppl. 2, Gamopet.: 67 (1927). —Mendonça in Contrib. Conhec. Fl. Angol., 1 Compositae: 153 (1943). —Robyns, Fl. Sperm. Parc Nat. Alb. **2**: 595 (1947). —Humbert, Fl. Madag. Comp., III: 886 (1963). —C. Jeffrey in Kew Bull. **18**: 481 (1966). —Boulos in Bot. Not. **126**: 164 (1973). —Hilliard, Comp. Natal: 624 (1977). —Lawalrée in Fl. Rwanda, Spermat. **3**: 692, fig. 217(1), (1985). —Lawalrée, Dethier & Gilissen in Fl. Afr. Centr., Compos., Cichorioideae: 30 (1986). TAB. **41** fig. D. Type from Europe.

    Sonchus oleraceus var. *asper* L., Sp. Pl.: 794 (1753). Type as above.

    Sonchus gigas Humbert, Fl. Madag. Comp., III: 887 (1963), quoad descr. lat. et typum, excl. descr. gall., non Boulos (1959) nom. non rite publ. —Boulos in Bot. Not. **126**: 170, figs. 11, 12 (1973). —Hilliard, Comp. Natal 624 (1977). —Lawalrée, Dethier & Gilissen in Fl. Afr. Centr., Compos., Cichorioideae: 32 (1986). Type: Zambia, Muckle Neuk, *Robinson* 904 (K, holotype).

    Sonchus friesii sensu L. Boulos in Bot. Not. **127**: 422 (1974) pro parte quoad specim. *Rogers* 8645.

An erect robust long-lived annual, or perennial herb 20–220 cm. tall. Stem simple below, sparsely branched above, stout leafy hollow, striately ribbed becoming angular below in larger plants, reddish-purple tinged, glabrous or sometimes ± densely glandular-setose above; setae c. 1 mm. long, patent. Leaves numerous appressed-ascending, the lowermost exauriculate and with petiole-like midribs, smaller and less dissected than the cauline leaves; cauline leaves up to c. 32 × 9 cm., smaller near the stem apex, oblanceolate or lorate-oblanceolate in outline, sinuate-dentate to runcinate-pinnatilobed, apices acute, leaf and lobe margins irregularly dentate with teeth spinulose and up to 6 mm. long, bases auriculate and semi-amplexicaul, the auricles rounded ± appressed and spinulose-dentate on the margins; upper leaves sessile, lanceolate, lobed to subentire, ± attenuate at the apices, semi-amplexicaul, grading into the synflorescence bracts. Capitula stalked, ± aggregated in corymbiform cymose clusters; stalks glabrous or glandular-setose. Involucres up to c. 14 mm. long, and c. 11 mm. in diam. when mature, very broadly cylindrical eventually spreading, densely white tomentose below, glabrescent, involucre and phyllaries at the base becoming swollen in fruiting capitula. Phyllaries many-seriate imbricate, glabrous except for the ciliate apices and sometimes large glandular setae on the midribs; the outermost phyllaries c. 2 mm. long, ovate, increasing ± uniformly to c. 14 mm. long inside, becoming narrowly lanceolate, the 2 innermost series subequal. Florets very numerous, corollas yellow, up to c. 10 mm. long, ligules c. 4 × 0.5–0.75 mm. Achenes reddish-brown 2.5–3.3 × 1–1.25 mm., broadly oblong-ellipsoid, strongly flattened with wing-like ribs on the margins and c. 3 narrow ribs on each face, smooth or minutely rugulose sometimes retrorsely hispidulous on the ribs. Pappus copious, white, c. 8 mm. long ± equalling or slightly exceeding the phyllaries, composed of barbellate slightly flattened setae intermixed with down-like hairs.

    Botswana. N: Maun, 19°29'S, 23°26'E, 9.vi.1975, *P.A. Smith* 1393 (SRGH). SE: Gaborone Dam, 1000 m., 30.ix.1975, *Mott* 741 (K; SRGH). SW: Ghanzi Pan, 14.5 km. E. of Ghanzi, 1000 m., 21.x.1969, *R.C. Brown* 6984 (K; PRE). **Zambia**. B: Sikelenge, Luampa R., 64 km. W. of Kaoma (Mankoya), 19.xi.1959, *Drummond & Cookson* 6621 (K; SRGH). N: Mbala, Lake Chila, 1650 m., 4.i.1952, *Richards* 237 (K). W: Mufulira, 20.v.1934, *Eyles* 8218 (K; SRGH). C: Mt. Makulu Res. Station, 19 km. S. of Lusaka, 27.vi.1956, *Angus* 1360 (K; SRGH). S: Namwala, Lochinvar Nat. Park, Hot Springs, 1000 m., 28.ix.1971, *van Lavieren, Sayer & Rees* 143 (SRGH). **Zimbabwe**. N: Makonde Distr., Mhangura (Mangula), 920 m., 1.i.1964, *Jacobsen* 2369 (PRE; SRGH). C: Gweru, Fletcher High School, 1400 m., 19.xi.1967, *Biegel* 2327 (SRGH). E: Chipinge, Sabi Valley Expt. Sta., x.1959, *Soane* 122 (SRGH). S: Chiredzi, Hippo Valley Estate, 16.x.1973, *Lonsdale* 328 (SRGH). **Malawi**. N: Mzuzu Distr., Viphya Range, 1500 m., 24.ii.1961, *Richards* 14452 (K). C: Dedza Distr., near Masawa, 12.vi.1973, *Salubeni* 1898 (MAL; SRGH). **Mozambique**. N: Lichinga (Vila Cabral), vi.1934 to i.1935, *Torre* 405 (LISC; COI). Z: Zambezia, 5.viii.1945, *Pimento* 160 (LISC; SRGH). M: Near Incanhini, 24.iii.1960, *Myre* 3839 (SRGH).

    Also in Eurasia, the rest of Africa and Madagascar. Widespread, but usually infrequent, as a ruderal or weed of cultivation often in permanently moist or swampy soil around pans and on river and lake margins. Occasionally also in miombo woodland.

    S. asper as treated here includes *S. gigas* Humbert and is perhaps best considered to be an aggregate species.

    Humbert in taking up Boulos's invalidly published name *Sonchus gigas* (proposed initially for an Egyptian plant) indicated as its holotype *Robinson* 904, thus applying the name to south tropical African specimens. These specimens cannot be satisfactorily distinguished from *S. asper* on morphological characters alone. *S. asper* is diploid (2n = 9) and at least some plants referred to *S. gigas* by Boulos are tetraploid, cryptic differences between them would therefore be expected.

    Tetraploid specimens previously assigned to *S. gigas* include: *Wickens* 1703 from Jebel Marra, Sudan, reported in Taxon **19**: 102 (1970) and *Ram* s.n. from Natal, South Africa, reported in Notes from Roy. Bot. Gard. Edin. **32**: 434, 436 (1973). The chromosome number of the *S. gigas* holotype specimen has not been established.

Humbert, Fl. Madag. Comp. III: 887 (1963) and Boulos in Bot. Not. **126**: 172 (1973) give pollen morphology as a further distinction between the two species; *S. asper* is reported as having tricolporate pollen and *S. gigas* a mixture of tricolporate and tetracolporate pollen.

9. **Sonchus oleraceus** L., Sp. Pl.: 794 (1753). —Oliv. & Hiern in F.T.A. **3**: 457 (1877) pro parte. —Harv. in Harv. & Sond., F.C. **3**: 528 (1865). —O. Hoffm. in Engl., Pflanzenw. Ost-Afr. **C**: 421 (1895). —Hiern, Cat. Afr. Pl. Welw. **1**, 3: 622 (1898). —Humbert, Comp. Madag.: 157 (1923). —R.E. Fr. in Acta. Hort. Berg. **8**: 95 (1924). —S. Moore in Journ. Bot. **65**, Suppl. 2, Gamopet.: 67 (1927). —Mendonça in Contrib. Conhec. Fl. Angol., 1 Compositae: 152 (1943). —Robyns, Fl. Sperm. Parc Nat. Alb. **2**: 595 (1947). —Sussenguth & Merxm. in Trans. Rhod. Sci. Assn. **43**: 71 (1951). —F.W. Andr., Fl. Pl. Sudan **3**: 50 (1956). —Wild, Common Rhod. Weeds: t. 75 (1955). —Adams in F.W.T.A., ed. 2, **2**: 296 (1963). —Humbert, Fl. Madag. Comp. III: 887 (1963). —C. Jeffrey in Kew Bull. **18**, 3: 481 (1966). —Merxm., Prodr. Fl. SW. Afr. 140: 4 (1967). —Boulos in Bot. Not. **126**: 155 (1973). —Agnew, Upland Kenya Wild Fls.: 498 (1974). —Hilliard, Comp. Natal: 623 (1977). —Lawalrée in Fl. Rwanda, Spermat. **3**: 692, fig. 212(5), (1985). —Lawalrée, Dethier & Gilissen in Fl. Afr. Centr., Compos., Cichorioideae: 32 (1986). TAB. **41** fig. C. Type from Europe.

A stout erect annual herb. Stem up to c. 150 cm. tall, simple or sparsely branched in the upper half, leafy hollow usually reddish-tinged often glaucous, stem and branches glabrous or sometimes ± densely glandular-setose and somewhat viscid, setae up to c. 1 mm. long patent purplish; branches up to c. 45 cm. long ascending. Lower leaves crowded, up to c. 28 × 9 cm., ± oblanceolate in outline, usually pinnately-lobed to coarsely runcinate-pinnatipartite, each with a broadly ovate-hastate apical lobe rounded to acute at the apex; margins usually coarsely and somewhat irregularly dentate with spinulose-acicular teeth; midrib narrowly winged, ± auriculate and semi-amplexicaul below; auricles ± sagittate-spreading, entire to dentate. Upper leaves sessile, smaller, lanceolate, ± deeply lobed to subentire, coarsely and sharply dentate on the margins, basal auricles usually large with acute spreading lobes. All leaves ± spathulate at first, often deeply laciniate in depauperate specimens. Capitula stalked, many in corymbiform cymes, stalks glabrous or glandular-setose; involucres up to c. 13 mm. long and 10 mm. in diam., very broadly cylindrical later spreading, densely white tomentose below soon glabrescent, involucre becoming swollen in fruiting capitula. Phyllaries many-seriate imbricate increasing in length from c. 3 mm. on the outside to c. 13 mm. inside, narrowly lanceolate, glaucous, glabrous except for ciliate apicies and occasionally 1–several large glandular-setae up to c. 2 mm. long along the midrib, the midribs below becoming swollen and corky. Florets numerous; corollas yellow up to c. 13 mm. long, tube slender and pubescent above, ligule c. 5.5 mm. long and strap-shaped. Achenes light to reddish-brown, up to c. 3 × 1 mm., narrowly obovoid and somewhat flattened, ± uniformly many-ribbed; ribs minutely muricate with numerous retrorse projections. Pappus white copious c. 7 mm. long, composed of barbellate setae intermixed with down-like hairs.

Botswana. N: Maun, 19°29'S, 23°26'E, 20.iii.1975, *P.A. Smith* 1292 (SRGH). SE: Gaborone Distr., Content Farm, 24°23'S, 25°57'E, 1050 m., 22.viii.1978, *Hansen* 3438 (C; GAB; K; SRGH). **Zambia**. W: Kitwe, 10.vii.1968, *Mutimushi* 2627 (K; NDO). C: Munali, 8 km. E. of Lusaka, 1300 m., 24.vii.1955, *King* 70 (K). S: Livingstone, Mulobezi Sawmills, 17.viii.1947, *Brenan* 7726 (K). **Zimbabwe**. N: Mazowe (Mazoe) Citrus Estates, 1200 m., 2.iii.1971, *Searle* 115 (K; SRGH). W: Bulawayo, 1380 m., v.1955, *Miller* 2843 (K; SRGH). C: Gweru R. bridge, 12.x.1974, *Biegel* 4650 (K; SRGH). E: Chipinge, Chibuye Project, 520 m., v.1958, *Davies* 2463 (K; SRGH). S: Mwenezi (Nuanetsi), Chilonga Irrigation Scheme, 18.ix.1970, *Taylor* 12 (K; SRGH). **Malawi**. N: Mzimba Distr., Mzuzu, Marymount, 1380 m., 20.iii.1974, *Pawek* 8228 (MO; SRGH). C: Dedza Distr., Chongoni For. School, 28.v.1963, *Banda* 508 (K; SRGH). S: Blantyre Distr., Limbe, 1140 m., 24.ii.1970, *Brummitt* 8736 (K). **Mozambique**. MS: SW. foot of Gorongosa Mt., c. 430 m., 9.vii.1969, *Leach & Cannell* 14260 (SRGH). GI: Outskirts of Inhambane, 23°54'S, 35°20'E, 20 m., ix.1935, *Gomes e Sousa* 1647 (K). M: Maputo, Jardim Tunduru (Jardim Vasco da Gama), 3.ix.1971, *Balsinhas* 1940 (K; LISC).

A native of Eurasia and N. Africa, now more or less cosmopolitan as a weed. A weed of gardens, irrigation schemes and roadsides.

Tribe 5. ARCTOTIDEAE Cass.

Arctotideae Cass. in Journ. Phys. **88**: 159 (1819). —Lewin in Feddes Repert., Beih. XI: 1–75 (1922). —Roessler in Mitt. Bot. Staatss. Münch. **3**: 72–500 (1959). —T. Norl., *Arctoteae*, Syst. Rev. in Heywood et al., Biol. & Chem. Comp.: 943–959, figs. 1–7 (1977).

Annual or perennial herbs, with or without stolons, sometimes scapose, sometimes suffrutescent with annual stems from woody rootstocks, or small to large subshrubs. Leaves alternate, or radical and ± rosulate. Capitula heterogamous and radiate or homogamous and discoid. Involucres campanulate, hemispherical, ovoid or subglobose; phyllaries usually many-seriate, free or fused at the base, margins and tips ± scarious or spinescent. Receptacle naked or fimbriate-alveolate (rarely paleate). Ray-florets female or neuter, styles filiform; style branches linear-lanceolate or elliptic; staminodes often present. Disk-florets hermaphrodite, or the inner functionally male; corolla limbs regularly 5-lobed, campanulate or gradually widening upwards; anthers entire or ± sagittate at the base, the auricles obtuse or acute rarely tailed, pollen spiny known to be lophate only in *Berkheya*; fertile styles terete or flattened, style branches thickened-connate to near the tips and with a collar of sweeping hairs or papillae at their base. Achenes glabrous or hairy, often ribbed; pappus of scales, coroniform or setiform, often 2-seriate, or wanting.

The tribe *Arctotideae* consists of 16 genera comprising some 200 species. It is Old World in distribution and centred in Africa. Differentiation of species and genera is greatest in South Africa, with many species endemic to the Cape region.

For *Berkheya*, *Hirpicium* and *Gazania* the preliminary work of H. Roessler in Kirkia **10**: 73–99 (1975) is gratefully acknowledged.

Key to the genera

1. Phyllaries ± free, apices obtuse (at least in inner phyllaries), scarious at the tips and on the margins - - - - - - - - - - - - - - - - - - - 2
  – Phyllaries connate at the base or higher, apices acute or acuminate, often spine tipped 4
2. Achene with 2 linear grooves or longitudinal chambers on one face; receptacle alveolae setose - - - - - - - - - - - - - - - **29. Arctotis**
  – Achene obscurely longitudinally ribbed, not developing longitudinal chambers on one face; receptacle alveolae ± fimbriate, not long-setose - - - - - - - - - 3
3. Achenes glabrous or pubescent with a basal tuft of short or long upward-pointing hairs; ray-florets female; pappus of overlapping scales - - - - **30. Haplocarpha**
  – Achenes densely woolly, without a basal coma of hairs; ray-florets neuter, ovary rudimentary; pappus ± fused, coroniform - - - - - - - - **31. Arctotheca**
4. Phyllaries connate at the base only, always spiny - - - - - **32. Berkheya**
  – Phyllaries connate to higher up forming a cup-shaped involucrum, mostly bristly, hardly spiny - - - - - - - - - - - - - - - - - 5
5. Pappus 1-seriate, or 2-seriate and dimorphic, the outer whorl of overlapping sessile scales, the inner whorl of much smaller stipitate scales; plants ± bristly-setose or glabrescent, rarely glandular-hairy - - - - - - - - - - - - **33. Hirpicium**
  – Pappus 2-seriate of subequal slender accrescent scales, outer whorl scales not overlapping, inner whorl scales ± similar in length and shape to the outer but with a short stalk; plants ± araneose or glabrous not hispid - - - - - - - - - - - - **34. Gazania**

29. ARCTOTIS L.

Arctotis L., Sp. Pl. **2**: 922 (1753); Gen. Pl. ed. 5: 394 (1754). —Beauverd in Bull. Soc. Bot. Genève, sér. 2, **7**: 41 (1915). —Lewin in Fedde, Repert., Beih. XI: 54 (1922). *Venidium* Less. in Linnaea **6**: 91 (1831).

Perennial or annual herbs, araneose-lanate or glandular, often acaulescent. Leaves alternate or radical, entire to lyrate-pinnatifid. Capitula heterogamous radiate, solitary, large and many-flowered. Phyllaries free, many-seriate, scarious-hyaline on the margins, the outer phyllaries usually with apical appendages. Receptacle flat, fimbrilliferous-alveolate with walls produced into setae. Ray-florets female, sometimes neuter, uniseriate; rays strap-shaped, 3-denticulate; staminodes sometimes present; style branches free and lanceolate. Disk-florets hermaphrodite, numerous, sometimes central florets functionally male; corollas infundibuliform, regularly 5-lobed; anther bases entire or shortly sagittate; style branches thickened-connate nearly to the apex. Achenes cylindric-turbinate, ribbed or winged, with the lateral wings folding to form 2 linear chambers on one side, sometimes transversally rugose on the other side, with or without a basal coma of long hairs, otherwise villous or glabrous; pappus c. 2-seriate of delicate hyaline scales, the inner pappus of 5–10 scales equalling the achene in length, the outer pappus smaller.

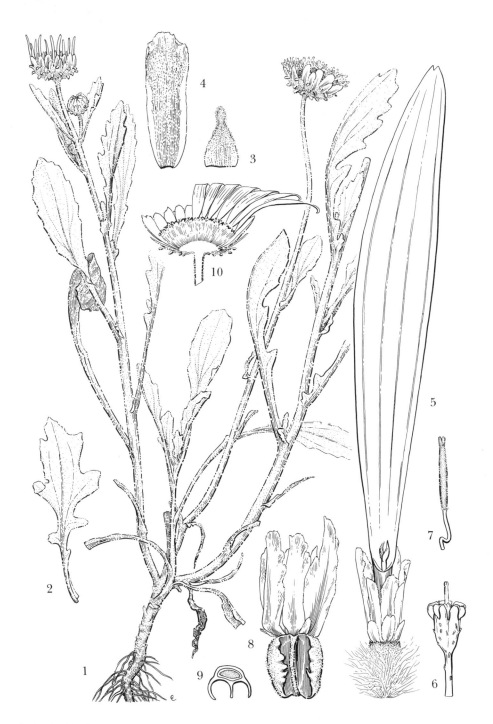

Tab. 42. ARCTOTIS VENUSTA. 1, habit (× ⅔), from *van Vuuren* 1050; 2, leaf (× ⅔); 3, outer phyllary (× 3); 4, inner phyllary (× 3); 5, ray-floret (× 5); 6, disk-floret corolla (× 5); 7, disk-floret style apex, branches connate from the style shaft apex, free above (× 10); 8, achene with pappus (× 5); 9, T/S through achene showing wing-like ribs (× 5), 2–9 from *Mott* 1089; 10, section through capitulum (× 1), from *Seydel* 3612. Drawn by Eleanor Catherine.

A genus of about 50 species, concentrated in the Cape Province of South Africa, one species extending to Namibia, Angola and Botswana. *Arctotis stoechadifolia* Berg., a perennial herb with yellow disk-florets, is cultivated as a garden ornamental in SW. Europe and Australia, and has become naturalised in sandy waste places in S. Portugal.

Arctotis venusta T. Norl. in Bot. Not. **118**: 406, figs. 1, 3 (1965). —Merxm. in Prodr. Fl. SW. Afr. 139: 27 (1966). TAB. **42.**Type from South Africa.
    Arctotis stoechadifolia sensu Harv. in Harv. & Sond., F.C. **3**: 454 (1865) pro parte. —Dinter in Fedde, Repert., Fasc. **15**: 341 (1919). —Lewin in Fedde, Repert., Beih. XI: 69, tab. 5, fig. III (1922), non Berg. (1767).
    Arctotis grandis sensu Wittmack, Gartenflora **49**: 557, fig. 71 (1900), non Thunb. (1799).
    Arctotis stoechadifolia var. *grandis* sensu Warren in Ann. Natal Mus. **6**: 171, fig. 1–7, 12 (1929), non Less. (1832).

An erect annual greyish-araneose herb, 8–60 cm. tall from a woody taproot. Stems short, few to many-branched from near the base, sometimes apparently acaulous with rosulate basal leaves; branches ascending, simple or branching above, leafy, stout, striate-sulcate, araneose to glabrescent, pithy. Leaves spathulate to narrowly oblanceolate in outline, expanded in the apical one half to one third and ± gently tapering into a linear petiole-like base, coarsely remotely serrate or sinuate-dentate to lyrate-pinnatilobed with 2–5 lobes on each side, obtuse at the apex, semi-amplexicaul to subauriculate at the base, thinly araneose; the expanded part strongly 3(5)-nerved from near the base; basal leaves of rosulate plants crowded, mostly 10–20 × 2–4 cm., ± spathulate; upper leaves more remote, mostly up to c. 9 cm. long, oblanceolate. Capitula many, solitary and terminal on the main and secondary branches. Involucres 12–15 mm. long and 12–20 mm. in diam., broadly cupuliform; the outer 2–3 series of phyllaries 2–4 mm. long, produced at the apex into a ± abruptly narrowed oblong-oblanceolate appendage 2–4 mm. long, often recurved, araneose outside; inner phyllaries up to c. 13 × 6 mm., oblong usually widening towards the apex, margins scarious-hyaline to broadly so about the apex, araneose-lanate and glandular on the back except for the margins and apex. Receptacle membranous-alveolate with the walls produced into setae to c. 4 mm. long. Ray-florets; rays white or pinkish inside and blue or purplish outside, 10–30 × 1.5–4 mm., strap-shaped, 3-denticulate at the apex; achenes c. 3 mm. long, ± turbinate, truncate at the apex, at first 3-ribbed, the ribs accrescent as the achene matures becoming wing-like with the lateral wings folding around to nearly meet the central wing and so form 2 open longitudinal chambers on one face of the achene, lateral wings pale dentate-lobed on the vertical margins; pubescent-lanate on the sides with a copious tuft of straight hairs arising from the achene base and exceeding the achene in length; pappus c. 2-seriate, of delicate overlapping narrowly oblong-oblanceolate hyaline scales to c. 4 mm. long, the outer series much smaller. Disk-florets very numerous; corollas blue or purplish, funnel-shaped, c. 5 mm. long; the outer 4–5 series of florets hermaphrodite producing achenes similar to those of the ray-florets; the central series ? functionally male with rudimentary ovaries.

    Botswana. SE: c. 156 km. W. of Lobatse on road to Molopo Farms in valley of fossil R. Sekhutane, c. 1000 m., 1.iii.1977, *Mott* 1089 (K; SRGH).
    Also in Namibia and South Africa (Cape Province, Orange Free State and Transvaal). In hot dry areas, usually in sandy soil, in dry river beds, vleis and disturbed soil on roadsides.
    Cultivated as a garden ornamental as the species or as hybrids of it in southern and eastern Africa and in Europe. It has been recorded as a garden escape in East Africa, and as a weed of cultivation in South Africa.

30. HAPLOCARPHA Less.

Haplocarpha Less. in Linnaea **6**: 90 (1831); Syn. Comp.: 36 (1832). —Beauverd in Bull. Soc. Bot. Genève, sér. 2, **7**: 50 (1915). —Lewin in Fedde, Repert., Beih. XI: 50 (1922).
    Landtia Less., Syn. Comp.: 37 (1832).

Scapose perennial herbs from woody, often rhizomatous rootstocks, occasionally stoloniferous. Leaves radical sessile elliptic, oblanceolate or subspathulate, entire, dentate or pinnatifid, usually white-felted on the undersurface. Scapes 1–many from the rootcrown, simple. Capitula heterogamous, radiate, solitary and terminal. Involucres broadly to narrowly campanulate; phyllaries free, few–many-seriate, scarious-hyaline on the margins. Receptacle shallowly fimbriate-alveolate. Ray-florets female, uniseriate; rays strap-shaped, 3-denticulate; style terete, branches free and lanceolate. Disk-florets numerous, hermaphrodite; corollas infundibuliform, deeply regularly 5-lobed; anther bases sagittate; style branches thickened-connate to near the tips, and free above as short blunt arms. Achenes turbinate, ribbed, with a basal coma of long or short hairs; pappus 2–several-seriate, occasionally reduced or wanting, the inner pappus of delicate hyaline lanceolate scales equalling the corolla tube in length, the outer pappus smaller.

An African genus of about 8 species. Subgenus *Haplocarpha* comprises 4 species, 3 of which are confined to the Cape Province, with one (*H. scaposa*) extending northwards as far as Angola, Zaire and Tanzania. Subgenus *Landtia* (Less.) Beauverd is represented by perhaps 3 species in the high mountains of East Africa, Ethiopia and Eritrea, plus one species (*H. nervosa*) which extends from Zimbabwe to the Cape Province.

Pappus scales longer than the achene; achene with a copious basal coma of long, silky hairs which
 exceed it in length; scapes mostly more than 10 cm. tall - - - - - 1. *scaposa*
Pappus scales and the basal coma of the achene both shorter than the achene; scapes up to c. 10 cm.
 tall - - - - - - - - - - - - - - 2. *nervosa*

1. **Haplocarpha scaposa** Harv. in Harv. & Sond., F.C. **3**: 465 (1865). —Oliv. & Hiern in F.T.A. **3**: 427 (1877). —Lewin in Fedde, Repert., Beih. XI: 52 (1922). —S. Moore in Journ. Bot. **65**, Suppl. II, Gamopet.: 62 (1927). —Mendonça, Contrib. Conhec. Fl. Angol., 2 Compositae: 129 (1943). —Martineau, Rhod. Wild Fls.: 91, plate 31 (1953). —Hilliard, Compos. Natal: 538, fig. 19 C (1977). —Lisowski, Aster. Fl. Afr. Centr. [in Fragm. Flor. Geobot. **36**, Suppl. 1]: 562, fig. 115 (1991). TAB. **43** fig. A. Types from South Africa.
 Haplocarpha thunbergii DC., Prodr. **6**: 494 (1838) non Less. (1832). Types from South Africa.
 Arctotis scaposa (Harv.) O. Hoffm. in Engl., Pflanzenw. Ost-Afr. **C**: 419 (1895). —Eyles in Trans. Roy. Soc. S. Afr. **5**: 520 (1916). Types as above.

A scapose perennial herb from a woody rootstock or rhizome; roots numerous thong-like. Leaves up to c. 12, rosulate suberect, mostly 5–20 × 1.5–6 cm. exceptionally to c. 30 cm. long, or 10 cm. wide, elliptic to oblanceolate, sometimes elongate-subspathulate narrowing from about the middle into a linear petiole-like base, apex obtuse to somewhat rounded, margins entire to sinuate-dentate, midrib composed of 5–7 distinct veins running parallel to diverge in pairs at ± regular intervals, each curving to the leaf apex, subprominent on both surfaces; upper surface finely pilose, lower surface densely white-felted. Scapes 1–4 from the rootcrown, up to c. 75 cm. tall, stout, white-araneose and also with scattered patent soft reddish-purple pilose hairs to c. 1 mm. long. Involucres 12–25 × 8–13 mm., broadly cupuliform. Phyllaries many-seriate, margins scarious-hyaline; outer phyllaries 4–6 mm. long, ± expanded in the apical half, or linear; the middle phyllaries 7–9 × 2–3.5 mm., oblong-lanceolate, rounded at the apex; the innermost to c. 12 × 1.5–2 mm., strap-shaped. Rays yellow, usually erect, 10–21 × 1.5–2.8 mm.; achenes 1.5–1.8 mm. long, turbinate, obscurely c. 8-ribbed, glabrous except for a basal coma of copious hairs longer than the achene; pappus c. 2-seriate, of c. 8 delicate overlapping narrowly lanceolate inner scales to c. 5 mm. long, and a few smaller outer scales. Disk-florets with yellow corollas c. 5.5 mm. long, achenes and pappus similar to those of the ray-florets.

 Zambia. N: Mbala Distr., Zombe Plain, c. 1524 m., 11.ix.1969, *Sanane* 921 (K). W: Mwinilunga Distr., c. 16 km. W. of Lunga R., c. 64 km. S. of the Boma, 12.viii.1930, *Milne-Redhead* 878 (K). C: c. 4.8 km. W. of Karubwe Siding, N. of Lusaka, 12.ix.1963, *Angus* 3741 (FHO; K). E: Nyika, 30.xii.1962, *Fanshawe* 7353 (K; NDO). **Zimbabwe**. W: Matobo, 29.xii.1947, *West* 2508 (K; SRGH). C: Harare, c. 1432 m., 8.v.1927, *Eyles* 4936 (K; SRGH). E: Chirinda Distr., near Umzilizwe (Umswilizwe) R. headwaters, 7.xi.1909, *Swynnerton* 1812 (K). **Malawi**. N: Nyika Plateau, Lake Kaulime, c. 2150 m., 24.x.1958, *Robson & Angus* 322 (K). C: Dedza Distr., Chongoni Forestry School, 18.i.1967, *Salubeni* 505 (K; SRGH). S: Blantyre Distr., Upper Hynde Dam, 2 km. N. of Limbe, c. 1170 m., 14.ii.1970, *Brummitt* 8560 (K; MAL). **Mozambique**. MS: Mossurize, prox. de Espungabera, 12.x.1943, *Torre* 6169 (LISC).
 Also in Angola, Zaire, E. Africa, Swaziland and South Africa (Transvaal, Natal, Orange Free State and Cape Province). A pyrophyte of submontane grassland, high rainfall grassland before the grasses appear, miombo and *Acacia* woodlands bordering dambos or on floodplains, usually in seasonally waterlogged or swampy soil of dambos or beside rivers and dams, often in black clay soils.

2. **Haplocarpha nervosa** (Thunb.) Beauverd in Bull. Soc. Bot. Genève, sér 2, **7**: 51 (1915). —Hilliard in Notes Roy. Bot. Gard. Edinb. **31**: 11 (1971); Compos. Natal: 538, fig. 19 B (1977). TAB. **43** fig. B. Type from South Africa.

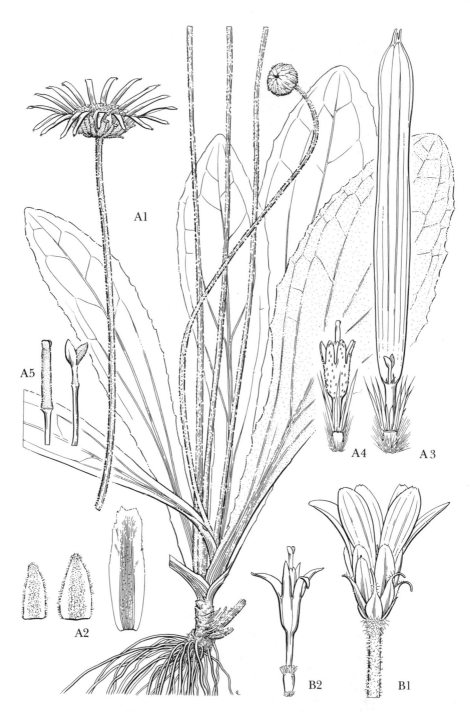

Tab. 43. A. —HAPLOCARPHA SCAPOSA. A1, habit (×⅔), from *Brummitt & Pawek* 11067; A2, outer, middle and inner phyllaries (× 3); A3, ray-floret (× 6); A4, disk-floret (× 6); A5, disk- and ray-floret style apices, branches connate from the style shaft apex, free above (× 8), A2–A5 from *Richards* 22595. B. —HAPLOCARPHA NERVOSA. B1, capitulum (× 2); B2, disk-floret (× 4), B1 & B2 from *Wild* 1509. Drawn by Eleanor Catherine.

Perdicium nervosum Thunb., Fl. Cap., ed. 2: 689 (1823). Type as above.
Leria nervosa (Thunb.) Spreng., Syst. Veg. **3**: 501 (1826). Type as above.
Landtia nervosa (Thunb.) Less., Syn. Comp.: 38 (1832). —Harv. in Harv. & Sond., F.C. **3**: 466 (1865). Type as above.
Landtia hirsuta Less., Syn. Comp.: 37 (1832). —Harv. in Harv. & Sond., F.C. **3**: 466 (1865). Type from South Africa.
Arctotis echinata DC., Prodr. **6**: 486 (1838). Type from South Africa.
Landtia media DC., Prodr. **6**: 495 (1838). Type from South Africa.
Haplocarpha hirsuta (Less.) Beauverd in Bull. Soc. Bot. Genève, sér 2, **7**: 51 (1915). Type as for *Landtia hirsuta*.
Haplocarpha serrata Lewin in Fedde, Repert., Beih. XI: 53 (1922). Type from South Africa.
Haplocarpha ovata Lewin in Fedde, Repert., Beih. XI: 54 (1922). Type from South Africa.

A mat-forming scapose perennial herb, stoloniferous; rootstock woody with numerous thong-like lateral roots. Leaves crowded-rosulate and prostrate, very variable, mostly 3–10 × 1–2.5 cm., exceptionally to c. 21 cm. long or 7.5 cm. wide, elliptic to oblanceolate in outline sometimes elongate-spathulate, gradually or abruptly narrowed from about the middle into a linear petiole-like base, apex subacute to rounded, base cuneate to long-attenuate becoming dilated and ± stem-clasping, margins entire or repand, irregularly toothed, subruncinate or subpinnatifid, midrib broad composed of numerous ± distinct veins running parallel to diverge in pairs at regular intervals, prominent beneath; upper surface glabrous or thinly pilose; lower surface densely white-felted. Scapes several from the rootcrown, up to c. 8 cm. tall but usually shorter than the leaves, simple, glabrous or white-araneose with ± obvious scattered patent soft reddish-purple pilose hairs to c. 1 mm. long. Involucres 8–11(14) × 6–9 mm., broadly obconic; phyllaries few-seriate, narrowly hyaline on the margins, glabrous occasionally pilose outside; outer phyllaries c. 6 mm. long, narrowly ovate, acute to attenuate at the apex; the inner phyllaries up to c. 10 × 3 mm. and oblong-lanceolate, rounded-mucronate at the apex. Ray-florets; rays yellow, spreading, 10–20 × 4–7 mm., narrowly-elliptic, narrowly 3-lobed at the apex; achenes 1–2.5 mm. long, turbinate, obscurely c. 8-ribbed, the ribs transversally rugose or muricate, glabrous except for a short delicate coma; pappus variable, sometimes wanting, 2–3-seriate, of delicate scarious overlapping narrowly lanceolate scales to 1 mm. long, the inner tapering to an attenuate apex, the outer much smaller and blunt. Disk-floret corollas yellow, up to c. 6 mm. long, achenes and pappus similar to those of the ray-florets.

Zimbabwe. E: Nyanga Distr., Mt. Inyangani, c. 2438 m., 5.ix.1954, *Wild* 4597 (K; SRGH).

Also in South Africa (Cape Province, Natal, Transvaal) and Lesotho. Known only from Mt. Inyangani in Zimbabwe where it occurs in moist submontane grassland, streamsides and marshy localities, at times partly submerged. Under wet conditions it forms extensive mats.

31. ARCTOTHECA Wendl.

Arctotheca Wendl., Hort. Herrenhus.: 8, t. 6 (1798). —Beauverd in Bull. Soc. Bot. Genève, sér. 2, **7**: 51 (1915). —Lewin in Fedde, Repert., Beih. XI: 48 (1922).
Cryptostemma R.Br. in Aiton, Hort. Kew. ed. 2, **5**: 141 (1813).
Microstephium Less. in Linnaea **6**: 92 (1831).

Perennial herbs, often white felted-lanate, stoloniferous. Stems erect to prostrate or creeping. Leaves alternate or subradical, lyrate-pinnatifid or pinnatilobed (bipinnately lobed) or spathulate. Capitula heterogamous, radiate, large, solitary, stalked. Involucres hemispheric; phyllaries many-seriate free imbricate, margins scarious. Receptacle deeply membranous-alveolate. Ray-florets neuter, rays yellow, strap-shaped, 3-denticulate. Disk-florets hermaphrodite; corollas tubular, regularly 5-lobed; anther bases sagittate; style terete, branches thickened-connate to near the apex and free above as short blunt arms. Achenes cylindric-turbinate, tomentose to sericeous-lanate, occasionally glabrous; pappus of small hyaline scales or coroniform, occasionally wanting.

A South African genus comprising 4 species, mainly from the Cape Province with one extending into southern Mozambique.
Arctotheca calendula (L.) Levyns is cultivated as a garden ornamental in Europe. It has become naturalised in the Mediterranean area, the SE. United States of America, Chile, Argentina, Australia and New Zealand.

Tab. 44. ARCTOTHECA POPULIFOLIA. 1, habit (× $\frac{2}{3}$), from *Acocks* 19789; 2, inner phyllary (× 3); 3, ray-floret (× 5); 4, disk-floret (× 5), 2–4 from *Retief* 273.01. Drawn by Eleanor Catherine.

Arctotheca populifolia (Bergius) T. Norl. in Aquilo **6**: 84, photos. 1 & 2 (1967). —Hilliard, Compos. Natal: 541 (1977). TAB. **44**. Type is a specimen from a plant cultivated in Holland.

Arctotis populifolia Bergius, Desc. Pl. Cap.: 323 (1767). Type as above.

Osteospermum niveum L.f., Suppl. Pl.: 386 (1782). Type from South Africa.

Microstephium niveum (L.f.) Less., Syn. Comp.: 55 (1832). —DC., Prodr. **6**: 496 (1838). —Harv. in Harv. & Sond., F.C. **3**: 468 (1865). Type as above.

Microstephium populifolium (Bergius) Druce in Rep. Bot. Exch. Cl. Brit. Isles **IV**, Report 1916: 636 (1917). Type as above.

Arctotheca nivea (L.f.) Lewin in Fedde, Repert., Beih. XI: 50 (1922). —Levyns in Adamson & Salter, Fl. Cap. Pen.: 828 (1950). Type as above.

A shortly trailing, tough-fleshy white-felted sometimes sticky perennial herb with a woody taproot. Stems herbaceous, robust, to c. 1 cm. in diam., prostrate to decumbent, or suberect to c. 30 cm. tall, branched, leafy, producing adventitious roots when buried in sand. Leaves thick-textured, mostly 6–20 cm. long, usually spathulate, expanded in the apical c.⅓ and abruptly narrowed below, sometimes ± lyrate; the apical part 2–7 × 2–7 cm., ± broadly ovate or broadly elliptic, sometimes lobed at the base, apex rounded to obtuse, base rounded-truncate to cordate or broadly cuneate, margins subentire to shallowly dentate with callose-tipped teeth, strongly 5–7-nerved from the base; the lower c.⅔ to c. 11 mm. wide, linear, sometimes widening into 2–many large lobes towards the apex, widening again near the base and somewhat conduplicate. Capitula few–many, solitary and terminal on simple or sometimes 1–2-branched axillary stalks up to c. 12 cm. long. Involucres 1–1.5 cm. in diam., hemispheric; phyllaries from c. 3 mm. long and broadly triangular outside to c. 10 mm. long and lanceolate-ovate inside, all white-felted to lanate on the back apart from scarious tips and margins. Receptacle deeply, membranous-walled alveolate. Ray-floret corollas yellow, rays 4–14 mm. long, strap-shaped, 3-denticulate apically; ray-floret achenes to c. 3 mm. long, rudimentary. Disk-floret corollas yellow, 3–4 mm. long, funnel-shaped with a short tube and a long campanulate limb with lobes c. 1 mm. long; disk-floret achenes c. 5 mm. long, laterally compressed-turbinate, wider than the corolla, finely lanate; pappus cupuliform consisting of a row of broad partly connate scales 0.5–1 mm. deep, membranous or chartaceous.

Mozambique M· Matutuíne, Ponta do Ouro Beach, 24.xii.1948, *Gomes e Sousa* 3904 (K); Inhaca Isl., Lighthouse N. foreshore, 30.ix.1957, *Mogg* 27072 (K).

Also in South Africa (Natal & Cape) and introduced in Australia. Early coloniser on shifting sand of foreshore dunes, forming patches.

32. **BERKHEYA** Ehrh.

Berkheya Ehrh., Beitr. **3**: 137 (1788), nom. conserv. —Roessler in Mitt. Bot. Staatss. Münch. **3**: 104 (1959); in Kirkia **10**: 74–86 (1975).

Crocodilodes Adans., Fam. Pl. **2**: 127 (1763).

Stobaea Thunb., Prodr. Pl. Cap. praefat. (1800); Nov. Gen.: 183 (1801); Fl. Cap. ed. Schultes: 620 (1823).

Perennial herbs sometimes suffrutescent with annual stems from woody rootstocks, or subshrubs. Indumentum felted-tomentose on leaf under surfaces, araneose-tomentose elsewhere, consisting of many-celled uniseriate soft hairs with extremely long filiform white apical cells, their bases tending to become rigid and persistent. Leaves alternate, sometimes crowded on the lower stem, ± spiny, subentire to dentate, pinnatifid or pinnatisect with spinescent teeth, sometimes decurrent on the stem as spiny wings. Capitula heterogamous and radiate or homogamous and discoid, solitary or subcorymbosely arranged. Phyllaries several-seriate, connate at the base, always spiny with spines sometimes slender and bristle-like. Receptacle with deep alveolae which totally enclose the achenes leaving only the pappus exserted; the alveolae margins entire or irregularly toothed or setose. Ray-florets uniseriate, neuter, corollas yellow sometimes white, always with staminodes, achenes wanting. Disk-florets hermaphrodite numerous; corollas yellow, infundibuliform, deeply 5-lobed; anther bases sagittate; style branches linear; achenes ± turbinate, ribbed, villous pubescent or glabrous; pappus ± 2-seriate, of numerous short or long scales, sometimes partly connate.

A genus of 74 species in Southern and tropical Africa, 11 species occurring in the Flora Zambesiaca area.

1. Leaf upper surface beset with straw-coloured bristle-like spines - - - - - 2
   - Leaf upper surface glabrescent or ± araneose or harshly pubescent, not bristly   4
2. Stem wingless - - - - - - - - - - - - - - 7. *setifera*
   - Stem winged - - - - - - - - - - - - - - - 3
3. Capitula radiate; wings continuous on the stem - - - - - - 4. *polyacantha*
   - Capitula discoid; wings discontinuous on the stem - - - - 8. *johnstoniana*
4. Stem winged - - - - - - - - - - - - - - - 6. *radula*
   - Stem wingless - - - - - - - - - - - - - - 5
5. Leaves with petioles 5–30 cm. long, lamina subentire abruptly narrowed and ± decurrent on
   the petiole - - - - - - - - - - - - - - 9. *speciosa*
   - Leaves sessile or subsessile, apparently petiolate if lamina pinnately divided but then petiole less
   than 5 cm. long - - - - - - - - - - - - - - 6
6. Leaves undivided, margins sub-entire or at most spinescent-denticulate, with long slender
   bristle-like mostly submarginal spines towards the base - - - - - - 7
   - Leaves toothed, lobed or pinnatifid, all teeth and lobes produced in spines; submarginal
   bristle-like spines absent - - - - - - - - - - - 8
7. Corolla lobes of disk-florets yellow or brownish (anther appendages often purplish), glandular-
   tuberculate outside particularly towards the apex; rays yellow - - - 11. *zeyheri*
   - Corolla lobes of disk-florets dark-purplish-black in the upper half, smooth calloso-glandular at
   the apex; rays creamy-yellow or whitish, often with black-tipped lobes   10. *rehmannii*
8. Capitula discoid - - - - - - - - - - - - - 2. *bipinnatifida*
   - Capitula radiate - - - - - - - - - - - - - - 9
9. Achenes glabrous; leaves lobed, especially the middle and lower cauline leaves, each lobe
   usually with 2 or more spine-tipped teeth; outer phyllaries 5–7 mm. broad   5. *robusta*
   - Achenes sericeous; leaves sharply toothed or serrate but never with toothed lobes; outer
   phyllaries mostly 1–2(3) mm. broad - - - - - - - - - - 10
10. Leaves serrate; phyllaries araneose or glabrescent outside, spines usually paired on the
   margins - - - - - - - - - - - - - - 3. *angolensis*
   - Leaves pinnatifid-dentate or coarsely dentate; outer phyllaries resembling the leaves, felted-
   tomentose outside, spines single or only weakly paired on the margins   1. *carlinopsis*

1. **Berkheya carlinopsis** Welw. ex O. Hoffm. in Bol. Soc. Brot. **13**: 34 (1896). O. Hoffm. in Warb.,
   Kunene-Samb. Exped., Baum: 425 (1903). —Mendonça, Contrib. Conhec. Fl. Angol., **1**
   Compositae: 131 (1943). —Roessler in Mitt. Bot. Staatss. Münch. **3**: 142 (1959); op. cit. **11**: 96
   (1973); in Kirkia **10**: 75 (1975). Type from Angola.
   Crocodilodes carlinopsis (Welw. ex O. Hoffm.) Hiern, Cat. Afr. Pl. Welw. **1**, 3: 608 (1898). Type as
   above.

A perennial herb or subshrub up to c. 1.5 m. tall. Stems branched, whitish araneose-
tomentose, or glabrescent, leafy. Leaves sessile, 3–6 cm. long, dentate to pinnatifid-
dentate; lamina (excluding the teeth or lobes) 2–3(4) mm. wide and linear or 5–15 mm.
wide and lanceolate; teeth 3–8 on each side, each tooth 2–6(10) mm. long, triangular or
linear and extended in a tawny spine 2–3 mm. long; margins of teeth and sinuses entire or
armed with smaller spines; upper surface smooth or somewhat scabrous, slightly to
densely araneose-tomentose or glabrescent; lower surface whitish felted-tomentose.
Capitula radiate, solitary and terminal on the branches, or subcorymbosely arranged,
2.5–5(+?) cm. in diam. including the rays. Phyllaries spreading, felted-tomentose outside,
subglabrous or glabrous inside, 10–20 × 1–3 mm., linear-lanceolate, spiny-acuminate,
ciliate-spinescent on the margins with spines 1–3 mm. long; the outermost phyllaries ±
leaf-like with small spine-tipped teeth; inner phyllaries smaller and less spinescent-
ciliate. Margins of the receptacular alveolae extended into straw-coloured bristles 1–2
mm. long. Achenes 1.5–3.5 mm. long, turbinate, 8–10-ribbed, strigose-sericeous,
glandular-viscid at the apex. Pappus scales 2-seriate, 1–1.5 mm. long, narrowly oblong,
acute or subobtuse, denticulate towards the apex.

Subsp. **sylvicola** (S. Moore) Roessler in Mitt. Bot. Staatss. Münch. **11**: 96 (1973). Type from Angola.
   Berkheya sylvicola S. Moore in Journ. Bot. **65**, Suppl. II: 63 (1927). Type as above.
   Berkheya carlinopsis var. *sylvicola* (S. Moore) Roessler in Mitt. Bot. Staatss. Münch. **3**: 143
   (1959). Type as above.

Leaf lamina 2–3(4) mm. wide, linear and ± pinnatifid-dentate.

Zambia. W: Solwezi, 25.vii.1964, *Fanshawe* 8843 (K; M; NDO).
Also in S. Angola. With grass in chipya woodland.
Berkheya spekeana Oliv. can be somewhat similar in appearance to this subspecies but is readily
distinguished by its leaves which are beset with scattered bristle-spines on the upper surface.
   Subsp. *sylvicola* (S. Moore) Roessler is distinguished from the other subspecies by its linear, 2–3(4)
mm. wide, and ± pinnatifid-dentate leaf blade, with the lobe length usually exceeding the blade

width. Subsp. *carlinopsis* is confined to SW. Angola and has dentately-lobed leaves with lanceolate blades 5–15 mm. wide and triangular teeth. Plants of this species from the Transvaal have been treated as subsp. *magalismontana* (Bolus) Roessler and are characterised mainly by the leaves bearing spines only at the apex of the teeth and not elsewhere on the margins as in the typical subspecies.

2. **Berkheya bipinnatifida** (Harv.) Roessler in Mitt. Bot. Staatss. Münch. **3**: 212 (1959); in Kirkia **10**: 79 (1975). —Hilliard, Compos. Natal: 562 (1977). Type from South Africa.

    Stobaea seminivea DC., Prodr. **6**: 518 (1838), non *Berkheya seminivea* Harv. & Sond., F.C. **3**: 507 (1865). Type from South Africa.

    Stobaea bipinnatifida Harv. in Harv. & Sond., F.C. **3**: 499 (1865). Type as for *Berkheya bipinnatifida*.

    Crocodilodes seminiveum (DC.) Kuntze, Rev. Gen. Pl. **1**: 333 (1891). Type as for *Stobaea seminivea*.

    Crocodilodes bipinnatifidum (Harv.) Kuntze, Rev. Gen. Pl. **1**: 332 (1891); Rev. Gen. **3**, 2: 143 (1898). Type as for *Berkheya bipinnatifida*.

A spinescent perennial herb 0.5–3 m. tall. Stems sparingly branched; branches ascending, leafy and like the stems thinly araneose at first. Leaves sessile, upper cauline leaves grading into the phyllaries, the lower- and mid-cauline leaves 5–30 × 3–18 cm., elliptic to oblong in outline, tapering and petiole-like, or not narrowed towards the base, pinnatilobed to pinnatifid sometimes somewhat bipinnatifid; lobes or segments narrowly triangular or linear-lanceolate and slightly to strongly dentate or cut, the lobes and teeth extended into spines 1–10 mm. long, the margins ± spinose-denticulate; the base briefly dilated and semi-amplexicaul with spiny auricles, or cordate; leaf upper surface thinly araneose becoming ± scabridulous when dry; leaf lower surface whitish felted-tomentose. Capitula homogenous and discoid, paniculately arranged, ± clustered at ends of branches, 1–2 cm. in diam. Phyllaries mostly 8–12 × 1–2.5 mm., the outer and inner ones shorter, all narrowly lanceolate and tapering to an apical spine, margins with 2–3 stout spines 2–4 mm. long on each side, these often with secondary patent spines at their bases (therefore paired), sometimes with smaller spines between the main spines, lamina at least in inner phyllaries spinescent-ciliate about the apex and glandular-hairy outside, or araneose. Receptacular alveolae margins paleaceous-setose, equalling or exceeding the pappus in length. Florets numerous; corollas white, c. 10 mm. long, the limb narrowly funnel-shaped and deeply lobed, achenes 2–3 mm. long, obconic-turbinate, 8–10-ribbed, ± sparsely hispidulous; pappus of c. 20 overlapping oblong scales 0.2–1.5 mm. long, obtuse or acute at the apex, fimbriate.

Subsp. **echinopsoides** (Bak.) Roessler in Mitt. Bot. Staatss. Münch. **3**: 214 (1959); in Kirkia **10**: 80 (1975). —Hilliard, Compos. Natal: 563 (1977). —Lisowski, Aster. Fl. Afr. Centr. [in Fragm. Flor. Geobot. **36**, Suppl. 1]: 565 (1991). Type: Malawi, Nyika Plateau, *Whyte* 186 (K, holotype).

    Berkheya echinopsoides Bak. in Bull. Misc. Inf., Kew **1898**: 155 (1898). Type as above.

Phyllary marginal spines usually with secondary patent spines at their bases and with several smaller spines between the main spines, lamina at least in inner phyllaries spinescent-ciliate about the apex and scattered glandular-hairy outside, never araneose. Corolla limb narrowly funnel-shaped and tapering into the corolla tube, the sinuses between the lobes not extending to base of limb.

Key to varieties

Leaves all narrowed and petiole-like to the base, briefly dilated and semi-amplexicaul
                                                 var. *echinopsoides*
Leaves, at least those on the upper stem and branches cordate, not tapering to the base
                                                     var. *cordata*

Var. **echinopsoides** (Bak.) G.V. Pope, stat. nov. Folia omnia basin angustata petioliformes, breviter dilatata atque semiamplexicaules. Basionym: *Berkheya echinopsoides* Bak. in Bull. Misc. Inf., Kew **1898**: 155 (1898).

Zambia. N: Mbala Distr., Saisi Valley, Kalambo Farm, 1800 m., 21.v.1952, *Richards* 1763 (K; SRGH). **Malawi**. N: Chitipa Distr., c. 40 km. SE. of Chisenga beyond Wenya, 1750 m., 21.iv.1976, *Pawek* 11182 (K; MAL; MO; SRGH).

Also in Tanzania and Zaire (mainly in the area from around the southern end of Lake Tanganyika to the northern end of Lake Malawi) and in Swaziland. In escarpment miombo and chipya woodland, in high altitude and submontane grassland and in dambos, often on rocky hillsides or on termitaria in dambo grassland.

Var. **cordata** G.V. Pope var. nov. A varietate echinopsoide foliis caulinis superioribus atque ramulinis basin cordatis nec attenuatis differt. Typus: Natal, 2732 Bb Ubombo, Manzengwenya, 28.xi.1969, *Moll* 4801 (K, holotype; NH).

Mozambique. M: Matutuíne, Ponta do Ouro para Zitunde, 19.vi.1981, *de Koning, Hiemstra & Nuvunga* 8835 (K; LMU); prox. de Zitunde, 17.xi.1944, *Mendonça* 2898 (LISC).

Also in adjacent northern Natal (Sordwana Bay and Ubombo area) and in Tanzania (from the Itigi area to Lake Victoria). Grassland, on sandy soils or black-cotton clays.

Subsp. *bipinnatifida* occurs in the eastern Cape Province and Natal in South Africa, on forest margins, often as a pioneer in disturbed forest on roadsides and clearings. It is morphologically very similar to subsp. *echinopsoides*, but may be distinguished by its phyllaries, corollas and pappus. The phyllaries of the typical subspecies are all ± araneose outside or glabrescent, the marginal spines are never paired and smaller spines are absent from between them. The corolla limb narrows abruptly into the tube with the sinuses between the lobes extending to the base of the limb.

3. **Berkheya angolensis** O. Hoffm. in Engl., Bot. Jahrb. **24**: 475 (ii.1898). —Mendonça, Contrib. Conhec. Fl. Angol., **1** Compositae: 131 (1943). —Roessler in Mitt. Bot. Staatss. Münch. **3**: 232 (1959). —Lisowski, Aster. Fl. Afr. Centr. [in Fragm. Flor. Geobot. **36**, Suppl. 1]: 567 (1991). Type from Angola.

    Crocodilodes andongensis Hiern, Cat. Afr. Pl. Welw. **1**, 3: 609 (xii.1898). Type from Angola.

    Crocodilodes angolensis (O. Hoffm.) Hiern, Cat. Afr. Pl. Welw. **1**, 3: 609 (1898). Type as for *Berkheya angolensis*.

    Berkheya andongensis (Hiern) K. Schum. in Just, Jahresb. **26**, 1: 375 (1900). Type as for *Crocodilodes andongensis*.

A slender perennial herb up to 1.5 m. tall, from a woody rootstock; roots thong-like. Stems annual erect simple, or branched in the upper part, thinly araneose, leafy. Leaves ± shortly petiolate, mostly 5–15 × 1–2(5) cm., linear-elliptic, tapering-acute above with a 1–3 mm. long apical spine, margins serrulate-dentate occasionally coarsely so, the teeth tipped with ± ascending bristle-spines 1–3 mm. long, also with 1–several smaller spines between the teeth; leaf upper surface very thinly araneose, becoming glabrescent or ± scabrous, the lower surface whitish felted-tomentose. Capitula radiate, few to many, solitary and terminal on stem and branches, oldest at the stem apex and overtopped by the younger capitula on lower branches, 5–7 cm. in diam. including the rays. Phyllaries thinly araneose outside, subglabrous inside, the inner often sparsely ciliate-pubescent, mostly 15–25 × 1.5–2.5 mm., outer and inner ones shorter, all linear-lorate tapering to an apical spine, margins of outermost phyllaries with up to c. 8 slender spines 3–6 mm. long on each side, these usually with a second patent spine at the base (phyllary spines paired), inner phyllaries with fewer and smaller spines. Receptacular alveolae margins with 1–2 mm. long spinescent setae not exceeding the achene in length. Rays yellow, mostly 12–25 × 2–3.5 mm., 4-lobed at the apex often unequally and deeply so with 1 or 2 sinuses to c. 12 mm. deep. Disk-floret corollas orange-yellow, c. 12 mm. long, narrowly funnel-shaped, glandular hairy outside especially towards the lobe apices; achenes 2.5–4 mm. long, turbinate, slightly 8–10-ribbed, hispid-strigose; pappus ± 2-seriate of overlapping narrowly obovate-oblong stramineous scales, denticulate-fimbriate about the apex.

Zambia. W: Mwinilunga Distr., SE. of Dobeka Bridge, 3.i.1938, *Milne-Redhead* 3934 (K; LISC; PRE).

Also in Angola and Zaire. *Brachystegia* and chipya woodland, on sandy soil.

4. **Berkheya polyacantha** Bak. in Bull. Misc. Inf., Kew **1898**: 156 (1898). —Brenan in Mem. N.Y. Bot. Gard. **8**: 487 (1954). Type: Malawi, Nyika Plateau, vii.1895, *Whyte* 150 (K, holotype).

    Berkheya parvifolia Bak. in Bull. Misc. Inf., Kew **1898**: 155 (1898). Syntypes: Malawi, Kondowe to Karonga, vii.1896, *Whyte* 357 (K); the other syntype from Tanzania.

    Berkheya echinacea subsp. *polyacantha* (Bak.) Roessler in Mitt. Bot. Staatss. Münch. **3**: 167 (1959); in Kirkia **10**: 76 (1975). —Brummitt in Wye Coll. Malawi Proj. Rep.: 52 (1973). Type as above.

A slender perennial herb up to c. 140 cm. tall, from a woody roostock; roots numerous, swelling to form fusiform root tubers. Stems annual 1–several erect simple or ± branched above, leafy, leaves decurrent on the stem in continuous wings; wings to 1.5 cm. wide, sinuate-dentate to coarsely dentate on the margins, spinescent as in leaf upper surfaces. Leaves ± crowded on the lower stem, sessile; lower leaves 12–22 × 2–5 cm., oblanceolate; upper cauline leaves mostly 3–12 × 0.5–2 cm., linear-oblong or lanceolate, acute; all leaves sinuate-dentate to coarsely dentate or sub-entire, upper surfaces and margins beset with patent stramineous bristle-like spines 2–7 mm. long, longest at the teeth apices, lower surfaces glabrous or thinly pilose-araneose. Capitula radiate, solitary and terminal, or several ± clustered on stems and branches, 4–6 cm. in diam. across the expanded rays. Phyllaries mostly 12–20 × 1–2 mm., the outer and innermost somewhat shorter, linear and spine-tipped; the outer phyllaries leaf-like and dentate-lobed with 2–4 bristle-like spines 3–5 mm. long on each tooth, often also spinescent on the margin between the teeth, often thinly araneose outside; the inner phyllaries ± unarmed. Receptacle alveolae paleaceous-setose on the margins, the setae to c. 2 mm. long. Rays yellow, mostly 15 × 1.5–4 mm. Disk-floret corollas yellow, c. 7 mm. long; achenes 1–2 mm. long, turbinate, ± densely sericeous-strigose; pappus scales ± biseriate, 1.5–2 mm. long, narrowly oblong-obovate.

Zambia. N: Mbala Distr., Nkali (Kali) Dambo, c. 1500 m., 10.iii.1955, *Richards* 4861 (EA; K). **Zimbabwe**. E: Nyanga Distr., Heights, c. 11 km. on Circular Drive from Pungwe Rest Huts, 1800 m., 18.i.1951, *Chase* 3573 (BM; LD; SRGH). **Malawi**. N: Nyika Plateau, 2300 m., 16.viii.1946, *Brass* 17246 (K; MO).
Also in Tanzania. Submontane grassland, often in marshy places.

Material from the Mbala area of Zambia represents a distinct variant which, while retaining the habit of this species with the stem lower leaves large and persistent, is distinguished by the upper stems being more densely leafy than is usual for the species, with the leaves ± appressed, overlapping and lorate.

5. **Berkheya robusta** Roessler in Mitt. Bot. Staatss. Münch. **3**: 260 (1959); in Kirkia **10**: 86 (1975). —Hilliard, Compos. Natal: 572 (1977). Type: Mozambique, Maputo, próx. de Catuane, 22.iv.1944, *Torre* 6495 (PRE, holotype; LISC).

A robust perennial herb 1–2 m. tall. Stem single erect stout shortly-branched above, leafy, coarsely pilose-araneose. Leaves sessile, 7–16(20) × 3–6(10) cm., elliptic-oblong in outline, pinnatilobed; leaf and lobe apices strongly spine-tipped; margins entire or ± dentate, spinescent; bases semi-amplexicaul with coarsely-dentate spiny auricles, or the lower leaves narrowed to petiole-like bases, not decurrent; lobes mostly 3–6 on each side, ± spaced, triangular and entire or somewhat rounded or irregular in outline and coarsely dentate, the teeth aristate-spinose with spines to c. 6 mm. long; leaf upper surface thinly araneose with scattered pilose hairs, becoming scabridulous when dry, the lower surface whitish or greyish felted-tomentose. Capitula radiate, numerous, 2–several clustered at the ends of short branches, radiate, c. 5–7 cm. in diam. across the expanded rays. Phyllaries mostly 14–22 × 4–8 mm., the outer and inner somewhat shorter, thinly araneose; the outer ± leaf-like, ovate-lanceolate and dentate to somewhat lobed with spine-tipped teeth, spines 2–4 mm. long and usually with a second patent spine at the base (phyllary spines paired), ciliate-spinescent between the spine-tipped teeth, succeeding inner phyllaries becoming linear-lanceolate with smaller spines. Receptacle alveolae margins extended into bristles 1–2 mm. long. Rays yellow, mostly 10–20 × 2–5 mm., shallowly 4-lobed at the apex. Disk-floret corollas yellow, 8–10 mm. long, lobes glandular-tuberculate at the apex; achenes 2–3 mm. long, very narrowly turbinate, 4–5-angular, tapering and ± grooved to the base, glabrous; pappus scales 1-seriate, ± connate at the base, c. 1 mm. long, oblong.

Mozambique. M: Namaacha, 29°59'S, 32°01'E, 600 m., ii.1931, *Gomes e Sousa* 414 (K).
Also in South Africa (Transvaal, Natal) and Swaziland. Wooded grassland often on rocky hillsides.

6. **Berkheya radula** (Harv.) De Wild., Ic. Sel. Hort. Then. **2**: 89 (1901). —Burtt Davy in Bull. Misc. Inf., Kew **1935**: 571 (1935). —Roessler in Mitt. Bot. Staatss. Münch. **3**: 244 (1959). Type from South Africa (Transvaal).
    Stobaea radula Harv. in Harv. & Sond., F.C. **3**: 491 (1865). Type as above.
    Crocodilodes radula (Harv.) Kuntze, Rev. Gen. Pl. **1**: 333 (1891). Type as above.
    Berkheya adlamii Hook. f. in Curtis, Bot. Mag. **123**: t. 7514 (1897) as "adlami". —Eyles in Trans. Roy. Soc. S. Afr. **5**: 520 (1916). Type a plant cultivated in the Kew Gardens.

A robust prickly perennial herb 25–100 cm. tall, from a woody taproot. Stems annual single erect, branched in the upper part, leafy, longitudinally striate, ± scattered pilose with hairs persistent and ± rigid when dry; the lower 4–8 leaves ± crowded and subrosulate, the cauline leaves decreasing in size and density towards the stem apex and decurrent in continuous wings extending the length of the stem; wings to 1 cm. wide, sinuately lobed to coarsely dentate and spinescent as in leaf margins, indumentum as for leaves. Leaves alternate, basal leaves 12–40 × 3–12 cm., oblanceolate to elliptic; upper cauline leaves narrowly oblong to linear, grading into synflorescence bracts; all leaves subobtuse to rounded and spine-tipped at the apex, repand-sinuate to sinuately lobed, the lobes semi-circular with sinuses 0.5–2 cm. deep, the lamina tapering from about the middle to a ± narrowly winged or lobed midrib, the margins of lobes and sinuses shallowly denticulate, the teeth aristate-spinose with spines to 8 mm. long, margins spinescent between the teeth, upper surface with scattered pilose hairs becoming ± scabrous as in stems, lower surface felted-tomentose. Capitula radiate, mostly 3–5 laxly arranged, 3–5 cm. in diam. across the expanded rays. Phyllaries mostly 12–20 × 1.5–3 mm., the outer and innermost somewhat shorter, all ± lorate and spine-tipped, sparsely thinly araneose often also with gland-tipped hairs or glabrescent; the outer phyllaries with 5–8 stout spines on each margin, 3–6 mm. long, these usually with a second patent spine at the base (paired phyllary spines), ciliate-spinescent between the spine-tipped teeth; the inner phyllaries with up to 10 smaller spines per side, often not paired. Receptacular alveolae margins with teeth c. 1 mm. long. Rays yellow, mostly 12–15 × 1.5–4 mm., 4-lobed at the apex and often unequally and deeply cleft with 1 or 2 lobes up to c. 8 mm. long. Disk-florets numerous; corollas yellow, 8–10 mm. long; achenes 3.5–4.5 mm. long, very narrowly turbinate, 4–5-angular, tapering and ± grooved towards the base, glabrous; pappus scales ± biseriate, overlapping, 1.5–2 mm. long, narrowly oblong, fimbriate-denticulate about the apex.

Botswana. SE: Farm Springfield, c. 3.2 km. S. of Lobatse, E. of railway, 17.i.1960, *Leach & Noel* 159 (K; SRGH). **Zimbabwe**. W: Esigodini (Essexvale), c. 1200 m., 15.i.1921, *Borle* 89 (K; M; PRE; SRGH). C: Charter, c. 8 km. N. of Lalapansi on Great Dyke, 17.i.1962, *Wild* 5622 (K; M; MO; PRE; SRGH). E: Makoni, Maidstone, 4.i.1931, *Norlindh & Weimarck* 4074 (K; LD). S: Zingedzi River Bridge on road between Mbrengwa (Belingwe) and West Nicholson, 4.v.1972, *Pope* 632 (K; SRGH).
Also in South Africa (Transvaal, Orange Free State and Cape Province). *Acacia* woodland and wooded grassland on flood-plains, river banks and vleis, usually on black alluvium, also on disturbed ground.

7. **Berkheya setifera** DC., Prodr. **6**: 507 (1838). —Harv. in Harv. & Sond., F.C. **3**: 509 (1865). —Roessler in Mitt. Bot. Staatss. Münch. **3**: 182 (1959); in Kirkia **10**: 77 (1975). Type from South Africa.
　　Crocodilodes setiferum (DC.) Kuntze, Rev. Gen. Pl. **1**: 333 (1891); op. cit. **3**, 2: 144 (1898). Type as above.
　　Berkheya setifera var. *tropica* S. Moore in Journ. Linn. Soc., Bot. **40**: 123 (1911). —Eyles in Trans. Roy. Soc. S. Afr. **5**: 521 (1916). Type: Zimbabwe, Gazaland, Lusitu River (hills), 1500 m., 20.ix.1906, *Swynnerton* 1869 (BM, holotype; K; MO).

A perennial herb up to c. 120 cm. tall, from a stout woody rootstock (? with root tubers). Stems annual single erect ± branched above, glabrous or thinly pilose-araneose. Leaves mostly basal with 1–several cauline leaves diminishing in size to the stem apex, not decurrent on the stem; basal leaves ascending-subrosulate, 10–45 × 4–12 cm., obovate to oblanceolate, apex rounded or subobtuse, base ± long-tapering becoming petiole-like, dilated and somewhat stem clasping, margins subentire to dentate or sinuate-dentate, upper surface and margins beset with patent stramineous spine-like bristles 2–7 mm. long, the longest at tooth apices, lower surface indumentum as for stem; cauline leaves few, becoming bract-like at the stem apex. Capitula radiate, solitary and terminal on the branches or several ± clustered, 3–6 cm. in diam. across the expanded rays. Phyllaries from c. 4 mm. long in the outer series to c. 12 mm. long in the inner series, mostly c. 1 mm. wide at the base and linear-tapering to a spinescent apex, spinescent-ciliate on the margins and pilose-araneose on the back; the inner ± unarmed. Receptacular alveolae margins with spinescent setae to c. 6 mm. long. Rays yellow, mostly 12–20 × 1.5–4 mm. Disk-florets numerous; limb yellow, c. 13 mm. long from the tubular lower half; achenes 2–3 mm. long, turbinate, 8–10-ribbed, hispid-strigose; pappus 2–3-seriate, of narrow paleaceous barbellate setae, 8–10 mm. long.

Zimbabwe. E: Chimanimani Distr., Rocklands, 1400 m., 8.x.1950, *Wild* 3539 (K; LD; LISC; S; SRGH). **Mozambique**. MS: Manica, Serra Zuira, Tsetserra, c. 6 km. from stables on road to Chimoio (Vila Pery), c. 1800 m., 3.iv.1966, *Torre & Correia* 15632 (LISC). M: Libombos, near Namaacha, Mt. Mpondium, 800 m., 22.ii.1955, *Exell, Mendonça & Wild* 491 (BM; LISC; SRGH).
Also in South Africa (Transvaal, Natal, Orange Free State and E. Cape Province), Swaziland and Lesotho. Submontane grassland or on the Libombo Mts. at lower altitudes in shrub savanna.

8. **Berkheya johnstoniana** Britten in Trans. Linn. Soc., Ser. 2, **4**: 22 (1894). —Brenan in Mem. N.Y. Bot. Gard. **8**: 487 (1954). —Roessler in Mitt. Bot. Staatss. Münch. **3**: 187 (1959); in Kirkia **10**: 78 (1975). Type: Malawi, Mt. Mulanje (Milanje), 1891, *Whyte* 7 (BM, holotype; G; K).

A perennial herb up to c. 80 cm. tall, from a stout creeping woody roostock. Stems annual single erect simple, thinly pilose-araneose, leafy with the leaves decurrent in discontinuous wings on the stem; wings usually not extending to the node below, absent from uppermost leaves, up to c. 5 mm. wide, coarsely dentate with indumentum as in leaves. Leaves crowded at the stem base, decreasing in size and density towards the stem apex, sessile; basal leaves ascending-subrosulate, 10–27 × 2–6 cm., narrowly oblong-oblanceolate in outline, apex acute, base ± long-tapering becoming petiole-like, somewhat stem clasping, margins pinnatilobed or pinnatifid with the lobes irregularly and often inconspicuously dentate, upper surface and margins beset with patent stramineous spine-like bristles 2–4 mm. long, the longest at tooth apices, lower surface white felted-tomentose; upper cauline leaves fewer, decreasing in size becoming linear and bract-like. Capitula discoid, 3–7, laxly to densely clustered at the stem apex, 2–2.5 cm. in diam. Phyllaries mostly 8–15 × 1–2 mm., the outer and innermost somewhat shorter, linear-tapering or subulate, spine-tipped; the outer somewhat leaf-like with 1–4 mm. long spine-like bristles on the margins as in leaves, thinly pilose-araneose on the backs or glabrescent; the inner spinescent-ciliate to -barbellate or unarmed. Receptacular alveolae margins with paleaceous setae to c. 6 mm. long. Florets numerous; corollas yellow, 10–16 mm. long; achenes 2–3 mm. long, turbinate, obscurely 8–10-ribbed, hispid-strigose; pappus scales 20–40, 2–3-seriate, of subulate paleaceous barbellate setae 8–10 mm. long.

Malawi. S: Mt. Mulanje, NW. slopes of Ruo Basin, c. 2130 m., 7.iv.1970, *Brummitt* 9705 (K).
Endemic on Mt. Mulanje. Submontane grassland.
Related to *Berkheya umbellata* DC. from E. Cape Province and Natal, but that species lacks the white felted tomentum on the leaf under surface.

9. **Berkheya speciosa** (DC.) O. Hoffm. in Ann. Nat. Hoffmus. Wien **24**: 314 (1910). —Roessler in Mitt. Bot. Staatss. Münch. **3**: 177 (1959). —Hilliard, Compos. Natal: 553 (1977). Type from South Africa.
 Stobaea speciosa DC., Prodr. **6**: 518 (1838) incl. var. *macroglossa* DC. —Harv. in Harv. & Sond., F.C. **3**: 492 (1865). Type as above.
 Crocodilodes speciosum (DC.) Kuntze, Rev. Gen. Pl. **1**: 333 (1891). Type as above.

A perennial herb up to c. 100 cm. tall, from a woody rhizome with fusiform root tubers. Stems annual single erect ± branched above, coarsely pilose. Leaves mostly basal with 1–several smaller cauline leaves; basal leaves ascending-subrosulate, petiolate. Petiole 2–30 cm. long, tapering-winged from the apex and spinescent sinuate-dentate, the wings reducing to setae 2–7 mm. long towards the petiole base. Leaf lamina mostly 6–30 × 3–18 cm., ovate to lanceolate, apex subobtuse to rounded, margins subentire to repand-sinuate rarely dentate, irregularly spinose with c. 2 mm. long spines interspersed with smaller marginal spines, base rounded to cordate, sometimes ± decurrent on the petiole; upper surface scattered pilose becoming rough with age; lower surface ± felted-tomentose. Capitula radiate, solitary and terminal on the branches or several ± clustered, up to c. 7 cm. in diam. across the expanded rays. Phyllaries up to c. 20 mm. long, linear-tapering from a c. 2 mm. wide base to an apical spine, spinescent-ciliate on the margins with spines up to c. 11 mm. long, pilose-araneose on the back. Receptacular alveolae margins fimbriate-setose. Rays yellow, mostly 15–25 × 2.5–4 mm. Disk-florets numerous; yellow, 9–12 mm. long, darkly glandular at the apex outside; achenes c. 2 mm. long, turbinate, 8–10-ribbed, glabrous or puberulous; pappus 2–3-seriate, of narrow ± spathulate overlapping scales to c. 3 mm. long.

Subsp. **speciosa**

Basal leaves broadly ovate or broadly elliptic, base rounded to cordate, petiole 5–30 cm. long.

Mozambique. M: Matutuíne, entre Zitundu e Manhoca, 29.xi.1979, *de Koning* 7714 (K; LMU).
Also in South Africa (Natal and E. Cape). Grassland often in moist localities.
Subsp. *lanceolata* Roessler with lanceolate basal leaves is recorded from the Transvaal Highveld.
Subsp. *ovata* Roessler with basal leaves ovate or elliptic and sessile or with petioles 2–3 cm. long, and solitary capitula is recorded from Natal and the E. Cape above c. 1525 m.

10. **Berkheya rehmannii** Thell. in Viert. Nat. Ges. Zürich **74**: 128 (1929). Type from South Africa (Transvaal).
 Berkheya zeyheri subsp. *rehmannii* (Thell.) Roessler in Mitt. Bot. Staatss. Münch. **3**: 221 (1959); in Kirkia **10**: 82 (1975). —Hilliard, Compos. Natal: 556 (1977). Type as above.

A perennial herb up to c. 100 cm. tall, from a woody rootstock with fusiform root tubers. Stems annual 1–several, strictly erect or sometimes ± decumbent, mostly simple, striately-ribbed, ± densely leafy with leaves smaller fewer and ± bract-like towards the apex, thinly araneose becoming scabridulous. Leaves sessile, 4–16 × 0.3–2.0 cm., elliptic-linear, tapering to an aristate-bristle 1–5 mm. long, margins spinose-denticulate with 2–10 submarginal bristles 5–15 mm. long on each side mainly in the basal half of the blade and there most closely spaced at the base or on the narrow auricle; leaf upper surface thinly araneose to glabrescent, ± scabridulous when dry; lower surface whitish felted-tomentose except on the pilose mid-nerve. Capitula radiate, solitary and terminal on stem or branches, 4–7 cm. in diam. across the expanded rays. Phyllaries spreading, becoming ± reflexed with age, 12–20 × 1 mm., the outer and innermost shorter, linear, tapering into a slender apical spine and with 3–8 bristles 3–6 mm. long spaced along each side, thinly araneose and glandular-hairy outside. Receptacular alveolae margins with short paleaceous setae c. 1 mm. long. Rays cream-coloured, white or pale-yellow, black-tipped, mostly 12–20 × 1.5–5 mm., lobes 4 dark-purple or black, glandular outside. Disk-floret corollas yellow, 7–11 mm. long, lobes shiny dark-purple or blackish in the upper part, smoothly calloso-glandular and sharply curved inwards at the apex; achenes remaining firmly in place for some time, 2–3 mm. long, obpyramidal-turbinate, 4-angular or -ribbed, sericeous, glandular-viscid at the apex, pappus 2-seriate of c. 20 overlapping oblanceolate shiny stramineous scales 2.5–3 mm. long.

Mozambique. M: Maputo Distr., Goba, prox. do rio Maiuana, 2.xi.1960, *Balsinhas* 161 (K; LISC); Goba, prox. da Fonte dos Libombos, 13.xii.1961, *Lemos & Balsinhas* 300 (K; LISC).
Also in South Africa (Transvaal and Natal) and Swaziland. Wooded and submontane grassland, often on hillsides.

11. **Berkheya zeyheri** Oliv. & Hiern in F.T.A. **3**: 429 (1877). —Eyles in Trans. Roy. Soc. S. Afr. **5**: 521 (1916). —Hopkins et al., Common Veld Fl.: 116 (1940). —Suesseng. & Merxm., Contrib. Fl. Marandellas Distr.: 63 (1951). —Roessler in Mitt. Bot. Staatss. Münch. **3**: 218 (1959); in Kirkia **10**: 81, tab. 1 (1975) pro parte excl. subsp. *rehmannii*. —Plowes & Drumm., Wild Fl. Rhod.: 140, pl. 182, 183 (1977). —Hilliard, Compos. Natal: 556 (1977) pro parte. TAB. **45**. Type from South Africa.
    Stobaea zeyheri Sond. & Harv. in Harv. & Sond., F.C. **3**: 496 (1865) nom. illegit., non (Less.) DC. (1838). Type from South Africa.
    Crocodilodes zeyheri (Oliv. & Hiern) Kuntze, Rev. Gen. Pl. **1**: 333 (1891). Type as above.
    Berkheya subulata sensu Eyles in Trans. Roy. Soc. S. Afr. **5**: 521 (1916).
    Berkheya insignis sensu Brenan in Mem. N.Y. Bot. Gard. **8**: 487 (1954).

A tufted perennial herb 10–90 cm. tall, from a large woody rootstock; roots numerous, each with a fusiform root tuber. Stems annual 1–several, strictly erect or sometimes ± decumbent, often reddish, mostly simple, striately-ribbed, ± densely leafy with leaves becoming smaller fewer and ± bract-like towards the apex, thinly araneose or glabrescent, sometimes scabridulous. Leaves sessile, 3–15 × 0.1–1.3 cm., linear or lanceolate, tapering to a bristle-like apical spine 1–5 mm. long, margins spinescent-denticulate with 2–15 submarginal bristles on each side, these bristles 5–20 mm. long and mainly in the basal half of the blade and there most closely spaced at the base or on the narrow auricle; leaf upper surface thinly araneose to glabrescent, ± scabridulous when dry; lower surface whitish-tomentose except on the mid-nerve. Capitula radiate, solitary and terminal on stem or branches, 4–6 cm. in diam. including the rays. Phyllaries spreading, ± reflexed with age, 12–20 × 1 mm., the outer and innermost shorter, linear, tapering into a slender apical spine and with 3–10 bristles 3–8 mm. long spaced along each side, thinly araneose and glandular-hairy outside. Receptacular alveolae deep with short paleaceous setae c. 1 mm. long on the margins. Rays yellow, mostly 15–20 × 1.5–4 mm., lobes glandular outside but not dark-coloured. Disk-floret corolla limbs yellow, 9–11 mm. long, lobe apices glandular-tuberculate and somewhat thickened calloso-glandular, not dark-purple, limb apex spherical in bud; achenes remaining firmly in place for some time, 1–2 mm. long, turbinate, densely sericeous, pappus 2-seriate of c. 20 overlapping oblanceolate shiny stramineous scales 2.5–3 mm. long.

Zambia. E: Nyika Plateau, within 90 m. of (Govt.) Rest-House, 23.xi.1955, *Lees* 49 (K; NDO). **Zimbabwe.** N: Mazowe, 1360 m., viii.1905, *Eyles* 178 (BM; SRGH). W: Bulawayo, i.1898, *Rand* 113 (BM). C: Gweru Distr., Old Dog Ranch, 12.i.1963, *Loveridge* 551 (K; M; MO; SRGH). E: c. 27 km. S. of Nyanga (Inyanga), 19.xi.1930, *Fries, Norlindh & Weimarck* 3072 (K; LD; LISU; M; SRGH). S: Mberengwa Distr., Buchwa Mt. 29.iv.1973, *Pope* 1016 (K; SRGH). **Malawi.** N: Nyika Plateau, by main road, 4 km. SW. from Rest-House, 2150 m., 22.x.1958, *Robson & Angus* 234 (BM; K; LISC; PRE;

Tab. 45. BERKHEYA ZEYHERI. 1, habit (×⅔), from *Stolz* 116 & from *Gilliland* 107; 2, ray-floret corolla (× 3); 3, disk-floret corolla (× 4); 4, style apex, branches connate from the style shaft apex, free above (× 8); 5, stamens (× 8); 6, achene with pappus (× 8); 7, pappus (× 4), 2–7 from *Zimmer* 245. Drawn by V. Goaman.

SRGH). C: Ntchisi, 1350 m., 9.ix.1946, *Brass* 17574 (K; MO; SRGH). S: Zomba, slopes of Zomba Mountain, 28.x.1963, *Salubeni* 121 (SRGH). **Mozambique**. N: serra de Ribáuè, Mepáluè Mt., c. 640 m., 23.i.1964, *Torre & Paiva* 10147 (LISC). T: Macanga, between Furancungo and Angónia, 29.ix.1942, *Mendonça* 536 (LISC). MS: Manica, Rotanda, prox. de Messambuzi, c. 1000 m., 17.xi.1965, *Torre & Correia* 12987 (LISC).

Also in Tanzania, South Africa (Transvaal and Natal) and Swaziland. A pyrophyte of submontane and plateau grassland, miombo and *Uapaca* woodlands, in dambos, on rocky outcrops or in shallow stony soils.

Leaf shape is variable in this species. Leaves are predominantly linear with revolute margins in plants from the plateau and submontane grasslands of Zimbabwe and the Transvaal. However, in Malawi and southern Tanzania the leaves can also be narrowly lanceolate and flat, often in plants from the same locality (eg. *Hilliard & Burtt* 4616 (K), with linear revolute leaves and *Pope et al.* 2268 (K), with wide flat leaves and spinescent margins, both collected on the Chambe cableway path on Mt. Mulanje, Malawi).

33. HIRPICIUM Cass.

Hirpicium Cass. in Bull. Soc. Philom.: 27 (1820); in Dict. Sci. Nat. **21**: 238 (1821); tom. cit. **29**: 448 (1823). —Roessler in Mitt. Bot. Staatss. Münch. **3**: 333 (1959); in Kirkia **10**: 86–95 (1975).
Berkheyopsis O. Hoffm. in Engl. & Prantl, Naturl. Pflanzenfam. IV, **5**: 311 (1892); in Bol. Soc. Brot. **10**: 179 (1892); in Bull. Herb. Boiss. **1**: 89 (1893).

Annual or perennial herbs, often suffrutescent with annual stems from a woody rootstock, or subshrubs. Stems and branches ± bristly-setose, sometimes glandular–hairy or glabrous. Leaves alternate, sometimes radical, densely bristly-setose, hispid or scabrous, sometimes glabrous with bristles on the margins only, the lower surfaces white-felted, rarely only glandular-hairy, linear ± oblanceolate oblong or obovate, entire or ± dentate sometimes pinnately lobed. Capitula heterogamous and radiate, stalked, or ± sessile and subtended by leaves, solitary and terminal on the stem and branches, sometimes ± racemose or corymbosely arranged. Phyllaries several-seriate, connate forming a ± obconic or cup-shaped involucre; free parts linear, strap-shaped or triangular-lanceolate, bristly or setose- to spinose-ciliate. Receptacle shallowly or deeply alveolate. Ray-florets uniseriate, neuter; corollas yellow with rays sometimes purplish outside, mostly without staminodes; rays erect strap-shaped, margins becoming inrolled; achenes wanting. Disk-florets hermaphrodite, numerous; corollas yellow infundibuliform; anther bases minutely sagittate; style branches linear; achenes ± ribbed, villous-sericeous; pappus scales 1- or 2-seriate, the outer large and overlapping, the inner when present smaller and stipitate.

A southern and tropical African genus of 12 species, 7 species occurring in the Flora Zambesiaca area.

1. Plants tufted; stems scapiform, simple; leaves mostly basal, filiform-linear; rootcrowns lanate
                                                    7. *angustifolium*
– Plants not tufted; stems leafy, simple or bushy and branching from near the base; leaves mostly cauline, not filiform; rootcrowns not lanate   -   -   -   -   -   -   -   -   2
2. Leaf lower surface glandular-hairy   -   -   -   -   -   -   -   -   - 1. *echinus*
– Leaf lower surface araneose-tomentose   -   -   -   -   -   -   -   -   -   3
3. Rays of the outer florets exceeding the involucre; involucre rounded at the base; pappus 2-seriate, the inner whorl of shorter stipitate scales; phyllaries spinose-ciliate; receptacle shallowly alveolate   -   -   -   -   -   -   -   -   -   -   -   -   -   4
– Rays mostly shorter than or ± equalling the involucre; involucre obconic, tapering to the base; pappus 1-seriate; phyllaries denticulate or hispid-ciliate on margins not spinose-ciliate; receptacle deeply alveolate   -   -   -   -   -   -   -   -   -   -   -   6
4. Outer phyllaries longer than the inner   -   -   -   -   -   - 3. *gazanioides*
– Outer phyllaries shorter than the inner   -   -   -   -   -   -   -   -   5
5. Pappus scales of the disk-florets 5–6 mm. long, the corolla usually ± equalling the pappus, or only the corolla lobes exceeding the pappus; annual herbs   4. *gorterioides* subsp. *gorterioides*
– Pappus scales of the disk-florets less than 4 mm. long, the corolla limb overtopping the pappus; perennial herbs   -   -   -   -   -   -   -   -   -   -   - 2. *bechuanense*
6. Outer phyllaries leaf-like, intergrading with the leaves; involucres more than 20 mm. long; perennial herbs   -   -   -   -   -   -   -   -   -   -   - 5. *antunesii*
– Outer phyllaries not leaf-like, capitula distinctly stalked; involucres less than 10 mm. long; annual herbs   -   -   -   -   -   -   -   -   -   -   - 6. *gracile*

1. **Hirpicium echinus** Less., Syn. Comp.: 55 (1832). —DC., Prodr. **6**: 502 (1838). —Roessler in Mitt. Bot. Staatss. Münch. **3**: 340 (1959). —Merxm., Prodr. Fl. SW. Afr. 139: 102 (1967). —Roessler in Kirkia **10**: 88 (1975). —Barnes & Turton, List Fl. Pl. Botswana at Nat. Mus., Sebele & Univ. Botswana: 34 (1986). Type from South Africa.

Gazania burchellii DC., Prodr. **6**: 514 (1838). Type from South Africa.

Berkheyopsis echinus (Less.) O. Hoffm. in Engl. & Prantl, Naturl. Pflanzenfam. IV, **5**: 311 (1892). —Leistner in Koedoe **2**: 172 (1959). Type as for *Hirpicium echinus*.

Berkheyopsis kuntzei O. Hoffm. in Kuntze, Rev. Gen. Pl. **3**, 2: 136 (1898). Type from South Africa.

A prickly caespitose or bushy, somewhat aromatic perennial herb up to c. 40 cm. tall, from a long slender woody taproot. Stems tufted and densely leafy from the rootcrown, or bushy and strongly branched from the base; sometimes in older plants stems procumbent woody and tufted-leafy at the nodes; branches decumbent to ascending, leafy, ± densely glandular-hairy, becoming woody at the base. Leaves sessile, 2–8 cm. long and 1–2 mm. broad, linear, usually with 1–3 linear lobes 1–10 mm. long on each side; leaf and lobe apices tipped with broad-based rigid white spinules 0.5–2 mm. long; margins ± revolute with up to c. 20 spaced white spine-like bristles 1–3 mm. long on each side; lamina hispid with scattered spinules up to 0.5 mm. long, particularly on the upper surface, glandular-hairy, not tomentose below. Capitula many, terminal on leafless glandular-hairy stalks 1.5–10 cm. long. Involucres 8–15 mm. in diam., the connate part obconic to hemispheric, glandular pubescent-pilose; outer phyllaries 2–4 × 0.5–1 mm., subulate and ± reflexed, spinescent-acuminate, spinose-ciliate; the inner c. 2 series 7–12 × 2–4 mm., ovate-lanceolate, long acuminate-spinose, with margins broadly hyaline, entire or ± spinosc-ciliate. Receptacle deeply alveolate. Ray-florets uniseriate, rays 10–15 mm. long, linear, yellow usually with a purplish transverse band. Disk-floret corollas yellow, c. 7 mm. long, deeply lobed; achenes 3–5 mm. long, ± stipitate-turbinate, densely long-sericeous with hairs equalling the pappus; pappus scales 2-seriate, the outer 2.5–6 mm. long, lanceolate-attenuate, hyaline overlapping; the inner c. 1 mm. long, oblong-ovate, shortly stipitate.

Botswana. SW: c. 11 km. N. of Union End, 15.iii.1969, *Rains & Yalala* 34 (K; SRGH).

Also in Namibia and South Africa (western Cape Province). Short grassland on sand, often on dunes.

2. **Hirpicium bechuanense** (S. Moore) Roessler in Mitt. Bot. Staatss. Münch. **3**: 343 (1959); in Kirkia **10**: 88, tab. 2 (1975). —Barnes & Turton, List Fl. Pl. Botswana at Nat. Mus., Sebele & Univ. Botswana: 34 (1986). TAB. **46**. Type: Botswana, Mahalapye, 31.i.1912, *Rogers* 6106 (BM, holotype; K; SRGH; Z).

Berkheyopsis bechuanensis S. Moore in Journ. Bot. **51**: 185 (1913). —Eyles in Trans. Roy. Soc. S. Afr. **5**: 520 (1916). —Martineau, Rhod. Wild Fl.: 91, pl. 33 (1954). —Burtt Davy in Bull. Misc. Inf., Kew **1935**: 571 (1935). Type as above.

Berkheyopsis rehmannii Thell. in Viert. Nat. Ges. Zürich **61**: 458 (1916). Syntypes: Botswana, Serowe, v.1904, *Blackbeard* 17 (GRA), and from the Transvaal.

Berkheyopsis brevisquama Mattf. in Notizbl. Bot. Gart. Berlin **8**: 180, 284 (1922). Syntypes: Botswana, Shashi River, ix.1896, *Klingberg* (S), and from the Transvaal.

A bushy ± scabrous perennial herb, branching from the base, 10–40 cm. tall from a woody taproot. Branches ascending, leafy, beset with scattered 0.5–2 mm. long patent white bristles, sometimes sparsely lanate-araneose, not glandular-hairy, becoming woody below. Leaves sessile, 1.5–6 × 0.2–1.0 cm., ± oblanceolate to linear, entire or irregularly and remotely toothed to lobed, leaf and lobe apices subacute white spinescent-mucronate, margins ± revolute with up to c. 8 patent white 2–4 mm. long bristles towards the base on each side; upper surface scabrous, very densely beset with conical white spinules; lower surface whitish lanate-tomentose except for the mid-nerve. Capitula terminal on branches, subsessile and usually subtended by 1–3 small leaves. Involucres 10–15 mm. in diam., the connate part broadly obconic to hemispheric, lanate-tomentellous to scabridulous or glabrescent; outer phyllaries 4–7 mm. long, strap-shaped, spine-tipped and with 2–5 patent white spine-like bristles 1–2 mm. long on each side, the innermost series 4–8 × 2–3 mm., lanceolate, barbellate and often spinose-ciliate on the margins, scabridulous outside. Receptacle shallowly alveolate. Rays yellow 8–20 × 1–3 mm., linear-oblanceolate. Disk-floret corollas yellow, 5–7 mm. long, deeply lobed; achenes 3–5 mm. long, tapering stipitate-turbinate, 10–12-ribbed, glandular, densely long-sericeous with hairs ± equalling the pappus; pappus scales 2-seriate, brownish hyaline, the outer ones 2.5–3.5 mm. long, oblong-elliptic, overlapping, the inner ones 2–2.5 mm. long, elliptic, stipitate.

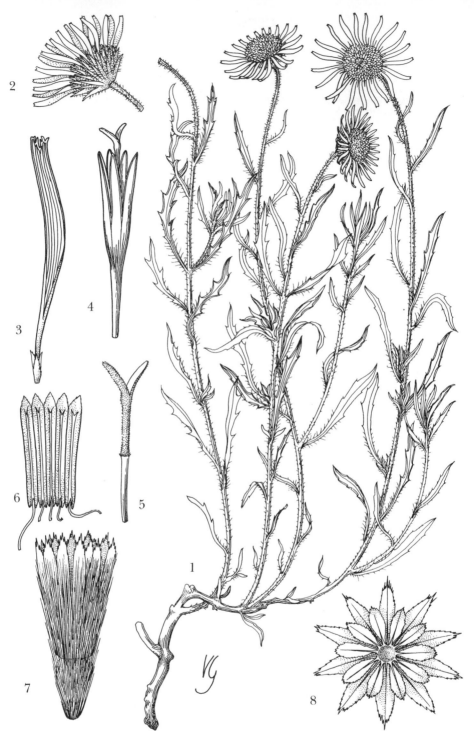

Tab. 46. **HIRPICIUM BECHUANENSE**. 1, habit (× $\frac{2}{3}$); 2, capitulum (× 1), 1 & 2 from *Van Son* 20728a & from *Cheeseman* 82; 3, ray-floret corolla and pappus (× 4); 4, disk-floret corolla (× 6); 5, style apex, branches connate from the style shaft apex, free above (× 8); 6, stamens (× 8); 7, achene with pappus (× 6); 8, pappus (× 4), 3–8 from *Cheeseman* 82. Drawn by V. Goaman.

Botswana. SW: Ghanzi, Eaton's Farm, TR 65, 30.iv.1969, *R.C. Brown* 6006 (K). SE: c. 20 km. beyond Molepolole on road to Letlhakeng (Letlaking), 15.ii.1960, *Wild* 4951 (K; LD; M; MO; PRE; SRGH). **Zimbabwe**. W: SW. Matopos, Maleme Valley, 10.i.1963, *Wild* 5978 (K; LISC; MO; SRGH). S: Beitbridge, 10.i.1961, *Leach* 10683 (K; M; MO; PRE; S; SRGH).

Also in South Africa (Transvaal and adjacent Cape Province). Low altitude mixed dry deciduous woodland and short grassland with scattered bushes, in sandy soils. Sometimes a weed of cultivation.

3. **Hirpicium gazanioides** (Harv.) Roessler in Mitt. Bot. Staatss. Münch. **3**: 345 (1959). —Merxm., Prodr. Fl. SW. Afr. 139: 102 (1967). —Roessler in Kirkia **10**: 90 (1975). —Barnes & Turton, List Fl. Pl. Botswana at Nat. Mus., Sebele & Univ. Botswana: 34 (1986). Type from Namibia.

Berkheya gazanioides Harv. in Harv. & Sond., F.C. **3**: 508 (1865). Type as above.

Gazania pechuelii Kuntze in Jahrb. Bot. Gart. Berlin **4**: 267 (1886). Type from Namibia.

Berkheya pechuelii (Kuntze) O. Hoffm. in Engl., Bot. Jahrb. **10**: 281 (1888). Type as for *Gazania pechuelii*.

Crocodilodes gazaniodes (Harv.) Kuntze, Rev. Gen. Pl. **1**: 332 (1891). Type as for *Hirpicium gazanioides*.

Berkheyopsis pechuelii (Kuntze) O. Hoffm. in Bol. Soc. Brot. **10**: 181 (1892); in Engl. & Prantl, Nat. Pflanzenfam. IV, **5**: 311 (1892). Type as for *Gazania pechuelii*.

Berkheyopsis pechuelii var. *glabrescens* Thell. in Viert. Nat. Ges. Zürich **61**: 460 (1916). Type from Namibia.

An erect annual herb, 5–40 cm. tall, branched from near the base. Branches leafy, hispid, beset with scattered 1–3 mm. long patent whitish bristles. Leaves 2–7.5 × 0.4–1.5 cm., elliptic to narrowly oblanceolate, tapering into a petiole-like base to c. 2 cm. long, entire or sometimes lobed, apex subobtuse mucronulate, margins ± revolute; upper surface scabrid-hispid, densely beset with conical whitish spinules which grade into scattered conical-based spine-like bristles up to 4 mm. long, the latter usually most numerous towards the leaf base; lower surface whitish or greyish lanate-tomentose except for the mid-nerve. Capitula terminal on branches, sessile and subtended by 3–many leaves which usually exceed the involucre in length. Involucres 15–30 mm. in diam., the connate part broadly obconic to hemispheric, mostly glabrous; outer phyllaries with free portions 10–20 × 0.5 mm., linear, bristle-tipped, pectinately ciliate on the margins with 10–20 or more bristles 2–3 mm. long on each side; innermost phyllaries with free portions 6–10 × 1–3 mm., narrowly triangular, acute to tapering acuminate, barbellate on the margins, membranous with a narrow dark-purple central part and broad margins. Receptacle shallowly alveolate. Rays yellow, 20–25 mm. long, linear-elliptic. Disk-floret corollas yellow, c. 8 mm. long, deeply lobed, thinly araneose outside especially on lobes; achenes 4–6 mm. long, narrowly obconic, 10–12-ribbed, glandular, densely long-sericeous; pappus scales 2-seriate, brownish hyaline, the outer ones 5–7 mm. long, oblong-lanceolate, overlapping, the inner ones 2.5–3 mm. long, oblong-ovate, stipitate.

Botswana. N: perhaps collected at Tioge River N. of Lake Ngami, 1855, *Wahlberg* (S). SW: c. 35 km. S. of Takatshwane Pan, 21.ii.1960, *Wild* 5106 (K; LD; M; MO; PRE; SRGH).

Also in Namibia and South Africa (Cape Province). Semi desert, usually with short grasses in dried-up pans, on sandy or rocky soil.

4. **Hirpicium gorterioides** (Oliv. & Hiern) Roessler in Mitt. Bot. Staatss. Münch. **3**: 348 (1959). —Merxm., Prodr. Fl. SW. Afr. 139: 102 (1967). —Roessler in Kirkia **10**: 92 (1975). —Barnes & Turton, List Fl. Pl. Botswana at Nat. Mus., Sebele & Univ. Botswana: 34 (1986). Type: "South Tropical Africa" (probably Botswana), *Baines* (K, holotype).

Berkheya gorterioides Oliv. & Hiern in F.T.A. **3**: 429 (1877). —Eyles in Trans. Roy. Soc. S. Afr. **5**: 521 (1916). Type as above.

Crocodilodes gorteriodes (Oliv. & Hiern) Kuntze, Rev. Gen. Pl. **1**: 332 (1891). Type as above.

Berkheyopsis angolensis O. Hoffm. in Bol. Soc. Brot. **10**: 180 (1892). Syntypes from Angola.

Berkheyopsis gorterioides (Oliv. & Hiern) Thell. in Viert. Nat. Ges. Zürich **61**: 461 (1916). Type as for *Hirpicium gorterioides*.

Berkheyopsis gorterioides var. *lobulata* Thell. in Viert. Nat. Ges. Zürich **61**: 461 (1916). Type from Namibia.

Berkheyopsis gossweileri S. Moore in Journ. Bot. **65**, Suppl. 2: 63 (1927). Type from Angola.

Berkheyopsis langii Bremek. & Oberm. in Ann. Transv. Mus. **16**: 442 (1935). Type: Botswana, Gemsbok, 1.v.1930, *van Son* in Tvl. Mus. 28723 (BM, holotype; PRE).

An annual scabridulous herb, 10–50 cm. tall. Branches erect or decumbent, leafy, coarsely hispid, beset with scattered 1–2 mm. long patent conical-based bristles, also thinly araneose towards the apex. Leaves mostly 2–8 × 0.5–1.5 cm., narrowly elliptic to oblanceolate, sometimes ± linear, tapering into a cuneate or petiole-like base to c. 1.5 cm. long, entire to irregularly serrate-dentate or lobed, apex subobtuse mucronulate, margins

± revolute; upper surface scabrid, densely beset with conical spinules which grade into scattered conical-based spinose bristles up to 3 mm. long, the latter usually most numerous towards the leaf base and becoming somewhat pectinate-ciliate to c. 4 mm. long on the margins near the base; lower surface whitish or greyish lanate-tomentose except for the midrib. Capitula terminal on branches, sessile and subtended by 3–several leaves which exceed the capitulum in length. Involucres 10–15 mm. in diam. increasing to c. 30 mm. when achenes are mature, the connate part broadly obconic to hemispheric, ± araneose-lanate; outer phyllaries with free portions 8–12 × c. 0.5 mm., linear, bristle-tipped, margins pectinately ciliate with 10–20 or more bristles 2–3 mm. long on each side; innermost phyllaries with free portions 6–13 mm. long, 2–3 mm. wide at the base, shortly triangular in the lower part and tapering-acuminate above, barbellate on the margins and sometimes pectinate-ciliate near the apex. Receptacle shallowly alveolate. Rays yellow, 10–25 × 2–5 mm., linear-elliptic. Disk-floret corollas yellow, to c. 6 mm. long, usually not or hardly overtopping the pappus scales of mature achenes, glabrous or somewhat puberulous not araneose; achenes 4–5 mm. long, narrowly obconic, 10–12-ribbed, glandular, densely long-sericeous, hairs shorter than pappus; pappus scales 2-seriate, brownish hyaline, the outer ones increasing to 5–6 mm. long (2–3 mm. long in subsp. *aizoides*), oblong-elliptic, overlapping, the inner ones c. 2.5 mm. long, oblong-ovate, stipitate.

Subsp. **gorterioides**

Botswana. N: Maun, near Post Office, ii.1967, *Lambrecht* 22 (K; MO; PRE; S; SRGH). SW: c. 14 km. S. of Ghanzi on Lobatse Road, 1.ii.1970, *Brown* 8270 (K; SRGH). SE: Boteti delta area, NE. of Mopipi, 900 m. 16.iv.1973, *Tyers* 4 (K; SRGH). **Zimbabwe**. W: Bulilima Mangwe Distr., Dombodema Mission Station, 1300 m., 18.ii.1972, *Norrgrann* 98 (K; SRGH).
    Also in Angola and Namibia. Low rainfall areas in short open grassland or wooded grassland with *Acacia* spp. or *Colophospermum mopane*, on margins of seasonally flooded pans or on floodplains.
    Subsp. *aizoides* (O. Hoffm.) Roessler occurs in southern Angola and is distinguished by its outer pappus scales being 2–3 mm. long and obovate-oblanceolate and by its leaves mostly less than 3 cm. long.
    Subsp. *schinzii* (O. Hoffm.) Roessler, from northern Namibia (Ovamboland), with outer pappus scales 3–6 mm. long is distinguished from the typical subspecies by its basally branching habit and the free part of the inner phyllaries less than 6 mm. long.

5. **Hirpicium antunesii** (O. Hoffm.) Roessler in Mitt. Bot. Staatss. Münch. **3**: 356 (1959); in Kirkia **10**: 93 (1975). —Lisowski, Aster. Fl. Afr. Centr. [in Fragm. Flor. Geobot. **36**, Suppl. 1]: 569, fig. 116 (1991). Type from Angola.
    Berkheya antunesii O. Hoffm. in Bol. Soc. Brot. **10**: 181 (1892). —Mendonça, Contrib. Conhec. Fl. Angol., **1** Compositae: 132 (1943). Type as above.
    Crocodilodes antunesii (O. Hoffm.) Hiern, Cat. Afr. Pl. Welw. **1**, 3: 608 (1898). Type as above.

A hispid-scabridulous perennial herb, 30–70 cm. tall from a swollen woody rootstock. Stems annual, erect sometimes decumbent, 1–several, simple or with a few stiff erect branches, densely coarsely hispid with bristles 1–3 mm. long, leafy. Leaves sessile, 3–14 × 0.3–0.8 cm., linear to linear-oblanceolate, acute to subacute, entire, margins revolute, bases tapering and ± petiole-like; upper surface densely beset with conical-based bristles up to 2 mm. long, these ± ciliate and up to c. 3.5 mm. long on the margins towards the base; lower surface whitish lanate-tomentose except for the midrib. Capitula terminal on stem and branches, subsessile and subtended by 3–many leaves which grade into the phyllaries; involucres 15–30 mm. in diam. and up to c. 30 mm. long, obconic, araneose-lanate. Phyllaries numerous, the outer series with free portions leaf-like, to c. 18 mm. long, linear to subulate, spine-tipped, margins hispid-ciliate, hispid on the midrib outside and on both surfaces towards the apex, araneose-lanate outside as in leaf undersurface; innermost series with the free part up to 12–17 mm. long, 3 mm. wide at the triangular base, subulate-acuminate in the apical half, membranous, barbellate-denticulate on the margins. Receptacle deeply alveolate. Rays yellow, often brownish-purple outside, 16–25 mm. long. Disk-floret corollas yellow, 6–8 mm. long, shallowly lobed; achenes 4–6 mm. long, narrowly turbinate, c. 10-ribbed, densely long-sericeous with hairs ± equalling the pappus in length; pappus scales c. 10 uniseriate, brownish-hyaline, overlapping, 5–6 mm. long, oblong-lanceolate, fimbriate at the apex.

Zambia. N: Mbala Distr., Chilongowelo Escarpment, 1500 m., 6.iv.1962, *Richards* 16273 (K; SRGH); Mporokoso Distr., c. 10 km. E. of Lumangwe Falls, Kalungwishi R., 14.iv.1989, *Pope, Radcl.-Sm. & Goyder* 2117 (BR; K; NDO). **Malawi**. C: Dedza, Chongoni Forest Boundary, 13.ii.1969, *Salubeni*

1262 (SRGH). **Mozambique**. N: Massangula, 15.v.1948, *Pedro & Pedrógão* 3512 (EA).

Also in Angola, Zaire and SW. Tanzania. High rainfall grassland and open *Brachystegia* woodland, in sandy or rocky soil.

6. **Hirpicium gracile** (O. Hoffm.) Roessler in Mitt. Bot. Staatss. Münch. **3**: 357 (1959); in Kirkia **10**: 93 (1975). —Lisowski, Aster. Fl. Afr. Centr. [in Fragm. Flor. Geobot. **36**, Suppl. 1]: 571, fig. 117 (1991). Type from Angola.

   Berkheya gracilis O. Hoffm. in Bol. Soc. Brot. **13**: 35 (1896). —Mendonça, Contrib. Conhec. Fl. Angol., 1 Compositae: 133 (1943). Type as above.

   Crocodilodes gracilis (O. Hoffm.) Hiern in Cat. Afr. Pl. Welw. **1**, 3: 609 (1898). Type as above.

   Athrixia diffusa Bak. in Bull. Misc. Inf., Kew **1898**: 152 (1898). Type: Malawi, Chitipa (Fort Hill), Nyasa-Tanganyika plateau, vii.1896, *Whyte* (K, holotype).

A slender, or somewhat bushy, wiry annual herb 10–90 cm. tall. Stems erect branched leafy, whitish thinly-araneose and hispid with ascending brownish or purple ± flattened scale-like bristles up to c. 1 mm. long; branches ascending 1–40 cm. long, leafy. Leaves sessile, mostly 1.5–7 × 0.1–0.5 cm., linear or linear-lanceolate, becoming bract-like on the upper stem and branches, occasionally the lower leaves to c. 8 mm. wide, entire, acute to subacute mucronulate, margins revolute ciliate towards the base usually with 4–10 bristles c. 1.5–3 mm. long on each side; upper surface scabrous with very short conical whitish spinules; lower surface whitish lanate-araneose except for the midrib. Capitula terminal, or ± racemose on simple stems and the longer branches, or ± laxly clustered. Involucres mostly 5–8 mm. in diam., the connate part 3–5 × 3–5 mm., obconic, araneose-lanate and with a few small scale-like bristles mainly on the midrib; outer phyllaries with their free portions 2–4 mm. long, subulate, minutely spinose-denticulate on the margins; inner phyllaries with their free portions 4–6 mm. long and c. 1.5 mm. wide at the base, triangular with a thick midrib continued into a pungent apex, margins barbellate-denticulate. Receptacle alveolae deep, membranous, exceeding the achene in length. Ray-florets mostly 2–6, seldom overtopping the involucre, 5–6 mm. long, rays pale yellow often purplish outside. Disk-floret corollas yellow, 2.5–3.5 mm. long; achenes 2–3 mm. long, obconic, glandular-viscid with glands in 10–12 broad longitudinal bands, sparsely whitish sericeous with hairs just exceeding the achene; pappus scales uniseriate, brownish hyaline, overlapping, 2–3.5 mm. long, ovate-lanceolate in outline, fimbriate-setose about the apex.

Zambia. B: 85 km. from Zambezi (Balovale) on Kabompo road, 26.iii.1961, *Drummond & Rutherford-Smith* 7397 (K; PRE; SRGH). N: 20 km. SE. of Kasama, 23.iii.1961, *Robinson* 4552 (EA; K; M; MO; SRGH). W: 6 km. W. of Solwezi, 18.iii.1961, *Drummond & Rutherford-Smith* 6992 (K; LISC; PRE; SRGH). C: Great North Road, 2 km. NE. of Serenje turnoff, 29.iii.1984, *Brummitt, Chisumpa & Nshingo* 16953 (K). S: c. 8 km. E. of Choma, c. 1300 m., 26.iii.1955, *Robinson* 1159 (K; SRGH). **Zimbabwe**. N: Gokwe Distr., Sengwa Res. Station, 20.iv.1976, *P. Guy* 2428 (K; SRGH). W: Matobo Distr., Besna Kobila Farm, c. 1500 m., 31.iii.1963, *Miller* 8430 (K; SRGH). C: Goromonzi, entrance to Ruwa Kennels, 23 km. E. of Harare, 15.iii.1969, *Biegel* 2909 (K; LISC; SRGH). E: Mutare Distr., Odzani River Valley, 1915, *Teague* 472 (BOL; K). **Malawi**. N: Nthalire, 23.v.1989, *Pope, Radcl.-Sm. & Goyder* 2322 (BR; K). C: Dedza Distr., Chongoni Forest, 17.iii.1969, *Salubeni* 1271 (K; SRGH).

Also in Angola, Zaire and Tanzania. In miombo, mixed deciduous woodland, mopane and Kalahari Sand woodlands, and in short grassland, on sandy soils or rocky outcrops often in shallow soil over granite slabs, sometimes a roadside weed.

7. **Hirpicium angustifolium** (O. Hoffm.) Roessler in Mitt. Bot. Staatss. Münch. **3**: 359 (1959). —C. Jeffrey in Curtis's Bot. Mag. 180, 2: 73, tab. 676 (1974). —Roessler in Kirkia **10**: 95 (1975). —Lisowski, Aster. Fl. Afr. Centr. [in Fragm. Flor. Geobot. **36**, Suppl. 1]: 573 (1991). Type from Tanzania (lectotype chosen by C. Jeffrey).

   Gazania angustifolia O. Hoffm. in Engl., Bot. Jahrb. **30**: 439 (1901). Type as above.

An erect tufted perennial herb 3–30 cm. tall, from a small rootstock; roots numerous, thong-like and somewhat fleshy; rootcrowns lanate. Stems annual, 1–3, purplish, usually simple, scapiform with 1–4 reduced leaves, ± araneose and also sparsely hispid with scattered brown or purplish weak bristles, especially towards the apex. Main leaves basal erect numerous, crowded, 10–30 cm. long, filiform-linear, entire, revolute except below, ciliate with bristles to c. 2.5 mm. long and most numerous near the leaf base, fewer or absent from the apical portion; upper surface sparsely scabridulous to glabrous, lower surface araneose-lanate except for the midrib. Cauline leaves 1–3 cm. long, otherwise similar to the basal leaves. Capitula solitary and terminal. Involucres 5–10 mm. in diam., obconic-campanulate; the connate part 3–7 × 4–7 mm., araneose, sometimes with a few small bristles. Outermost phyllaries often free, extending briefly onto the capitulum stalk,

the free portion of the fused outer phyllaries to c. 7 mm. long, triangular-subulate and ±
spinose-ciliate on the margins or entire; free portion of inner phyllaries to c. 8 mm. long,
c. 1.5 mm. wide at the base, narrowly triangular with a thick purplish midrib continued
into a pungent apex, margins hyaline barbellate-denticulate. Receptacle shallowly
alveolate. Rays 15–20 mm. long, yellow, often brown or purplish outside, erect, flat and
oblanceolate or margins inrolled. Disk-floret corollas yellow, 4.5–5.5 mm. long; achenes
c. 2 mm. long, obconic, somewhat viscid-glandular, densely sericeous with hairs equalling
or exceeding the pappus; pappus scales uniseriate, brownish-hyaline, overlapping, 2.5–3
mm. long, oblong, fimbriate about the apex.

Zambia. N: Mbala Distr., Zombe Plain, Lombwe drainage, c. 457 m. 11.x.1966, *Richards* 21516
(K). W: Chingola Distr., Luano For., 21.viii.1969, *Mutimushi* 3535 (K; NDO). **Malawi**. N: Chisenga,
foot of Mafinga Mts., 1850 m., 13.xi.1958, *Robson & Fanshawe* 613 (K; LISC; SRGH). C: Chongoni
For. Res., Dedza, 20.vii.1967, *Salubeni* 781 (SRGH).
Also in Tanzania and Zaire. A pyrophyte, in burnt seasonally flooded grassland.

34. GAZANIA Gaertn.

Gazania Gaertn., Fruct. Sem. Pl. **2**: 451 (1791), nom. conserv. —Roessler in Mitt. Bot.
Staatss. Münch. **3**: 364 (1959); in Kirkia **10**: 95–99 (1975).
Meridiana Hill., Veg. Syst. **2**: 121 (1761) nom. rejic.

Perennial herbs with woody rootstocks, (rarely annual herbs or dwarf subshrubs with
woody branches). Plants scapose, or with suberect to decumbent leafy stems. Leaves basal
and crowded, or alternate, araneose or glabrescent on the upper surface, white felted-
araneose beneath, ± linear to oblanceolate or obovate, entire or pinnatilobed to
pinnatifid. Capitula heterogamous, radiate, solitary on scapes or terminal on the stem
and branches. Phyllaries connate below into a deeply cup-shaped ± campanulate
involucre, sometimes ± obconic; free parts in 2–3 rows inserted on the rim of the connate
part, with none or few scattered on the fused portion. Receptacle ± shallowly alveolate.
Ray-florets uniseriate, neuter, without staminodes; corollas mostly yellow with rays
sometimes whitish or orange inside and greenish or reddish-striped outside, sometimes
with a dark spot at the base, strap-shaped ascending-erect; achenes wanting. Disk-florets
hermaphrodite, numerous, corollas yellow, infundibuliform; anther bases sagittate,
shortly mucronate-acuminate; style branches linear; achenes narrowly turbinate, villous
with hairs exceeding or ± equalling the pappus; pappus of biseriate subequal narrow
delicate accrescent scales, at first difficult to distinguish from the ovary hairs, the outer
pappus scales sessile, the inner scales shortly stipitate.

A South African genus of 17 species, of which *G. krebsiana* subsp. *serrulata* is widespread from the
Cape Province to Angola and Tanzania, and *Gazania rigens* var. *uniflora* extends into southern
Mozambique.
Gazania hybrids are often cultivated as garden ornamentals.

Stems stout, decumbent to suberect; leaves cauline, leathery, mostly more than 10 mm. wide; plants
   forming mats colonising coastal sand dunes   -   -   -   -   -   -   - 1. *rigens*
Stems absent or much abbreviated, plants scapose; leaves crowded and basal, not leathery, mostly
   linear-oblanceolate and less than 10 mm. wide; plants tufted pyrophytes of grasslands
                                                                      2. *krebsiana*

1. **Gazania rigens** (L.) Gaertn., Fruct. Sem. Pl. **2**: 451, t. 173 (1791) quoad syn., non quoad icon. et
   descript. —Roessler in Mitt. Bot. Staatss. Münch. **3**: 370 (1959); in Kirkia **10**: 96 (1975). Type a
   cultivated plant.
   Gorteria rigens L., Amoen. Acad. **6**: 105 (1763); Sp. Pl. ed. 2: 1284 (1763). Type as above.

A shortly trailing perennial herb from a woody taproot. Stems 1–several, usually
decumbent, often ± stoloniferous up to c. 40 cm. long and 4–6 mm. in diam., or suberect to
c. 30 cm. tall, simple or branched, ± uniformly leafy, glabrous or ± araneose. Leaves
leathery, mostly 3–11(15) × 0.4–2.3 cm., lanceolate or obovate-lanceolate, tapering-
attenuate petiole-like below and stem-clasping at the base, mostly simple sometimes
pinnatisect with 1–2 lobes on each side, apices acute or subobtuse, margins somewhat
revolute, entire; upper surface glabrous or thinly araneose; lower surface whitish felted-
tomentose except for the midrib. Capitula usually solitary and mostly axillary on upper

stem and branches; capitula stalks 4–15 cm. long, naked or sometimes bracteate, glabrous or araneose. Involucres 10–15 × 8–15 mm., narrowly campanulate, connate in the lower two-thirds. Phyllaries several-seriate and longest towards the inside, free parts 5–10 mm. long, the outer usually strap-shaped and often felted-tomentose outside, the inner triangular and ± attenuate at the apex, glabrous. Receptacle alveolae walls c. 0.5 mm. high, membranous. Rays yellow, ascending-erect, 14–25 mm. long, linear-oblanceolate. Disk-florets numerous; corollas yellow, 5–6 mm. long; achenes 3–5 mm. long, narrowly subfusiform-turbinate, obscurely 10–12-ribbed, densely long-sericeous with the hairs hiding the pappus; pappus 2-seriate of subequal accrescent scale-like setae 2–5 mm. long, the setae developing a somewhat thickened stalk-like base and narrowly hyaline lateral wings which taper into a fine apex, the outer series briefly connate at the base, the inner free.

Var. **uniflora** (L.f.) Roessler in Mitt. Bot. Staatss. Münch. **3**: 371 (1959); in Kirkia **10**: 97 (1975). —Hilliard, Compos. Natal: 543 (1977). Type from South Africa.
  Gorteria uniflora L.f., Suppl.: 382 (1782). Type as above.
  Gazania uniflora (L.f.) Sims, Bot. Mag. **48**: t. 2270 (1821). Type as above.

Leaf upper surface, stem, capitulum stalk and involucre ± glabrous. Capitula 2.5–4 cm. diam. including the rays. Connate part of involucre campanulate, c. 7–8(10) mm. broad. Rays without an eye-spot near the base.

Mozambique. M: Matutuíne, Ponta do Ouro xii.1978, *de Koning* 7339 (K; LMU); Inhaca Island, Ponto Abril, 30.ix.1957, *Mogg* 27671 (K); Bela Vista, Zitundo, Ponta do Ouro, 2.x.1968, *Balsinhas* 1330 (LISC).
Also on the east coast of South Africa from Natal to Knysna in the Cape Province. Cultivated in Kenya. Coastal dune coloniser forming extensive mats.
Var. *rigens* is only known in cultivation. It is distinguished by its large capitula (4–8 cm. in diam.), and by its rays which are yellow or orange and have an eye-spot near the base. Var. *leucolaena* (DC.) Roessler, differs from var. *uniflora* by being ± araneose on all parts. It occurs with var. *uniflora* along the eastern coast of the Cape Province.

2. **Gazania krebsiana** Less., Syn. Comp.: 44 (1832). —Roessler in Mitt. Bot. Staatss. Münch. **3**: 401 (1959). —Harv. in Harv. & Sond., F.C. **3**: 475 (1865) incl. var. *hispidula* Harv. —Eyles in Trans. Roy. Soc. S. Afr. **5**: 520 (1916). —Hilliard, Compos. Natal: 545 (1977). Type from South Africa.

A tufted scapose perennial herb 6–30 cm. tall; rootstock woody with 1–several abbreviated apical underground branches, or stolons bearing leaf tufts, roots numerous fibrous to thong-like. Leaves few to numerous, densely crowded, all undivided or some pinnatifid, 3–30 × 0.15–0.8 cm., filiform to linear or linear-oblanceolate, narrowly tapering below before widening briefly and ± stem sheathing at the base, acute to subacute, margins ± strongly revolute and barbellate to pilose-ciliate; upper surface glabrescent or sometimes with a few short scattered bristles; lower surface whitish felted-tomentose except for the midrib. Scapes 1–many, 3–18 cm. tall, simple, tomentellous soon glabrescent, naked or with a few linear bracts. Capitula solitary; involucres 10–15 × 6–10 mm., narrowly campanulate to ± obconic, connate for half or more of the length, free parts of the phyllaries 2–3-seriate at the rim of the fused portion, absent from the sides or with a few scattered there; free parts of outer phyllaries strap-shaped, mostly 3–6 mm. long but up to c. 15 mm. long when arising from near the base of the involucre, usually spinose-denticulate on the margins; free parts of inner phyllaries ± triangular, 3–8 mm. long, obtuse or acuminate, the innermost usually with broadly membranous margins. Receptacle shallowly alveolate. Rays yellow, often with a greenish or reddish band outside, ascending-erect, 14–28 mm. long, linear-oblanceolate or oblanceolate. Disk-florets numerous; corollas yellow, 6–10 mm. long, ± shallowly-lobed; achenes 2.5–4 mm. long, narrowly subfusiform-turbinate, obscurely c. 10-ribbed, glandular, densely long-sericeous with the hairs hiding the pappus; pappus 2-seriate of subequal accrescent scale-like setae 4–6 mm. long, the setae developing a somewhat thickened stalk-like base and narrowly hyaline lateral wings which taper into a fine apex, the outer series briefly connate at the base, the inner free.

Subsp. **serrulata** (DC.) Roessler in Mitt. Bot. Staatss. Münch. **3**: 408 (1959); in Kirkia **10**: 97, tab. 3 (1975). —Merxm., Prodr. Fl. SW. Afr. 139: 73 (1967). —Hilliard, Compos. Natal: 545 (1977). TAB. **47**. Type from South Africa

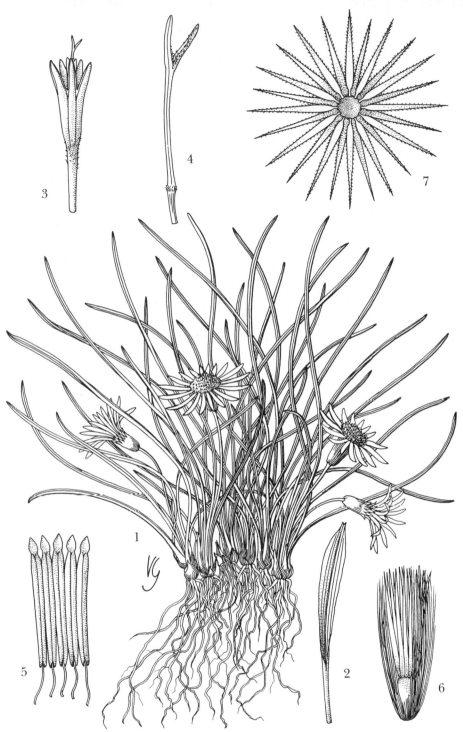

Tab. 47. GAZANIA KREBSIANA subsp. SERRULATA. 1, habit (× ⅔), from *Gibbs* 75; 2, ray-floret corolla (× 2); 3, disk-floret corolla (× 4); 4, style apex, branches connate from the style shaft, free above (× 8); 5, stamens (× 8); 6, achene with pappus (× 4); 7, pappus (× 4), 2–7 from *Eyles* 184. Drawn by V. Goaman.

Gazania serrulata DC., Prodr. **6**: 512 (1837). —Harv. in Harv. & Sond., F.C. **3**: 475 (1865). —Oliv. & Hiern in F.T.A. **3**: 428 (1877). —O. Hoffm. in Bol. Soc. Brot. **10**: 179 (1893); in Engl., Pflanzenw. Ost-Afr. **C**: 419 (1895). —Mendonça, Contrib. Conhec. Fl. Angol., **1** Compositae: 130 (1943). —Suesseng. & Merxm., Contrib. Fl. Marandellas Distr.: 65 (1951). —Barnes & Turton, List Fl. Pl. Botswana at Nat. Mus., Sebele & Univ. Botswana: 34 (1986). Type as above.

Gazania schinzii O. Hoffm. in Bull. Herb. Boiss. **2**: 215 (1894). Type from Namibia.

Leaves all undivided. Inner phyllaries obtuse or subacute, the free part 3–5 mm. long.

Botswana. SE: c. 10 km. N. of Ramatlabama, 25°30'S, 25°33'E, 12.xi.1977, *O.J. Hansen* 3278 (C; GAB; K; PRE; SRGH). **Zambia**. C: Mkushi, 24.ix.1957, *Fanshawe* 3717 (K; NDO). **Zimbabwe**. N: Mwami (Miami), c. 1360 m., 4.x.1946, *Wild* 1290 (K; SRGH). W: Hwange Nat. Park, c. 17 km. SSW. of Main Camp on Dapi Pan road, 20.xii.1968, *Rushworth* 1374 (K; SRGH). C: Marondera, Pasture Res. Station, 18.ix.1931, *Rattray* 419 (SRGH). E: Nyanga Village, 29.x.1930, *Fries, Norlindh & Weimarck* 2373a (COI; K; LD; LISU; SRGH). S: Great Zimbabwe, 1100 m., viii.1929, *Godman* 117 (BM). **Malawi**. N: Viphya Plateau, Kamunga Rock, c. 59 km. SW. of Mzuzu, c. 1767 m., 24.ix.1972, *Pawek* 5799 (K). C: Nkhota Kota Distr., Chenga Hill, 1600 m., 9.ix.1946, *Brass* 17590 (BM; K; MO; PRE; SRGH). S: Tung Station, Limbe, 17.viii.1950, *Jackson* 109 (K). **Mozambique**. N: Lichinga (Vila Cabral), 15.xi.1934, *Torre* 407 (COI; LISC). T: Macanga, c. 20 km. de Furancungo, estrada para Angónia, 29.ix.1942, *Mendonça* 545 (LISC). MS: NW. of Musapa Mt., 6.ix.1957, *Chase* 6705 (K; SRGH). GI: Inhambane, between Panda and Manele, 15.x.1968, *Balsinhas* 1388 (LISC; PRE). M: Matutuíne, Ponta do Ouro, 20.xi.1944, *Mendonça* 2928 (LISC).

Also in Tanzania, Angola, South Africa (Transvaal, Natal, Orange Free State and Cape Province), Swaziland and Lesotho. A widespread pyrophyte of short grassland from coastal to submontane habitats, often in dambos in miombo woodlands.

Subsp. *krebsiana* occurs in the Orange Free State, Natal, the Cape Province and Lesotho and may be distinguished by its acuminate inner phyllaries 4–8 mm. long. Subsp. *arctotoides* (Less.) Roessler, from the Orange Free State and Cape Province has some or all the leaves pinnately divided and the inner phyllaries obtuse.

INDEX TO BOTANICAL NAMES

257

Family				Family			
Chenopodiaceae	133	9(1)	1988	Irvingiaceae	42	2(1)	1963
Chrysobalanaceae	63	4	1978	Ixonanthaceae	33	2(1)	1963
Colchicaceae	189	—	—	Juncaceae	197	—	—
Combretaceae	73	4	1978	Juncaginaceae	206	—	—
Commelinaceae	195	—	—	Labiatae	129	—	—
Compositae	97	—	—	Lauraceae	145	—	—
Connaraceae	60	2(2)	1966	Lecythidaceae			
Convolvulaceae	117	8(1)	1987	see Barringtoniaceae	76	4	1978
Cornaceae	93	4	1978	Leeaceae	55	2(2)	1966
Costaceae	165	—	—	Leguminosae,			
Crassulaceae	67	7(1)	1983	Caesalpinioideae	61	—	—
Cruciferae	13	1(1)	1960	Leguminosae,			
Cucurbitaceae	86	4	1978	Mimosoideae	61	3(1)	1970
Cuscutaceae	118	8(1)	1987	Leguminosae,			
Cyperaceae	213	—	—	Papilionoideae	61	—	—
Dichapetalaceae	47	2(1)	1963	Lemnaceae	202	—	—
Dilleniaceae	5	1(1)	1960	Lentibulariaceae	122	8(3)	1988
Dioscoreaceae	172	—	—	Liliaceae	190	—	—
Dipsacaceae	96	7(1)	1983	Limnocharitaceae		—	—
Dipterocarpaceae	27	1(2)	1961	Linaceae	32	2(1)	1963
Droseraceae	68	4	1978	Lobeliaceae	101	7(1)	1983
Ebenaceae	107	7(1)	1983	Loganiaceae	109	7(1)	1983
Elatinaceae	24	1(2)	1961	Loranthaceae	149	—	—
Ericaceae	102	7(1)	1983	Lythraceae	78	4	1978
Eriocaulaceae	211	—	—	Malpighiaceae	35	2(1)	1963
Eriospermaceae	180	—	—	Malvaceae	28	1(2)	1961
Erythroxylaceae	34	2(1)	1963	Marantaceae	166	—	—
Escalloniaceae	64a	7(1)	1983	Mayacaceae	194	—	—
Euphorbiaceae	153		—	Melastomataceae	77	4	1978
Flacourtiaceae	18	1(1)	1960	Meliaceae	46	2(1)	1963
Flagellariaceae	196	—	—	Melianthaceae	57	2(2)	1966
Fumariaceae	12	1(1)	1960	Menispermaceae	7	1(1)	1960
Gentianaceae	113	7(4)	1990	Menyanthaceae	114	7(4)	1990
Geraniaceae	37	2(1)	1963	Mesembryanthemaceae	89c	4	1978
Gesneriaceae	123	8(3)	1988	Mimosaceae			
Gisekiaceae				see Leguminosae	61	3(1)	1970
see Molluginaceae	89b	4	1978	Molluginaceae	89b	4	1978
Goodeniaceae	98	7(1)	1983	Monimiaceae	144	—	—
Gramineae				Montiniaceae	65	4	1978
tribes 1–18	162	10(1)	1971	Moraceae	156	9(6)	1991
Gramineae				Musaceae	167	—	—
tribes 24–26	162	10(3)	1989	Myricaceae	158	—	—
Guttiferae	25	1(2)	1961	Myristicaceae	143	—	—
Haloragaceae	71	4	1978	Myrothamnaceae	69	4	1978
Hamamelidaceae	70	4	1978	Myrsinaceae	105	7(1)	1983
Hernandiaceae	146	—	—	Myrtaceae	74	4	1978
Heteropyxidaceae	75	4	1978	Nyctaginaceae	131	9(1)	1988
Hyacinthaceae	185	—	—	Nymphaeaceae	10	1(1)	1960
Hydnoraceae	140	—	—	Ochnaceae	44	2(1)	1963
Hydrocharitaceae	205	—	—	Olacaceae	48	2(1)	1963
Hydrophyllaceae	115	7(4)	1990	Oleaceae	108	7(1)	1983
Hydrostachyaceae	138	—	—	Oliniaceae	79	4	1978
Hypericaceae				Onagraceae	81	4	1978
see Guttiferae	25	1(2)	1961	Opiliaceae	49	2(1)	1963
Hypoxidaceae	178	—	—	Orchidaceae	163	—	—
Icacinaceae	50	2(1)	1963	Orobanchaceae			
Illecebraceae	22	1(2)	1961	see Scrophulariaceae	120	8(2)	1990
Iridaceae	191	—	—	Oxalidaceae	38	2(1)	1963

Family	No.			Family	No.		
Palmae	198	—	—	Santalaceae	150	—	—
Pandanaceae	199	—	—	Sapindaceae	56	2(2)	1966
Papavaraceae	11	1(1)	1960	Sapotaceae	106	7(1)	1983
Papilionaceae				Scrophulariaceae	120	8(2)	1990
see Leguminosae	61	—	—	Selaginaceae			
Passifloraceae	84	4	1978	see Scrophulariaceae	120	8(2)	1990
Pedaliaceae	125	8(3)	1988	Simaroubaceae	41	2(1)	1963
Periplocaceae				Smilacaceae	174	—	—
see Asclepiadaceae	112	—	—	Solanaceae	119	—	—
Philesiaceae	175	—	—	Sonneratiaceae	80	4	1978
Phormiaceae	181	—	—	Sphenocleaceae	100	7(1)	1983
Phytolaccaceae	135	9(1)	1988	Sterculiaceae	30	1(2)	1961
Piperaceae	142	—	—	Strelitziaceae	169	—	—
Pittosporaceae	19	1(1)	1960	Taccaceae	173	—	—
Plantaginaceae	130	9(1)	1988	Tecophilaeaceae	179	—	—
Plumbaginaceae	103	7(1)	1983	Tetragoniaceae	89d	4	1978
Podostemaceae	137	—	—	Theaceae	26	1(2)	1961
Polygalaceae	20	1(1)	1960	Thymelaeaceae	148	—	—
Polygonaceae	136	—	—	Tiliaceae	31	2(1)	1963
Pontederiaceae	192	—	—	Trapaceae	82	4	1978
Portulacaceae	23	1(2)	1961	Turneraceae	83	4	1978
Potamogetonaceae	208	—	—	Typhaceae	200	—	—
Primulaceae	104	7(1)	1983	Ulmaceae	154	9(6)	1991
Proteaceae	147	—	—	Umbelliferae	90	4	1978
Ptaeroxylaceae	58	2(2)	1966	Urticaceae	157	9(6)	1991
Rafflesiaceae	139	—	—	Vacciniaceae			
Ranunculaceae	4	1(1)	1960	see Ericaceae	102	7(1)	1983
Resedaceae	15	1(1)	1960	Vahliaceae	64	4	1978
Restionaceae	212	—	—	Valerianaceae	95	7(1)	1983
Rhamnaceae	53	2(2)	1966	Velloziaceae	171	—	—
Rhizophoraceae	72	4	1978	Verbenaceae	128	—	—
Rosaceae	62	4	1978	Violaceae	16	1(1)	1960
Rubiaceae				Viscaceae	149	—	—
(Rubioideae)	94	5(1)	1989	Vitaceae	54	2(2)	1966
Rubiaceae				Xyridaceae	193	—	—
(Cinchonoideae,				Zannichelliaceae	209	—	—
Antirheoideae)	94	—	—	Zingiberaceae	164	—	—
Rutaceae	40	2(1)	1963	Zosteraceae	210	—	—
Salicaceae	160	9(6)	1991	Zygophyllaceae	36	2(1)	1963
Salvadoraceae	110	7(1)	1983				

Obtainable from main sales Agent:
Natural History Museum,
Publications,
London SW7 5BD

Also available through:

Instituto de Investigação Científica Tropical,
Rua da Junqueira 86,
P-1300 LISBOA, Portugal

or from your bookseller